Mission Engineering and Integration

James D. Moreland, Jr.

Mission Engineering and Integration

Approach and Principles Leading to Integrated Solutions

Springer

James D. Moreland, Jr.
MEI Innovative Solutions Inc
King George, VA, USA

ISBN 978-3-032-32131-2 ISBN 978-3-032-32132-9 (eBook)
https://doi.org/10.1007/978-3-032-32132-9

This Springer imprint is published by the registered company Springer Nature Switzerland AG
The registered company address is: Gewerbestrasse 11, 6330 Cham, Switzerland

Foreword

In *Mission Engineering and Integration*, Dr. James Moreland, Jr. has provided a thorough and comprehensive introduction to all the complexities associated with modern defense systems acquisition as well as prescriptions for how to manage those complexities. This is a long overdue contribution to the field of defense acquisition.

Throughout my 50-plus year career developing, operating, and directing defense systems of all types and for each military service, I've had to respond to questions about why it seems so difficult to deliver these products on cost and schedule. I was serving as the Under Secretary of Defense for Acquisition, Technology and Logistics when Leon Panetta became Secretary of Defense. At our first meeting, he asked me why we couldn't acquire the F-35 fighter like we had the Mine-Resistant Ambush Protective Vehicles, or MRAPs. My answer was one word, "Complexity." Dr. Moreland's book takes the reader through all the many dimensions of complexity associated not just with a single weapon system, but with the acquisition of integrated mission capabilities. These integrated capabilities have to perform their missions successfully against threats designed specifically to defeat them, and they have to do so reliably over their operational life. Dr. Moreland deals with all of these problems and many more in this comprehensive work.

Today, we face challenging and sophisticated threats, especially China. The presumption of US dominance is no longer valid. It is critical that we make sound decisions about our modernization investments and manage those investments efficiently. When I came back to the Pentagon in 2010 after a 15-year stint in industry, I was shocked to see defense system requirements being set by senior leader intuition without supporting data or analysis. In the ensuing years, I kept a sign outside my office quoting G. Edward Deming, "In God We Trust, All Others Bring Data." Throughout *Mission Engineering and Integration*, Dr. Moreland stresses the need for data and analysis, and the methods to generate or acquire data and use it to support decisions, especially decisions about the specifications or requirements for defense system-of-systems.

As this book also makes very clear, acquisition management itself is incredibly complex and not for amateurs. One of my favorite quotes on defense acquisition is from General Mark Welsh, the former Air Force Chief of Staff. When Mark was asked to take on a more direct role as a Service Chief in managing acquisition programs, his answer was to the effect that he was not interested in having that responsibility. To paraphrase his response, "That's hard and I'm not qualified." Mark had spent one assignment tour working in the acquisition world. He knew what he was talking about.

Anyone who does wish to understand or become more qualified in defense acquisition would be well served to read and absorb the material covered in *Mission Engineering and Integration*.

Honorable Frank Kendall
26th United States Secretary of the Air Force and Under Secretary of Defense for Acquisition, Technology, and Logistics (AT&L)
Washington, DC, USA

Preface

To ensure the technological superiority of our nation, critical and prudent technological investments must provide increased effectiveness under affordable conditions to acquire safe and secure capabilities. It is imperative that our nation balances tactical and strategic investments across all accounts, ensures research, development, science and technology investments target the correct missions, and expeditiously transition technologies and capabilities to end users. The basic concept of Mission Engineering and Integration (MEI) looks across operational mission threads using an Effects Web Framework to see how systems work together in an operationally relevant environment. Through the rigor of experiments, testing, data-driven assessments, and acquisition oversight, all national industrial sectors are looking for game-changing innovations that integrate technology and systems to provide affordable solutions.

These end-to-end mission assessments must focus on the integration and interoperability across multiple systems evaluated in a representative operational environment to ensure accuracy in determining operational capability gaps and to lead to better acquisition decisions. Data from an authoritative source of truth should inform the process to increase reliability and decrease uncertainty with the solution space not limited to material solutions but also consider the broader spectrum of doctrine, organization, training, materiel, leadership and education, personnel, facilities, and policy domains to gain a holistic perspective. Understanding system-of-systems (SoS) at the proper level of detail reduces the risk of fielding fragmented mission engineering threads, resulting in fielded capabilities across system boundaries, operational domains, and within individual institutions. In addition, MEI considers the alternative mission engineering threads forming a mission engineering web to provide mission resiliency and redundancy in executing the mission threads while reducing unnecessary redundancies. This provides an accountable governance structure at the mission level ensuring the national investments across the critical institutions reflect the way the operational plans must be executed with integration and interoperability factors in mind.

This *Mission Engineering and Integration: Approach and Principles to Integrated Solutions* book starts with chapters defining integration, interoperability, and complex SoS, then leads into the seminal chapter on the MEI approach. This approach tracks the development of mission essential tasks following a critical path method and operational planning process. Further, it identifies mission success criteria essential to the development of mission engineering threads/webs. These mission engineering threads/webs can be continuously monitored using a mission capability portfolio management process to manage the execution health based on operational objectives set by the stakeholders. The implementation of MEI comes with challenges associated with data availability and collection, governance structure, system maturity levels across SoS, stakeholder coordination across national institutional boundaries, and workforce/tool development to name a few. However, the Effects Web Framework provides a translation mechanism between operational execution and engineering of complex SoS leading to advanced capabilities. This MEI multidisciplinary and interdisciplinary approach combines rigorous analyses, coordination, and collaboration to produce holistic views across domains and functionalities. Our dependencies on continuous network and system connectivity has become embedded in our societal makeup, permeating all aspects of our systems, networks, and business operations as we attempt to deliver and emphasize the importance of developing and maintaining operational cohesion across mission threads and objectives.

I was honored to receive the Department of the Navy's highest civilian service award, the Distinguished Civilian Service Award, in 2014 from the Secretary of the Navy and the Chief of Naval Operations for my pioneering, trailblazing, and leadership efforts in the development of MEI which started in 2010 as the Integration and Interoperability (I&I) Initiative within the Department of the Navy. The citation of this award read as follows:

> Dr. Moreland's considerable and widely recognized ability to link policy with technical details in a rigorously disciplined systems engineering fashion is exceptionally rare and valued by the Navy. His leadership on a range of naval, joint, and coalition efforts has demonstrated a unique capability to develop and implement innovative strategic concepts to shape future requirements, organizational strategy, and operational doctrine. Most recently, Dr. Moreland demonstrated this expertise by implementing a structured mission engineering approach for the Vice Chief of Naval Operations' Integration and Interoperability Activity. His exceptional professionalism, personal initiative, technical competence, and steadfast devotion to duty reflect great credit upon him and were in keeping with the highest traditions of the United States Naval Service.

The call for action for this important effort was led by the Chief of Naval Operations (ADM (ret) Gary Roughead), Fleet Forces Commander (ADM (ret) John Harvey), and Naval Air Systems Commander (VADM (ret) David Architzel), with a driving force from VADM (ret) David Dunaway. I was fortunate to be called upon by ADM (ret) Gary Roughead to lead this effort across the Naval Systems Commands with the summary of the effort best captured in a news release quote from VADM (ret) David Dunaway,

> "Naval Aviation leadership has embraced a forward-thinking innovative concept to realign 'stove-piped' programs to a 'system of systems' perspective," explained Vice Adm. David Dunaway, Commander, Naval Air Systems Command. "Often referred to as I&I, this per-

spective requires us to explicitly link our technical expertise and solutions with operational tactics from the very start of acquisition. Simply put, Jim's robust and expert leadership has helped Naval Aviation move our I&I initiative forward by delivering measurable results."

This trailblazing work led me to develop curricula, teach courses, and serve as a doctoral research advisor for MEI at Old Dominion University beginning in 2014 and at Virginia Tech starting in 2022. These efforts aim to embed this new approach into the education of the next generation of scientists and engineers. This academic expansion is ongoing, with several other institutions, including military academies, expressing interest in incorporating MEI into their programs. Additionally, I am currently establishing MEI credentials at the Defense Acquisition University and the University of New South Wales in Australia. On a parallel training path, I have developed and delivered numerous MEI workshops both domestically and internationally, a practice I continue to this day. Japan and Australia have been the primary international partners for these focused workshops, with additional inquiries from other nations. Domestically, the workshops have been conducted across all branches of the US military, ensuring widespread adoption and integration of MEI principles within national defense operations.

On the Department of Defense (DoD) organizational front, I delivered this innovative approach in September 2014 as the thought leader and pioneer upon entering the Senior Executive Service ranks in the position of Executive Director, Naval Warfare within the Office of the Under Secretary of Defense for Acquisition, Technology, and Logistics (OUSD AT&L). This push towards a capability- and mission-based approach played a major role in the restructuring of the OUSD AT&L into two smaller entities in February 2018: OUSD for Research and Engineering (OUSD R&E) and OUSD for Acquisition and Sustainment (OUSD A&S). This change, driven by Congress and reflected in the 2016 and 2017 National Defense Authorization Acts, aimed to enhance the focus on MEI to understand mission execution needs, SoS capabilities, and integration of systems. Within the newly formed OUSD A&S, MEI played a key role in the immediate transitioning of the Deputy Assistant Secretary of Defense (DASD) for Tactical Warfare Systems to DASD for Platform and Weapons Portfolio Management and in establishing the Acquisition Integration and Interoperability (AI2) Office in January 2023. Both offices are dedicated to building a robust acquisition infrastructure to deliver joint, SoS capabilities while managing resources across capability portfolios. Additionally, to further institutionalize MEI within the DoD, OUSD R&E established the new DASD for Mission Integration Office to provide mission-based inputs for the requirements process, enhancing joint capability integration and experimentation. This office also supports the development and refinement of joint warfighting concepts and operational plans in direct alignment with the MEI approach.

It took approximately 5 years to introduce the MEI approach across the military services due to the culture change required as well as the education and momentum to move forward. This activity was performed as additional duty until formalized with the introduction of Mission Integration Management (MIM) in the National Defense Authorization Act of 2017 (NDAA 2017), Section 855. From its 2014 introduction to 2025, the NDAA 2017, Section 855, and other policies have guided the implementation of MEI as outlined below:

- **NDAA 2017, Section 855:** Focuses on enhancing MIM using advanced data, modeling, and simulation capabilities following a MEI approach. It emphasizes improving the DoD's ability to integrate and align resources, strategies, and operations across missions. By leveraging these tools, the DoD aims to strengthen coordination and decision-making processes, thereby ensuring that mission objectives are achieved efficiently and effectively. The section also requires periodic assessments and reporting to Congress to foster accountability and ensure MIM remains adaptive to evolving defense needs and technological advancements.
- **DoD Directive (DoDD) 5000.01:** Provides the foundational framework for the Defense Acquisition System, emphasizing the efficient and effective development, procurement, and sustainment of defense capabilities. In the context of MEI, the directive highlights the importance of an SoS approach, integrating engineering disciplines to design, analyze, and optimize mission-level solutions. It prioritizes interoperability, scalability, and the alignment of capabilities with operational objectives to ensure systems work cohesively to achieve mission success. By embedding MEI principles throughout the life cycle of defense acquisition, DoDD 5000.01 fosters a holistic approach to addressing complex operational challenges and delivering resilient, adaptable, and mission-ready solutions.
- **DoD Instruction (DoDI) 5000.88, Engineering of Defense Systems:** Establishes policy and guidance for applying systems engineering across the life cycle of defense acquisition programs. In the context of MEI, this instruction emphasizes the importance of designing and managing systems to achieve integrated mission outcomes. It highlights the role of MEI in analyzing complex mission threads, identifying interdependencies, and ensuring interoperability across SoS. DoDI 5000.88 promotes the use of rigorous technical processes, digital engineering tools, and data-driven analysis to optimize mission effectiveness, reduce risk, and support decision-making. By embedding MEI into systems engineering practices, it ensures that capabilities are developed and delivered in alignment with overarching mission objectives and operational needs.
- **DoD Directive (DoDD) 7045.20, Capability Portfolio Management (CPM):** Establishes policies for managing groups of related capabilities to ensure they align with strategic priorities and deliver maximum value. In the context of MEI, this directive supports the integration and coordination of capabilities across systems and portfolios to achieve mission-level objectives. It emphasizes the importance of a holistic, SoS approach to analyze interdependencies, reduce redundancies, and ensure that capabilities are interoperable and effective within mission environments. By aligning capability investments with operational requirements, DoDD 7045.20 enables MEI to drive informed decision-making, resource optimization, and the delivery of cohesive, mission-ready solutions across the DoD.
- **DoD Directive (DoDD) 5135.02, Under Secretary of Defense for Acquisition and Sustainment (USD(A&S)):** Defines the roles and responsibilities of the

USD(A&S) in managing the acquisition and sustainment of defense capabilities. In the context of MEI, this directive supports the integration of engineering and acquisition efforts to ensure that systems and capabilities align with mission requirements. It emphasizes the application of MEI principles, such as an SoS approach, to ensure interoperability, life cycle management, and operational readiness. By fostering collaboration between engineering, acquisition, and sustainment activities, DoDD 5135.02 enables the delivery of cohesive, mission-focused solutions that address the DoD's complex operational challenges.

- **DoD Directive (DoDD) 5137.02, Under Secretary of Defense for Research and Engineering (USD(R&E)) Charter:** Outlines the responsibilities and authority of the USD(R&E) in overseeing the research, development, and engineering of advanced technologies to maintain the DoD's technological advantage. In the context of MEI, this directive assigns the USD(R&E) a central role in fostering the integration of innovative technologies and engineering practices to address mission-level challenges. It emphasizes advancing SoS engineering, enhancing interoperability, and ensuring that emerging technologies are aligned with mission requirements. By promoting collaboration across domains and leveraging MEI principles, DoDD 5137.02 ensures the effective design, development, and deployment of capabilities that support complex and evolving operational needs.
- **Chairman of the Joint Chiefs of Staff Instruction (CJCSI) 5123.01J, Charter of the Joint Requirements Oversight Council (JROC) and Implementation of the Joint Capabilities Integration and Development System (JCIDS):** Establishes guidance for identifying, prioritizing, and validating joint military capability requirements. In the context of MEI, this instruction emphasizes the importance of aligning capabilities with mission objectives through an SoS approach. It integrates MEI principles by ensuring capability development considers mission-level performance, interoperability, and operational effectiveness across joint and coalition environments. CJCSI 5123.01J supports rigorous analysis and collaboration to identify gaps, evaluate alternatives, and ensure that developed capabilities are fully integrated and optimized to meet the demands of complex missions.

The formalization of MEI within the DoD (converting to Department of War (DoW) in 2025) reached a new level of maturity in 2025 with the enactment of the *Supporting the Production of Essential and Emerging Defense (SPEED) Act*, and its implementation guidance in the August 2025 Secretary of Defense memorandum titled, *Reforming the Joint Requirements Process to Accelerate Fielding of Warfighting Capabilities*. Section 203 of the *SPEED Act* establishes the MEI Activity (MEIA) as a DoW-wide construct jointly governed by the USD(R&E) and USD(A&S). The MEIA institutionalizes the MEI framework as the DoW central mechanism for linking science and technology, research, engineering, and acquisition to operational mission outcomes. The MEIA is tasked to conduct MEI analyses across key operational problems (KOPs), refine capability requirements, and execute structured experimentation campaigns that bring together military service laboratories, defense agencies, and industry innovators. Through this mechanism, the DoW now integrates mission-level analysis, experimentation, and acquisition planning into a single, iterative process to accelerate the delivery of joint, interoperable capabilities to the end user. Collectively, the *SPEED Act* and its implementing memorandum retire the legacy JCIDS and replace it with a mission-based, engineering-driven governance model. MEI serves as the guiding principle of defense innovation by ensuring that future capability development is driven by operational missions, grounded in authoritative data, and focused on delivering unified outcomes.

James D. Moreland, Jr.

Acknowledgments

The development of this book, *Mission Engineering and Integration: Approach and Principles Leading to Integrated Solutions*, has been a deeply meaningful effort, shaped by the steady support, encouragement, and contributions of many individuals. It also represents a deliberate step in capturing the pioneering concepts, thought leadership, and many years of personal dedication that guided the creation and evolution of the Mission Engineering and Integration (MEI) discipline. In bringing this work into written form, the aim is to provide future generations of engineers, scientists, operators, and decision-makers across any area of expertise with enduring academic knowledge. In addition, this book offers a practical construct for implementing MEI that future practitioners and leaders can rely on when confronting the growing complexities of system-of-systems challenges.

First and foremost, I extend my heartfelt gratitude to my wife, Angela Clift Moreland, whose unwavering support, patience, and belief in this work made every stage of the journey possible. Her encouragement and understanding have been my foundation throughout long days, late nights, and my continuous commitment to the objective of advancing MEI as a national capability.

A special acknowledgment goes to Ms. Margie Stevens, whose meticulous attention to detail, and exceptional technical editing and writing skills served as the backbone of the editorial process. As technical editor, her careful review, thoughtful feedback, and dedicated effort ensured the material reflects the consistency and clarity essential to a discipline like MEI.

I am also grateful to Ms. Merritt Hepinstall, whose creative talent and exceptional skill as a graphic artist elevated the visual quality of this volume.

To all who contributed their insight, encouragement, and expertise, I offer my sincere appreciation. This book is as much a product of your support as it is of the ideas and principles it seeks to advance.

Competing Interests The author has no competing interests to declare that are relevant to the content of this manuscript.

Contents

List of Figures

List of Tables

Acronyms

3D	Three-dimensional
A&S	Acquisition and Sustainment
A2AD	Anti-access and Area Denial
AAF	Adaptive Acquisition Framework
ADM	Admiral
AI	Artificial Intelligence
AI2	Acquisition Integration and Interoperability
AoA	Analysis of Alternatives
API	Application Programming Interfaces
APT	Advanced Persistent Threats
A_o	Operational Availability
AT&L	Acquisition Technology and Logistics
BoE	Basis of Estimate
BOM	Bill of Materials
BSC	Balanced Scorecard
C2	Command and Control
CAS	Complex Adaptive System
CASCADE	Complex Adaptive System Composition and Design Environment
CBS	Cost Breakdown Structure
CDF	Cumulative Distribution Function
CDR	Critical Design Review
CER	Cost Estimating Relationship
CES	Cost Estimating Structure
CI	Configuration Item
CIL	Capability Immersion Layer
CJCS	Chairman Joint Chiefs of Staff
CKC	Cyber Kill Chain
CM	Configuration Management
CO	Contractor Owned
CONEMP	Concept of Employment
CONOP	Concept of Operation

COTS	Commercial-off-the-Shelf
COU	Course of Action
CPG	Contingency Planning Guidance
CPR	Cost Performance Report
CRL	Capability Readiness Level
CT	Category Theory
CT	Contractor Test
CTA	Cognitive Task Analysis
CTAWC	Cognitive Task Analysis and Workload Classification
CY	Current Year
DAO	Defense Acquisition University
DARPA	Defense Advanced Research Projects Agency
DASD	Deputy Assistant Secretary of Defense
DE	Digital Engineering
DevSecOps	Development, Security, and Operations
DL	Deep Learning
DLP	Data Loss Prevention
DMO	Distributed Maritime Operations
DNN	Deep Neural Networks
DoD	Department of Defense
DoDAF	Department of Defense Architecture Framework
DoE	Design of Experiment
DoN	Department of the Navy
DOTMLPF-P	Doctrine, Organization, Training, Material, Leadership, Personnel, Facilities, and Policy
DT	Developmental Test
DT	Digital Thread
E2EL	End-to-End Lifecycle
EDR	Endpoint Detection and Response
EEG	Electroencephalogram
EMS	Electromagnetic Spectrum
ERM	Enterprise-wide Risk Management
EVM	Earned Value Management
EW	Electronic Warfare
F2T2EA	Find, Fix, Track, Target, Engage, and Assess
FDNA	Functional Dependency Network Analysis
FFBD	Functional Flow Block Diagram
GA	Genetic Algorithms
GOMS	Goals, Operators, Methods, and Selection
GOTS	Government-of-the-Shelf
GPS	Global Positioning System
GR&A	Ground Rules and Assumptions
GRA	Government reference architecture
GRM	Government Reference Model
GUI	Graphical User Interface

HAMAS	Ḥarakah al-Muqāwamah al-ʾIslāmiyyah, meaning "Islamic Resistance Movement"
HFE	Human Factors Engineering
HFM	High-Fidelity Model
HSI	Human Systems Integration
HITL	Human-in-the-Loop
HoQ	House of Quality
HPT	High Performance Team
I&I	Integration and Interoperability
I&T	Integration and Test
ICAM	Identity, Credential, and Access Management
ICF	Integrated Capability framework
ICTB	Integrated Capabilities Technical Baseline
IDS	Intrusion Detection Systems
IP	Intellectual Property
IP	Internet Protocol
IPT	Integrated Product Team
IRL	Integration Readiness Level
ISR	Intelligence, Surveillance and Reconnaissance
IT	Information Technology
JCDL	Joint Capability Demonstration Layer
JCIDS	Joint Capability and Integration Development System
JROC	Joint Requirements Oversight Council
JSCP	Joint Strategic Capabilities Plan
JTC	Joint Test Concept
JTF	Joint Task Force
KPI	Key Performance Indicator
KPP	Key Performance Parameter
KSA	Key System Attributes
LAS	London Ambulance Service
LCC	Lifecycle Cost
LCCE	Lifecycle Cost Estimate
LFM	Low-Fidelity Model
LFT	Live Fire Test
LLM	Large Language Model
LNCH	Launch Scoring
LVC	Live, Virtual, and Constructive
M&S	Modeling and Simulation
MBD	Model-Based Design
MBE	Model-Based Engineering
MBM	Model-Based Manufacturing
MBS	Model-Based Sustainment
MBSE	Model-Based Systems Engineering
MDAO	Multidisciplinary Analysis and Optimization

MDO	Multi-domain Operations
MDSD	Model-Driven Software Development
ME	Mission Engineering
MEI	Mission Engineering and Integration
MER	Manpower Estimate Report
MET	Mission Essential Task
METL	Mission Essential Task List
MFHM	Multi-fidelity Hierarchical Model
MFSM	Multi-fidelity Surrogate Models
ML	Machine Learning
MOA	Memoranda of Agreement
MoE	Measure of Effectiveness
MoP	Measure of Performance
MoS	Measure of Success
MOSA	Modular Open Systems Approach
MRO	Maintenance, Repair, and Operations
MT	Mission Thread
MTB	Mission Technical Baseline
MTBF	Mean Time Between Failure
MVP	Minimum Viable Product
NASA	National Aeronautics and Space Administration
NATO	North Atlantic Treaty Organization
NAVAIR	Naval Air Systems Command
NAVSEA	Naval Sea Systems Command
NDAA	National Defense Authorization Act
O&S	Operations and Sustainment
OODA	Observe, Orient, Decide, and Act
OPTEMPO	Operational Tempo
OQE	Objective Quality Evidence
OSA	Open Systems Architecture
OSD	Office of the Secretary of Defense
OT	Operational Test
OUSD	Office of the Undersecretary of Defense
OV	Operational View
OV-1	High-Level Operational Concept
OWL	Web Ontology Language
PDR	Preliminary Design Review
PDCEA	Plan, Detect, Control, Engage, Assess
PERT	Project Evaluation and Review Technique
PLCCE	Program Lifecycle Cost Estimate
PMESII	Political, Military, Economic, Social, Information, and Infrastructure
PMR	Prevent, Mitigate, and Recover
PNT	Position, Navigation and Timing
PPBE	Planning, Programming, Budgeting, and Execution

QFD	Quality Function Deployment
QoS	Quality of Service
R&E	Research and Engineering
RDF	Resource Description Framework
RL	Reinforcement Learning
RoE	Rules of Engagement
RoI	Return on Investment
ROM	Rough Order of Magnitude
RPA	Robotic Process Automation
RY	Real Year
SBD	Set-Based Design
SDREN	Secret Defense Research and Engineering Network
SE	Systems Engineering
SE&I	Systems Engineering and Integration
SMART	Specific, Measurable, Achievable, Relevant, and Time-bound
SoS	System-of-Systems
SPL	System Performance Layer
SYSCOM	Systems Command
T&E	Test and Evaluation
TACSIT	Tactical Situation
TDP	Technical Data Package
TENA	Tactical Data Link Integration Exerciser
TEWA	Threat Evaluation and Weapon Assignment
TRL	Technology Readiness Level
TTP	Tactics, Techniques, and Procedures
U.S.	United States
UAF	Unified Architecture Framework
UFE	Unallocated Future Expenses
UJTL	Universal Joint Task List
VADM	Vice Admiral
VM	Virtual Machine
VVA	Verification, Validation, and Accreditation
W3C	World-Wide Web Consortium
Wasm	WebAssembly
WBS	Work Breakdown Structure
WOSA	Weapon Systems Open Architecture
ZTA	Zero Trust Architecture

Chapter 1
The Need for a New Approach

1.1 Evolving Landscape

Welcome to the world of Mission Engineering and Integration (MEI), a dynamic and interdisciplinary field where innovation, precision, and strategic thinking converge to shape the success of complex endeavors. This book embarks on a journey through the foundational principles, methodologies, and practical applications that define the MEI approach. The landscape of MEI is continually evolving, driven by advancements in technology, changes in mission objectives, and the increasing complexity of the environments in which missions are conducted. Several key trends and factors contribute to the evolving landscape of MEI:

Advancements in Technology

- **Impact**: Rapid advancements in technology, including artificial intelligence, robotics, sensors, communication systems, and materials are transforming the capabilities of mission systems. Mission engineers leverage these technologies to design more sophisticated and capable systems.

Increased Autonomy

- **Impact**: There is a growing trend toward the development and deployment of autonomous systems in missions. MEI is adapting to incorporate higher levels of autonomy, enabling systems to operate independently and make real-time decisions in dynamic environments.

Interconnected and Collaborative Systems

- **Impact**: Many modern missions involve the coordination of multiple interconnected systems working across delivery domains known in the military as Multi-Domain Operations (MDO). MDO refer to military strategies and activities that involve the integration and synchronization of operations across

J. D. Moreland, Jr., *Mission Engineering and Integration*,
https://doi.org/10.1007/978-3-032-32132-9_1

multiple domains. The term "domain" in this context refers to the different environments or realms in which activities take place. To illustrate, the traditional warfighting domains include:

Land: Involving operations on the ground, often characterized by infantry, armor, and other land-based forces.
Air: Encompassing operations in the air, involving aircraft, drones, and other aerial assets.
Sea: Involving naval and maritime operations, including activities on and under the water.
Space: Focused on operations in space, including satellite communications, reconnaissance, and other space-based activities.
Cyber: Encompassing operations in the cyber domain, including cyber warfare, electronic warfare, and information operations.

MDO recognize that modern conflicts are not confined to one specific domain but are often characterized by their cross-domain nature. The integration and coordination of activities across these domains are essential for achieving strategic objectives and maintaining a competitive advantage. The concept acknowledges the interconnectedness of the different operational environments and the need for joint and collaborative efforts to address evolving threats. MEI is evolving to address the challenges of designing, integrating, and managing collaborative systems that work together synergistically to achieve common objectives.

Integration of Commercial Solutions

- **Impact**: The integration of commercial-off-the-shelf (COTS) solutions and components is becoming more prevalent in MEI activities. This trend allows for faster development cycles, cost savings, and the incorporation of cutting-edge technologies.

Data-Centric Approaches

- **Impact**: The increasing importance of data in decision-making is influencing MEI approaches. Mission Engineers are adopting data-centric strategies, including data analytics, machine learning, and real-time data processing, to extract valuable insights from the vast amount of data generated during missions.

Miniaturization and CubeSat Technologies

- **Impact**: The trend toward miniaturization and the use of CubeSat technologies is influencing MEI across all industrial sectors to include space exploration. Smaller, more cost-effective systems like satellites enable more frequent and distributed missions, changing the traditional paradigm of large, singular missions.

Adaptive and Resilient Systems

- **Impact**: MEI is placing a greater emphasis on designing adaptive and resilient systems that can respond to unexpected challenges and uncertainties. This includes the development of systems that can autonomously adapt their behavior based on changing mission conditions.

Focus on Sustainability

- **Impact**: There is a growing awareness of the environmental impact of missions, especially in space exploration. MEI is evolving to incorporate sustainable practices, including the development of reusable systems and the consideration of end-of-life disposal.

Cybersecurity Considerations

- **Impact**: With the increasing reliance on connected systems and the integration of digital technologies, cybersecurity has become a critical aspect of MEI. Ensuring the security of mission-critical data and systems is a priority.

Global Collaboration

- **Impact**: MEI is increasingly becoming a collaborative effort on a global scale. International partnerships and collaborations bring together expertise and resources from different countries to address complex challenges and conduct joint/coalition missions. When implementing MEI, a common framework like the Effects Web Framework standardizes ontologies and taxonomies so results can be integrated and compared in a more seamless information exchange. Organizations often develop or tailor their own approaches for their specific needs and mission requirements but doing so creates challenges to collaboration and interoperability across efforts. At the writing of this book, multiple Allied Partners are seeking training for MEI to gain appreciation and understanding so that integration and interoperability can be addressed across forces to leverage combined capabilities.

Regulatory Compliance

- **Impact**: There is an increasing focus on regulatory compliance in MEI, especially in fields like space exploration. Engineers must navigate international and national regulations to ensure missions adhere to legal and ethical standards.

Public and Private Sector Collaboration

- **Impact**: Collaborations between public agencies and private sector companies are becoming more common. MEI must adapt to the dynamics of public-private partnerships, considering both governmental and commercial objectives.

Focus on Inclusivity and Diversity

- **Impact**: More and more, MEI teams are recognizing the importance of diversity and inclusion. Embracing a diverse range of perspectives contributes to innovative problem-solving and robust decision-making.

Ethical Considerations

- **Impact**: As technology advances, MEI is grappling with ethical considerations related to the use of autonomous systems, data privacy, and the potential impact of missions on local and global ecosystems.

Mission engineers play a pivotal role in navigating these challenges and driving innovation in the design and execution of missions. Some driving themes behind the continued adoption of MEI are:

- A new emphasis on acquisition priorities will involve maximizing the use of existing resources and continually integrating relevant commercial capabilities.
- Rather than the next greatest platform or system, decision leaders will need to better connect existing capabilities and not merely rely on enhancing exquisite capabilities.
- Over time, reliance on exquisite platforms will be replaced with incorporation of a more mixed use of advanced platforms, upgrades to existing platforms, and expendable and autonomous platforms. The goal is to achieve lower capability and cost but produce significantly higher numbers.
- The arrival of artificial intelligence/machine learning (AI/ML) capabilities and algorithmic strategies will enable smart systems that understand their mission and can adapt their courses of action or connectivity based on dynamic events.
- Finally, stakeholders will need to acquire capability at a speed that matches the evolutionary speed of threat and market advancements and technology development ("speed of relevancy"). Moreover, mission systems must be open to continuous upgrades and made resilient to adopting threats, such as digital effects across the electromagnetic spectrum.

Transforming our way of doing business toward digital transformation, model-based engineering, commercial methods, and focusing on integrated solutions are essential and urgent for mission and competitive success.

1.2 Importance of Integration in Complex Missions

Integration plays a crucial role in complex missions across various domains, including defense, space exploration, healthcare, business, and more, and is highlighted through several key aspects:

Interoperability: Complex missions often involve multiple systems, technologies, and entities. Integration ensures that these diverse components can work together seamlessly, which is essential for mission success. It allows different systems to communicate, share data, and collaborate effectively.

Efficiency and Optimization: Integration helps optimize the use of resources, both human and technological. By coordinating various elements of a mission, organizations can reduce redundancy, streamline processes, and make better use of available resources, thereby leading to improved efficiency and cost-effectiveness.

Enhanced Decision-Making: Integrated systems provide a comprehensive view of the mission landscape, allowing decision-makers to access real-time data and insights from different components. This enables informed decision-making, which is particularly crucial in dynamic and rapidly changing mission environments.

Risk Reduction: Complex missions often involve inherent risks. Integration promotes a more comprehensive understanding of potential risks and their interdependencies. By addressing and mitigating risks proactively through integrated planning and execution, organizations can enhance mission safety and success.

Adaptability and Flexibility: Integrated systems are more adaptable to changes and unexpected challenges. In dynamic mission environments, the ability to adjust strategies, reroute resources, and respond to emerging situations is critical. Integration facilitates agility and flexibility in adapting to evolving mission requirements.

Holistic View of the Mission: Integration provides a holistic view of the entire mission rather than focusing on isolated components. This comprehensive perspective allows for better coordination, monitoring, and control of the mission as a whole, leading to improved overall performance.

Communication and Collaboration: In a complex mission, effective communication and collaboration among different teams, departments, and stakeholders are paramount. Integrated systems enable seamless communication channels, fostering collaboration and coordination among diverse entities involved in the mission.

Data Fusion and Analysis: Integration of data from various sources facilitates data fusion and analysis, leading to a more accurate and comprehensive understanding of the mission environment based on an authoritative source of data with complete trust. This is particularly important in missions that require data-driven insights for planning, execution, and post-mission analysis.

Customer Satisfaction: In commercial missions, integration contributes to a smoother customer experience. Whether delivering products or services, an integrated approach ensures that customer expectations are met efficiently, leading to higher satisfaction levels.

Mission Success: Ultimately, the overarching importance of integration in complex missions lies in achieving mission success. A well-integrated mission ensures that all systems work cohesively toward common objectives, increasing the likelihood of meeting mission goals and objectives toward a desired end state.

In summary, integration is a critical enabler for managing the complexity of modern missions. It provides a structure for coordination, collaboration, and optimization across diverse elements to enhance overall mission effectiveness and success.

1.3 Purpose

The purpose of this book is to provide a vision, guidance, and a structured approach for understanding and addressing integration and interoperability gaps within complex adaptive system-of-systems (SoS). In an era where the *Internet of Things* is transforming industries, the need to share information seamlessly across systems and industrial sectors has become paramount. Yet, limitations in data-sharing capabilities often hinder collaboration and operational efficiency. Drawing on valuable lessons from military operations, where interoperability ensures seamless information exchange among joint and coalition forces, this book highlights the critical importance of creating robust communication networks. By bridging gaps between industrial domains and enabling the clear distribution of relevant information, this book aims to equip global entities with the tools to enhance efficiency, improve decision-making, and ensure the safety and well-being of all. Ultimately, it envisions a future where a capable and unified communication fabric empowers end users and decision-makers to fully realize the potential of a globally connected network.

1.4 Scope

This book explores the broad challenges of integration and interoperability across various industrial sectors, offering a comprehensive perspective on their implications for MEI. It serves as a guide for understanding operational gaps associated with these challenges, introducing a structured MEI approach to address them effectively. This book lays a solid foundation by defining integration and interoperability and delving into their underlying principles and philosophies. Furthermore, it identifies critical enablers necessary for successful MEI, ranging from architectural design considerations to the adoption of enabling technologies. By bridging theory and application, this book equips readers with the tools and insights needed to tackle complex integration and interoperability issues in diverse mission-critical environments.

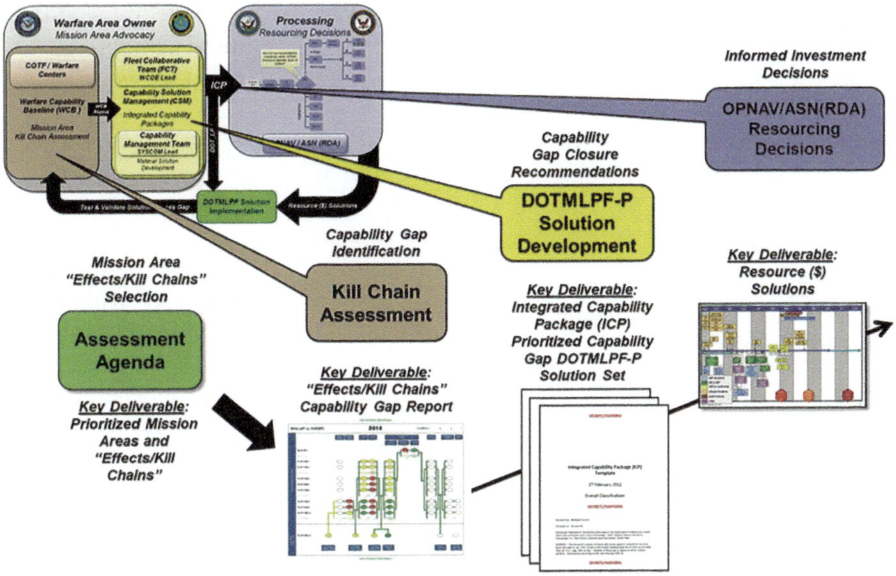

Fig. 1.1 Navy integration and interoperability initiative methodology

1.5 Evolution of Mission Engineering and Integration: A US Navy Initiative

MEI has undergone a significant evolution over the past decade, shaped by advancements in technology and the changing needs of military operations. The roots of modern MEI within the Navy can be traced back to the *Integration and Interoperability (I&I) Initiative* in 2010, which focused on addressing challenges in ensuring systems across platforms could work together seamlessly to achieve mission objectives [ANDERSON11]. However, the concept of mission engineering (ME) predates this initiative, with its earliest applications seen at the National Aeronautics and Space Administration (NASA). NASA utilized ME to design and execute highly complex, single-purpose missions such as space exploration and satellite launches, focusing on the detailed integration of systems for achieving precise outcomes. In contrast, the Department of Defense (DoD) faces the challenge of executing multiple overlapping and concurrent missions, requiring MEI to account for the interoperability and adaptability of systems across diverse operational environments. Adaptability is the capability to dynamically adjust system architectures, operational concepts, and decision-making processes in response to evolving mission needs, threats, and constraints while maintaining effectiveness and resilience. This difference in scope and scale has driven the evolution of MEI within the DoD toward a more dynamic and mission-centric approach.

The activities for the *I&I Initiative* are illustrated in Fig. 1.1 driven by the following original MEI definition:

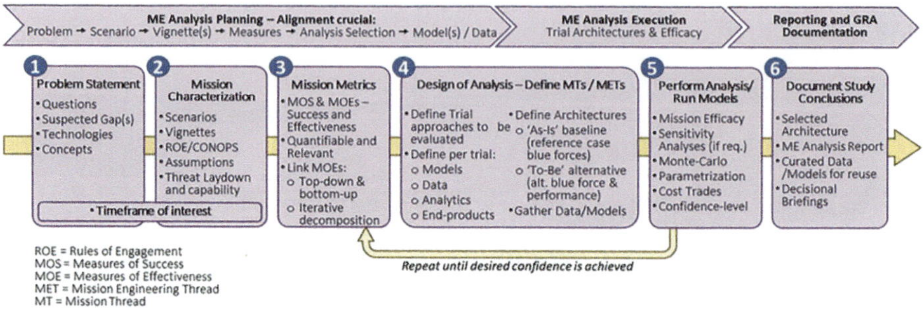

Fig. 1.2 *MEG v1.0* methodology

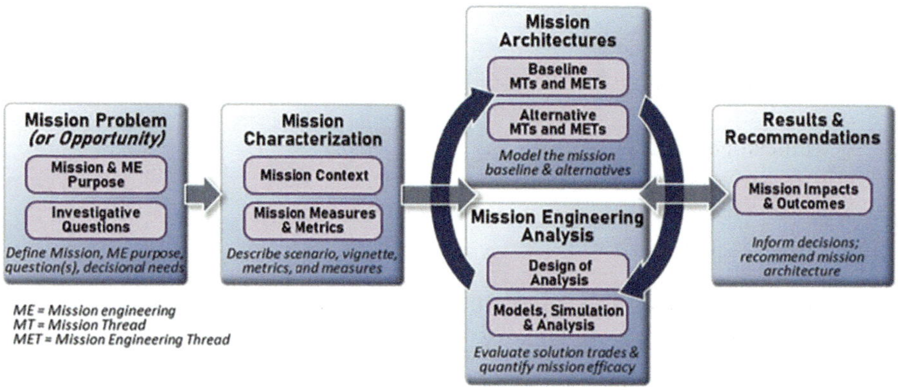

Fig. 1.3 *MEG v2.0* methodology

> The planning, analyzing, organizing, and integrating current and emerging operational concepts for the purpose of evolving the end-to-end operational architecture and capability attributes, across the DOTMPLF-P spectrum, including anticipated blue force (BLUFOR) and opposing force (OPFOR) behaviors, that are needed to inform the communities of interest involved in fulfilling mission needs statements. [MORELAND19]

The corresponding process followed a structured sequence of steps to assess kill chains which will be introduced in Chap. 8. Building on the foundational principles established by the Navy's *I&I Initiative*, MEI has since expanded into a comprehensive framework employed across the DoD starting under the author's trailblazing development and thought leadership in September 2014. This progression started with the *MEI Guidebook* in November 2019 to the *Mission Engineering Guide* (*MEG*) *v1.0* in November 2020 to the current *MEG v2.0* document released in October 2023 [MORELAND26]. Figures 1.2 and 1.3 show the general methodology for *MEG v1.0* and *MEG v2.0*. The definition for ME in *MEG v1.0* is the following:

> The deliberate planning, analyzing, organizing, and integrating of current and emerging operational and system capabilities to achieve desired warfighting mission effects. [MEG1.0]

Similar to the *MEG v1.0* definition, the following definition for ME can be found in *MEG v2.0*:

> An interdisciplinary process encompassing the entire technical effort to analyze, design, and integrate current and emerging operational needs and capabilities to achieve desired mission outcomes. [MEG2.0]

The original *MEI Guidebook* by the author provided a detailed process to guide this innovative approach with an effects web framework to incorporate a common framework for understanding across multiple stakeholders. This prescriptive process for MEI was generated based on military service leaders' input due to the in-depth technical rigor required and the unfamiliar territory it navigated within the OSD environment, where instructions and directives were typically written in a general fashion without excessive prescription. While this book will focus on the *MEI Guidebook* approach to be covered in detail in Chap. 8, it directly aligns with the later versions of *MEG v1.0* and *MEG v2.0* which are built upon the foundational work in the *MEI Guidebook* but with less emphasis on detailed execution of the process.

Key milestones in the evolution of MEI include the adoption of advanced architecture frameworks and digital engineering principles. Initially, the *C4ISR Architecture Framework* laid the foundation by establishing a structured methodology to describe and analyze command, control, communications, computers, intelligence, surveillance, and reconnaissance (C4ISR) systems. This framework evolved into the *Department of Defense Architecture Framework* (*DoDAF*) *v2.0*, which expanded its scope to include SoS integration, capability alignment, fit-for-purpose approach focused on authoritative source of data, and traceability to mission outcomes. More recently, the adoption of the *Unified Architecture Framework* (*UAF*) has further advanced MEI by enabling a holistic, model-based approach that integrates operational, systems, and capability viewpoints into a single, cohesive framework. Together, these frameworks have provided the necessary tools to align mission architectures with operational needs while ensuring consistency and interoperability across systems and SoS. Combined with digital modeling and simulation tools, these frameworks have enabled the DoD to create robust, data-driven mission models that simulate end-to-end execution, generating actionable insights for capability development, acquisition, and operational decision-making.

From a policy and statute perspective moving from a blank slate during the introduction of MEI to DoD in 2014 to today, MEI within the DoD (transitioning from the DoD to the Department of War (DoW) in the 2025 timeframe) is driven by a robust set of statutes and policies, beginning with the National Defense Authorization Act (NDAA) for Fiscal Year 2017, specifically *Section 855* (*10 USC 2358*). This specific section of *NDAA 2017* mandates the establishment of Mission Integration Management (MIM) activities for each mission area, designed to provide essential mission-based inputs into key defense processes such as requirements management, concept assessments, design options, budgeting, and program management, as outlined in *DoD Directive* (*DoDD*) *5000.01* (The Defense Acquisition System). A central component of MIM is the conduct of SoS analysis, ensuring the integration and interoperability of systems to meet critical mission requirements. The analysis framework aims to identify operational gaps and develop capabilities that enhance the warfighter's ability to execute mission essential tasks and mission threads. Additionally, *DoD Instruction (DoDI) 5000.88* emphasizes the integration of systems engineering best practices, focusing on a disciplined approach to balance cost, schedule, performance, and risk throughout the acquisition lifecycle. Supporting policies, such as *DoDD 7045.20* (Capability Portfolio Management) and *DoDD 5137.02* (USD(R&E) Charter), further guide the identification of options to close capability gaps and synchronize development efforts, ensuring that the DoD invests in appropriate capabilities to address mission needs effectively. *DoDD 5135.02* (USD(A&S) Charter) and *CJCSI 5123.01J* (Joint Requirements Oversight Council Charter) contribute to the integration of MEI practices, while the role of the USD(R&E) ensures the coordination and alignment of MIM activities with broader defense strategies. The *Supporting the Production of Essential and Emerging Defense (SPEED) Act* and the *Reforming the Joint Requirements Process to Accelerate Fielding of Warfighting Capabilities Memorandum* by the Secretary and Deputy Secretary of Defense dated 20 August 2025, further strengthens this framework by establishing the Mission Engineering and Integration Activity (MEIA) and the Requirements and Resourcing Alignment Board (RRAB). These governance activities help to institutionalize mission-based analysis and resource alignment with a focus on streamlining processes while canceling the legacy Joint Capabilities Integration and Development System (JCIDS). Collectively, these statutes, policies, reforms, and organizational transformations reflect a broader strategic evolution, ensuring that MEI remains central to national security governance, readiness, and the systematic, data-driven optimization of mission capabilities.

Today, MEI is recognized as a critical component of operational and capability planning and capability development while spreading into other industrial sectors and across the international landscape. By focusing on mission outcomes rather than individual systems, it enables stakeholders to identify and address gaps, assess the integration of new technologies, and optimize mission architectures. This mission-centric approach ensures that operational forces are equipped with interoperable and adaptable solutions, ultimately enhancing their ability to execute critical missions in complex and dynamic environments.

1.6 Overview of Book Structure

From the historical roots of MEI to the cutting-edge technologies shaping the future, this first edition of *Mission Engineering and Integration* delves into the heart of mission success. The reader navigates through the intricate landscape of system and mission architectures, interoperability, and data-driven decision-making, unraveling the complexities that Mission Engineers face. Through real-world case studies and a forward-looking lens on emerging trends, this book offers a comprehensive guide for both novices and seasoned professionals. Together, we will unlock the secrets

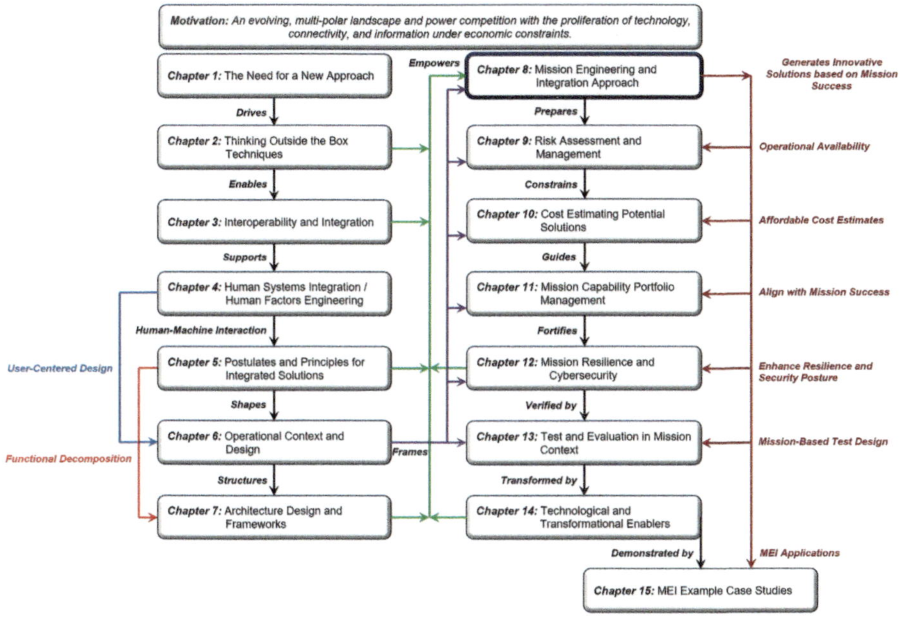

Fig. 1.4 Moving through the chapters

behind the orchestration of successful missions, where strategic planning meets adaptability, and where the pursuit of excellence is grounded in the principles of MEI to the integrated solutions results. Figure 1.4 depicts how this book will describe this new integrated solution approach focused on mission-based design and its implications.

Chapter 1, *The Need for a New Approach*, provided a glimpse of the evolving landscape with an emphasis on the integration of complex SoS required for mission execution toward national security operations.

Chapter 2, *Thinking Outside the Box Techniques*, delves into the significance of system theory, critical thinking, and innovation in shaping problem statements and devising potential solutions for intricate challenges within the context of applying the MEI approach. System theory provides a framework to comprehend the interconnectedness and interdependencies within an SoS, enabling holistic analysis of complex problems and identification of potential solutions. This theoretical framework leads to the need for critical thinking to question assumptions, evaluate evidence, and consider alternative perspectives, thus facilitating informed decision-making based on evidence and reason. Innovation is crucial for devising creative and effective solutions to intricate problems, often involving unconventional approaches and integration of existing technologies or methodologies in novel ways, leading to breakthrough advancements in MEI.

Chapter 3, *Integration and Interoperability*, explores key factors in ensuring effective coordination and communication across systems and operators. The chapter begins by examining historical challenges that have hindered seamless integration and the justification for adopting new approaches. It then outlines the philosophy of integration, emphasizing the role of enabling technologies such as modular open systems architecture and open standards. The discussion extends to interoperability, detailing essential enablers like waveforms, networks, gateways, semantics, and ontologies, as well as the importance of trust, security, and data-driven architectures. Finally, the chapter addresses the significance of global interoperability and the role of international standards in enhancing coordination among allied and partner nations.

Chapter 4, *Human Systems Integration/Human Factors Engineering*, discusses the significance of human factors and human systems integration within the context of MEI. Human Systems Integration (HSI) and Human Factors Engineering (HFE) are both critical for optimizing human performance and ensuring effective interaction between personnel and technology. The chapter begins with an overview of HSI, including its application to commercial-off-the-shelf products and the development of HSI-based requirements. It then explores HFE, addressing human abilities, concerns, and requirements, as well as practical applications in system design. A key aspect of the discussion is human-machine teaming, which examines how humans and automated systems can collaborate effectively to enhance mission effectiveness and operational outcomes. The chapter also delves into cognitive workload and human performance, outlining methods for assessing performance through behavioral and physiological measures. Together, these topics provide a comprehensive understanding of how human-centered design principles can enhance operational effectiveness and safety.

Chapter 5, *Postulates and Principles for Integrated Solutions*, explores the foundational principles that support MEI, emphasizing their importance in addressing complex operational challenges. The chapter begins by defining key concepts and presenting a systems engineering model that serves as a framework for integration. It then discusses the role of modeling and simulation (M&S), along with model-based systems engineering (MBSE), in enhancing decision-making and optimizing mission success. A central focus is the articulation of postulates and practical principles that guide the application of MEI in real-world scenarios. The chapter also highlights the necessity of adopting an MEI approach, detailing critical focus areas such as mission success measures, SoS analyses, design flexibility, and multidisciplinary analysis and optimization (MDAO). By integrating these elements, the chapter underscores the imperative of improving decision-making speed and effectiveness to maintain a competitive advantage in dynamic environments.

Chapter 6, *Operational Context and Design*, examines the principles of force design and Command and Control (C2) within a dynamic operational environment. The chapter emphasizes the role of continuous Threat Evaluation and Weapon Assignment (TEWA) analyses in shaping adaptive and resilient force structures. By leveraging universal mission command and multi-domain operations, this approach seeks to enhance force adaptability while creating complexity and uncertainty for adversaries. The chapter explores the philosophy behind operational planning, the competitive threat landscape, and the integration of TEWA processes with C2. It also delves into intelligence operations, detailing information processing, intelligence products, and the development of realistic operational scenarios. Additionally, the chapter discusses human intelligence modeling, governance structures, and strategies for effective decision-making across organizational frameworks. Through these elements, *Operational Context and Design* outlines a framework for enhancing operational effectiveness in complex and contested environments.

Chapter 7, *Architecture Design and Frameworks*, explores key principles and structures essential for designing robust system and mission architectures. It begins with an overview of architecture design and the role of reference architectures in standardizing solutions. The chapter categorizes architecture into different levels, including enterprise and mission architectures, and examines various architecture frameworks that provide structured methodologies. It introduces the *Integrated Capability Framework* (*ICF*), which aligns governance, mission models, and system/platform models to enhance mission success. The chapter further details methods for decomposing mission success criteria and aligning system and mission objectives. Finally, it discusses emerging trends in architecture tools, highlighting future advancements in architectural design and integration.

Chapter 8, *Mission Engineering and Integration Approach*, serves as the premier chapter of the book, introducing a structured framework for developing integrated solutions that align with mission success criteria. This chapter establishes the foundational concepts of MEI and explores the relationship between MEI and both systems and SoS engineering, as well as its implications for DoD acquisition processes. Focusing on mission decomposition techniques is key, and includes identifying mission essential tasks, and structuring mission engineering threads or webs using common frameworks such as the Effects Web Framework. Additionally, the chapter examines how MEI can help anticipate and address potential future national security challenges. Finally, it outlines a systematic 10-step approach for conducting MEI, culminating in the development of effective Concepts of Operations (CONOPS) and mission engineering threads or webs for specific missions. Through these elements, *Mission Engineering and Integration Approach* provides the foundation for applying MEI principles to complex operational environments, ensuring a cohesive and adaptable strategy for mission success.

Chapter 9, *Risk Assessment and Management*, delves into the critical processes of identifying, assessing, and mitigating risks within the framework of MEI. The chapter examines risk attitudes and introduces methodologies for effective risk management, emphasizing strategies such as risk identification, mitigation through burndown activities, and the safeguards necessary to maintain operational availability. A comprehensive risk and issue management framework is outlined, which includes a multi-step process: planning, identification, analysis, mitigation or correction, and continuous monitoring. The chapter also explores various categories of risk, particularly those relevant to mission execution, and presents approaches for developing risk measures. Additionally, it highlights the role of uncertainty in risk management and introduces the emerging field of digital risk management, emphasizing its relevance for the future of operational risk assessment. Through these strategies, the chapter provides a systematic approach to managing risks and ensuring the resilience and success of complex missions.

Chapter 10, *Cost Estimating Potential Solutions*, provides a comprehensive look at cost estimating processes essential for effective decision-making in MEI. This chapter covers various methods and tools used to estimate costs at different stages of the program lifecycle, from initial rough-order-of-magnitude (ROM) estimates to detailed budget quality assessments. Key topics include the Life Cycle Cost Estimate (LCCE), cost estimating classifications, and specific approaches such as Class A and Class B cost estimates for new construction and conversion projects. The chapter addresses feasibility cost estimates and ROM estimates and outlines the step-by-step cost estimating process across three phases: project definition, cost methodology, and cost estimating tasks. Additionally, the chapter discusses cost-risk and uncertainty, examining methodologies used for assessing and managing cost-related risks, as well as the critical distinction between risk and uncertainty in cost estimation.

Chapter 11, *Mission Capability Portfolio Management*, delves into the development and application of capability portfolio management in the context of MEI. The chapter emphasizes the importance of continuously monitoring and assessing the health of mission portfolios to ensure alignment with strategic objectives. By systematically evaluating performance metrics, capability effectiveness, and identifying areas for improvement, portfolio management enables organizations to realign or adjust capabilities to meet mission goals. The chapter also introduces the concept of elevating the trade space to the SoS level, promoting a holistic approach that considers the interdependencies and interactions among diverse systems. This approach prevents the suboptimization of individual systems by ensuring that decisions made for one system take into account their broader impact on the entire SoS. As a result, this integrated approach helps achieve more effective and efficient mission outcomes, enhancing the overall performance and success of mission portfolios.

Chapter 12, *Mission Resilience and Cybersecurity*, addresses a crucial aspect of MEI by focusing on the importance of mission resilience in the face of cyber threats. As modern missions rely increasingly on interconnected systems, ensuring the robustness and continuity of mission activities becomes dependent on effective cybersecurity measures. This chapter explores the science behind cybersecurity, detailing how to build resilience against cyber threats through robust defense techniques, such as Zero Trust Architecture and the Cyber Kill Chain. It also examines the policies that drive cybersecurity initiatives and the integration of risk management strategies to enhance cyber resiliency. By highlighting the critical role of cybersecurity in safeguarding mission success, this chapter underscores the need for proactive measures to protect against evolving cyber threats and ensure operational continuity.

Chapter 13, *Test and Evaluation for Mission Resilience*, examines the critical role of Test and Evaluation (T&E) in ensuring mission success within complex SoS. This chapter emphasizes the importance of rigorous T&E to verify the effectiveness and reliability of mission capabilities across diverse operational environments. A key focus is placed on incorporating Design of Experiments (DoE) methodology and the evolving *T&E as Continuum*, which adapts to dynamic operational needs. It explores the shift toward a Joint Test Concept (JTC), enabling integrated assessments across multiple systems and domains. Additionally, the chapter highlights the application of mission-based T&E, defined by mission success criteria and supported by key performance metrics, such as Measures of Success (MoS), Measures of Effectiveness (MoEs), and Measures of Performance (MoPs). Through a combination of M&S, integrated testing approaches, and specialized cybersecurity evaluation techniques, this chapter provides a comprehensive understanding of how T&E processes can be leveraged to ensure resilient, effective, and mission-ready capabilities.

Chapter 14, *Technological and Transformational Enablers*, in conclusion, focuses on the key technologies and innovations that enhance the MEI approach. This chapter examines how advancements such as artificial intelligence (AI), machine learning (ML), digital engineering (DE), and blockchain contribute to improving decision-making, operational effectiveness, and overall mission resilience. It highlights how these enablers help forces present a dynamic range of challenges to adversaries, making it more difficult for them to succeed. The chapter discusses the application of AI/ML in MEI, the strategic use of DE to support integration and transformation, the role of blockchain in secure data management, and the unifying mathematical framework of category theory. Additionally, it explores the importance of authoritative data sources, data security, and the adoption of cutting-edge tools to ensure the continuous evolution and effectiveness of the MEI framework. Ultimately, the chapter emphasizes how these transformative technologies enable forces to stay ahead of threats and rapidly adapt to changing operational needs.

Chapter 15, *Mission Engineering and Integration Example Case Studies*, illustrates how the MEI 10-step process is applied in real-world settings through three practical examples: the Hurricane Katrina Search and Rescue mission, NASA's Moon to Mars initiative, and the Ohio Health Information Exchange known as *CliniSync*. These case studies highlight how MEI principles, analytical methods, and mission-focused frameworks can be translated into concrete actions that support mission success across humanitarian response, national space exploration, and statewide healthcare integration. By examining each case through mission decomposition, task identification, mission thread development, and final evaluation, readers gain a clear understanding of how structured MEI approaches improve decision-making and integrated solution design. Including these diverse examples is essential because they demonstrate the broad applicability of MEI, show its relevance across multiple sectors, and reinforce how theory becomes practice when addressing complex, multi-stakeholder mission challenges. Together, they provide a powerful foundation for understanding how the MEI approach delivers coherence, clarity, and measurable mission outcomes in a wide range of operational environments.

The MEI approach is applicable across various industrial sectors where SoS are required to execute objectives through mission essential tasks. Therefore, this book has been written using general terms like "effects web" instead of defense-specific terms like "kill chain." This broader terminology ensures the concepts and methodologies presented are relevant and accessible to a wide range of industries and applications by examining specific contexts that reflect their differing objectives, environments, and operational paradigms. Across industrial sectors, the detection phase focuses on identifying inefficiencies, bottlenecks, threats, or anomalies using sensors, Internet-of-Things devices, data analytics, and AI to ensure real-time awareness of system health. Decision-making involves analyzing data with machine learning and predictive analytics to optimize resource allocation and improve workflows. Solutions are then implemented through automation and advanced manufacturing techniques, enhancing operational efficiency and output quality. Lastly, continuous assessment through monitoring and feedback ensures sustained operational excellence. This book and its accompanying research are built on past DoD experiences that have demonstrated improved effectiveness and efficiency in fielding integrated and interoperable SoS using an MEI approach or similar mission-based concepts.

As we venture into Chap. 2, multiple techniques are considered for the creation of innovative ideas where the convergence of complex SoS demands a multifaceted approach to problem-solving. As operational missions evolve to encompass interconnected networks, technologies, and stakeholders, the need for dynamic thinking becomes paramount. This next chapter will uncover how fostering a mindset that embraces critical analysis, imaginative exploration, and inventive solutions is not only advantageous but imperative for effectively addressing the complexities inherent in MEI. Through the lens of critical, creative, and innovative thinking, the aim is to illuminate the pathways to success in designing and integrating comprehensive solutions that meet the ever-evolving demands of operational mission execution.

References

1. Anderson, S. E., Baron, N. T., Garrett, R. K., Jr., & Moreland, J. D., Jr. (2011). Managing the interstitials, a system of systems framework suited for the ballistic missile defense system. *Systems Engineering, 14*(1), 87–109.
2. Department of Defense. (2020). *Mission engineering guide v1.0*. Office of the Under Secretary of Defense for Research and Engineering. Retrieved from https://ac.cto.mil/wp-content/uploads/2020/12/MEG-v40_20201130_shm.pdf
3. Department of Defense. (2023). *Mission engineering guide v2.0*. Office of the Under Secretary of Defense for Research and Engineering. Retrieved from https://ac.cto.mil/wp-content/uploads/2023/11/MEG_2_Oct2023.pdf
4. Moreland Jr, J (2019). *Mission engineering and integration guide*. Retrieved from https://www.ise.vt.edu/content/dam/ise_vt_edu/adjunct/MEI%20Guidebook_Nov%202019.pdf
5. Moreland, J., Ambler, N., LaMar, C., Nix, M., & Sawyer, C., (2026). Accelerating Mission Engineering: Evidence of a Stratified Landscape and a Path Toward Unified Practice. *Systems Engineering*, e70076. https://doi.org/10.1002/sys.70076

Chapter 2
Thinking Outside the Box Techniques

2.1 The Cognitive and Strategic Shift to Decision Superiority

Foundational advanced thinking techniques like systems theory, critical thinking, creative thinking, and innovation serve as driving forces behind Mission Engineering and Integration (MEI) solutions, fostering out-of-the-box thinking. Systems theory provides a framework for understanding complex interdependencies and optimizing system performance. Critical thinking enables rigorous analysis and evaluation of problems, leading to insightful solutions. Creative thinking encourages novel idea generation and exploration of unconventional approaches. These techniques drive innovation that leads to groundbreaking solutions by leveraging new technologies and methodologies. Together, these techniques empower MEI practitioners to devise state-of-the-art and effective solutions that address complex mission challenges with creativity and ingenuity.

To maintain a competitive edge in an increasingly complex landscape, organizations must adopt a new approach that prioritizes adaptability, understands and incorporates mission success criteria, examines system-of-systems (SoS) functionality and dependencies, increases decision speed, and includes uncertainty factors. Relying on well-known strategies and operational models that competitors have already adopted limits innovation and erodes long-term advantages. Just as the US military leveraged emerging technologies and novel concepts during the Cold War to counter numerical disadvantages, today's competitive environment demands a shift from traditional, large-scale strategies to more dynamic and distributed models. For survival, industry leaders must utilize integrated SoS approaches and indirect tactics to exploit vulnerabilities. Similarly, in business and technological competition, dominant players leverage data-driven strategies, automation, and agile methodologies to outmaneuver rivals. A decision-centric approach, enabled by artificial intelligence (AI), automation, and decentralized execution, can provide a strategic advantage by allowing organizations to operate with greater flexibility,

J. D. Moreland, Jr., *Mission Engineering and Integration*,
https://doi.org/10.1007/978-3-032-32132-9_2

impose multiple challenges on competitors, and sustain a prolonged lead in innovation. By focusing on rapid adaptation and leveraging emerging technologies for distributed decision-making, organizations can transition from a sustained competitive depletion strategy to one that ensures enduring strategic dominance.

This new approach to competition is characterized by adaptability, composability, and decision speed, shifting away from rigid, monolithic strategies toward a more dynamic and distributed model with the ability to analyze and construct complex mission engineering webs. By decomposing large, multi-functional units into a greater number of smaller, specialized elements, organizations increase operational flexibility and create a more unpredictable environment for competitors. This modular and reconfigurable strategy allows rapid integration of new technologies, enhancing agility and innovation. Additionally, it enables more precise resource allocation, reducing inefficiencies and ensuring that competitive actions are tailored to specific challenges. By presenting multiple simultaneous challenges to adversaries, this approach raises the complexity of competitive decision-making, making it harder for opponents to predict strategies and counter them effectively. The implementation of AI and autonomous systems further accelerates decision cycles, allowing for real-time adaptation and disruption of competitors' strategic orientation. This strategy seeks to overwhelm adversaries by forcing them into a state of confusion and reactive decision-making by ensuring that organizations maintain a first-mover advantage by outpacing rivals. The execution and strategic evolution of this strategy requires the MEI approach to decompose complex missions and their respective mission engineering webs to understand and build effective SoS capabilities.

The Observe-Orient-Decide-Act (OODA) loop, developed by military theorist John Boyd, is a foundational concept in decision-centric operations, emphasizing the importance of rapid and adaptive decision-making in competitive environments. Boyd broke down the decision process into four key phases: **observation**, which involves gathering information about both adversary and friendly forces; **orientation**, where the gathered data is analyzed to determine the adversary's intentions and possible actions; **decision**, in which a course of action (COA) is developed and chosen; and **action**, where the selected COA is executed. Boyd argued that the key to gaining a decisive advantage is not merely acting faster, but disrupting the adversary's ability to orient, thereby slowing and ultimately collapsing their decision cycle, see Fig. 2.1.

This focus on disrupting orientation is particularly relevant today, as modern sensor networks make it increasingly difficult to prevent adversaries from observing military or competitor movements. By embracing a decision-centric approach, US forces create ambiguity and complexity, making it harder for adversaries to determine targets, intentions, and operational strategies. Implementing this concept fully requires shifts in force structure, command and control (C2) processes, and operational system design, moving away from static architectures and toward dynamic, reconfigurable effects webs that maximize adaptability and unpredictability in combat. Chapter 6 provides more detail on John Boyd's foundational concept of the OODA loop.

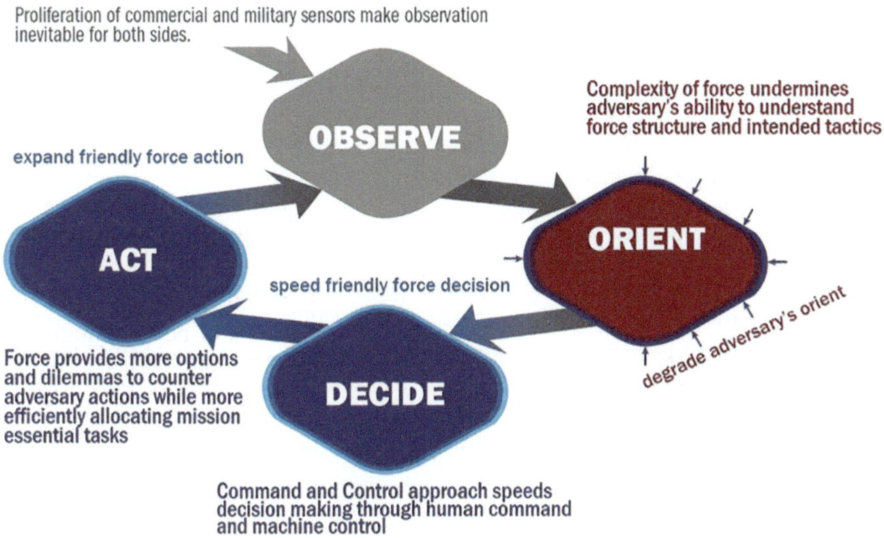

Fig. 2.1 Impacting the OODA Loop in new approach

2.2 Systems Theory

Systems theory provides a holistic understanding of the interconnectedness and interdependencies within complex systems, enabling MEI practitioners to analyze problems from a systemic perspective. By considering the broader system dynamics and feedback loops, practitioners can better identify root causes, anticipate potential impacts, and design more robust and resilient solutions. This interdisciplinary field of study is scalable and explores the relationships among systems within larger, more intricate SoS. Regardless of the specific discipline, its central tenet is that the collective whole surpasses the mere sum of its individual parts [EISNER05]. In essence, when analyzing how individual systems interact to impact the overarching complex SoS, certain traits of the whole cannot be readily understood by examining any single component in isolation. Systems theory endeavors to elucidate and formulate hypotheses concerning characteristics that emerge within complex SoS, ones that seemingly cannot be observed within any isolated system. This phenomenon is termed emergent behavior, signifying that a complex SoS possesses qualities that its individual components do not exhibit independently.

Certain disciplines, such as engineering, exhibit a wide range of systems theory. Collectively, its concepts provide a basis for understanding systems not merely as collections of components, but as interacting structures whose relationships, feedback mechanisms, adaptation, and behavior must be considered at the system level [EISNER13]. Key terms and concepts of systems theory as they pertain to engineering, psychology, and sociology include:

- **System**: A structure comprising interconnected or interdependent parts.

- **Complex SoS**: The overarching SoS capability composed of smaller, individual systems.
- **Homeostasis**: The state of stable conditions within a system or SoS, with these entities consistently moving toward equilibrium.
- **Adaptation**: The tendency of a system or SoS to adjust in response to new environmental factors to ensure its protection.
- **Feedback loop**: The occurrence of a system's or SoS's outputs impacting its inputs and causing the system to cyclically influence itself [FORRESTER68].

The origins of modern systems theory can be traced back to the post-World War II era, coinciding with significant technological advancements of that time. As human interactions with emerging technologies increased, a necessity arose for a deeper comprehension of human behavior in relation to the surrounding mechanisms, whether machine-driven or otherwise. This demand spurred the development of signal detection theory, driven by the imperative to devise more efficient radar and sonar systems, ensuring military personnel could effectively discern between diverse signals. The initial impetus for a comprehensive general systems theory emerged from the research efforts of Ludwig von Bertalanffy during the 1940s and 1950s. His pioneering work culminated in the publication of "General System Theory: Foundations, Development, Applications" [BERTALANFFY68], marking a significant milestone in the formalization and dissemination of systems theory principles. The aim of his book was to delineate fundamental laws that apply across a broad spectrum of scientific domains. Through his lens, Ludwig von Bertalanffy perceived the cyclical interplay among individual components within a complex SoS, wherein each element both influences and is influenced by the interdependencies within the SoS, offering valuable insights across diverse contexts. Building upon this premise, Ludwig von Bertalanffy posited the existence of universal principles or guidelines that could harmonize disparate fields, particularly the natural and social sciences, fostering a more cohesive educational framework. He succinctly summarized systems theory as follows:

> *General system theory, therefore, is a comprehensive study of entirety. ...The essence of the somewhat enigmatic phrase, 'The whole is more than the sum of its parts,' lies in the fact that the constituent attributes cannot be fully explained solely by analyzing the individual elements. As a result, the characteristics of the complex system manifest as novel or emergent.* [BERTALANFFY68]

This holistic approach, combined with the empathetic and creative aspects of design thinking and the analytical rigor of critical thinking, empowers MEI practitioners to develop comprehensive and effective solutions that address the multifaceted challenges of mission-critical systems. The principles of systems theory provide an important foundation for systems engineering, translating systems thinking into structured methods for defining, designing, integrating, and evaluating successful systems [EISNER11]. For MEI, this perspective extends beyond the performance of individual systems to consider the different sources of complexity, how systems interact within the operational environment, and how these interactions collectively contribute to mission outcomes, requiring the Mission Engineer to understand the nature of the complexity when evaluating the overall SoS [SHEARD11]. Systems thinking also provides a practical means for addressing real-world problems by considering the relationships, interactions, and broader context that influence system behavior [MONAT17]. A critical first step in performing the MEI approach is the accurate framing of the complex problem domain. Systems thinking emphasizes that complex problem situations should be examined within their broader context, including the perspectives, interests, and interactions of the stakeholders involved [CHECKLAND99]. Finding the appropriate methodology ensures the operational analyses and engineering design progresses effectively. One powerful tool identified is the "6 + 1 Questions" framework [KEATING15]. This methodology proves essential for mission engineers aiming to incorporate the operational context into SoS Engineering (SoSE) and helps formulate the mission capabilities and requirements from the operational concepts developed within the defined context. The key to avoiding analytical or design misalignments lies in using a structured approach and the 6 + 1 questions delineated below.

The first question to address is: "**What is the problem or need?**" Defining the problem or need is foundational to the analytical process. Failing to accurately articulate the problem leads the mission engineer down a misguided path, similar to a pyramid erected on a poorly constructed base. The mission engineer must ensure the problem is clearly defined to establish a strong foundation for the rest of the effort.

Next, the researcher must ask: "**What are the interests, values, and management strategy for stakeholders?**" Understanding the stakeholders' interests and values ensures that the mission engineer is designed to provide relevant data that is of practical use to the involved parties. The identification of stakeholders also enables the mission engineer to involve the right individuals on the MEI team, ensuring that the analytical effort serves its purpose.

The third question is: "**What is the relevant context?**" Context refers to the broader set of circumstances influencing the design, deployment, operation, and maintenance of complex systems or SoS. The mission engineer must define the boundaries of this context, which may include factors like human behavior, social and economic conditions, political influences, technical aspects, or policies. A clear understanding of the context ensures that the analytical process is appropriately grounded.

The fourth question is: "**What constitutes success or failure of the complex system, SoS, and the complex system endeavor?**" Success or failure needs to be defined early in the MEI analytical process to evaluate whether the objectives have been met. This includes not only analytic performance but also how well the results are applied in practice. By setting clear benchmarks for mission success, the mission engineer can measure the value and impact of the MEI approach.

The fifth question is: "**Is the approach sufficiently compatible to proceed with the complex system or SoS endeavor**?" This question ensures that the MEI approach aligns with the problem context, the mission objectives, and the worldview of the stakeholders. Compatibility between the MEI approach and the operational environment is essential to producing valid and actionable results. If the approach is not suitable, the mission engineer may need to reconsider or adjust the methodology.

The sixth question is: "**What are the likely systemic failure modes for the complex system or SoS as well as the complex system endeavor?**" Acknowledging

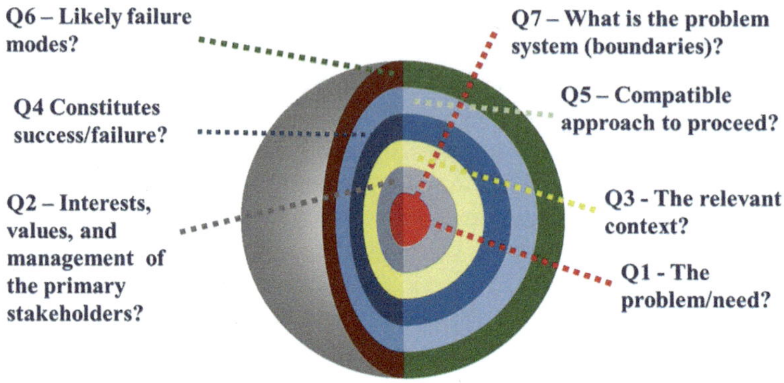

Fig. 2.2 6 + 1 Important systemic questions for MEI approach

the potential failure modes, whether in the system itself (e.g., faulty communication) or in the MEI (e.g., flawed data collection methods), helps the mission engineer anticipate challenges and refine the approach. Failure is not inherently negative; it often provides insights that guide the MEI effort in a more fruitful direction.

Finally, the "+1" question asks: "**What is the complex system or SoS, including boundaries and environment?**" This question prompts the mission engineer to define the systems' boundaries, the elements they comprise, and their environment. By addressing the interrelationships within and across the systems and understanding how external factors may influence their operation, the mission engineer is better prepared to identify key systems, components, and areas for exploration in the MEI effort.

Once these 6 + 1 questions are thoroughly explored as depicted in Fig. 2.2, the mission engineer has a solid framework to begin addressing the complex problems of MEI. A clear, structured approach not only clarifies the problem but ensures that the MEI effort remains on track, offering valuable insights into the mission capabilities and requirements. This methodology prevents the MEI approach from deviating from its original goals, thus providing a robust foundation for tackling complex engineering challenges.

2.2.1 Complexity

Complexity, when viewed from the perspective of an SoS, embodies the intricate interplay and interdependence among the constituent systems within a larger interconnected network. It encompasses the nuanced relationships, emergent phenomena, and dynamic interactions that emerge from the collective behavior of the individual systems. This perspective is consistent with Weaver's concept of organized complexity, in which numerous interrelated factors interact in ways that require consideration of the collective system rather than its individual elements in isolation [WEAVER48]. Within an SoS, complexity arises from the interconnectedness of systems, where changes in one system can propagate throughout the entire network, leading to feedback loops and nonlinear relationships. Moreover, the emergent properties that manifest at the level of the SoS—which cannot be fully understood by examining the individual systems in isolation—contribute to its complexity. This dynamic nature of SoS, coupled with their heterogeneous composition comprising diverse systems with varying structures and functions, introduces challenges in terms of interoperability, coordination, and alignment of goals. Additionally, the uncertainty and risk inherent in SoS operations further amplify their complexity, requiring adaptive and resilient management approaches.

An example could be the development of a national missile defense system, which integrates multiple, interconnected systems such as radar, satellite surveillance, communication networks, and interceptor missiles. Each component may function well individually, but their collective behavior introduces uncertainties due to interactions, environmental variables, and evolving threats. By embracing flexible designs, continuous testing, and the ability to rapidly adjust to new information, uncertainties in such a complex SoS can be managed effectively using a combination of approaches to include modeling and simulation for risk assessment, redundancy and fail-safes, incremental testing and iteration, and agile system upgrades.

Therefore, complexity in an SoS perspective encompasses the multifaceted dynamics, emergent properties, heterogeneity, and uncertainties inherent in the interconnected network of systems, emphasizing the need for comprehensive understanding and effective management strategies to navigate and harness their potential.

2.2.2 Balancing Complexity and Scale

A paradox arises from the contrast between the apparent simplicity of each mission essential task when viewed from the perspective of individual systems and the inherent complexity it entails within the broader context of SoS integration necessary for mission execution. To reconcile this seeming paradox, the MEI practitioner must acknowledge that the complexity of a system is contingent upon the level of detail employed in its description, thus varying with scale. This entropy signifies the complexity at the smallest scale, yet a comprehensive understanding of a system necessitates grasping its complexity across multiple scales. A system's complexity profile illustrates its complexity relative to scale. Although scale is often measured in length, it essentially reflects the number of coordinated components contributing to a behavior, with physical length serving as a proxy.

Figure 2.3 depicts how random, coherent, and correlated systems exhibit distinct complexity profiles. Random systems display peak complexity at the smallest scale, characterized by the finest granularity and highest level of detail. However, as the scale increases and individual component behaviors converge, the complexity diminishes rapidly. In contrast, coherent systems maintain consistent complexity across scales; the overall behavior of the system also encompasses the behavior of its components. For coherent systems, all components exhibit the same behavior; for

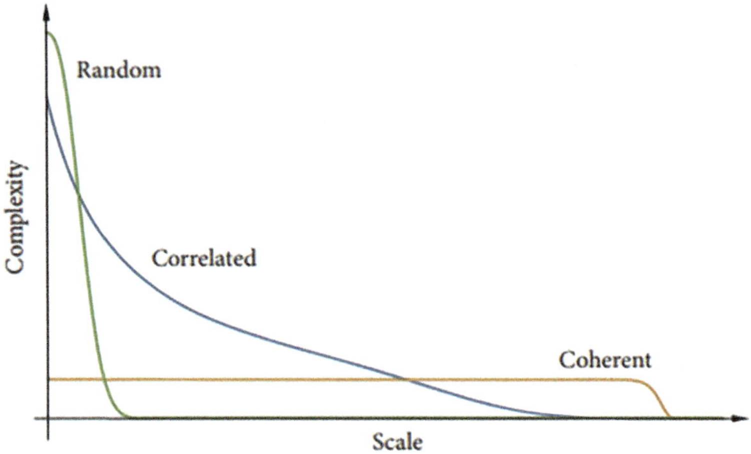

Fig. 2.3 System types regarding complexity

example, the behavior (location, orientation, and velocity) of one part of the system completely determines the behavior of the other parts. Notably, complexity tends either to increase or remain constant as the scale decreases, as examining a system in finer detail yields more information while still encompassing the entire system. Correlated systems, on the other hand, manifest various behaviors across different scales, resulting in a gradual increase in complexity with greater levels of detail. Each level of detail expands the length of the description, illustrating the multiscale structure and gradually increasing complexity characteristic of complex systems.

2.2.3 *Mathematical Framework for Complexity*

A significant gap exists in contemporary mathematical discourse, particularly concerning complex systems theory. Complex systems theory deals with systems composed of many interacting components, where the behavior of the system as a whole cannot be easily predicted from the behavior of its individual parts. While this area of study has gained considerable attention in various disciplines, including physics, engineering, biology, economics, and sociology, there remains a lack of a unified mathematical framework to fully understand and analyze these systems. One primary mathematical issue raised by complex systems theory is its challenge to standard analytical mathematical dynamics. Traditional analytical mathematical approaches, which often rely on linear equations and deterministic models, struggle to capture the nonlinear and emergent behaviors characteristic of complex systems. Addressing these challenges requires new mathematical tools and methodologies.

Without a coherent mathematical framework, researchers are left with a fragmented collection of specific models and mathematical approaches, leading to a lack of cohesion and synergy in research efforts. Moreover, the absence of a unified mathematical underpinning for complex systems theory poses a significant obstacle to interdisciplinary collaboration. Without a common language and framework, researchers from different fields may struggle to communicate and integrate their findings effectively. This results in a simulation challenge for future research. By acknowledging and confronting the mathematical challenges posed by complex systems theory, researchers can pave the way for the development of new mathematical tools and frameworks that better capture the dynamics of complex systems, including self-organized emergence and other key features. Addressing these challenges is crucial for advancing our understanding of complex systems for MEI and addressing real-world problems in various domains.

Functional Dependency Network Analysis (FDNA) is a powerful method for analyzing and understanding complex, interdependent systems, particularly within an SoS architecture. In the context of performing mission essential tasks for mission execution, FDNA helps to identify, model, and assess the relationships, dependencies, and impacts among different system components, revealing potential risks and optimizing performance. The step-by-step process for FDNA is:

1. **Define the SoS and Mission Essential Tasks (METs)**
   - **SoS Identification**: Identify the various systems involved in the mission, like command and control, communication, intelligence, logistics.
   - **Mission Essential Tasks**: List all critical tasks needed to achieve the mission goals. METs may include tasks such as surveillance, reconnaissance, target acquisition, and secure communications.
2. **Identify Functional Dependencies**
   - **Analyze Inter-System Dependencies**: For each system and MET, identify dependencies and interdependencies. For example, communications may rely on satellites, while intelligence gathering could depend on drones and data-analytics systems.
   - **Determine Input-Output Relationships**: Identify what each system requires as input (resources, data, etc.) from other systems and what output it provides to others. These dependencies are crucial for mapping functional interactions.
3. **Construct the Functional Dependency Network**
   - **Network Nodes**: Represent each system and MET as nodes in a network.
   - **Directed Edges for Dependencies**: Draw directed edges between nodes to show dependencies, with the direction indicating the flow of dependency (from provider to consumer).
   - **Dependency Strength and Type**: Label each edge with the strength or type of dependency (e.g., critical, moderate, or minor; data, resource, or communication dependency).
4. **Analyze Dependency Strength and Vulnerabilities**
   - **Assess Node and Edge Strength**: Evaluate how critical each node (system or MET) and edge (dependency) is to mission success. This can be quantified based on factors like failure probability, resilience, or operational importance.

- **Identify Vulnerabilities**: Nodes or edges with high dependency and low resilience represent potential vulnerabilities. For example, a single-point dependency like a centralized server for data processing might be a risk if it lacks redundancy.

5. **Evaluate System Performance Under Different Conditions**

   - **Simulate Scenarios**: Model various scenarios to assess how the SoS responds to stressors (e.g., node failure, cyber-attack, or resource scarcity). FDNA helps simulate the impact of these disruptions by visualizing how failure in one system affects others.
   - **Impact on METs**: Analyze how disruptions affect the METs. This can reveal how resilient the SoS is in achieving mission objectives even if certain dependencies fail.

6. **Optimize and Enhance SoS Resilience**

   - **Redesign or Introduce Redundancies**: For critical dependencies, explore options for redundancy (e.g., additional backup systems) or redesign interdependencies to reduce the risk of single-point failures.
   - **Adjust Resource Allocation**: Prioritize resources toward systems or tasks with high criticality to ensure they have the necessary robustness to withstand operational stresses.
   - **Develop Contingency Plans**: Use the insights gained from FDNA to create response strategies for potential failures in the SoS.

7. **Continuous Monitoring and Updating**
   - **Ongoing Analysis**: Regularly update the FDNA model with real-time data, adjusting for any changes in the SoS structure, mission objectives, or technology upgrades.
   - **Track System Performance**: Use FDNA for ongoing monitoring during missions to detect dependency shifts or emerging vulnerabilities, allowing for real-time adjustments if necessary.

   Practical Application Example: In a mission scenario, imagine that a reconnaissance task relies on both an Unmanned Air Vehicle system for data collection and a satellite for communication. FDNA can model this dependency network to reveal that if the satellite experiences a disruption, the entire reconnaissance task is at risk. By identifying this in advance, planners might add an alternative communication relay or prioritize satellite protection measures to mitigate this vulnerability.

Benefits of FDNA for SoS in Mission Planning

- **Enhanced Resilience**: By understanding dependencies and potential failure points, FDNA aids in building a more resilient SoS for mission-critical tasks.
- **Improved Decision-Making**: FDNA provides actionable insights into where to allocate resources and focus risk mitigation efforts.
- **Proactive Risk Management**: By simulating various operational conditions, FDNA helps preemptively identify and address vulnerabilities before mission execution.

2.2.3.1 Mathematical Formulation

Using FDNA with MoEs and MoPs as success criteria to develop the logical architecture of the SoS, Bayesian Networks for modeling dependencies combined with Genetic Algorithms (GAs) and Pareto fronts for multi-objective optimization can be used to evaluate the effectiveness. Pareto optimization allows the engineer to evaluate trade-offs between competing objectives (e.g., maximizing mission effectiveness and minimizing risk or resource cost). The generic step-by-step mathematical formulation and solution approach follows:

1. **Define Key Variables and Sets**

   - **Systems and Tasks**:
     - $S = \{S_1, S_2, \ldots, S_n\}$: Set of systems in the SoS.
     - $T = \{T_1, T_2, \ldots, T_m\}$: Set of Mission Essential Tasks (METs).
   - **Dependencies**:
     - $D \subset S \times S$: Set of directed dependencies between systems, where $(S_i, S_j) \in D$ indicates that S_j depends on S_i.

2. **Define MoEs and MoPs**

   - **MOEs**: Measures of Effectiveness evaluate the achievement of each mission task T_k based on relevant systems' performance.

$$\mathrm{MoE}_{Tk} \in [0,1]$$

   - **MOPs**: Measures of Performance assess each system's functional performance in executing its role.

$$\mathrm{MoP}_{Si} \in [0,1]$$

3. **Bayesian Network Model for Dependencies**
   Bayesian Networks are used to capture probabilistic dependencies between systems:

   - **Conditional Probability $P(\mathbf{MoP}_{Sj}|\mathbf{MoP}_{Si})$**: Expresses the probability of S_j's performance given the performance of $S_{i,}$ modeling uncertainty in dependencies.
   - **Bayesian Update Rule**: With new evidence, the probability distribution of $\mathrm{MoP}_{\mathrm{Sj}}$ is updated via Bayes' theorem:

$$P\left(\mathrm{MoP}_{Sj} | \mathrm{MoP}_{Si}\right) = \left[P\left(\mathrm{MoP}_{Si} | \mathrm{MoP}_{Sj}\right) \cdot P\left(\mathrm{MoP}_{Sj}\right)\right] / P\left(\mathrm{MoP}_{Si}\right)$$

This enables propagation of changes across the network, updating downstream performance probabilities and reflecting impacts on system-level dependencies.

4. **Aggregating System Performance into Task Effectiveness (MoEs)**
   For each mission task T_k, the effectiveness MoE_{Tk} is calculated based on the probabilistic performance of relevant systems:

   - **MOE Calculation**:
   $\text{MoE}_{Tk} = g_k(\alpha_{ik} \cdot \text{MoP}_{Si})$ where $g_k(\cdot)$ aggregates the impact of system performance on T_k and α_{ik} denotes the degree to which S_i impacts T_k.
   - **Expected MoE**: Using Bayesian probabilities, the expected effectiveness of task T_k is:

$$E\left[\text{MoE}_{Tk}\right] = \sum_{\text{MoP}_{Si}} P\left(\text{MoP}_{Si}\right) \cdot \alpha_{ik} \cdot \text{MoP}_{Si}$$

5. **Define Multi-Objective Optimization Problem with Pareto Fronts**
   The goal is to optimize the configuration of the SoS with respect to multiple objectives, such as maximizing mission effectiveness and minimizing system dependency risks or resource usage. Pareto fronts help identify optimal trade-offs between conflicting objectives.

   - **Objective Functions**:

     Maximize Overall Mission Effectiveness: $f_1(\text{Chromosome}) = \sum_{k=1}^{m} E[\text{MoE}_{Tk}]$

     Minimize Dependency Risk: $f_2(\text{Chromosome}) = \sum_{(S_i, S_j) \in D} \text{Risk}_{ij}\, w_{ij}$

     Minimize Resource Cost (if resources are constrained): $f_3(\text{Chromosome}) = \sum_{i=1}^{n} \text{Cost}_{Si}$

     Here, w_{ij} is the weight of the dependency between S_i and S_j, representing how crucial it is, and Risk_{ij} indicates the risk factor for each dependency.

   - **Pareto Front**: A Pareto front is the set of all non-dominated solutions, where a solution is non-dominated if no other solution improves one objective without worsening another. The Pareto front allows us to analyze trade-offs and choose configurations that balance multiple objectives.

6. **Optimization Using Genetic Algorithms with Pareto Front Analysis**

   (a) **Chromosome Representation**:

   - Each chromosome represents a configuration of system dependencies, resource allocations, and potential redundancies.
   - Genes represent dependency weights w_{ij}, resource allocations, or redundancies in the system.

(b) **Fitness Function with Multi-Objective Optimization**:

- Since multiple objectives are identified, a weighted fitness function can be used: $F(\text{Chromosome}) = w_1 f_1(\text{Chromosome}) - w_2 f_2(\text{Chromosome}) - w_3 f_3(\text{Chromosome})$. Alternatively, a genetic algorithm can be used which maintains a Pareto front of non-dominated solutions at each generation.

(c) **Genetic Algorithm Operations**:

- Selection: Use Pareto-based selection to prioritize chromosomes contributing to a well-distributed Pareto front.
- Crossover and Mutation: Apply crossover to mix parent configurations and mutation to explore alternative dependency setups.
- Pareto Front Evaluation: After each generation, compute the Pareto front and retain solutions that represent different trade-offs across the objectives.

(d) **Convergence and Solution**:

- Run the genetic algorithm until convergence (fitness scores plateau) or for a maximum number of generations.
- The resulting Pareto front provides configurations balancing mission effectiveness, risk, and resource use, allowing decision-makers to select based on the SoS mission priorities.

Solution Approach Steps

1. Build the Bayesian Network to represent dependencies and update system performance based on evidence.
2. Initialize the genetic algorithm with a population of random configurations.
3. Evaluate fitness using Pareto-based multi-objective criteria.
4. Select, crossover, and mutate to generate new solutions.
5. Compute the Pareto Front at each generation and store non-dominated solutions.
6. Converge optimal solutions by iterating over generations, yielding a set of Pareto-optimal configurations.

This combined approach using Bayesian Networks, Genetic Algorithms, and Pareto fronts provides a robust framework for balancing mission effectiveness, risk, and resource constraints within complex SoS configurations and offers a flexible, multi-objective optimization process for mission-critical scenarios.

2.3 Critical Thinking

Critical thinking refers to the ability to objectively analyze, evaluate, and synthesize information to form reasoned judgments or decisions. It involves actively questioning assumptions, considering evidence, and assessing the validity of arguments or claims. Critical thinking enables individuals to think logically, identify biases or fallacies, and make informed decisions based on evidence and reasoning rather than on emotion or outdated assumption(s). It is a valuable skill in various contexts, including problem-solving, decision-making, and effective communication.

Critical thinking in the context of MEI involves the systematic evaluation of information, evidence, and assumptions to make reasoned decisions and solve complex problems effectively. It requires MEI practitioners to analyze and interpret data, assess the validity of arguments, and identify potential biases or logical fallacies. Critical thinkers approach MEI with skepticism and intellectual rigor, questioning assumptions and exploring alternative perspectives to arrive at authoritative conclusions. By applying critical thinking skills, MEI professionals can mitigate risks, optimize resource allocation, and enhance the effectiveness of mission planning and execution.

Applying critical thinking entails a structured approach to problem-solving that involves several key steps and competencies. First, individuals must identify the problem or question at hand with clarity and precision, demonstrating a thorough understanding of the issue. Second, they gather relevant information from credible sources, critically evaluating its credibility, relevance, and significance. Analyzing this information involves identifying patterns, assumptions, biases, and logical fallacies, leading to a comprehensive assessment of the problem's complexities. Next, individuals generate multiple alternatives or solutions and evaluate them based on various criteria such as feasibility, effectiveness, and ethical considerations. Making informed decisions requires logical reasoning and evidence-based judgment, culminating in the selection of the most suitable course of action. Effective communication of the decision-making process, rationale, and conclusions is essential to ensure transparency and understanding among stakeholders. Finally, individuals reflect on the process, identifying areas for improvement to refine their critical thinking skills for future challenges.

Competencies in critical thinking include analytical skills for objective examination, problem-solving skills for identifying and exploring solutions, open-mindedness to consider diverse perspectives, skepticism toward information, logical reasoning, creative thinking for innovative approaches, and effective communication to facilitate collaboration and understanding. By honing these competencies and following the steps of critical thinking, individuals can navigate complex issues with clarity and confidence, contributing to effective problem-solving in various contexts.

2.4 Creative Thinking

Creative thinking is the ability to generate new ideas, concepts, or solutions by exploring different perspectives, making connections between seemingly unrelated concepts, and challenging conventional thinking patterns. This critical thinking approach serves as a fundamental component for driving the generation of innovative solutions to complex problems leading to design thinking and specific outcomes. Critical thinking extends beyond design thinking by encompassing a broader evaluation of assumptions, logic, and underlying principles, allowing for deeper analysis and problem-solving that challenges established frameworks and explores alternative perspectives. Design thinking encourages individuals and teams to approach problems with openness, curiosity, and a willingness to experiment. Creative thinking within design thinking involves brainstorming, free-generation of ideas, and prototyping to explore a wide range of possibilities and iterate on potential solutions until the optimal outcome is achieved. Creative thinking techniques enable teams to break free from traditional constraints, think "outside the box," and develop innovative approaches that effectively address user needs. By bringing together diverse perspectives and skill sets, creative thinking simulates the dynamic conditions typically found in emerging economies, where rapid adaptation and innovation are necessary for survival. It involves immersing teams in real-world situations where they must gather subjective data and insights directly from end users to best understand operational environments. This ethnographic research technique seeks to understand the beliefs, values, behaviors, and social dynamics of a group through direct observation and participation in their daily life to gain understanding of unfamiliar environments, communities, and systems.

Imposing constrained timelines using creative thinking involves leveraging time pressure as a catalyst for innovation and problem-solving. Setting strict deadlines forces teams to think creatively and efficiently to generate solutions within limited timeframes. This approach encourages out-of-the-box thinking, promotes rapid prototyping, and fosters experimenting to quickly iterate and refine ideas. It mirrors the fast-paced environments entrepreneurs face in emerging markets. This approach reflects the reality that operator needs and market dynamics evolve rapidly, necessitating an iterative approach to problem-solving. By honing skills in rapidly assessing complex situations and promptly prototyping ideas, creative design thinkers develop the ability to navigate uncertainty effectively. They learn to transform open-ended project mandates into tangible opportunities through practical experimentation and iterative refinement. Benefits of creative design thinking as an agile, imaginative approach are: uncovering hidden patterns, meeting user needs, making breakthrough discoveries, performing effective problem-solving, and navigating uncertainty.

2.5 Innovation

Innovation is the process of developing a new or significantly altered service, product, process, structure, or policy. It involves the creation of something novel or the substantial modification of existing elements. Adoption, on the other hand, occurs when individuals or organizations recognize, embrace, and effectively utilize an innovation. Change agents, including innovators, facilitators, leaders, and users, play a crucial role in this process. They identify, develop or acquire, integrate, and leverage the value of the innovation, driving its acceptance and implementation within their respective contexts.

Complete innovation adoption occurs when change agents successfully identify, develop or acquire, integrate, and leverage the value of a new or significantly changed service, product, process, structure, or policy. However, legacy sectors, characterized by entrenched structures and established norms, often resist or block innovations that challenge their existing models. Without external pressures or compelling reasons to change, organizations may be inclined to innovate within the confines of their established roles and interests, thereby protecting the status quo. While incremental and architectural innovations may find relatively smoother integration into existing systems and processes, radical and disruptive innovations often encounter resistance within legacy sectors. The acquisition system is designed to accommodate gradual changes and improvements; however, struggles may persist in adapting to innovations that fundamentally alter established paradigms. Effectively introducing innovative concepts into legacy sectors, therefore, presents a complex challenge that requires navigating institutional resistance, addressing vested interests, and fostering a culture conducive to change.

Innovation is a multifaceted process that often requires careful design and facilitation to unlock its full potential. By working through various methods and frameworks, the MEI practitioner/promotion team can expand thinking and exploration in a rich array of available choices to help navigate the complexities of innovation, identify unique opportunities, and develop creative solutions to pressing challenges. Whether leveraging design thinking, adopting agile methodologies, or embracing interdisciplinary collaboration, each approach offers valuable insights and techniques for driving innovation forward. Ultimately, by embracing diverse perspectives and actively engaging with the innovation process, the practitioner harnesses the power of creativity to achieve meaningful and impactful outcomes. The following are some drivers that provide a structured approach for innovation:

- ***Intelligence drivers***: The rapidly changing and often unpredictable nature of today's world requires us to leverage artificial intelligence (AI) and cognitive capabilities to make sense of complex and evolving patterns. Intelligence drivers play a crucial role in helping us interpret and navigate the vast amounts of information and data that surround us. AI serves as a powerful tool for processing and analyzing large volumes of data, identifying trends, and uncovering hidden patterns that may not be immediately apparent to human observers. By harnessing AI technologies, we can augment our ability to understand and interpret complex phenomena, enabling us to adapt and respond effectively to changing circumstances. Additionally, cognition plays a key role in innovation at both the individual and organizational levels. The ability to process information, apply knowledge, and explore new concepts is essential for generating novel ideas and solutions. Fostering cognitive flexibility and creativity reveals new sources of innovation and promotes breakthrough insights that drive progress and growth in an increasingly dynamic and interconnected environment.
- ***Technology drivers***: The landscape of collaboration is undergoing profound changes, driven by the need to tap into external networks and diverse perspectives for innovation and growth. Traditional notions of guarding internal

intellectual property are giving way to a more open and transparent approach. Organizations are increasingly embracing external collaboration to drive creativity and innovation. Technology plays a crucial role in facilitating this shift toward open collaboration. Organizations must revisit their technology infrastructure to accommodate the diverse opinions and perspectives that come with external collaboration. This digital transformation may involve implementing new platforms and tools that enable seamless communication, data sharing, open thinking, and collaboration across physical and virtual spaces. By designing technology solutions that prioritize openness, transparency, and inclusivity, organizations can create environments that foster innovation and drive progress.

- ***People drivers***: The heart of innovation beats within people, yet often actions fail to align with this fundamental truth. To unlock the full potential of the workforce, organization team leaders need to understand what drives individuals to excel and reach new heights. It is about striking the right balance between fostering creativity, nurturing talent, embracing diverse perspectives, encouraging open dialogue, and prioritizing relationship building. Moreover, it is about recognizing employees for their invaluable contributions and treating them as valuable assets. Instead of viewing them as disposable resources, management must invest in their growth and development, offer meaningful incentives, and reward them for driving innovation. Retaining experienced talent within organizations is becoming increasingly challenging, underscoring the importance of valuing and retaining employees. Empowering them with autonomy and authority rather than burdening them with restrictions is key to nurturing a culture of innovation.
- ***End User drivers***: Innovation aimed at delivering more value with fewer resources is particularly relevant in today's context by gaining a deep understanding of the end users' genuine needs. The landscape of marketing, for example, is undergoing a profound transformation as end user preferences evolve from a focus on price and volume to a demand for increased value and personalized engagement. In this shifting landscape, understanding the unmet needs of end users and translating them into innovative solutions has become more challenging than ever. As organizations expand globally, there is a growing necessity for reverse engineering innovation involving the alignment of products or services with the specific needs of diverse markets, especially in times of economic constraint.
- ***Rational drivers***: To achieve consistent innovation results requires a more structured, robust, and rational approach to innovation. This entails several key elements, including a heightened focus on innovation's strategic architecture, effective management of systems that facilitate innovation from conception to commercialization, and the design of innovation processes aimed at reducing time-to-market, enhancing productivity, and facilitating successful commercialization. Innovation strategic architecture involves systematic alignment of innovation efforts with overarching business goals and objectives. By ensuring that innovation initiatives are strategically aligned, organizations can prioritize projects that have the greatest potential to drive long-term value and competitive advantage.

- ***Environmental drivers***: The culture and climate within which organizations operate can either foster or hinder innovation. To promote innovation effectively, organizations must recognize the need to enhance cooperation across diverse aspects of innovation activity and engage with external partners on third-party platforms to tap into a broader range of perspectives and expertise. Instead of confining innovation efforts to siloed research and development efforts, organizations should consider the entire value chain and leverage diverse networks across the organization. However, there is a delicate balance to strike between achieving quick results, which are often necessary for immediate success, and establishing longer term structures that support sustained innovation on a broader scale. Innovation is increasingly driving the development of sustainable products, processes, and organizational designs, with environmental considerations becoming increasingly important.

Achieving an optimal balance between technological advancements and operational concepts is essential for fully realizing the potential of military or other competitive innovations. While new technologies can provide incremental improvements to existing capabilities, their true impact is often determined by how they are integrated into operational strategies. Historically, technological breakthroughs have been most effective when paired with complementary shifts in operational concepts. In the commercial sector, innovations such as small hard drives, initially inadequate for mainframes, found groundbreaking applications in portable music players, demonstrating that the value of a technology often depends on its use case rather than its standalone performance. Military innovation follows a similar pattern. Sustaining technologies enhance existing operational methods, while disruptive technologies introduce entirely new ways of achieving military objectives. The most transformative advances occur when revolutionary operational concepts are combined with disruptive technologies. However, implementing such innovations is challenging, as they often require a fundamental rethinking of success metrics, force structures, and doctrine. Balancing technological and operational perspectives ensures that emerging capabilities are not merely integrated into legacy approaches, but instead drive new paradigms of competition effectiveness. Figure 2.4 illustrates some predicted end-state boundaries when combining disruptive technologies and operational concepts with either evolutionary refinements or revolutionary shifts.

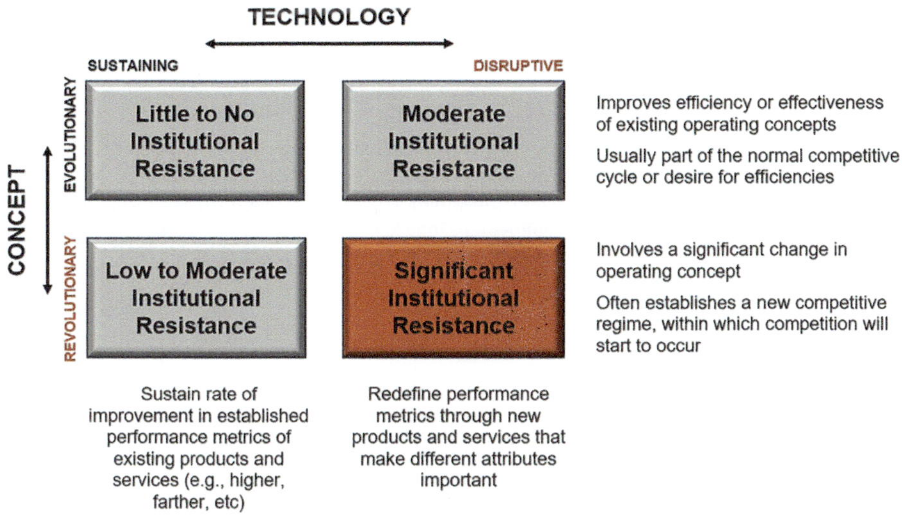

Fig. 2.4 Intersection of technological and operational perspectives

As we explored the foundations of **critical and creative thinking**, we recognize their essential role in problem-solving, innovation, and decision-making. These cognitive approaches enable us to break down challenges, envision new possibilities, and develop solutions that drive progress. However, effective thinking must be paired with seamless **integration and interoperability** to ensure successful real-world implementation for complex systems. Moving forward into the next chapter, we will define these concepts, examining how diverse systems, technologies, and processes must work together cohesively to achieve shared objectives. The ability to connect, communicate, and function as a unified whole is critical in today's interconnected environments, where success depends on more than just innovation; it depends on collaboration and compatibility.

References

1. Von Bertalanffy, L. (1968). *General system theory: Foundations, development, applications*. George Braziller.
2. Checkland, P. (1999). *Systems thinking, systems practice*. John Wiley.
3. Eisner, H. (2005). *Managing complex systems: Thinking outside the box*. John Wiley.
4. Eisner, H. (2011). *Systems engineering: Building successful systems*. Morgan & Claypool.
5. Eisner, H. (2013). *Topics in systems*. Mercury Learning and Education.
6. Forrester, G. (1968). *Principles of systems* (2nd ed.). Pegasus Communications.

7. Keating, C., Pyne, J., & Bradley, J. M. (2015). Advancing enquiry for system-of-systems engineering: Paradox, critical thinking and critical systems enquiry. *Int. J. Syst. Syst. Eng., 6*(4), 348–349.
8. Monat, J., & Gannon, T. (2017). *Using systems thinking to solve real-world problems*. Systems Engineering Program, Worcester Polytechnic Institute.
9. Sheard, S. A., & Mostashari, A. (2011). Complexity types: From science to Systems Engineering. In *Proceedings of the 21st Annual of the International Council on Systems Engineering (INCOSE) International Symposium, Denver, Colorado, USA, 20–23 June 2011.*
10. Weaver, W. (1948). Science and complexity. *American Science, 36*, 536–544.

Chapter 3
Integration and Interoperability

3.1 Historical Challenges

Integration and interoperability challenges continue to exist, and in some cases, connectivity is a widespread problem that continues to grow with the dependencies and reliance of an interconnected world. No longer are satellite communications, Global Positioning Systems (GPS), and satellite images exclusively controlled and used by the military. They have become tools of everyday life to millions around the world including many in underprivileged Third World nations. The innovations spurred from opening GPS to the commercial market, the creation of the internet, and the implications of Moore's Law are factors in the rise of globalization. From the way we communicate and make business transactions to the way operatives conduct and coordinate riots and revolutions, technology has changed the way societies interact.

The geopolitical landscape continues to evolve. A growing global population, with the majority of growth occurring in developing nations, continues to migrate to urban centers. Many of these centers are in the world's littorals in areas whose governments historically have been unable to meet the basic needs of their populations. Many future projections predict a complex future security environment in which technology continues to be distributed among a flattened and ever more interconnected world, one which will inevitably include high-security, anti-access technologies. Projections foretell a blurring of organizational and societal boundaries, the haves and have nots, super-empowered individuals, and operations in increasingly dense urban environments in and among the people requiring greater fidelity information and precision systems.

Bad actors and catastrophic events threaten normal operations through development of disruptive capabilities of commercial-off-the-shelf (COTS) technologies. These disruptive events are unpredictable and cause tremendous impact on the health and security of peaceful nations. Interoperability is not a "nice to have" but an

J. D. Moreland, Jr., *Mission Engineering and Integration*,
https://doi.org/10.1007/978-3-032-32132-9_3

essential requirement in order to maintain an advantage. It is imperative now to achieve interoperability parity with legacy systems as well as through the insertion of new innovative technologies. Much of the technology required is currently available, and in some cases, is already installed in legacy systems because of the many years of addressing and funding these issues when working across complex system-of-systems (SoS). But where all these systems meet is in the objective area, and therein lies the challenge. Responsible competencies fail to achieve alignment across the SoS in the design phase, resulting in gaps during acquisition execution and fielded systems. The following interoperability examples provide real-life examples of the complexity and challenges that lie ahead in achieving interoperability across SoS.

On October 26, 1983, soldiers from the 82nd Airborne Division engaged in a firefight on the southern part of the island of Grenada during Operation URGENT FURY. The story has been told many times, sometimes wildly embellished, and even portrayed in the movie Heartbreak Ridge. An AC-130 gunship was on station overhead, but the airborne soldiers could not reach the gunship on their frequency. A soldier made a commercial phone call to Fort Bragg, NC, whereupon one of the watch officers relayed to the AC-130 via satellite communications what frequency they should use to communicate with the soldiers. The crew of the AC-130 came up on the frequency, and the soldiers received the close air support they needed. That episode was used as a metaphor for the Department of Defense's (DoD) lack of interoperability among the military services. Among the outcomes was the Goldwater-Nichols Act of 1986 [GOLDWATER86] which reorganized the US military by strengthening joint operations, enhancing the authority of combatant commanders, and improving inter-service coordination, reducing bureaucratic rivalries.

Another famous interoperability failure is the London Ambulance Services computer-aided dispatch system. This system was designed to partially automate the manual processing of emergency calls. An emergency call should be answered by a London Ambulance Service employee, who should ask for the location of the emergency. Based on this information, the system should determine the ambulances closest to the location of the emergency. An ambulance close to the emergency location and ready for service should then be dispatched to the emergency. However, in the 1966 incident, inappropriate user interfaces for the communication devices in the ambulances caused partly incorrect and insufficient delivery of emergency information. In addition, the communication devices did not take into consideration the existing radio blackout spots in the communication network, and therefore, experienced communication failures in the area. As a result, ambulances that were not ready for service were sent to the location of an emergency, and other ambulances were delayed in responding due to denied, disconnected, intermittent, and limited bandwidth and communication channels. These communication challenges caused intolerable delays and endangered people's lives.

Fast forward to 2005 when Hurricane Katrina made landfall on the Gulf Coast creating a historical disaster. New Orleans flooded causing the death of more than 1800 people and $100 billion in property damage. The storm's damage was intensified by many failures to include substantial interoperability issues in disaster preparedness and response systems. The 2006 bipartisan House report on the disaster, "A Failure of Initiative," [CONGRESS06] stated, "a complete breakdown in communications that paralyzed command and control and made situational awareness murky at best." Response agencies failed to communicate due to equipment failures and a lack of system interoperability. These problems occurred despite focused investments in federal, state, and local governments for emergency communication systems since the beginning of the Cold War. Specifically, the Shreveport, La., Fire Department's radio system was found to be useless in the New Orleans area although their systems could communicate with police, emergency medical services, and 50 other agencies in their home region. When in New Orleans, this police force used an alternative, incompatible system and therefore could not talk to the Shreveport squad. In another instance, the destruction of infrastructure and lack of interoperable communication systems caused a reliance on paper relays or face-to-face meetings to exchange critical information between the Mississippi National Guard and other first responders along the Gulf Coast. As a result of these emergency response issues, calls for assistance were delayed and response times slowed. Despite the severity of communication failures uncovered after Katrina, emergency personnel from 50 public safety agencies in Virginia, Maryland, the District of Columbia, and the federal government responded to the September 11, 2001, attack on the Pentagon with similar challenges. The responders' incompatible systems made communication attempts futile; therefore, responders had to rely on runners to shuttle messages from commander to commander.

These failure events highlight the interoperability and connectivity challenges that still exist today. Given the state of commercial communications technology, any outside observer would surmise that the most advanced nation in the world would not experience the persistent and existing challenges in exchanging information among end users going into harm's way. The reality is our industrial sectors are still working diligently to find paths to allow groups and their respective SoS to access and exchange information in an effective and safe manner.

3.2 Justification for Change

Globalization concepts all visualize the use of international expertise that will conduct operations across the full range of functional necessities to support the health and security of all nations. All populations and individuals will need to conduct stand-alone operations while maintaining the capability to partner in larger combined operations. To meet these challenges and maintain an advantage over an ever-expanding set of functional demands, all end users will need to thoroughly understand and exploit current and future technologies to create an architecture that supports independent operations yet seamlessly connects to the combined global network. This network ensures that all relevant end users and their supporting elements can contribute to and access critical information thereby maintaining true end-to-end capability fundamental to a robust future global network.

To stay ahead of global competition, stakeholders need to accelerate network-centric and distributed operational capabilities by continuing their legacy practices of building the next bigger, faster, more powerful, more sustainable version of legacy systems and technologies. Until recently, the acquisition structure allowed capabilities to be developed in isolation and based on a system architecture built simply for individual system requirements compliance. As a result, deployed systems fell short of the expected operational utility or suboptimization of capability. This is amplified by competitors' ability to out-innovate at an accelerating pace. Aided by widespread global availability of very advanced technologies, global competitors have degraded the US' dominant advantage. In the DoD, emerging policies are now requiring acquisition approaches that ensure individual systems will integrate into the larger mission enterprise architecture. In addition, the current and emergent threats require faster development of new capabilities, increased readiness, and interoperability at reduced cost. Through the implementation of MEI, the DoD is moving to a more holistic approach to address the challenges associated with fielding, integrating, training, and readying of integrated capabilities.

A new paradigm is needed that values effectiveness over dominance. Effectiveness may be thought of as the ability to deliver a desired effect at will through many dynamic operational conditions. System performance must be employed such that it is not only highly effective at any time but also provides a visible demonstration of resilient and continuous operations under varying risks.

3.3 Integration Philosophy

Integration is the process of cohesively merging multiple applications or systems into a unified framework that enhances overall operational effectiveness. Systems integration involves interfacing system elements and enabling their interactions so that they collectively provide the functionality required to accomplish system objectives. [MADNI14] Successful integration therefore requires consideration not only of individual system performance, but also of the interfaces, interactions, dependencies, and behaviors that emerge when system elements operate together. In the context of MEI, this process extends beyond simply connecting systems; it requires a structured approach to ensure that independent and functionally distinct systems interoperate seamlessly to deliver enhanced mission capability. Integration is particularly critical in SoS environments, where the convergence of multiple platforms, sensors, networks, and decision-support tools must create a synchronized and responsive operational ecosystem.

A key example of integration can be seen in how organizations consolidate multiple technologies, often developed separately, into a single application designed to address specific mission needs. This mirrors the commercial sector, where companies integrate acquired technologies to improve efficiency and functionality. However, integrating multiple systems poses significant challenges, particularly when these systems are developed by different suppliers, each with varying design philosophies, levels of technological maturity, and proprietary constraints. This issue is especially pronounced in complex operational environments, such as defense, healthcare, and aerospace, where systems must support diverse functions

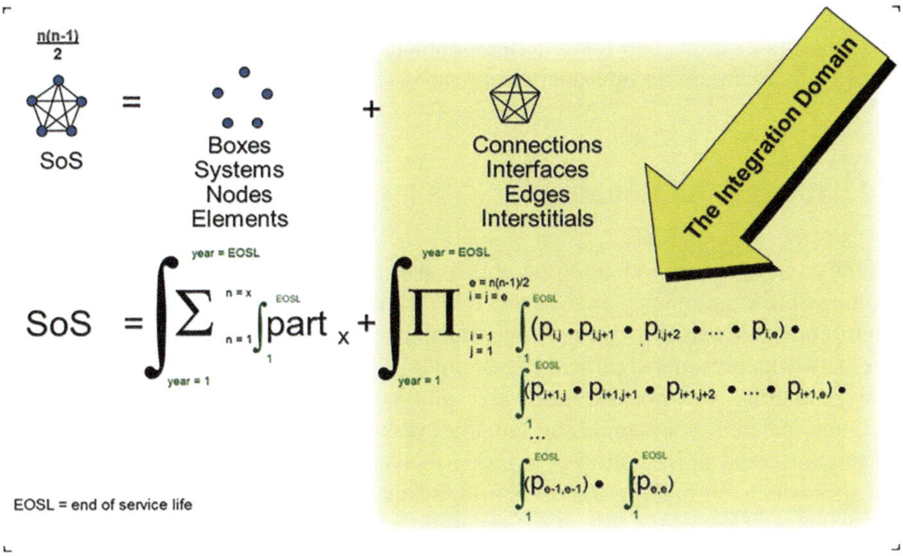

Fig. 3.1 Integration of the individual systems and networks

such as data analytics, connectivity, communications, and workflow management. The diversity of these functional areas makes it impractical to create a "one-size-fits-all" solution without sacrificing quality, performance, or security. Instead, integration efforts must focus on customized, modular approaches that allow specialized components to interoperate without compromising their individual strengths.

Beyond technical challenges, integration efforts can be hindered by proprietary architectures that restrict open connectivity and data sharing. A highly integrated application environment that relies on closed, vendor-specific methods can impede innovation, particularly when new suppliers attempt to introduce disruptive technologies that require seamless interaction with legacy systems. Vendor lock-in due to proprietary designs create barriers to interoperability, limiting the ability of organizations to incorporate emerging technologies rapidly and cost-effectively. Traditional approaches have often focused on deploying standalone applications or systems to fulfill distinct enterprise or mission functions. However, the increasing complexity of modern operational challenges demands more cohesive, interoperable, and cost-efficient solutions that enable real-time information sharing and cross-domain collaboration.

Given that traditional capability development methods emphasize acquiring individual systems, there remains a critical need to elevate integration as a primary design and budget consideration, see Fig. 3.1. This requires establishing well-defined interfaces, standardized data exchange protocols, and open architectures to enable the fluid exchange of mission-critical information across diverse operational environments. The integration domain must not be treated as an afterthought but as a foundational enabler of mission success, ensuring that systems function harmoniously within a broader enterprise framework. By embedding integration principles into the early stages of system development and leveraging open systems approaches (OSA), organizations can foster agility, enhance adaptability, and reduce lifecycle costs while maintaining interoperability across evolving mission landscapes.

3.4 Integration Enablers

Effective integration and interoperability within an SoS are critical to ensuring seamless information exchange across diverse networks, platforms, and operational environments. Designing systems that transfer real-time data across multiple disparate networks presents significant challenges, particularly in merging communication protocols, waveforms, hardware, and software. While overcoming these technical barriers is essential, an equally pressing challenge lies in the collation, distribution, and presentation of mission-relevant data to end users in a way that enhances decision-making. As users bridge networks and gain access to an increasing volume of data, the risk of information overload becomes a key concern. To address this, MEI principles must guide the development of decision-support tools that filter, prioritize, and present only the most mission essential information. A critical enabler in this effort is the design of Graphical User Interfaces (GUIs) that optimize information visualization, ensuring users can quickly assess the operational environment and act accordingly. These interfaces must not only standardize how information is formatted and displayed but also account for human systems integration (HSI) factors to enhance cognitive processing and reduce decision fatigue.

Interoperability capability gaps in mission-critical systems can be categorized into two primary groups: intra-system interoperability gaps and inter-system interoperability gaps. Intra-system gaps occur within an individual system or agency, often stemming from proprietary architectures, legacy constraints, or inconsistent data formats. Addressing these gaps requires adherence to open systems architectures (OSA) and modular design principles, ensuring that internal components can be rapidly updated or reconfigured to align with evolving mission needs. Inter-system gaps, on the other hand, arise between individual systems operating within a larger SoS, where disparate networks, security policies, and data standards hinder seamless information exchange. Effective integration at this level requires a common data framework, cross-domain solutions, and alignment with enterprise-level information-sharing strategies. The ability to fuse and interpret data across multiple operational layers ensures that operators have access to a cohesive and accurate operational picture.

To mitigate these interoperability gaps, developers must adopt an iterative, mission-driven approach that continuously assesses system performance and refines integration strategies over time. MEI enablers, such as model-based systems engineering (MBSE), digital engineering, and agile development, provide structured frameworks for identifying, testing, and resolving these gaps in a risk-tolerant and economically sustainable manner. By embedding these integration enablers into the system design process, organizations can enhance operational effectiveness, reduce integration costs, and improve the agility of mission-critical systems in dynamic environments.

3.4.1 Modular Open Systems Architecture

The rapid evolution of technology necessitates a paradigm shift in how future network architectures are designed and implemented. Traditional models relied on long-term investments in proprietary hardware and software, which were effective in their time but have proven costly and unsustainable amid continuous technological advancements. The rigidity and high cost of upgrades in this past approach hinder the ability to adapt to emerging mission needs. The Modular Open Systems Approach (MOSA) applies modular design principles and open interfaces to facilitate interoperability, technology insertion, competition, and system evolution throughout the acquisition life cycle [OUSDAS23]. By reducing dependence on proprietary designs and interfaces, MOSA can mitigate vendor lock-in and provide greater flexibility to integrate new or upgraded capabilities as technologies and mission needs evolve. From an MEI perspective, these characteristics enable constituent systems and capabilities to evolve without requiring redesign of the entire SoS, supporting more rapid adaptation of mission threads to changing operational requirements. Given this reality, MEI must embrace a MOSA approach, enabling flexible, scalable, and rapidly upgradable systems that can evolve alongside operational demands.

To effectively support mission objectives, MOSA must facilitate the seamless integration of new capabilities, technologies, and platforms into existing SoS environments. This requires an intentional shift toward open architectures that utilize advanced programming languages, open standards, and modular designs. By doing so, organizations can plan for obsolescence rather than react to it, ensuring that new systems can be incorporated without the prohibitive time and cost constraints of legacy models. Interoperability, flexibility, modularity, and portability must be treated as essential nonfunctional requirements. This ensures that diverse systems, developed at different times with different technologies, can communicate and function cohesively within a broader mission ecosystem.

The adoption of MOSA is driving a fundamental change in procurement strategies. Traditionally, acquisition programs relied on systems integrators to design and manage interoperability between systems. However, a new end user-driven approach is emerging, wherein government agencies and operational stakeholders define the reusable architecture that will shape future SoS acquisitions. This architectural ownership is a deliberate effort to catalyze open competition, shifting the market away from vendor lock-in and toward a government reference architecture that promotes true interoperability and innovation. This approach recognizes that systems vary in their need for openness. Instead, interoperability should be tailored to ensure effective integration across different architectures, data models, and operational lifecycles. The pivotal technical goal driving this shift is interoperability, elevating it from a secondary design consideration to a primary performance metric. Instead of viewing open architecture as a checkbox requirement, organizations must ensure that MOSA principles actively reduce lifecycle costs, enhance adaptability, and shorten response times between user-identified needs and system implementation.

A truly integrated open architecture is not merely about adopting open standards or modular components; it is about seamlessly combining these elements within a cohesive open infrastructure. While many system integrators claim compliance with open-architecture principles, the real value of MOSA is realized only when systems are designed from the outset to support long-term adaptability, affordability, and mission-driven innovation. By embedding MOSA principles within MEI, organizations can ensure that future mission systems are resilient, scalable, and capable of evolving alongside the ever-changing operational landscape.

3.4.2 Open Standards

Building an SoS capability based on open standards is crucial for achieving interoperability among systems, SoS, and applications that are developed and procured at different times. Open standards enable procurement agencies to define the architecture requirements, instead of being constrained by potentially dominant suppliers who dictate system designs. Historically, although open architecture initiatives were intended to achieve interoperability, they often fall short in practice, with systems struggling to communicate seamlessly. The key to resolving this issue and ensuring true interoperability lies in controlling the data flow across systems. Given the complexity of SoS, a data modeling approach aligned with the MEI approach is essential. While technologies can be mapped using defined concepts of employment, understanding the fundamental information flow within the SoS is often not straightforward. In the past, this understanding has been inferred during the integration process, often leading to escalating costs as systems scale and complexity increases over time. As each new system is integrated, the challenge grows, and the resulting costs can become prohibitive.

A more effective way to define an SoS architecture is to focus on its data. This approach necessitates a data-centric middleware open standard that provides the foundation for effective integration. Without such a standard, the procurement process risks creating vendor lock-in, where a supplier's technology becomes embedded in the system supply chain, forcing future procurements to rely on that supplier. To avoid this, the open standard for SoS integration must mandate both a wire protocol and a programming interface, ensuring interoperability across different technologies. While these technical foundations set the stage for interoperability, they are still insufficient on their own. Communication and connectivity may be standardized, but semantic interoperability, or the understanding of the meaning of the data being exchanged, remains a challenge. This issue can be addressed by developing a system data dictionary that specifies both the content and context of the data being communicated across the SoS. Unlike a wire protocol, which defines the format of the data, a system data dictionary serves as a full specification of the data's meaning, ensuring that it can be correctly interpreted and exchanged across diverse technologies.

In addition to defining the structure and meaning of the data, the data dictionary also includes metadata, which defines the semantic information associated with each piece of data. This metadata provides context for the data, enabling it to be re-contextualized by the systems using it. For example, in the case of track data (e.g., the location of an air vehicle), the data dictionary would clarify the units of measurement (sea level or above ground, kilometers or miles), whether negative values are valid, and other critical context such as the data's origin, age, and reliability. These context elements contribute to the semantic understanding of the data, helping systems assess the veracity and utility of the information. For instance, if one system reports the location of an air vehicle at a certain altitude, the data dictionary ensures that all systems in the SoS understand whether that altitude is measured from sea level or above ground. Moreover, it answers important questions like, "Who is providing this data?" and "How old is the data?" These questions, although subtle, impact the validity and relevance of the information for mission-critical decisions. Without such context, systems can produce errors such as duplicated tracks, incomplete or conflicting data, and inefficient use of computing resources.

When systems work on a common semantic data model, rather than simply exchanging syntactic messages, they can achieve true interoperability. This shared data model provides a common understanding of the information exchanged across the SoS, enabling systems to work together effectively and make informed decisions. While some open system programs focus on portability, it is important to distinguish portability from interoperability. Although portability allows applications or systems to run on different platforms, it does not guarantee that new capabilities or features will integrate smoothly. As systems evolve and new requirements are introduced, portability may not prevent the cascading impact of changes through the SoS, resulting in costly updates and integration challenges. True interoperability ensures that new capabilities can be introduced without disrupting existing systems or creating inefficiencies.

3.5 Interoperability Philosophy

Interoperability refers to the ability of systems to transfer and share information seamlessly, regardless of hardware, software, or communication protocols. In the context of MEI, interoperability is crucial because it enables diverse systems and SoS to work together effectively, even when they are developed at different times, by different organizations, or with varying technologies. Despite its importance, the term "interoperability" has become contentious within many industrial sectors, often misused, misunderstood, or conflated with other related concepts. To address this confusion, it is essential to first establish a clear and consistent understanding of what interoperability truly involves.

One common issue is the frequent misuse of terms that are often assumed to be interchangeable, but actually describe distinct concepts. For example, terms like compatibility, integration, and data sharing are sometimes mistaken for interoperability, yet they each have specific meanings in the context of MEI. Although integration and interoperability are closely related, they are not synonymous. Integration generally focuses on combining system elements to operate as part of a larger whole, whereas interoperability, emphasizes the ability of systems to exchange information and effectively use the information exchanged. Consequently, connecting systems or enabling information exchange does not by itself ensure interoperability, particularly when differences exist in data representations, semantics, context, interfaces, and system constraints [POLLOCK01]. Recognizing these differences is crucial for defining precise requirements and ensuring that the systems and SoS being developed can effectively work together. The lack of a unified set of definitions for these related terms, as illustrated in Table 3.1 with definitions drawn from various sources like military directives and industry standards, highlights the ongoing challenge in the field. This lack of clarity underscores the historical struggle within the industry to consistently explain and apply the concept of interoperability. In the context of MEI, this lack of consistency can lead to inefficiencies and misaligned expectations, making it essential to establish a shared understanding of interoperability and its implications for system design, procurement, and integration.

Interoperability traverses all industrial sectors, domains, hierarchical structures, and operational functions. Although each sector, domain, and function supports distinctive missions, all are interdependent. The shared awareness provided by connected nodes provides the speed and focus we celebrate in our strategies and standard operating procedures. Of all the consistent patterns discerned in past events, two concepts of universal significance in generating global dominance are speed and focus. Speed refers to the rapidity of action and applies to both time and space. When considering time, speed is associated with tempo, which is the consistent ability to operate quickly. In terms of distance, speed refers to the ability to move rapidly over space. Both forms are genuine sources of performing effective operations across complex SoS irrespective of the specific industrial sector or domain. In other words, speed is a force multiplier when trying to execute functions in a competitive environment or market. In markets, it is relative speed that matters rather than absolute speed, leading us to the concept of Speed of Relevance. Superior speed enables stakeholders to seize the initiative and dictate the terms of action that are most relevant to the adversary or competitor, forcing them to react rather than act proactively.

By emphasizing interoperability, the industrial sectors seek to align the markets and manufacturers more closely with the operation of open commercial and government products. The most important market benefit sought is open market competition for system and subsystem supply. Procurement organizations seek to foster investment ahead of specified requirement, driving differentiation through cost and innovation. Most important, the goal is to build agility into the procurement process so that the end users are better served with more capability, delivered more rapidly, with increased update cycles to stay at the forefront of technological competition.

Table 3.1 Related interoperability definitions

Term	Technical definition	Relation to interoperability and commercial context
Interoperability	The ability of systems, units, or end users to provide services to and accept services from other systems, units, or end users, and use the services exchanged to enable them to operate effectively together	A fundamental prerequisite for an open competitive supply chain. Defines an in-service system capability as much as an initial development capability. enables integrators to connect multiple components developed by different parties without change
Integratability	To be able to form, coordinate, or blend into a functioning or unified whole. To incorporate into a larger, functioning or unified whole	Makes no claims as to the system interoperability. In extremis, any software system is integratable — at a cost. Does not imply any in-service system attribute
Replaceability	One system or actor taking the place of another especially as a substitute or successor	While a replaceable system is an asset, it does not imply that the replaced system can be enhanced or altered in any way. More likely, it has to remain identical in functionality
Interchangeability	To put each of (two systems) in the place of the other, or to be used in place of each other	An improvement on replaceability because the systems are likely to behave differently based upon the SoS the system is placed into. But this system context usually must be pre-determined and fixed ahead of development time
Extensibility	The ability to add new systems, components, and capabilities to an SoS	There is no limit to the domino effect of change requests needed across the rest of the SoS to integrate the new system. Nor does it imply that the new system can meaningfully exchange information with any other system already in the SoS
Componentization	A software package, service or module that encapsulates a set of related functions that communicate via defined interfaces	A valuable building block in software architectures, but its interfaces are not necessarily openly defined for interoperability and can still end up delivering stove-piped systems
Modularity	Clarifies the functional blocks of a system or SoS, separating capability into modules	Improves maintainability but makes no claims for interoperability as interfaces can be closed and proprietary

The core technology enabling tenet of this approach is to develop a common SoS architecture that satisfies the systems nonfunctional requirements. But commonality alone is insufficient to reap the agility and commercial benefits sought because it does not necessarily facilitate the integration of new innovative capabilities with older "common" functionality or explicitly identify the legitimate points of variation.

To achieve a fully interoperable SoS architecture, the developer needs to satisfy three levels of interoperability:

- **Technical Interoperability**—Bits and bytes are exchanged in an unambiguous manner via a set of standardized communication protocols.
- **Syntactic Interoperability**—A common data format is defined for the unambiguous sharing of information.
- **Semantic Interoperability**—The meaning of data is exchanged through a common information model and the meaning of information is unambiguously defined and shared.

Note, semantic interoperability is not common practice today. The focus on the data semantics will facilitate the drive toward interoperability. However, even if all these capabilities in an SoS are enabled, there is no guarantee that an integrated open architecture will result. One more key step is required that is to delegate the syntactic and semantic interoperability to a software infrastructure layer that can be common across every system. Thus, interoperability can now be sustained at the SoS level rather than solely between two or more interoperating applications or systems. Achieving this requires a shift in perspective, where the focus moves from traditional SoS communication to a broader, more holistic approach that ensures all components within the SoS can effectively interact. To facilitate this, the key to associating interoperability with data and information flows at the system level is through a data-centric design approach. This approach ensures that all systems within the SoS share a common understanding of the data, enabling seamless exchange and integration. If a system cannot interpret and process the SoS data language, meaning the standardized data models, formats, and semantics used across the entire network, true interoperability is not realized. Therefore, it is essential for each system within the SoS to be capable of speaking the same data language to ensure effective communication and collaboration across diverse technologies and operational environments.

In the context of MEI, the relationships between scalability, performance, and reliability are essential trade-offs that systems and mission engineers must navigate to optimize one attribute at the expense of the others. Traditionally, achieving higher performance or scalability would often come at the cost of reliability, and vice versa. However, in modern SoS, such compromises are no longer acceptable, especially in competitive markets where operational continuity and resilience are critical. Within an SoS, the failure of any individual system or subsystem should not lead to a failure of the entire SoS. This is a fundamental principle that has driven system developers to implement well-architected middleware solutions, ensuring no single point of failure exists within the architecture. Unlike traditional client-server models, where a server failure can cause all connected clients to fail or suffer reduced functionality, a true peer-to-peer middleware implementation prevents this issue by distributing responsibility across the network.

Moreover, the middleware must support the SoS by enabling the ability to sustain a data-driven state across the network. This means that the system's state, in terms of data flows, is continuously maintained as data moves between systems. Achieving scalability is largely about decoupling systems and ensuring each system can communicate with the broader network independently, using a data bus. Rather than systems interacting directly with each other, they communicate through the data bus, either placing data onto it or pulling it off as necessary. This structure allows systems to interact based on their data needs, ensuring flexibility and scalability. To maintain coherence and continuity, it is essential to implement a mechanism for managing "data in motion," which refers to the dynamic flow of data throughout the SoS. This mechanism also plays a crucial role in maintaining the state of the SoS and ensuring semantic interoperability across the systems, which is vital for aligning data exchanges with the operational context.

Incorporating Quality of Service (QoS) into systems and across SoS capabilities is critical to ensure that performance, reliability, and scalability are achieved without compromise. QoS parameters such as bandwidth, latency, and packet loss need to be actively managed to guarantee that data flows meet mission-critical requirements. QoS management enables systems to operate at optimal levels, even in dynamic and resource-constrained environments, and ensures that critical data is delivered within the required parameters to support timely decision-making and mission success. Even with these inherent scalability and reliability features, performance cannot be compromised. The open standards middleware must be capable of handling large-scale communications across diverse communication channels, including high-frequency radios, ethernet, satellite links, or proprietary connections. This ensures that systems can effectively exchange information, regardless of the scale or complexity of the SoS, while maintaining the high performance needed for mission-critical operations.

3.6 Interoperability Enablers

When thinking in terms of information sharing, one will see that each node distributing information requires four technological components: a sensor to gather information; a processor to convert that information into usable data; an interface to display and interact with that data; and a transceiver with the requisite antennas or apertures to transmit and receive data.

Interoperability serves to enhance situational awareness by providing information collected by sensors throughout the global environment and presenting the information in a usable way to both the actors and decision-makers. Nodes throughout the global environment populate the network with information which can then be displayed across the global network. For instance, integrated sensors onboard aircraft detect and display threats to a pilot and then offboard that information to the network. Other systems used for navigation provide the operator with situational awareness about the operator's specific location and transmit that precise location information on the network. The transmission includes pieces of information such as asset status, maintenance information, and fuel states to provide advance notice and real-time awareness across the air space. The collation of data from multiple networks and merging the information provides one coherent picture to all participants. This common understanding and representation of the environment provides a shared awareness to all network users, a greater understanding of the global environment, and serves as a force multiplier for the decision-makers.

The interoperability continuum includes information producers, users, and the data transportation mechanisms which allow information to flow across the global enterprise. It provides both consistent situational awareness and effective decision-making for mission-critical assets. The capabilities and supporting technologies are critical in their own right; however, the integration across the SoS for mission thread execution is required to maximize their capabilities and to facilitate interoperability. While the focus of this chapter is on capabilities, specific material and non-material solutions serve as examples of system components necessary to achieve interoperability. These critical enablers are the waveforms, networks, and message protocols that enable information exchanges required to support operations, techniques, and procedures.

In this section, we examine a number of factors that must be considered when trying to achieve interoperability in modern software-intensive SoS. Of the many complicating factors that exist, the following are exemplars:

- The movement of critical information across transport medium
- The problem of semantics
- The difference between bounded and unbounded software systems
- The need for trust, trustworthiness, and security in software systems

While each factor is discussed separately, significant interrelationships become readily apparent. The first and second factor, that of moving information across transport medium and semantics, permeates every other aspect of interoperability. Then, after noting that a key difference between bounded and unbounded systems is that unbounded boundaries are far more fluid. Fourth, even conjecturing about unbounded systems necessitates considering how security and trust can operate in such a context.

3.6.1 Waveforms, Networks, and Gateways

In the context of MEI, the terms "waveforms," "networks," and "gateways" play a critical role in ensuring seamless communication across diverse systems and platforms. The term waveforms refers to the modulation techniques applied to the electromagnetic wave by the transmitter. These modulations, including amplitude, frequency, phase, and polarization, are fundamental to how data is transmitted over various mediums. The choice of waveform directly impacts the performance, range, and security of communications in complex mission environments, where systems must reliably exchange data across various platforms, including satellites, radios, and other communication devices. Different waveforms may be required to support unique operational needs, from high-bandwidth data transfer to low-latency command and control communications.

Networks, on the other hand, refer to the infrastructure that connects two or more computer systems, enabling the sharing of information, messages, and files. In the mission context, networks can operate over both wired connections, such as cables, switches, and routers, as well as wireless connections, such as radio frequencies or satellite links. The design and optimization of these networks are vital for maintaining operational continuity across a wide range of mission-critical activities. As systems and platforms become more integrated, ensuring that these networks can handle varying demands for bandwidth, reliability, and security is paramount. Furthermore, networks must be able to accommodate both centralized and distributed architectures, where data flow needs to be efficiently routed and processed across geographically dispersed locations.

Gateways serve as the crucial interface between different systems and networks, enabling interoperability by receiving various protocols and waveforms, processing and translating them, and forwarding usable information to operational data systems. In a distributed, multi-network environment, gateways act as translators, converting data from one communication protocol to another, ensuring that systems with incompatible communication standards can still interact. For instance, a system using a proprietary waveform might need to communicate with a different system that uses an open standard. In such cases, a gateway is responsible for converting the data and ensuring it can be properly processed by the receiving system. Given that the global enterprise often employs multiple protocols and waveforms and operates in a distributed manner across vast geographic areas, the need for scalable, organic, plug-and-play gateways becomes even more critical. These gateways must be flexible and able to dynamically integrate with new systems and technologies without requiring extensive reconfiguration or downtime. This plug-and-play capability enables rapid adaptation to evolving mission requirements across diverse operational environments, including air, land, sea, and space domains. To support mission success in complex, high-stakes environments, the integration of waveforms, networks, and gateways must be meticulously designed to ensure robust, secure, and scalable communication, with the ability to dynamically reconfigure and integrate various systems, protocols, and waveforms.

3.6.2 Semantics

Achieving interoperability in complex systems requires not only aligning at the physical and protocol levels but also ensuring systems can consistently interpret the meaning of the data exchanged. This concept, known as semantic interoperability, involves systems sharing a common understanding of the data's meaning in context, not just its raw format. For example, if one system sends the number "5" to another, its meaning depends on both systems agreeing whether it represents a high-priority risk, the fifth day of the week, or some other interpretation. Without this shared understanding, the data is rendered meaningless. Furthermore, in dynamic environments, this alignment must also be maintained as systems evolve or new systems are added. The interpretation of data may shift over time, requiring ongoing efforts to ensure that the meaning remains consistent and relevant across all participating systems.

Universal semantic agreements across diverse systems remain a significant challenge. In dynamic SoS, where new systems are frequently integrated, and legacy systems with varying data formats and protocols exist, the complexity of aligning meanings grows. Establishing universal standards for semantic meaning is particularly difficult due to the varying contexts in which systems operate. While lower level communication standards (e.g., hardware, basic protocols) can resolve many interoperability challenges, semantic meaning often depends on context, cultural nuances, and the specific needs of each system, making universal agreements difficult to attain. The challenge is compounded when integrating legacy systems, as their data formats and operational logic may not align with newer, standardized approaches, often creating friction when systems need to exchange meaningful data.

The ambiguity inherent in human language plays a critical role in these challenges. Misunderstandings can occur even in straightforward human communication, and this ambiguity is replicated in software systems. Human cognitive interpretations of language vary, and when these interpretations are not explicitly defined in a system's design, the system-generated representations can lead to different understandings of the same data. In contrast, machine interpretation of language is often rigid and context-sensitive, amplifying the potential for misinterpretation and complicating system interoperability. One promising approach to overcoming these challenges is semantic technology, which enables systems to dynamically infer meaning and reasoning from data. This technology is especially useful in two key areas for MEI applications: artificial intelligence (AI) with automated reasoning and flexible, transparent data management. AI systems can use semantic technology to infer new facts from existing data, particularly when dealing with complex datasets that are difficult for humans to rationalize. AI also benefits from different reasoning mechanisms such as deductive reasoning, which derives conclusions from established premises; inductive reasoning, which makes generalizations based on specific observations; and abductive reasoning, which infers the most likely explanation for a set of observations. These types of reasoning are essential when data is incomplete or inconsistent, enabling systems to fill gaps or resolve conflicts in data interpretation. Moreover, in evolving systems where data sources and applications are not fully known upfront, semantic technology offers a flexible solution that can adapt to changes in underlying data storage formats without requiring complete re-engineering. This adaptability allows systems to continue functioning effectively even as new data types and relationships emerge.

In recent years, large language models (LLMs) and large knowledge models (LKMs) have become essential tools in supporting semantic interoperability. These models, trained on vast datasets, are capable of understanding and interpreting complex language constructs, often in ways that go beyond traditional machine learning models. By leveraging LLMs, systems can process and infer meaning from unstructured data (such as text or speech) and generate contextually appropriate responses or insights. LLMs provide the capability to understand nuances in language, including context, tone, and ambiguity, which is often a challenge in traditional semantic processing approaches. Furthermore, large knowledge models (LKMs), which organize information into structured knowledge graphs, provide an additional layer of semantic understanding by defining relationships between concepts and entities. These models allow systems to infer relationships and discover hidden patterns within data by tapping into vast stores of accumulated knowledge. LKMs enable systems to dynamically reason across different domains and integrate new knowledge as it becomes available, facilitating continuous adaptation in evolving environments. The integration of LLMs and LKMs into semantic technology helps to overcome the limitations of static ontologies by enabling more flexible, real-time reasoning across systems with diverse data formats and protocols.

Semantic technology also enhances scalability by providing a storage-agnostic approach that integrates diverse data sources and formats, including structured, semi-structured, and unstructured data. By using canonical forms and semantic reasoning, systems can infer relationships between data points based on a small set of rules, which reduces the complexity of managing large data volumes. This capability allows for more efficient data analysis and supports interoperability across systems with different data formats and storage systems. Furthermore, semantic technology helps overcome information silos, enabling diverse systems to collaborate despite variations in data representations. A key component of using semantic technologies is the use of ontologies, which define a shared understanding of concepts and relationships within a domain. Ontologies provide a framework for semantic alignment by establishing a formal vocabulary or taxonomy that ensures systems have a common understanding of data and its relationships. This shared understanding enables systems to infer new relationships or conclusions based on established rules, facilitating dynamic reasoning and knowledge sharing. Ontologies play a crucial role in ensuring that systems can work together, despite differences in how data is represented or stored.

3.6.3 Ontologies

An ontology defines the terms and relationships among terms that represent an area of knowledge. In software engineering, computer-readable ontologies are growing in importance for defining basic concepts within a domain. If multiple-domain applications are developed utilizing shared ontology, or if their distinct ontologies can be related, then the applications can have a common understanding of data, and semantic interoperability is enhanced. In addition, ontologies can be developed that relate information across domains, opening up new possibilities for interoperability.

While offering promise for enhanced semantic interoperability by helping developers to locate relevant descriptions and allowing computers to infer relationships and properties, ontologies are hard to define well because:

- Few people are expert in the representation of knowledge, and these experts are rarely the domain experts building ontologies. As a result, ontologies are often poorly constructed and hard to maintain.
- Consensus building is a hard task. It can be made more difficult by the scope and diversity of the organization and domain(s), by the existence of legacy applications that encourage advocates to fight for their solutions, and by widely different intended uses of the ontology.

In addition, the long-term evolution of ontologies is a complex task. Because applications are strongly coupled to specific ontology versions, evolution will be constrained to maintain upward compatibility unless mappings between versions are provided. Several ontology languages have been developed, but interest is now focused on the widely used Web Ontology Language (OWL). OWL was standardized by the World Wide Web Consortium (W3C). The W3C OWL is a semantic web language designed to represent rich and complex knowledge about things, groups of things, and relations between things to support interoperation across systems and organizational boundaries by providing shared meaning to data. A number of tools exist to create and manipulate OWL ontologies or to reason about OWL ontologies. OWL builds on other web standards to define ontologies that can be distributed across the Web. A growing number of tools support OWL, and hundreds of domain ontologies representing commercial, government, military, and academic interests have been developed.

Merging ontologies or mapping between them is a current research topic. While a variety of approaches and tools are under investigation, almost all require significant human intervention, normally by an individual or group of individuals familiar with the ontologies to be joined. Two additional and contradictory problems with ontology evolution have been noted: diffusion (or mission creep) and enforced orthodoxy. The latter refers to avoiding change and hindering new ways of thinking, particularly regarding revolutionary paradigm shifts to an ontology. This can occur even when change is needed because the new thoughts that can be constructed are limited by the language that is used.

Although ontologies are recognized for their expressiveness and flexibility, most semantic technologies use them exclusively to capture data sources and do not employ rules to capture system dynamics, behavior, or policies. However, only the combination of ontologies and rules provides the full power of semantic technologies as a basis to implement applications and adaptable that grow with the needs of the underlying user needs or system requirements. The main objective of semantic technologies is to provide a conceptual and logical foundation to how information is modeled and interrelated. As such, semantic technologies provide an infrastructure on which to build applications because they have been proven effective as a means to integrate disparate information, especially in applications that have traditionally used taxonomies or controlled vocabularies.

3.6.4 Semantic Interoperability Approach

Detecting and addressing semantic interoperability problems between disparate, heterogeneous systems is possible only if knowledge about the native semantics of those systems has been captured. To tackle this issue, it is essential to explicitly express the knowledge that is currently implicit in message structures. Instead of defining a field's meaning in a paper-based specification document and assuming consistent interpretation of its values, the meaning must be represented as a machine-processable digital artifact. This artifact not only defines the field and its values but also illustrates how it relates to other similar artifacts. It captures the knowledge about the field within a message and describes what the value means and how it relates to other information beyond the message's context. This knowledge, stored separately, can then be used to interpret and process message instances.

Capturing such knowledge employs formalized structures known as ontologies, defined within the knowledge engineering community as explicit, formal specifications of shared conceptualizations. A significant benefit of using these formalized structures is that a suitably constructed software program can understand the semantics of the represented knowledge and process it appropriately, without needing the semantics implicitly encoded in the software program's procedures. This means that data about real-world phenomena, or conceptualizations, are bound to self-describing metadata. Such a construct allows for a fundamentally different kind of software program, one not tailored exclusively to handle only one possible meaning of data, but to correctly extract and process the meaning of any specific data instance.

Many terms used in natural languages have several distinct meanings. When such ambiguous terms specify the information content shared among different systems, different assumptions may be implicitly encoded in software procedures themselves, leading to semantic interoperability problems. An ontology lexicon provides a constrained semantic interpretation of these terms and formal definitions, a process known as ontological commitment. This involves mapping ontology terms to their intended meanings, specifying precisely which meaning is intended. The computing world recognizes ontologies as an emerging mechanism for addressing semantic interoperability in information systems. This aligns with the acknowledged importance of semantic understanding and interoperability for organizations and their systems to successfully and competitively provide their services. By specifying conceptualizations in terms of an "agreement" on meaning between involved parties, an ontology becomes a reification of a knowledge agreement.

When integrated into heterogeneous systems, ontologies offer several benefits. They serve as repositories for metadata needed to ensure correct communication between different systems, especially those operating with incompatible data models—a common scenario in coalition operations. While data exchange itself is challenging and requires significant effort, understanding and processing data as intended by the provider is even more challenging without established message formats. By providing meaning to concepts, attributes, and attribute values used in respective data models, ontologies shift communication from merely exchanging data to a communication between systems' ontology components, leveraging their flexibility and adaptability.

When it comes to data exchange between systems using different message formats, external semantics encoded as ontologies can help determine how the data should be interpreted and facilitate the construction of mappings between ontologies for translating messages. In an operational context, the urgent need to exchange data between multiple systems is often constrained by limited time. Analyzing existing data structures and message formats, reverse engineering-intended results from message exchanges, and constructing ontologies for a system is time-consuming. This makes constructing mappings for data exchange between systems on short notice challenging, if not impossible, and can significantly delay operational readiness. Formal representations of semantics through ontologies are a crucial step toward automated information interoperability. However, beyond the use of ontologies and related tools, a common information exchange methodology is also necessary. For this purpose, ontologies must become integral to the lifecycle of information systems and the recommended implementation strategy. The boundary conditions, necessary methods, and rules for their application should be part of a framework that supports semantic interoperability, providing a scalable solution for cooperation.

3.6.5 Trust, Trustworthiness, and Security

Trust, trustworthiness, and security play a critical role in achieving interoperability. All depend on deep, rich, and shared semantic understanding. Discussion of these notions is also dependent on our initial definition, where we stress that interoperability can be considered only in a given context.

In the envisioned landscape of unbounded SoS, trust emerges as a foundational element vital for seamless collaboration among interoperating entities. Within this dynamic environment, each participant in an interaction must possess, cultivate, or perceive a degree of confidence in the reliability and competencies of their counterparts. This trust is far from binary; rather, it spans a spectrum, influenced by contextual nuances and subject to temporal shifts. One entity may be deemed trustworthy in specific scenarios while raising doubts in others. Moreover, the trust bestowed upon a party is not static; it evolves over time, contingent upon ongoing actions and outcomes. Importantly, the absence of trust or outright distrust, though not devoid of challenges, poses fewer risks than misplaced trust. It is preferable to acknowledge the limitations of trust in a particular party than to mistakenly place reliance where it is unwarranted, thereby leading to systemic vulnerabilities and potential failures. Thus, in the pursuit of effective collaboration within unbounded SoS, nurturing and managing trust relationships with sensitivity to context and time dynamics becomes paramount.

Trustworthiness is intricately linked to the genuine performance and reliability of the end-to-end service delivered, particularly within the framework of an SoS. Misplaced trust occurs when one party holds the perception that another party is trustworthy, yet the latter fails to uphold this trustworthiness in practice. Unlike trust, which can vary in degrees and evolve over time, trustworthiness tends to exhibit a more binary nature within a specific context and instance. At any given moment, the end-to-end capability is deemed trustworthy or not, based on its actual performance and fulfillment of expectations. However, trustworthiness remains subject to contextual and temporal influences, wherein its assessment may fluctuate across different scenarios and over time. Therefore, ensuring the alignment between perceived trust and actual trustworthiness is crucial for fostering reliable and resilient interactions within complex SoS environments.

Security encompasses several key principles, including confidentiality, integrity, and availability, all essential for safeguarding information within systems. Confidentiality ensures that information is accessible only to authorized individuals, maintaining its secrecy from unauthorized parties. Integrity ensures that information remains unaltered and free from corruption, whether due to intentional tampering or accidental errors. Availability ensures that information is consistently accessible and not subject to disruptions or unauthorized deletion. Addressing security concerns typically involves a combination of security policies and technologies:

- Security policies outline risks, threats, and guidelines for system management and legitimate usage, as well as protocols for responding to security breaches.
- Security technologies minimize intrusions, detect unauthorized access, and limit damage. Examples include one-time password systems, firewalls, monitoring tools, security analysis software, and encryption methods.

While security primarily focuses on preventing unauthorized access, corruption, and disruptions, trust and trustworthiness encompass broader considerations related to the reliability, predictability, and consistency of system behavior. It is conceivable for a highly secure and error-free entity, operating within a secure network, to lack trustworthiness in a specific context or for a particular need. The distinction lies in the fact that trust involves more than just ensuring secure interactions but requires confidence that the information exchanged is appropriate for the given context and timeframe. While security mechanisms are essential for safeguarding interactions, they alone are insufficient for fostering trust and constructing trustworthy capabilities between components. In traditional system development, ensuring trustworthiness often involves close collaboration with component providers, rigorous modeling, and comprehensive testing of end-to-end capabilities. By working closely with providers and thoroughly evaluating system behavior under various conditions, stakeholders can establish trust in the reliability and suitability of components for their intended purposes. This holistic approach goes beyond mere security measures for it emphasizes the importance of consistent performance and alignment with expected outcomes. Ultimately, establishing trust between interoperating systems requires not only secure interactions but also a deep understanding of system behavior, context, and the reliability of information exchanged.

3.6.6 Data-Driven Architectures

To achieve SoS-level interoperability, procurement agencies are recognizing that the SoS state needs to be maintained in the architecture infrastructure and not within applications or in a specific system. In fact, the entire SoS state must be made explicit within the integrated open architecture by every connected functional system or application. This decouples not only the communication between systems, but also the state information. Multiple guiding principles to achieving data-led interoperability follow:

- All data intended for exchange is rigorously defined with clear semantics, thoroughly described, documented, and readily accessible.
- Data exists within the SoS independently of any specific application or function and is managed by the infrastructure itself. In this approach, the infrastructure serves as the custodian of data, ensuring its availability, integrity, and accessibility across the entire ecosystem of interconnected systems and applications.
- A centralized data repository which may be fully distributed and serverless in nature must be established in accordance with open standards. This repository serves as the single authoritative source of state information within the SoS architecture.

Adhering to these principles instead of relying on inter-system or inter-application communication, any system or application can access state information directly from the software infrastructure. This shift reduces coupling between components and eliminates isolated data silos. A prevalent example of this approach is the widespread use of databases as central repositories, effectively serving as a "cache" for state data. Databases enable client applications to efficiently retrieve, display, and modify data and its state without requiring constant system-wide connectivity. While this method is highly effective for managing data-at-rest, it falls short in the context of distributed, non-homogeneous real-time systems. Centralizing state information in such scenarios is impractical and unfeasible. Instead, a fully distributed, real-time bus system capable of maintaining the state of data-in-motion ensures seamless communication and responsiveness across the SoS landscape. The following approaches play a critical role in unlocking the full potential of data assets and driving business value in the global data-driven landscape.

3.6.6.1 Seamless Data Aggregation

The data architecture enables enterprise-wide, seamless data access, and persistence to support data aggregation across diverse data domains. Data producers use analytics to access data across data domains and classification levels to produce a refined aggregated data product and tag the aggregated data product with classification and other access attributes to allow user/analytic access at various levels. This seamless aggregation provides comprehensive awareness and leverages related data that informs end users. These data combine into an entity state product, with each data element tagged for source, access, and processing provenance. Seamless data aggregation requires that data producers and consumers use data management services and concepts for:

- Data access by local and remote analytics to the best data sources, regardless of physical data location.
- Data query/retrieval/update across security enclaves to access the best data sources for analytic processing.
- Metadata tagging to record data pedigree, provenance, and lineage, to support future use of the aggregated data product.
- Data archiving to ensure change tracking and access to prior versions of data to support multiple hypotheses, backtracking, and forensic analysis of past, current, future data products and all supporting data.
- Realizing an Extensible Object Model for consistent use of data between different data domains.

Seamless data aggregation supports tasking, collection, processing, exploitation, and dissemination prior to acquiring a complete understanding of the end state and its characteristics. Seamless data aggregation provides multi-information aggregations to support analysis methods designed to discover unknown objects, which are then resolved and added to the baseline.

3.6.6.2 Location of Agnostic Data

Location-agnostic data enables authorized users and systems to discover, identify, access, and retrieve data without requiring knowledge of where the user or data is physically or logically located. Data can be distributed, replicated, or cached across enterprise nodes according to mission needs, operational conditions, and the attributes and access privileges of the requesting person or non-person entity. This approach supports users operating in denied, disconnected, intermittent, and limited-bandwidth (D-DIL) environments by enabling need-based data movement, local availability, and synchronization when connectivity permits. Location-agnostic data relies on persistent identification and metadata to separate the identity and meaning of data from its physical storage location, application, platform, or network. The use of globally unique identifiers enables the same data entities to be consistently identified across distributed enterprise nodes. When combined with authoritative data sources, metadata, access controls, version management, and synchronization mechanisms, this approach helps maintain data consistency, provenance, integrity, and trust across the enterprise. For MEI, location-agnostic data improves interoperability and information sharing across heterogeneous systems and enables mission participants to access the trusted data required to support timely decisions and mission outcomes.

3.6.6.3 Secure Data Access

The secure data-access characteristics enable access to co-mingled data of differing security classifications. All data in the data architecture is tagged with cryptographically bound security tags. Secure data access uses these tags to access data in support of analytics, dissemination, and display. Secure data access enables producers and consumers to access the most consistent, value-added products in the knowledge base at the security level for which they are cleared. Thus, access to the data is controlled, monitored, and managed. Secure data access requires data producers and consumers to use data management services and concepts for:

- Authentication and authorization services
- Cross-security domain data transfer services
- Data query and selection services
- Lifecycle data management services
- Cybersecurity audit and logging services

3.6.6.4 Right Data, Right Time, Right Location

The data architecture requires defining local and enterprise-level mission needs including specified QoS for data dissemination. The "right data at the right time at the right location" characteristic depends on maintaining local node cache, providing consistent data dissemination methods and patterns, and interfacing with prioritized communications. In denied, disconnected, intermittent, and limited-bandwidth environments, the right data may not be the most up-to-date data, but the best available data at the time. This concept requires data producers and consumers to use data management services and concepts for:

- Selection of specifically profiled data (mission profiles, analytic queries, support to activity modeling)
- Distribution of data on the local node and across the enterprise
- Access to local data
- Management of the local node cache
- Options negotiation for prioritized communication

Enterprise data architecture enables big data analysis to facilitate the discovery of the right data and supports activity-based data analysis and exploitation.

3.6.6.5 Trusted Data

The trusted-data characteristic emphasizes that all data include comprehensive metadata to document data source, pedigree, provenance, and lineage to support data assessment and application to missions. Insight into data trust is essential for machine and human consumption of the data. Data trustworthiness requires data producers and consumers to use data management services and implement key data strategy concepts to include:

- Metadata tagging to record data pedigree, provenance, and lineage
- Use-defined patterns and practices for domain-specific data extensions to the Extensible Object Model, including assignment of metadata, types, and terms
- Alignment with and extension of the controlled vocabulary to avoid semantic overlap
- Open use specifications and standards for data access and dissemination and assignment of identifiers

3.6.6.6 Data at the Speed of Relevancy

Operational timelines require the data architecture to monitor and dynamically refine data processing and access timelines in support of mission requirements. End-to-end operational timelines are decomposed into time budgets allocated to specific performance budgets for specific processes and activities. Support to operational timelines requires data producers and consumers to use data management services and implement key data strategies to include:

- Minimize, automate, or eliminate the need for data mediation by using extensible object models and vocabularies to support analysis across data domains
- Expose data and analytic results back into the enterprise with complete metadata tagging (security, characteristics, pedigree, provenance, and lineage) to allow data dissemination services to meet required performance budgets
- Use dissemination and cache management services to support streamlined data sharing of objects and their relationships and to prioritize use of available communications capabilities
- Create metadata to manage and monitor analytics' ability to meet performance timelines

3.7 Global Interoperability and Standards

Discussions of modernization and mission-centric development frequently converge on the importance of modular open standards. Certainly, adaptive global interoperability is a critical need for all tenets of modernization, and global standards are widely accepted as the best if not the only means of achieving this degree of interoperability. While well-meaning with many positive attributes, recent developments indicate that newer technical approaches will enable other means of achieving global interoperability. Although global common standards play a critical role in achieving global interoperability, strong cautions against widespread imposition of global common standards are advised. Alternatives are offered in the paragraphs that follow.

Fig. 3.2 Global standards do not necessarily provide interoperability

Global standards and open systems architectures are fundamentally good ideas that solve many legacy problems of system and SoS designs. Traditional stovepiped or vertically integrated systems use locally defined, often proprietary interfaces between systems and subsystems, or possibly no interfaces at all. Any modification of these systems requires a great amount of manual design work. Usually, the owner of the system alone has access to design details and system experience necessary to make these modifications. This legacy approach leads to systems that cannot readily be adapted, are slave to the incumbent system integrator, and fundamentally not interoperable. Neither this inefficient approach nor mandatory global standards guarantee an interoperable product as illustrated in Fig. 3.2.

In contrast, if global standards exist, systems can be updated and combined very efficiently, provided they all comply with the standard. When open systems do work, they provide many advantages:

- Enables rapid tailoring of a system to a mission need
- Promises wider access to innovation
- Addresses obsolescence risk
- Drives risk-prudent competition, providing a constant battle rhythm of competition by leveling the playing field and reducing barriers to market

But what happens when one wants to compose capabilities that span beyond a single system or domain as when executing a mission? Global interoperability then becomes a driver. Consider the toy example in Fig. 3.3 where open system standards show rapid creation of new capabilities within given systems or domains. While each type of toy provides the flexibility promised by open standards, there is no easy way to create something out of a combination of the different toys; there is no global interoperability.

A solution often proposed is moving to a more global standard—a "universal translator." If the scope of the standard can be expanded to cover the full range of

Fig. 3.3 Open systems standards promote rapid development of new capabilities

Fig. 3.4 Universal translators are not the answer

interfaces at a minimum, interoperability is achieved, correct? While appealing in principle as a logical extension of open standards, this approach quickly fails when the diversity of the computing ecosystem grows too large. Therefore, as depicted in Fig. 3.4, a universal translator is not the answer as it is inefficient, does not adapt, and in the longer term actually stifles innovation and modernization.

Consider some commercial examples. Global standards work well in cloud computing. A common virtual machine (VM) standard enables universal plug-and-play into a cloud because there is relatively little variation between different hardware and operating system choices. A VM universal translator can be relatively simple while spanning the entire cloud ecosystem. The internet is a more complex example. On the one hand, a global standard, Internet Protocol (IP), exists. However, IP describes only one layer of the internet, how a network is defined between computer nodes. Nothing else about the internet and its usage adheres to a global standard. Indeed, there is an almost uncountable number of configurations of physical data links and applications and even ways that data are routed running on the internet, and the number is continuously growing. This adaptability and innovation came about precisely because there was not a single global standard.

So why don't global standards work? The simple reason is complexity. If only a few systems or functions are involved, creating a standard is fairly simple. However, as the number of systems and functions grows, the complexity of a standard grows geometrically. This challenge is compounded by requirements processes, which strive for predictability and performance. Before a standard can be defined, all potential users must predict and define its uses for some extended time, likely to be several years. If a standard is meant to be universal across the global environment, that list of requirements could be quite large. Once the process of developing the technical specification begins, the superset of requirements makes the resulting specification intractable; therefore, some characteristics or functions need to be sacrificed. This begins a negotiation process of what functions each stakeholder is willing to live without. As a result, the standards definition process itself can take many years.

Once a standard is agreed upon, the goal of creating efficiency, agility, and innovation is often not realized. First, acceptance is difficult independent of policy. There are certainly political and business motivations that work against acceptance; there may even be legitimate reasons for resistance. The standards creation process often is not driven by the developers who will ultimately use the standards sitting at the table as they are being defined. As a result, it may not really suit the needs of their system, and even when it is suitable, there is time and effort associated with learning and adapting to it. Often, driven by the compromises that need to be made in the requirements process and negotiation, it may not be an ideal technical solution. All of this leads to developers not using the standard directly but instead learning how to translate from their existing interfaces to the standard. This increases the probability of success and provides some opportunity to integrate different things together, but it does not fulfill the full desire for innovation and certainly is not very efficient.

Another problem arises that stifles innovation further. If new capabilities fit within the scope of the standard, opportunities for competition and innovation arise for a period of time. But if a new capability falls outside the standard definition, the standard becomes a barrier to innovation. Because of the time involved in defining and managing a global standard, opportunities to modify it are limited. The new capability will likely have to wait longer for the next standard version than if the laborious process of designing a custom interface were used. Furthermore, it may not be possible to make every change needed in a standard because of backward compatibility. If a change is not compatible with everything contained in prior versions, it requires every system using the standard to make an update creating an enormous barrier to making modifications to standards.

So how can global interoperability be achieved if global standards are not the answer? Just mandating interoperability as a policy statement is not sufficient. It is too difficult to define in a way that is meaningful and cannot be resisted or misinterpreted. Instead, there is a technology solution: if a system simply has well-defined, exposed interfaces, there are tools being developed that can enable on-the-fly, tailored interfaces with other systems. Interoperability is thereby achieved without a common standard.

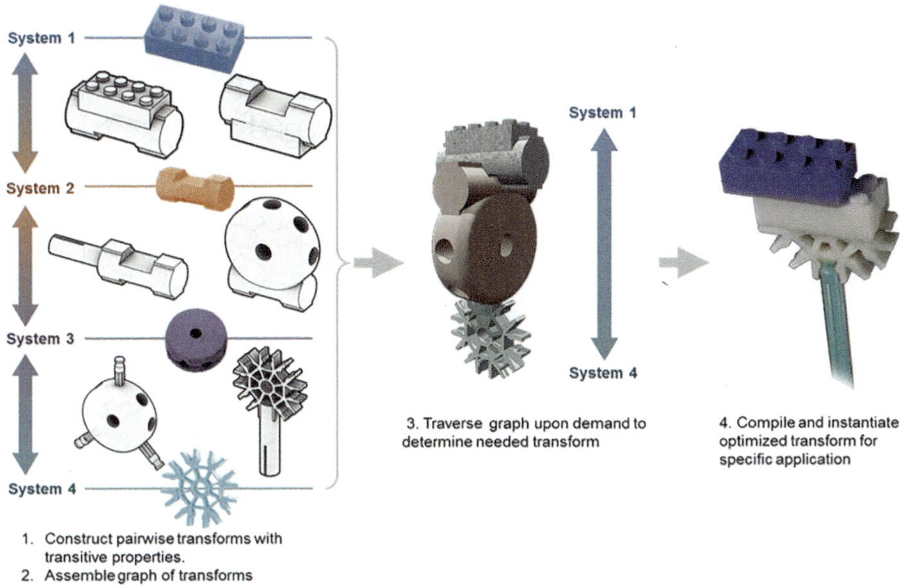

Fig. 3.5 Toy examples enabling creation of custom interfaces

Figure 3.5 illustrates a toy incompatibility to demonstrate this ability to combine different systems into an integrated SoS with interoperability to create an effective product.

Suppose a pair of systems (e.g., a Lego and a Lincoln Log) have already been integrated once in the past, or they are part of an ecosystem that has its own local standard. A specific translator now exists for those two systems. Likewise, suppose other systems also have been integrated in the same pairwise manner. In the case of the toy problem, one translator takes Lincoln Log to Tinker Toy, and another Tinker Toy to K'Nex. Now a universal translator tool can provide the means of defining all of these pairwise translators as mathematical transformations. Suppose an effects web builder wants to use two systems within a mission engineering thread that have never integrated before (e.g., integrating a Lego to a C'Nex), which do not share a common standard, but have been integrated with other systems. The proposed software tool automatically discovers a set of existing transformations that can connect the two new systems and automatically generates a new custom interface. This software process does not eliminate interface definitions, manual system engineering, or even localized standards, rather it provides a pathway to creating on-the-fly global interoperability between any two systems without the existence of a shared global standard. Table 3.2 shows implications and opportunities that arise from a future state that achieves global interoperability without global standards. It will be disruptive to the industrial base in the short term; however, this movement presents great business opportunities in the long-term over and above the game-changing opportunities to promote global well-being.

Table 3.2 Implications and opportunities of global interoperability

Implications	Challenges	Opportunities
Reduced importance of massive monolithic development programs More, smaller, focused elemental acquisition programs	Forecasting uncertainty	More chances to win; no "bet the company" competitions
	Burden on business development/capture teams	Increased opportunities for innovation, invigorated engineering workforce
		Operations efficiency; level utilization of human and physical capital
Industry has leadership role in exposing interfaces and fostering ecosystem for interoperability	Need to self-organize and cooperate on interfaces	People building the systems get to define natural interfaces
		Interfaces can be defined to protect intellectual property and the ability to differentiate
More commercial-like acquisition of new tools for integration support	Industry accepts more up-front risk	Potentially greater financial rewards (e.g., higher margins)
		Renewed ability to innovate; not hindered waiting for requirements

Term	Technical Definition	Relation to Interoperability & Commercial Context
Portability	The ability of something, e.g.,a software application, to be readily moved from one environment to another, usually due to a common platform	While this aids re-use of the application, it has no association with interoperability of the application with other applications in the environment it has been moved into. It only facilitates integratability with the platform or SoS
Open System	Provides for "some" level of system capability that exhibits interoperability, portability and use of open standards	There is no standard to define the level of openness, nor its relationship to interoperability

There is not necessarily a complete answer to this global interoperability problem, but this proposed approach does provide an example of the benefits of customization and standards may both be achieved. Indeed, one may consider a software translator package and the protocol for capturing interface specifications itself a standard, but it is not a standard as commonly used by the design community today, i.e., an exact interface specification. Regardless of what it is designated, these integration tools can create a pathway to a future global state in which design of complex effects webs and mission engineering threads and distributed architectures can be made in the field at operational planning timescales to satisfy the speed of relevance. Achieving global interoperability without global standards is the key to achieving continuous speed.

This chapter explored the challenges and enablers of **integration and interoperability**, emphasizing the need for seamless system interactions through modular open systems architecture, open standards, and semantic interoperability. However, mission success depends not only on technical integration but also on how effectively humans interact with these systems. Chapter 4 shifts focus to **Human Systems Integration (HSI) and Human Factors Engineering (HFE)**, examining how human capabilities, limitations, and cognitive processes influence system design and performance. As systems grow more complex, ensuring intuitive human-machine interaction becomes essential for reducing cognitive workload and enhancing decision-making. This next chapter explores human-centered design principles to optimize system usability, efficiency, and mission effectiveness.

References

1. Madni, A. M., & Sievers, M. (2014). Systems integration: Key perspectives, experiences, and challenges. *Systems Engineering, 17*(1), 37–51.
2. Office of the Under Secretary of Defense for Acquisition & Sustainment. (2023). *Open systems approach to weapon system acquisition*. U.S. Department of Defense.
3. Pollock, J. T. (2001). The big issue: Interoperability vs. integration. *Enterprise Application Integration Journal, 3*(10), 20–22.
4. U.S. Congress, Goldwater-Nichols Department of Defense Reorganization Act of 1986, Pub. L. No. 99–433, 100 Stat. 992, October 1, 1986.
5. U.S. House of Representatives, Select Bipartisan Committee to Investigate the Preparation for and Response to Hurricane Katrina, A Failure of Initiative: Final Report of the Select Bipartisan Committee to Investigate the Preparation for and Response to Hurricane Katrina, H. Rep. No. 109–377, 109th Congress, 2nd Session, U.S. Government Printing Office, Washington, DC, 2006.

Chapter 4
Human Systems Integration/Human Factors Engineering

4.1 Human Systems Integration

Human Systems Integration (HSI) is an interdisciplinary and comprehensive management and technical process that emphasizes integrating human considerations into system acquisition and development processes. The goal of HSI is to enhance human-system design, reduce lifecycle ownership costs, and optimize total system performance. HSI encompasses various domains, including manpower, personnel, training, human factors engineering, safety, health, habitability, and survivability. These activities are considered concurrently and integrated with all other systems engineering (SE) design activities, ensuring that human factors are thoroughly addressed throughout the system lifecycle. By incorporating HSI principles, organizations can create systems that are more user-friendly, efficient, and effective while also minimizing risks and costs associated with human factors issues.

Three key concepts define an effective HSI program. First, it is essential to recognize that systems comprise hardware, software, and human personnel, all operating within a surrounding environment. Often, acquisition programs for systems (Program of Record) overlook the human capacity or requirements as part of the system, leading to poor task allocation among hardware, software, and human users or supporters. To ensure optimal task allocation, it is critical to consider the human element early in the system development process. Second, successful HSI depends on integrating the functional HSI domains into acquisition planning efforts, such as participation in program High Performance Teams (HPTs) and Integrated Product Teams (IPTs). HSI domains often function independently due to the distribution of expertise within an organization. In a "stovepipe" approach, each discipline struggles to assert its importance in the acquisition process. Proper HSI implementation through participation in HPTs and IPTs integrates these domains, leveraging their interdependencies to achieve an optimal design. This integration allows for effective HSI through trade-offs and collaboration, providing a common basis for informed

J. D. Moreland, Jr., *Mission Engineering and Integration*,
https://doi.org/10.1007/978-3-032-32132-9_4

decision-making. The results of these efforts should be reflected as updates to system requirements and documents. Finally, HSI must be considered early in the requirements development phase of system design and acquisition to maximize return on investment (RoI) and system performance. HSI requirements should be developed alongside capability-based requirements through functional analyses. These HSI requirements will drive HSI metrics and embed HSI considerations within the system design. Once a system is designed, addressing HSI oversights can become very expensive, highlighting the importance of early integration.

HSI incorporates several functional areas, referred to as domains, which are critical to its implementation. Those domains are listed in Table 4.1 and cover a broad spectrum of human factors and system interactions. HSI personnel, who provide integrated domain oversight, implement HSI processes and practices and ensure the involvement of HSI domains throughout the system's engineering lifecycle. Effective HSI implementation hinges on the comprehensive integration and oversight of these domains.

While there may be overlap among those responsible for overall HSI domain integration and specific domain expertise, the individuals tasked with providing consolidated HSI input depend heavily on discipline experts within each HSI domain. Functional implementation of HSI is based on regular and frequent communication, coordination, and integration across the HSI domains to provide comprehensive human-system expertise. This approach ensures that all aspects of human interaction with systems are considered and optimized from the earliest stages of system development through to operation and maintenance.

4.1.1 HSI for Commercial-Off-the-Shelf Products

Program managers are encouraged to use Commercial Off-The-Shelf (COTS) products to save time and money considering rapid development and transition as well as meeting affordability limits. However, a common misperception is that using COTS eliminates the need for HSI. In fact, incorporating COTS content increases the need for HSI assessment and risk mitigation. The more extensive the use of off-the-shelf content—whether COTS or Government Off-The-Shelf (GOTS)—the more crucial it is for HSI to inform equipment selection decisions. Without proper HSI evaluation, inexpensive COTS equipment can lead to high human-related costs and poor overall system performance. To avoid these risks, the HSI team must identify all human-related issues before selecting products, preferably during market research of available commercial products or those used by other agencies and programs. Optimally, usable systems are achieved through the successful integration of equipment with human components. Early testing and evaluation are crucial to ensure proper system usability. Measures of Effectiveness (MoE) and Measures of Performance (MoP) should consider the end user as part of the system, going beyond models and simulations to ensure the system is not only suitable but also operable and maintainable. Insights gained from usability testing and evaluation can

Table 4.1 HSI functional domains

Domain	Definition	Examples of expertise
Human Factors Engineering (HFE)	Designing hardware and software to optimize human well-being, overall system safety, performance, and operability involves focusing on human capabilities and limitations as they affect and are affected by system design across various mission environments and conditions (nominal, contingency, and emergency). This approach ensures robust integration of all humans interacting with a system throughout its life cycle. HFE solutions are guided by three core principles: system demands must be compatible with human capabilities and limitations; systems should enable the utilization of human capabilities in non-routine and unpredictable situations; and systems must be designed to tolerate and recover from human errors	Task analysis, human performance measures (workload, usability, situation awareness), HFE Design (biomechanics, crew functions, habitat architecture), Human in-the-loop (HITL) Evaluation, Human Error Analysis, Human-system Interface, Systems Design, and HFE Analysis
Operations Resour ces	Operations planning and execution require careful consideration of operability and human effectiveness. These factors drive the design and development phases of systems, ensuring they meet the needs and capabilities of their human operators. Effective planning involves making trade-offs in function allocation, determining the appropriate balance between human tasks and automated processes, and establishing the level of autonomy that systems should possess	Operations process design involves creating efficient workflows to ensure optimal human/machine resource allocation. This includes planning and coordinating mission operations, conducting resource modeling, and performing complexity analysis to manage mission intricacies
Maintainability and Suppor tability	Designing to simplify maintenance and optimize human resources, spares, consumables, and logistics is crucial for operational missions where time, access, and distance are limited. This approach ensures maintenance tasks are straightforward, reducing the need for extensive training and minimizing errors. Optimizing human resources involves efficient allocation and utilization of end users' skills and time	Maintenance and housekeeping during operations are critical aspects of operations. Additionally, sustainability and logistics play key roles in ensuring the smooth functioning of operations
Habitability and Environment	Consideration of both external and internal environments are crucial for human habitats, encompassing exposure to natural elements and factors related to living and working conditions. These considerations are essential to sustain t morale, safety, health, and performance of the user population and directly impacts personnel effectiveness	Environmental health considerations encompass various domains, including radiation health, toxicology, nutrition, acoustics, and architecture
Safety	Safety factors are paramount to ensure that mission activities are executed with minimal risk to personnel. Mission success entails not only achieving objectives but also the safe return of operators and safety of support personnel throughout the mission	Safety analysis, reliability assessments, and quality assurance are crucial components of ensuring the survivability and human rating of systems. Additionally, hazard analysis plays a vital role in identifying and mitigating potential risks

(continued)

Table 4.1 (continued)

Domain	Definition	Examples of Expertise
Training	Developing a training program aims to streamline the resources needed to equip personnel with the necessary knowledge, skills, and abilities to effectively operate, maintain, and support the system	Instructional Design, Training, Facility Development

significantly inform the development of system training. Furthermore, usability evaluations should continue through fielding and periodically throughout the acquisition lifecycle to ensure ongoing benefits in real-world applications and settings.

Many factors should be considered in assessing the suitability of COTS equipment for operations. Since COTS devices are primarily designed for varying applications and operational conditions, engineers must often conduct comparative assessments between candidates using prioritized criteria and the input of experienced domain experts and stakeholders. From a human safety perspective, engineers should inspect the item to ensure proper workmanship and the absence of prohibited parts or materials. Early testing of the COTS item is advised to verify its expected function, quality, and to protect the project schedule if issues arise. Engineers should work closely with domain experts and adhere to the certifying authority's regulations for COTS use on the project/platform. All COTS items should be identified in design reviews and documentation to ensure visibility to regulating authorities. Additionally, the engineer is wise to assess and document the risks associated with using COTS products for the project.

4.1.2 Developing HSI-Based Requirements

HSI requirements are a critical tool for impacting system design and performance, despite their cost and schedule implications. These requirements ensure that human factors are adequately considered during system design. HSI requirements are developed, integrated, interpreted, and verified with support from HSI personnel, SE personnel, and discipline experts across each HSI domain. While some HSI requirements originate from standards, others are derived from the Concept of Operations through functional analysis of the mission, scope, relevant HSI domains, and human risk mitigation. Each standard statement essentially defines an acceptable risk level in a specific area. These standards are translated into program HSI requirements managed by the HSI team to ensure traceability back to specific organizational standards while providing usable information for the project/platform teams in SE, design, and operations. These requirements are sometimes very specific and quantitative due to the constraints imposed by human capabilities and limitations, such as maintaining health and performance. However, they also allow as much design and operational trade space as practical while ensuring human protection. Thus, HSI requirements help balance the need for system performance and usability with the necessity of safeguarding human well-being. By thoroughly integrating HSI requirements into the design process, organizations can ensure that systems are optimized for human use and performance, yet safe from potential human-related risks.

HSI practitioners use metrics to estimate and track the effectiveness of HSI implementation. Employing quantitative metrics whenever possible allows HSI requirements to be measurable, which is crucial for effective implementation within the SE lifecycle. To be effective, HSI metrics must be objective, verifiable, and validated against system goals. Such metrics can reveal trends and identify areas requiring trade studies. Key categories of metrics for characterizing the success of HSI efforts include cost, schedule, risk, and performance. Operational resources, personnel skill levels, and training are key metrics often used to measure a program/project's HSI success (or failure). Program-specific metrics may be developed to track high-visibility elements of system development that pose significant risks to budget, schedule, or mission success. These metrics need to be defined and tracked in an HSI Plan and reported during system design reviews, maturing over the project lifecycle to improve their significance and merit, and showing trends in relation to program goals over time. Key Performance Parameters (KPPs) are metrics derived from overarching program goals and are applied to a particular scope and operational context. Depending on the nature of the program risk area, an HSI practitioner characterizes the risk using one or more measurement methodologies as highlighted in Chap. 9, Risk Assessment and Management. MoEs and MoPs can be used to establish and manage technical margins, thereby reducing development risk and increasing the probability of mission success.

4.2 Human Factors Engineering

Human Factors Engineering (HFE) involves integrating human characteristics into the definition, design, development, and evaluation of systems to optimize human-system performance in operational conditions. HFE focuses primarily on designing human-machine interfaces that align with the physical, cognitive, and sensory abilities of the user population. Specifically, considerations in this domain encompass ergonomics, psychological principles, human behaviors, and characteristics such as sensory, perceptual, mental, and physical attributes as they relate to the design, development, test, and evaluation of systems. Considering these factors, HFE aims to enhance usability, safety, and efficiency, ultimately improving the overall effectiveness of systems and technologies.

The Program Manager employs HME in system design to achieve the following requirements for effectiveness:

- Optimize human-system performance under operational conditions
- Provide effective training
- Ensure systems can be operated, maintained, and supported by specific personnel
- Guarantee systems are suitable (habitable and safe with minimal safety and occupational health hazards) and survivable for both the crew and equipment

Integrating human characteristics and limitations into system design, development, and evaluation criteria supports the goal of optimizing human performance during operations, maintenance, support, and training activities. The HME effort should focus on developing or improving all human interfaces of the system to ensure the required effectiveness of human performance during system operation, maintenance, support, control, and transport. It should also aim to minimize economical demands upon personnel resources, skills, training, and costs. To achieve these goals, the HFE effort must be well integrated with other HSI domains and supported by HSI practitioners, user advocates, and other functional leads (such as acquisition program function leads) when necessary. This collaborative approach ensures a comprehensive consideration of human factors throughout the system's lifecycle, leading to optimized human-system interaction and overall system performance.

4.2.1 Human Abilities, Concerns, and Requirements

Human factors encompass the cognitive, physical, and sensory abilities required by end users to perform system operational, maintenance, and support tasks. Human factors engineers contribute to the acquisition process by ensuring that the Program Manager designs systems that leverage the abilities of the user population without exceeding them. The HFE community integrates the human characteristics of the user population into the system definition, design, development, and evaluation processes. The goal is to optimize human-machine performance for the operation, maintenance, and sustainment of the system. HSI ensures that systems are user-friendly, efficient, and effective, thereby enhancing overall system performance and user satisfaction. Human-centered design concerns include but are not limited to:

- ***System Elements***—Hardware and software elements designed to enable and facilitate effective and safe human performance. These include user-friendly controls, clear displays, ergonomic workstations, well-organized worksites, accessible entrances and exits, legible labels and markings, robust structures, stable steps and ladders, strategically placed handholds, and provisions for easy maintenance.
- ***Mission Essential Tasks***—Assigning functions and tasks and deciding whether they are performed by humans or automated systems based on human capabilities and limitations.
- ***Physiological***—Human characteristics including factors such as muscle strength and endurance in various positions, visual acuity, tolerance to temperature extremes, the range of human hearing frequencies, and sensitivity to vibration.

- ***Information Collection/Situational Awareness***—Data, information, and its attributes that equip humans with knowledge, comprehension, and awareness of events in both the tactical environment and the system itself.
- ***Cognitive***—Guidelines for decisions, systems supporting decision-making, methods to maintain situational awareness, mental models of the tactical environment, methods for generating knowledge, cognitive abilities, attitudes, and aids for memory.
- ***Environmental Factors***—Natural and constructed environments, controls over environmental factors, and facility designs that promote optimal human performance.
- ***Operational Factors***—Components of a system that facilitate its successful operation, including procedures, documentation, workload management, and job aids.
- ***Organizational Culture***—Aspects such as job design, management structure, command authority, policies, and regulations that influence behavior.
- ***Communication***—Providing facilities for enhancing team performance, fostering cooperation, promoting collaboration, and communicating among team members and with other personnel.
- ***Ergonomics***—Ensuring that system interfaces and controls are designed to fit the physical and cognitive capabilities of users, minimizing strain and maximizing comfort and efficiency.
- ***Usability***—Designing systems that are intuitive and easy to use, allowing users to achieve their goals with minimal effort and error.
- ***Human-Machine Interaction***—Creating interfaces that facilitate effective and efficient interaction between users and systems, considering factors such as feedback, control, and automation.

Cognitive requirements pertain to the human ability to assess and process information efficiently. These requirements are usually defined in terms of response times and are set to prevent excessive cognitive workload. Operations involving numerous complex tasks within a short timeframe can lead to cognitive overload and safety risks. Capability documents should outline whether human presence in operations is necessary, such as requirements for "human in control," "manual override," or "fully autonomous operations." Physical requirements are typically defined by anthropometric measurements, strength considerations, and weight limitations. These requirements are crucial for ensuring human performance, safety, and occupational health within a system. To guarantee that users can effectively operate, maintain, and support the system, the program office should specify requirements tailored to the user population. Sensory requirements are generally described in terms of visual, olfactory (smell), or auditory (hearing) factors. For instance, systems might need to function in loud environments caused by system discharges or operate under conditions of low visibility, such as on a cloudy, moonless night without additional lighting. Specific sensory capabilities, like visual acuity, may restrict the suitability of certain roles for individuals with particular sensory profiles.

4.2.2 Application of Human Factors Engineering

To fully leverage its benefits, HFE should be integrated into the design process from the outset of the acquisition phases as part of conceptual development and early solution design. HFE should draw support from other HSI domains and SE processes. Early decisions regarding human-machine interfaces in the design phase optimize human performance and overall total system performance. HFE involvement should persist through each subsequent acquisition phase, balancing trade-offs informed by inputs from other HSI domains, hardware and software designs, and adaptations. HFE practitioners bring expertise in design criteria, analysis, modeling tools, and measurement methods. Incorporating the input from HFE practitioners is vital to making sure the program office systems are operationally suitable, safe, survivable, effective, user-friendly, and cost-effective.

In any system acquisition process, it is crucial to acknowledge the varying competencies (skills and knowledge) required by different users. HFE processes help identify these competencies necessary for performing tasks effectively. They assess whether requirements and other attributes align with what users can realistically provide, highlighting potential training or operational challenges arising from any gaps. HFE tools and techniques are instrumental in identifying the attributes of diverse user classes and levels, guiding the development of tailored information products, training programs, systems, and other supportive aids. While understanding the information processing and network-centric requirements of the system is paramount, considering how data should be formatted and displayed to users to prevent cognitive overload is also crucial. This applies both to the system being designed and its interfaces with other systems to form the system-of-systems (SoS) capabilities. The system must avoid imposing excessive workload or stress on these interfaces.

4.3 Human-Machine Teaming

The assumption that the difficulty of automating a task correlates with the amount of human mental effort associated with it is flawed. A more useful approach is to consider how readily an activity can be codified when evaluating its automation potential. Tasks with clear rules, metrics, and recognition features are more likely to be optimized for machine execution. This has led to surprising outcomes, where roles traditionally deemed challenging and high-paying, involving data sorting or deterministic analysis, are now automated. Tasks that humans struggle to comprehend will present significant challenges in terms of codification and automation.

Computer algorithms excel at certain tasks, such as sorting and searching through large amounts of structured data (e.g., text and document processing, enterprise information, and genetics), conducting deterministic analysis (e.g., counting, classifying, and game playing), and executing predictable mechanical interactions (e.g., manufacturing, flying, and driving). However, they face limitations when it comes to understanding complex unstructured data (e.g., images, acoustics, and environmental context), performing non-deterministic analysis (e.g., road scene understanding or predicting human behavior), and executing dexterous actions (e.g., fine manipulation requiring touch and pressure feedback or handling deformable objects). Despite these challenges, machines are increasingly surpassing humans in some of these demanding tasks, such as image recognition. Machines do not experience concentration lapses or fatigue, assuming they have access to a constant power supply.

The challenges faced by computer algorithms stem from their struggle with uncertainty and ambiguity in both data interpretation and decision-making processes. Humans tend to outperform machines in understanding context, a trend likely to persist for the foreseeable future. Machines often falter when exercising nuanced judgment in complex or ambiguous contexts, which can moderate decision outcomes. Additionally, because machines are programmed or trained using established datasets relevant to a specific task or problem, encountering new or divergent situations can lead to failure. In contrast, human adaptability to new situations tends to be far superior, with even imperfect responses often proving more functional. Humans are able to employ mental substitutions or approximations from familiar skills or tasks to arrive at solutions, a capability that machines struggle to replicate outside specific narrow confines. Furthermore, AI technologies can suffer from catastrophic forgetting, wherein previous algorithm optimizations or task skills are lost when trained on new tasks and datasets.

4.4 Addressing Cognitive Workload and Human Performance

Because cognitive tasks are unobservable and complex, they are not particularly well understood. Tasks of varying complexity can result in different levels of cognitive workload in humans. Cognitive workload is a latent construct that describes the effort required by working memory to perform a cognitive task. This workload can be measured through various methods, including self-reporting, performance measures (such as task accuracy and timing), behavioral observations, and neurophysiological measures. Neurophysiological measures encompass pupil response, heart rate variability, core temperature, and electroencephalogram (EEG) tests.

Cognitive workload can influence human performance in various ways. Increased cognitive and visual workload can affect attention and the ability to detect information in the peripheral visual field, demonstrating that workload can influence an operator's ability to perceive and respond to relevant stimuli [VANWINSUM18]. In human-automation interactions, cognitive load and task complexity can also influence susceptibility to automation bias, affecting how operators evaluate and respond to automated information or recommendations [LYELL18]. Excessive cognitive workload is associated with poor performance and increased errors, while moderate levels of elevated workload have been linked to enhanced performance. Understanding the operators' workload levels in complex systems is crucial for optimizing human performance. Developing a method to classify tasks based on cognitive workload provides a foundation for such evaluations. Before classifying tasks, however, the program engineer must decompose the high-level system goal into meaningfully distinct sub-tasks. This decomposition helps identify and evaluate the cognitive workload associated with each task, ultimately aiding in system optimization for better human performance.

Task analysis is the process of breaking down a skill, movement, or cognitive process into smaller sub-tasks that a system operator must perform to accomplish the system's overarching goals. This process is valuable because it allows the systems engineer to identify and independently evaluate or modify task subcomponents. Task analysis is a fundamental tool in human factors research and has been widely applied in various fields, including product design, instructional design and training, function allocation, and error and workload assessment. The focus of this section is on performing task analysis specifically on cognitive tasks, known as Cognitive Task Analysis (CTA). Applied Cognitive Task Analysis (ACTA) provides a practitioner-oriented approach for identifying and understanding the cognitive demands associated with complex tasks, including the knowledge, judgments, decisions, and cues used by experienced operators [MILITELLO98].

CTA focuses on the underlying mental framework, thought processes, and knowledge required to perform a task. It helps identify hidden and ineffective cognitive strategies as well as tasks that induce high cognitive demand. Additionally, CTA can serve as a baseline model for task optimization to enhance human performance. Researchers have identified over 100 varieties of CTA, but most approaches share five common steps:

1. Collect preliminary knowledge
2. Identify knowledge representation
3. Apply focused knowledge elicitation methods
4. Analyze and verify acquired data
5. Format results for the intended application

Approaches are typically classified by the knowledge elicitation method, generally including observation and interviews, process training, and conceptual techniques. CTA has been applied across various domains, such as autonomous vehicle display design, electronic health record design, instructional design, and medical training program development. The following steps introduce a CTA methodology to predict operator cognitive workload, entitled Cognitive Task Analysis and Workload Classification (CTAWC):

1. **Traditional Task Analysis**: The task or process of interest is decomposed into basic actions that the user performs to achieve the end goal. This is done iteratively with the assistance of users and expert stakeholders to ensure accuracy and comprehensiveness.

2. **Cognitive Task Analysis (CTA)**: For each traditional task identified in the previous step, the required cognitive tasks are defined using cognitive and psychomotor taxonomies. These tasks are gradually decomposed to a satisfactory level of detail, ensuring all cognitive requirements are captured.
3. **Experimental Validation**: A controlled experiment is designed to simulate the tasks and measure cognitive workload for the classified tasks of interest. The experimental workload data are then statistically compared with theoretical workload predictions to validate the methodology.

This structured approach facilitates a detailed understanding and classification of cognitive workload, which helps to optimize tasks to enhance operator performance. Optimizing a system for maximum human performance necessitates a deep understanding of the cognitive tasks required of the user. CTA provides a structured process for isolating the distinct mental processes performed by system operators. This framework has potential applications across any system where unobservable human actions play a significant role in system operation.

Future research should focus on validating this approach in different settings and explore the discriminative power of taxonomy classifications. For the most part, the approach differentiates between "high" and "low" workload; however, demonstrating that individual levels within defined taxonomies can discriminate among varying degrees of cognitive workload could provide additional validation. Effective and safe system operation depends on accurate models of operator workload, which serve as a foundation for system optimization. Having an approach for creating and validating such models paves the way for improved human performance in complex systems.

4.4.1 *Performance Measures*

Some workload measurement techniques focus on objectively measuring aspects of the operator's performance. These performance measures of workload share a common characteristic: they solely consider the task being performed or the work produced by the human operator. By evaluating metrics such as task accuracy, completion time, and error rates, these techniques provide an objective assessment of how workload influences performance. This focus on measurable outcomes helps in understanding the relationship between cognitive workload and operator efficiency, enabling targeted improvements to enhance overall system performance.

4.4.1.1 Measuring Speed and Accuracy

The simplest performance measurement technique involves assessing the speed and/or accuracy with which an operator completes a task. This approach operates under the assumption that as workload increases, the operator's performance is likely to degrade. Measuring speed and accuracy can be as straightforward as using a stopwatch to record the time taken to complete a task and noting the operator's success rate. One advantage of this method is its simplicity; observers only need to observe the operator's performance to determine workload acceptability. However, there are drawbacks to this approach. It is relatively insensitive to the operator's state or condition during task execution. Factors such as feeling overworked or underperforming are not considered in performance assessments. Additionally, prolonged task performance may lead to fatigue, boredom, or other undesirable conditions, which can impact performance. For instance, a low-workload task may initially yield excellent performance, but performance may deteriorate as the operator becomes bored. Similarly, operators may initially perform well under high-workload conditions but experience fatigue over time.

4.4.1.2 Measuring Activity

Another performance measure, which is more sensitive to the state of the operator, focuses on quantifying the actions necessary for task completion. This technique involves cataloging and counting the number of steps or actions the operator takes to execute the task. A high number of steps suggests high workload, whereas fewer steps to complete a task implies a low workload. Actions considered in this technique may include control inputs, verbal responses, mental calculations, decisions, and visual searches. One advantage of this approach is its simplicity; observers only record the user's actions and tally them once the task is finished.

A basic disadvantage becomes evident in that this technique does not directly address the concept of workload as commonly understood. The number of steps required for task completion does not necessarily correlate with the operator's perceived workload. Furthermore, this method overlooks skill discrepancies between operators. While one operator may effortlessly complete a task that challenges another, the action-counting technique may not differentiate between their levels of effort or workload.

4.4.1.3 Task Analysis

A variation of tabulating operators' actions involves enumerating the procedural steps required to complete a task without directly observing operators during task performance. One primary advantage of this task analysis technique is its independence from testing with human operators. However, several disadvantages exist. First, cataloging procedural steps can be time-consuming. Second, the technique overlooks differences in skill levels among operators. Note, however, that some variability reduction can be achieved if analysts work within a common model or architecture, such as GOMS (Goals, Operators, Methods, and Selection rules). Finally, this approach assumes that all operators will follow the same procedure to complete the task, which may not always be the case in practice.

4.4.2 *Physiological Measures*

Physiological measures of workload aim to correlate physiological changes with varying levels of workload. Initially, researchers sought physiological measures as objective indicators of workload, independent of assumptions about how individuals perceive workload or subjective ratings from operators. The ideal physiological measure would serve as a "tell-tale" sign of workload experience. Despite extensive investigation into various physiological measures, none has conclusively captured the multifaceted concept of workload. While these measures offer valuable insights into physiological responses to workload, they have yet to provide a universally applicable and definitive indicator of workload perception.

4.4.2.1 Heart Rate

Heart rate stands as one of the simplest, longstanding physiological measures of workload. It can be easily monitored using devices akin to those employed during exercise. Monitors sample and record heart rate at regular intervals, typically once per second, storing the data for subsequent analysis on personal computers. While numerous researchers have investigated the relationship between heart rate and human task performance, as well as its correlation with other workload measures, findings have been varied. Heart rate appears to correlate well with physical activity and demonstrates a modest correlation with mental activity.

4.4.2.2 Heart Rate Variability

Heart rate variability represents a slightly more advanced measure of workload. It quantifies the variations in time intervals between heartbeats, irrespective of the heart rate itself. Monitoring heart rate variability requires more advanced equipment compared to simple heart rate monitoring. Several researchers have reported success in correlating heart rate variability with other indicators of mental workload.

4.4.2.3 Evoked Potentials

More advanced physiological recording techniques involve measuring changes in electrical potential in response to visual and auditory stimuli or capturing images of the brain while operators engage in tasks. These techniques, while highly sophisticated and informative, are also costly and specialized.

The versatility of the human mind is crucial to our adaptability, but as man-machine teaming becomes more significant, it is essential to integrate human capabilities into future systems. Various technologies have the potential to enhance both physical and cognitive human performance. Prosthetics and neuroprosthetics

This chapter focused on **Human Systems Integration (HSI) and Human Factors Engineering (HFE)**, emphasizing how human capabilities, cognitive workload, and performance measures influence system design and operational effectiveness. However, integrating human considerations into complex systems requires a structured approach grounded in fundamental engineering principles. The next chapter transitions to **postulates and principles for integrated solutions**, introducing key systems engineering models, modeling and simulation techniques, and MEI focus areas that provide a foundation for designing and evaluating integrated solutions. These postulates and principles establish a structured framework to guide decision-making, ensuring that both technical and human factors are addressed in system development. This next chapter will explore how these principles shape mission success by promoting adaptability, flexibility, and a multidisciplinary approach to system design.

(brain-controlled mechanical devices), including exoskeletons and telexistence, can help overcome physical limitations. Augmented reality and real-time language translation aid in how we visualize information and learn. However, we must balance these advantages with ethical and legal considerations, aware that many potential adversaries may not be as restrained. In the future, addressing threats may require us to reconsider current legal boundaries or risk losing a decisive advantage to adversaries.

References

1. Lyell, D., Magrabi, F., & Coiera, E. (2018). The effect of cognitive load and task complexity on automation bias in electronic prescribing. *Human Factors: The Journal of the Human Factors and Ergonomics Society, 60*, 1008–1021.
2. Militello, L. G., & Hutton, R. J. B. (1998). Applied Cognitive Task Analysis (ACTA): A practitioner's toolkit for understanding cognitive task demands. *Ergonomics, 41*, 1618–1641.
3. Van Winsum, W. (2018). The effects of cognitive and visual workload on peripheral detection in the detection response task. *Human Factors: The Journal of the Human Factors and Ergonomics Society, 60*, 855–869.

Chapter 5
Postulates and Principles for Integrated Solutions

5.1 Background and Definitions

This chapter outlines the concepts and principles setting the foundation for the comprehensive Mission Engineering and Integration (MEI) framework as grounded in traditional disciplines and aims for an integrated solution as the desired end state. These connections draw from insights from prior work in systems theory and operational planning techniques. While previous theories in systems engineering (SE) have provided valuable concepts, none has presented a definitive theory for executing SE at the intricate system-of-systems (SoS) level while adequately incorporating operational context. The reader will do well to read more from past experts. Ludwig von Bertalanffy's [BERTALANFFY68] general systems theory concepts, rooted in empirical analysis, sought commonalities (i.e., isomorphism) among various system types. W. R. Ashby [ASHBY58] took a deductive approach, starting from the set of "all conceivable systems" and identifying their shared features. Jay W. Forrester [FORRESTER68] approached systems theory with a system dynamics perspective, examining how systems evolve over time, with broad applicability across all system types. Robert Rosen [ROSEN71] contributed to dynamical system theory for biological systems, offering an additional angle to the intricate challenge. MEI builds upon these theories but presents a distinctive approach to representing a complex SoS within specific operational conditions.

The resilience established via interconnected nodes eliminates dependence on a single point of failure, along with enhanced operational clarity derived from a shared understanding of the environment and a streamlined mission thread. This collectively contributes to an accelerated decision-making cycle (e.g., Observe-Orient-Decide-Act (OODA) loop) for end users. This capability enables faster reception, processing, and response to information than competitors to yield a decisive edge in conflict. Rapid processing and response times are crucial for shaping the tempo of an operational mission. The interconnection within our fast-paced

J. D. Moreland, Jr., *Mission Engineering and Integration*,
https://doi.org/10.1007/978-3-032-32132-9_5

societal fabric is permeating our systems, networks, space assets, and organizations. An illustration of this complexity is evident in determining optimal investments for the development of space assets and safeguarding positioning, navigation, and timing applications. This demands a systematic approach to identify key interfaces and intersection points across numerous systems in a multi-domain, multi-functional environment. A nuanced understanding of SoS at this level enables the mitigation of risks associated with deploying flawed mission engineering threads and, consequently, operational capabilities. Moreover, this MEI approach facilitates a focus on resilient alternatives for executing mission threads while minimizing unnecessary redundancy within and across industrial sectors.

MEI is an interdisciplinary field of engineering and engineering management that is systematic, quantifiable, and iterative. It combines the structure of SE and the tactical insights of operational planning to achieve mission wholeness. Social and physical complexities across organizations and systems lead us to the necessity to embrace this complexity leading to a competitive advantage and is the core emphasis of MEI.

5.2 Systems Engineering Model

The discipline of SE employs various models to guide the design, development, and management of complex systems. The Waterfall Model follows a linear, sequential approach, progressing through phases like requirements, design, implementation, testing, and maintenance, making it suitable for projects with well-defined requirements. The V-Model, which expands on Waterfall by emphasizing verification and validation and aligning design and testing activities, will be the model focused on within this book. In contrast, the Iterative Model allows for repeated cycles of development and refinement, adapting to changes and uncertainties. The Spiral Model combines iterative processes with risk management, making it ideal for projects requiring continuous assessment and risk mitigation. Agile-based SE integrates agile principles, emphasizing flexibility, collaboration, and incremental delivery, often through sprints or iterations. The Model-Based Systems Engineering (MBSE) approach uses digital models to represent and analyze system components and their interactions, fostering collaboration, and reducing errors. Each model serves specific needs based on the project's complexity, uncertainty, and stakeholder requirements.

All technical projects must have documented SE processes. This documentation must include a description of the SE approach and specific engineering activities. As previously mentioned, there are multiple acceptable development approaches (e.g., Spiral, Evolutionary, Agile, Rapid, Prototyping). Neither a specific approach nor a set of engineering activities is mandated. However, when defining, documenting, or validating a SE process, the following common SE dimensions should be considered:

- **Understanding the Problem Space**: The set of activities associated with defining the system objective, capabilities, and requirements.
- **Understanding the Solution Space**: The set of activities associated with the development of the architecture, design, and requirements implementation. *It is understood that Understanding the Problem Space and the Solution Space is an iterative process with dependencies and overlap of activities.*
- **Systems Analysis and Control**: The set of activities that should be considered throughout all phases of the development effort (e.g., Configuration Management, Risk Management, Safety, Anti-Tamper, Environmental, Human System Integration (HSI), Information Assurance, Logistics).
- **Integration and Interoperability**: The set of activities associated with integrating all the components and verifying and validating that system performance and mission capability are achieved.
- **Operational Performance**: The set of activities associated with deploying the system, ensuring the system meets the user needs, and continuing to support the system through the end of the lifecycle. Objective Quality Evidence (OQE) should be used during project reviews to demonstrate that the project has addressed the SE process considerations provided in these SE dimensions.

Each project's documented SE process must include a diagram illustrating the SE approach and activities. Figure 5.1 provides an example of how the development of SE approaches and activities using the SE V-Model can be captured in a diagram. This generic figure does NOT imply that all technical projects must utilize this exact type of V-based approach. It is understood that:

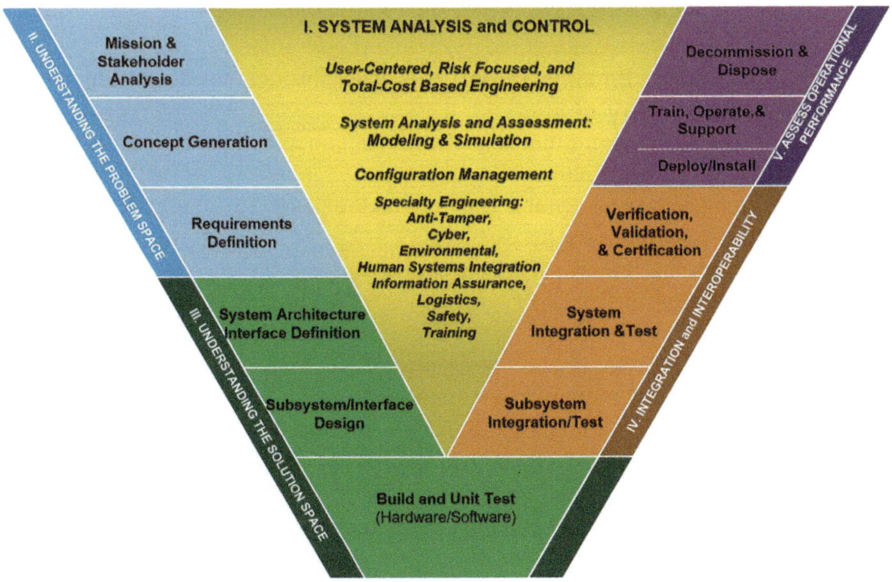

Fig. 5.1 Generic SE V-Model activities

(a) There are multiple different development approaches.
(b) SE is a highly iterative process. The activities shown in Fig. 5.1 may not necessarily be executed in a sequential or waterfall method.
(c) The time and level of detail allocated to each SE phase and activity are project dependent.

Regardless of the SE approach, all SE processes should address the SE activities shown in Fig. 5.1.

I. **System Analysis and Control Activities:**
   (a) Systems Analysis that includes well-defined and documented:
   - Alternative approaches
   - Analysis and decision criteria
   - Quantitative and qualitative analysis methods
   - Data sources and collection method
   - Decisions that are based on and aligned with contributions to mission capabilities
   - Total Lifecycle planning and cost, where lifecycle includes concept development, system design and production, deployment, operations, maintenance and support, decommission, and disposal

   (b) Technical Planning that includes well-defined and documented:
   - Risk-focused, user-centered, and total Lifecycle Cost (LCC)-based SE processes applied throughout all phases of SE processes
   - Technical scope that incorporates inputs from appropriate subject-matter experts and addresses, through all the project phases, the applicable specialty engineering activities (Safety, Anti-Tamper, Environmental, HSI, Information Assurance, Logistics, etc.)
   - Methodology to collect, analyze, and report on data
   - Metrics-based quality control processes throughout all the engineering phases and project lifecycle

   (c) Technical Assessment that includes well-defined and documented:
   - Technical Performance Metrics
   - Entrance and exit criteria for technical reviews
   - Methods of periodic assessment
   - Appropriate levels of review that have been incorporated into the SE process
   - Methodology for conducting failure investigations
   - Methodology for documenting results and sharing lessons learned
   - Models and simulations

   (d) Risk Management that includes:
   - Addressing cost, schedule, technical performance; and all specialty engineering (safety, environmental, etc.) risks
   - Documented risk management plans and strategies that begin at project initiation and are continued throughout all lifecycle phases
   - Documented risk management processes with risk mitigation strategies that are approved by the stakeholders

- Methodology for regularly tracking and reassessing cost, programmatic and technical risk(s) through risk closure
- Identification and tracking of residual (accepted) risks across all phases of the project
- Regular stakeholder review of risk status

(e) Configuration Management (CM) that includes:

- Documented CM plans and strategies that begin at project initiation
- Documented CM processes to include change control
- CM processes that have been accepted by the stakeholders
- Configuration Items (CIs) that have been clearly identified and baselined to an appropriate level
- A method to continuously identify and track CIs across all phases of the project
- Configuration control and change tracking within the Requirements Development and Management process
- Updating all impacted programmatic and technical artifacts reflecting accepted changes
- An effective configuration audit system that:
  - Ensures the functional and performance attributes of the system and subsystems (Functional Audit) are achieved
  - Ensures the system and subsystems are built and installed in accordance with the requirements of its documentation (Physical Audit)

II. **Understanding the Problem Space Activities:**

(a) Mission and Stakeholder Analysis that:

- Identifies and engages operational and acquisition stakeholders
- Determines all applicable mission areas
- Documents a strategy for mission validation

(b) Concept Generation, Requirements Definition, and Management that includes:

- Clearly defined and documented system capabilities and requirements that address mandates and constraints
- Inputs and review by all project stakeholders
- Mission-level requirements that are traceable to system-level requirements
- System-level requirements that are traceable to the subsystem and component-level requirements
- Identification and traceability of integration dependencies and relationships between requirements within the system and from higher level operational (mission) requirements
- Identification of the trade space boundary defined by the requirements that must be satisfied to meet the mission-level capabilities and those requirements that can be optimized through the program's system-level design
- Assessment of potential requirements changes for programmatic and technical impact

III. **Understanding the Solution Space Activities:**
   (a) System Architecture and Interface Management that includes:
      - Formal documentation that is CM controlled for the system architecture
      - Functional analysis performed at the system, subsystem, and component level
      - Documentation of all internal and external interfaces, as well as the system data exchanges associated with each
      - Determination of Integration Readiness Level (IRL) for each interface
      - Architecture and interface designs that have been presented to and understood by the stakeholders
      - Methodology to ensure complete interface evaluation for compliance for all interfaces and directions
      - A process across the lifecycle for reviewing and processing interface changes
      - Establishment of Memoranda of Agreement (MOA) with external programs that control or influence interfaces implemented by the system
   (b) Design Solution that:
      - Documents the system design, including how that design satisfies all system requirements, operational use cases, including specialty engineering requirements (Safety, Anti-Tamper, Environmental, HSI, Information Assurance, Logistics, etc.)
      - Allocates functions addressed explicitly throughout the design process
      - Analyses and documents design alternatives at all system levels
      - Provides well-defined design alternatives and selected design that consider impacts on cost, schedule, and technical performance
      - Designs and implements individual CIs that satisfy all requirements, including specialty engineering requirements
      - Maintains the design artifacts that are consistent with the actual implementation components
      - Considers test impacts and required objective quality evidence articles

IV. **Assess Integration and Interoperability Activities:**
   (a) Test Planning that:
      - Identifies methods to validate requirements including technical, performance, environmental, and operational that lead to system certification
      - Identifies relevant systems outside of the project boundary used in accomplishing higher level mission objectives
      - Includes mission-focused test design and artifacts
      - Identifies early opportunities to perform mission-based testing to include potential for preproduction activities
   (b) Identification of requirements for:
      - Facilities
      - Models and simulations
      - Automated test and analysis tools and methods
      - Networking and information assurance

- Safety
- Documentation
- Personnel
- Selection of mature, validated and verified facilities, models, simulations, and test tools
- Prioritized test objectives traced to performance metrics

(c) Test Execution that includes:

- System configurations that are operationally representative
- Pretest verification of test configuration and facilities
- Configuration-managed equipment under test
- Test harness comprised of relevant systems (virtual, operational representative, and/or operational components)
- Clearly documented and accessible data and results
- Generation of objective quality evidence

(d) Verification, Validation, and Certification must be conducted in accordance with the project's governing policies and instructions.

V. **Assess Operational Performance (Capability) Activities:**

(a) Deployment, Operation, and Support that includes:

- Early identification of and communication with the key personnel and authorities for system delivery and installation acceptance
- Formal configuration control
- All artifacts required to support safe and effective operations
- Authorization and certification prior to system use
- Process for obtaining feedback from operators and stakeholders
- A documented process to adjudicate reported problems, requested capabilities and enhancements
- Designation of organizations with specified roles and responsibilities to provide in-service maintenance and updates

(b) Decommission and Disposal should include:

- A documented and approved (by designated authority) process that addresses the security, safety, environmental, legal, and all other specialty engineering requirements
- A documented and approved (by designated authority) process that addresses the potential exposure of the systems to hazardous materials and instructions for subsequent safe handling and decontamination
- Consideration throughout all the lifecycle phases

Traditional SE was developed to manage the design and development of monolithic single systems with self-contained, well-defined systems comprising clear boundaries. This approach involves breaking down the system into subsystems, designing each part individually, and ensuring that when the components are integrated, the entire system works as expected. For many decades, this method worked well for relatively straightforward systems, such as individual aircraft, vehicles, or manufacturing machines. In these monolithic systems, requirements were typically stable, interactions between subsystems were well-understood, and the design process could be largely linear. Systems engineers could focus on optimizing the performance of the entire system within a controlled environment with known inputs and predictable outputs. However, today's engineers increasingly deal with complex SoS consisting of networks of interdependent systems that operate together to achieve larger objectives. Examples of these include smart cities, global transportation networks, modern military defense systems, and advanced healthcare ecosystems. These SoS are composed of multiple interconnected, often independently developed systems, each with its own lifecycle, objectives, and management structures. The traditional SE principles remain necessary for the disciplined engineering of individual systems and their integration; however, the traditional SE approach alone is no longer sufficient to address the complexity of modern SoS for several reasons:

1. **High Interconnectivity and Dynamic Interactions**: In SoS, interactions between individual systems are highly complex and often dynamic. Unlike monolithic systems, where interactions between subsystems are usually predefined, SoS involve multiple systems that may be interacting in real time, with unpredictable consequences. For example, in a smart transportation system, traffic signals, autonomous vehicles, and city-wide communication systems must interact seamlessly, often adapting to real-time traffic conditions that require continuous coordination and flexibility.
2. **Independent Management and Development Cycles**: In SoS, each system may have its own independent development cycle, operating under different stakeholders, timelines, and goals. These systems may evolve independently and unpredictably, which is different from the centralized, controlled development process in monolithic systems. Engineering these systems requires new approaches to coordination, where the focus is not on controlling every system but managing the interfaces and ensuring interoperability.
3. **Emergent Behavior**: In complex SoS, the overall behavior of the system cannot be predicted solely by understanding each individual system. Emergent behavior, defined as the unanticipated outcomes resulting from the interaction of different systems, poses a significant challenge. For instance, in a networked power grid, the failure of a single node could cascade, leading to widespread blackouts that are difficult to anticipate without sophisticated modeling of interdependencies.
4. **Scale and Scope**: The sheer scale of SoS is beyond the scope of traditional methods. Systems such as national air traffic control networks or global communications systems span across different organizations, geographic regions, and regulatory environments, making it nearly impossible to treat them as single, monolithic entities. These systems must also accommodate a broader range of disciplines, from cybersecurity to environmental impact assessments, all while ensuring long-term resilience.
5. **Adaptability and Flexibility**: Modern SoS must be adaptable to evolving conditions, whether driven by new technologies, changing user needs, or external factors like regulations and environmental shifts. Traditional SE processes,

which often assume static requirements and solutions, struggle to accommodate this level of flexibility. Instead, new methods are required to continuously monitor and adjust the behavior of interconnected systems in real time.

Kurtz and Snowden distinguish among ordered, complicated, complex, and chaotic problem contexts and emphasize that the appropriate decision-making approach depends on the nature of the environment [KURTZ03]. To address the unique challenges posed by SoS, SE has evolved to incorporate new principles and practices. Architectural frameworks, MBSE, and agile methodologies are now used to handle the complexity and unpredictability of SoS. These approaches emphasize collaboration across systems, continuous integration, and the adaptability to emergent behaviors. Additionally, modern SE for SoS incorporates risk management, resilience engineering, and uncertainty quantification as core elements, ensuring that even in the face of uncertainty, systems can maintain functionality and adapt to new circumstances. SoS engineers are also researching and exploring advanced techniques leveraging the use of digital twins. Using digital twins involves virtual modeling for physical systems that can be used for real-time monitoring and predictive maintenance across the system network.

The new paradigm of SE focuses on enabling communication, coordination, and continuous improvement across multiple systems to ensure overall mission success in increasingly complex environments. In addition, this new paradigm within the SE discipline is advanced even further with a strong emphasis on mission execution and success through MEI. In the representation of the SoS for MEI, the complexity of operational missions often necessitates the use of multiple SE V-Models to represent the SoS required for their execution. Each V-Model encapsulates the lifecycle of an individual system, from defining requirements and architecture to verification and validation. By interlinking these V-Models (discussed later in Chap. 8), engineers can capture the interactions and interdependencies between various systems that collectively achieve mission objectives. This approach allows for concurrent development and integration of systems while maintaining alignment with overarching mission goals.

5.3 Modeling and Simulation

Modeling and simulation (M&S) are essential SE tools that support decision-making, analysis, and training across multiple functional disciplines. Models, simulations, and their associated data and artifacts are developed and utilized within a controlled engineering ecosystem to ensure consistency, integration, and reusability throughout a system's lifecycle. By providing a comprehensive and dynamic view of a project, these tools enhance efficiency, confidence, and collaboration among stakeholders. A model is defined as a physical, mathematical, or logical representation of a system, entity, phenomenon, or process, while a simulation implements a model over time to analyze system behavior. These tools are instrumental in exploring concepts, evaluating system characteristics, opening trade spaces, and facilitating informed decisions. Simulations also enable the study of stochastic interactions between systems, individuals, and environments that static models cannot achieve. For example, DoD M&S are supporting multi-domain warfare analysis and training by integrating diverse systems and domains into a unified simulation environment. Reusing existing models, simulations, and results wherever practical enhances cost-effectiveness and efficiency while supporting requirements development, prototyping, algorithm development, testing, and training. However, reusing models for purposes beyond their original intent may require additional engineering to address differences in assumptions, constraints, and compatibility.

Effective integration of simulations requires addressing touchpoints, including networks and computing environments, connectivity protocols, and compatibility of simulation models. While manual integration processes are currently common, they are time-consuming, costly, and prone to locking in variable features like weather and terrain. A more efficient approach is to federate existing M&S using interoperability standards, creating systems of interacting M&S that align with shared objectives. To maximize the value of M&S, their development and use must be carefully planned and managed. Key considerations include scoping activities to meet user expectations, ensuring relevance to the application domain and lifecycle phase, and maintaining consistency and interoperability of digital artifacts. M&S tools should be integrated into technical planning, appropriately resourced, and included as part of the authoritative source of truth. M&S should be developed, documented, and managed throughout the lifecycle of an effort. Proper documentation, including metadata and mapping to widely supported standards, ensures reusability and facilitates collaboration. These tools should also be included in the technical data package (TDP) and referenced in technical reviews to provide baseline information for subsequent phases and other projects. By coordinating and controlling M&S activities, systems engineers ensure alignment with architecture and design objectives, enabling informed decision-making and enhanced project outcomes.

5.4 Model-Based Systems Engineering

Model-Based Systems Engineering (MBSE) is a model-centric view and element of digital engineering. It incorporates a systemic and integrated use of models to represent, design, define, and analyze systems and components with business objectives. Using MBSE implements a means to describe and drive implementation of products using models across disciplines to support lifecycle activities from concept through disposal. The critical need for digital transformation is recognized across the engineering community and considered the preferred process by which the government acquires, develops, fields, and sustains future systems allowing the United States to keep pace with the ever-changing and increasingly complex competitive landscape. MBSE offers a solution for such a transformation within DoD acquisitions and development lifecycle processes. A fundamental MBSE tenet is that design truth is captured using digital artifacts within one or more descriptive and/or analytical models. In the future state, the descriptive and/or analytical models are digitally federated and linked to accurately represent the entire system or SoS. This holistic approach to information-access management includes use descriptions, exchange mechanisms, and data protection (both cyber and proprietary). Information protection provides a structured process to determine data protection requirements by program phase and product type.

As systems become increasingly complex, the tools and processes used to acquire, design, and field them must also increase in sophistication and flexibility to realize optimal solutions on schedule and at cost. This drives the speed of transition from document-centric to model-centric. The limitation with documents is the lack of a straightforward path to traceability and verification between all the detailed technical system-design specifications and the pages and pages of words contained in the review artifacts. In addition, mistakes and errors are more likely in these situations because the reviewers assume that the solution exists and is sound based on a thorough quantitative design analysis compared against properly defined requirements. One promising proposal is to develop a "document-centric to model-centric" artifact transition plan defined and implemented at the beginning of the MBSE collaborative concept of operations transformation. Specifically, this digital transformation plan is one where documents that once defined operational scenarios, requirements, and system design gradually are phased out and replaced with a single authoritative source of truth in the form of interconnected models. The goal of such a transition would be to relieve the burden of human consumption and review of large amounts of data and instead institute a more streamlined and machine-driven process through the continuous validation of traceable digital artifacts.

The digital engineering ecosystem allows the engineer to make critical and consistent design decisions early in a product's development while design freedom and trade space are at a maximum. Using models improves the rate at which engineers can transfer design knowledge while simultaneously reducing design ambiguity. Progression of time warrants decision support when dealing with large volumes of information because knowledge is increasing while interdependent attributes are integrated. Therefore, adoption of MBSE with the use of digital artifacts supports trade-offs among multiple conflicting goals and constraints most profoundly within the technical debt for system design and technology maturity where risk and uncertainty are high as depicted in Fig. 5.2.

MBSE moves the engineering discipline from its current state of stove-piped models and data sources toward a future state linking data sources and models across a product's lifecycle and provides the authoritative source in a digital ecosystem. These models need accurate and high-quality data that can be exchanged and transferred across models and organizational boundaries. It is critical to build metrics into the MBSE digital ecosystem to continuously monitor and modify this environment over time based on evolving needs and requirements. Having metrics to characterize the performance as the digital ecosystem is implemented and is crucial to optimize continuous improvements. Five potential key metrics to evaluate and monitor the MBSE environment are:

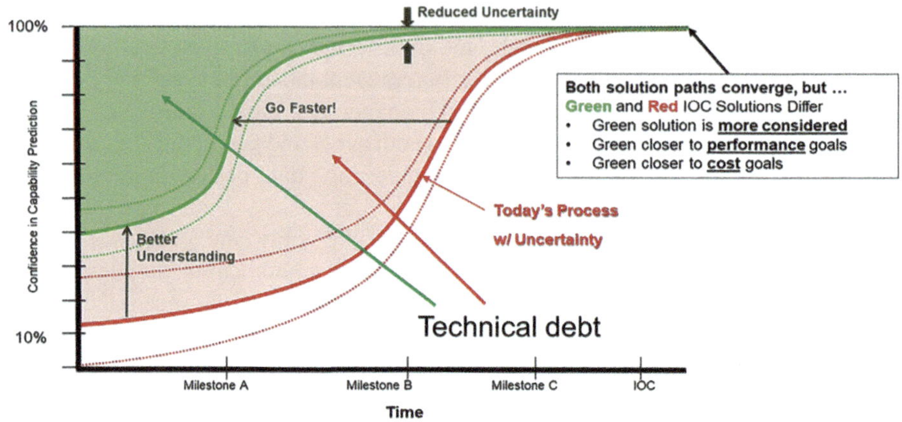

Fig. 5.2 Building capabilities at the speed of relevancy

1. Complexity of the models via number of modeling artifacts, objects added, as well as their interconnections.
2. The clarity and ease of understanding of the models to key stakeholders, subject-matter experts. and/or decision-makers.
3. Number of system defects/inconsistencies identified and mitigated via the interconnected models and the traceability of the digital thread.
4. Customer satisfaction with the model-centric environment and agile-based evaluations.
5. Time to generate and complete trade studies and new operational scenarios in the MBSE environment.

5.5 Mission Engineering and Integration Focus Areas

MEI is comprised of two focus areas: Systems Engineering Integration (physical) and Operational Context (social). These two thrusts encompass four components: Mission Context, Systems Integration, Organizational Structure, and Regulations as illustrated in Fig. 5.3.

MEI encompasses the physical, logical, and social (behavioral) dimensions of SoS within the framework of operational mission threads. A comprehensive understanding of these dimensions is imperative in the development of integrated solutions. Figure 5.4 depicts the traditional system architecture for single systems with multiple subsystems. Here, subsystems exchange vital information through the edges connecting the nodes, providing inputs, outputs, and feedback for control mechanisms. The software responsible for this logic requires insight into the systems' states to influence the intended control, which is intricately linked with the physical systems. This complexity is heightened in the context of SoS tasked with

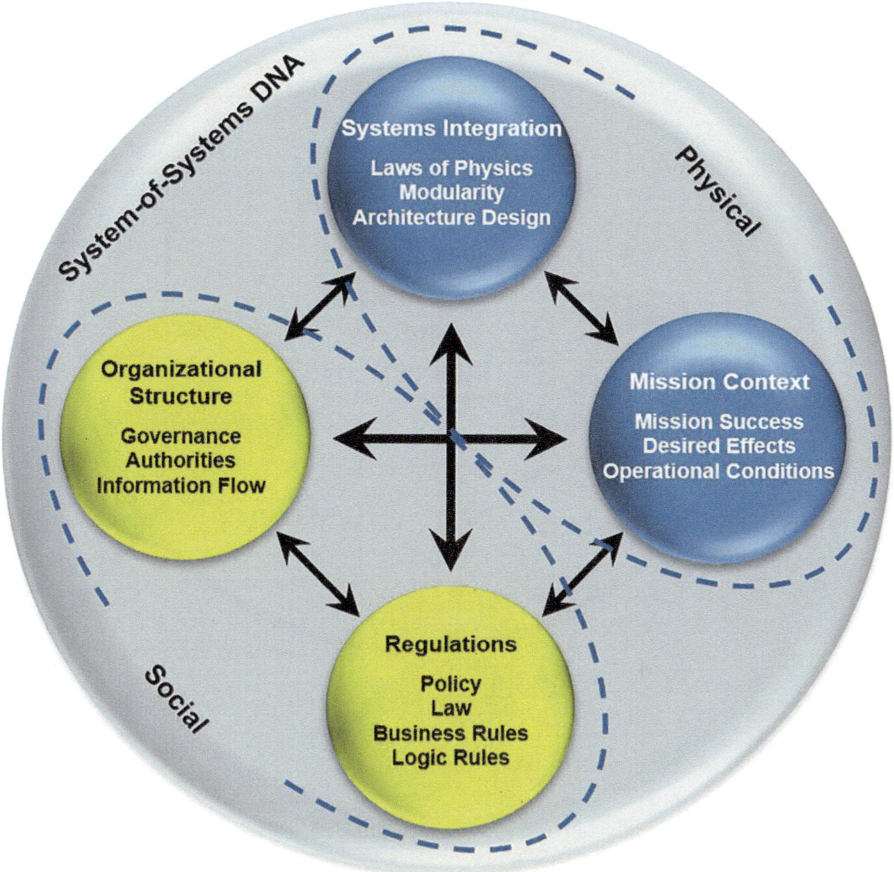

Fig. 5.3 Mission engineering and integration focus areas

Fig. 5.4 Traditional single system configuration

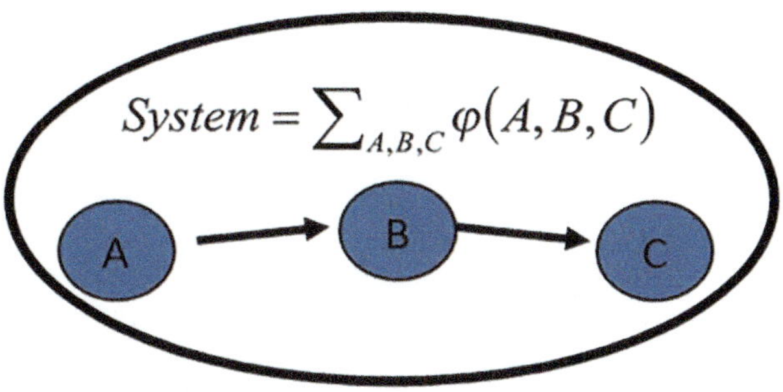

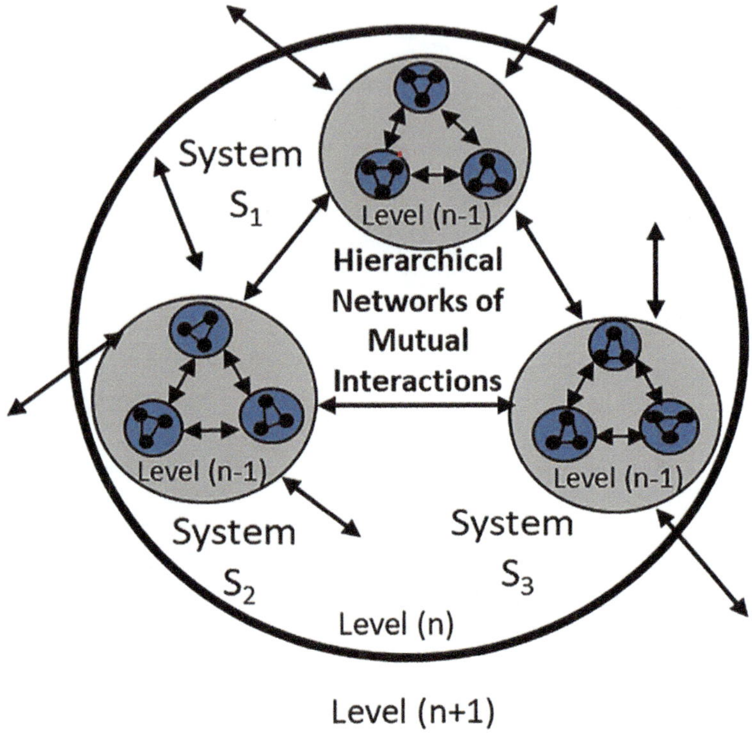

Fig. 5.5 Complex system-of-systems configuration

executing mission threads across system boundaries within mission engineering threads/webs as illustrated in Fig. 5.5.

This complex SoS must operate and interact under dynamic operational conditions such as environmental, political, social, and across constituent systems at the system boundaries. Additionally, the system integration region encompasses HSI aspects, wherein the physical and logical functional design must account for human physiology and psychology (see discussion in Chap. 4). This establishes a connection between the user, operator, maintainer, and manufacturer with the system structure serving as a bridge to the social systems involved in building, operating, and utilizing the systems. In addition, this is where the physical integration of systems forming the SoS as well as the integration of technologies into systems both work to evolve the capability and effectiveness. Integration means very different things to different communities and cultures. The design engineer and architect see integration through the lens of Newton's Laws involving hardware components and physical materials. The information systems designer sees the integration through the lens of Moore's Law with a focus on computer components and logical software. The challenge with complex SoS is the intersection of these domains through architectural models and process control that must bring both sides together into a safe, effective, and affordable operational capability.

Mission context impacts both the physical/logical aspects of systems and the social aspects, residing within the physical domain of the foundational structure for systems and SoS. The social connection to the mission context occurs at the intersection of the physical and social domains and drives the physical/logical choices made for the systems by emphasizing or amplifying these design choices. The social aspects constitute a significant focus outlined by the Organizational Structure and Regulations regions, addressing the exchange of information across organizational boundaries and the regulatory frameworks governing operations. This brings in the aspects of sociology where human behaviors and leadership drive organizational structure and operations. Efficient information flow is a crucial element in the design and operation of an integrated system. MEI ensures that the organizational structure facilitates the required information flow among system disciplines and guarantees that the design effectively captures this flow, highlighting integration and interoperability across systems. Concern arises when gaps, barriers, and organizational reservoirs hinder the flow of information within the organization.

Regulations in the form of policies, laws, business rules, and logic rules are generally social influences on the systems. This element of the social domain certainly influences the physical/logical aspects of the systems but is included with the social aspects of the systems due to their social and behavioral considerations. The final piece of this MEI structure consists of the edges between these regions which represent the flow of data to inform the connective tissue between all regions.

5.6 Postulates and Principles

The formulation of a set of fundamental postulates and principles for MEI was inspired by an examination of Ludwig Boltzmann's work and his postulates on gas distributions, illustrating an early approach to characterizing interactions in complex systems. This section introduces seven postulates leading to the derivation of four practical principles for MEI implementation. These postulates delineate the scope of SE, identifying the facets of systems and the influences pertinent to systems engineers. The resulting MEI principles lay the foundation for a comprehensive mathematical framework in SE. Serving as an extension of the postulates, these principles are listed subsequently, detailing the application of the MEI discipline.

5.6.1 Postulates

A postulate is an assumption accepted without requiring proof to be true, real, or essential. In the context of MEI, postulates establish the foundational assumptions for the discipline. These assumptions are then elaborated upon by a set of principles detailed in Sect. 5.6.2.

Postulate 1: MEI is product-specific and context-dependent

Description: Engineering activities and the integration of systems' physical, logical, social, and cognitive relationships (operational context) are propelled by the product (i.e., SoS) and its operational environment. Foundational to the specific product or SoS is a profound understanding of the mission or purpose, articulated through the product's desired objectives and mission success criteria. This encompasses the system aspects necessary for effective and safe operation, thus extending consideration throughout the entire product lifecycle.

Evidence: Widespread customizing of MEI strongly reinforces this postulate. MEI must be consistent with SE and operational mission activities and the systems being developed or operated. As developers and implementers of this new innovative MEI approach, many organizations are applying unique processes to produce mission engineering threads. Moreover, it is evident through direct engagement with many industrial sectors, military services, and academia both domestic and international that most are not following a standard approach.

Implications: This postulate asserts that the application of MEI should be structured according to the characteristics of the SoS under development or operation. When applied to an operational need, the MEI approach is expected to differ in emphasis and application depending on the unique nature of the product, its environment, and its mission context. Nevertheless, a shared framework is essential to ensure a fair comparison among multiple systems with similar functionality.

Postulate 2: The MEI domain consists of multiple systems, their interactions among themselves, and their interactions with the operational environment

Description: In terms of physical, logical, and structural characteristics, an SoS is not a singular mechanical, electrical, or chemical entity. Instead, it constitutes a compilation of interacting heterogeneous systems. MEI is focused on the amalgamation of these diverse systems, varying in physical and logical types, to form an optimally balanced functional whole that can achieve mission goals. This postulate specifically tackles the integration and interoperability facets of MEI from an SoS perspective. Postulate 3 further delves into the discipline of systems integration, adopting an interdisciplinary approach.

Evidence: The distinct engineering disciplines excel in developing their specific functions. Nevertheless, when these functions are integrated with one another and with the environment, the interrelationships play a crucial role in shaping the ultimate performance of the SoS. This includes the emergence of properties that may not be evident in the functions of individual systems. Consequently, while the engineering of individual functions is well-handled, the integration and interoperability of these engineering functions render them effective and purposeful as an SoS. The domain of MEI encompasses the set of these integrated relationships and solutions within a specific operational context.

Implications: The mission engineer is dedicated to understanding the interactions within the SoS, adopting a role distinct from that of a design engineer who concentrates on details. As a knowledgeable integrator, the mission engineer acknowledges that SoS interactions, along with their connections to the SoS environment, hold significant influence over the design and potentially shape system responses in unforeseen ways when combined. It is imperative for the mission engineer to possess the ability to anticipate and effectively manage these responses.

Postulate 3: The role of MEI is to proficiently integrate various engineering disciplines

Description: The MEI discipline constitutes its own engineering discipline, yet it is intricately connected with and not independent from other engineering and social disciplines. MEI aims to seamlessly integrate and incorporate these diverse engineering and social disciplines in a sophisticated manner and generate an effective and operational SoS across the lifecycle of its numerous constituent systems.

Evidence: The development of any intricate SoS is the result of the interaction among numerous engineering disciplines, with various social aspects exerting influence on the integration process. These engineering disciplines collaborate in an integrated manner, both formally and informally, to initiate these SoS capabilities.

Implications: The MEI approach centers on the interaction of various disciplines, working toward a fundamental comprehension of each discipline and a detailed understanding of their interactions. This involves incorporating organizational integration aspects, including Standard Operating Procedures. The mission engineer needs to be aware of the organizational and sociological influences on the development and operations of the SoS. Additionally, the mission engineer actively "engineers" these relationships to ensure effective integration.

Postulate 4: MEI influences and is influenced by organizational structure and culture

Description: MEI does not concentrate solely on the technical aspects of the SoS. Likewise, the organization's structure influences the engineering of individual systems and the resulting SoS. These factors also have a notable impact on the organizational culture.

Evidence: Organizational mirroring is exemplified by instances where the organization aligns with system functions.

Implications: The mission engineer needs to be aware of the organizational culture, interactions, and their potential influence on the design of individual systems and the resulting SoS. Understanding how information flows through the organization, undergoes filtering and interpretation, and is integrated into system and SoS design or operational procedures is crucial. Collaborating with project management and line management, the mission engineer should actively address issues related to organizational information flow and culture. This collaboration enhances the effectiveness of both the individual system and the SoS, particularly across organizational boundaries.

Postulate 5: MEI influences and is influenced by budget, schedule, policy, and law

Description: Every project is subject to overarching constraints that extend beyond physical and environmental boundaries. Typically, projects face limitations in terms of budget and schedule. In addition, all systems must adhere to established organizational and government policies and laws. The resulting SoS is required to function within these rule sets. Policies and laws impose additional constraints on budgets, schedules, and technical solutions. Collectively, these factors provide the context in which the mission engineer develops and operates the system and resulting. Moreover, the choices made in system and SoS design also influence these factors. Government policy and law reflect legislators' understanding of what systems and SoS can achieve in fulfilling their intentions. Similarly, corporate/company policy is shaped by the types of systems the organization chooses to develop and the relationships between these entities concerning the resulting SoS.

Evidence: Every project operates within these constraints; infinite budgets or schedules do not exist. The application of policy and law is pervasive in our systems and the resulting SoS. Similarly, corporate/company budgets and schedules are determined by executives' understanding of the financial resources and timeframes needed to develop a system or SoS. The executives' understanding is reflected in budget and schedule allocations, encompassing both total funding and a timeframe perspective.

Implications: Establishing constraints is driven by social choices. Individuals make decisions to set budget limits, schedule limits, policies, and laws, whether at the national or organizational level. Consequently, physical and logical solutions navigating these constraints are linked to social choice theory. The mission engineer makes choices as informed by an understanding of the abilities of systems and SoS to fulfill the intentions of government and corporate/company executives. This comprehension shapes budget and schedule allocations, as well as the formulation of policies. Likewise, the available budget, anticipated duration, existing policy, and law can influence choices in the development of a system or SoS.

Postulate 6: MEI spans the entire system lifecycle

Description: MEI extends beyond the development phase and remains integral throughout the entire lifecycle of a system, encompassing operation, decommissioning, and disposal. Organizational relationships and objectives evolve as the system and SoS advance through these phases. Nevertheless, MEI consistently integrates the functions of systems and various disciplines across all stages of the SoS lifecycle. Additionally, MEI carries the responsibility for the operation of the system and resulting SoS, including managing various changes and upgrades to system capabilities.

Evidence: MEI is intricate during the development phases, particularly for individual systems, although it is more comprehensible than the integration and interoperability of multiple systems. In the operational phases, MEI remains crucial as systems undergo maintenance upgrades, adapt to new applications, and experience obsolescence-driven redesigns. Later during decommissioning and disposal, MEI is again indispensable for ensuring the proper decoupling of systems and adherence to policies and laws governing system disposals.

Implications: As systems progress through their individual lifecycles, the role of MEI evolves. There is a shift from development to operations in terms of the scope of changes and organizational responsibility. MEI takes on the responsibility for integrating and ensuring interoperability of systems, including managing changes and upgrades to the systems. The baseline operational systems then become the environment where changes during the operational phase take place. The organization undergoes significant changes as systems transition from development to operations, resulting in different organizational relationships, needs, and potentially a distinct organizational culture. MEI must address these factors as they impact the systems. Another organizational change and a shift in culture occur during decommissioning and disposal, where MEI continues to play a crucial role.

Postulate 7: Understanding of the systems' evolution as the development and operation of these systems progresses

Postulate 7 Corollary: Understanding of the systems degrades during operations if the understanding of the systems is not maintained

Description: A more profound understanding of individual systems and the SoS is acquired as the systems advance through development and operations. During the development phase, the need for detailed decisions increases, and as comprehension deepens, more informed decisions can be made. However, it is essential to acknowledge that understanding of the systems could potentially regress, particularly if organizational changes occur (as per ***Postulate 4***). Factors such as the inactivity of an organizational element, loss of experience due to retirements of key individuals, or the closure of suppliers may contribute to a decline in understanding.

Evidence: The continuous increase in the depth of understanding is evident in the development of any individual system or SoS as the design matures. This progression is reflected in the technical assessment process, where systems advance from concept review to requirements review, design review, and finally, acceptance review. Abundant lessons are learned during the operations phase of any system or SoS. This profound understanding of the systems and practical application drives commercial product upgrades or the introduction of new models.

A good example from the automotive industry could be Tesla's profound understanding of electric vehicle technology and its practical application which drives the continuous upgrades to their existing car models. For instance, Tesla frequently releases software updates to improve battery performance, autopilot features, and energy efficiency. Additionally, this knowledge has led to the introduction of new models, like the Tesla Model Y, which addresses consumer demand for a more affordable electric SUV while incorporating cutting-edge technology from previous models.

On the contrary, regression in the understanding of systems can be observed in some lifecycle extension activities. When the understanding of systems is not maintained, the foundation of system and SoS specifications becomes unclear. This leads to instances where some systems do not perform in accordance with their specifications manifesting as either underperformance or overperformance. Furthermore, operational procedures may lose their basis, making it challenging to determine when systems should be maintained as they age or retired.

Implications: Requirements evolve as the design of individual systems and the SoS progresses. While mission requirements are defined at the project's outset, systems requirements cannot be predetermined but instead flow down from these mission requirements to capture operational context. They are shaped by design choices and become progressively clearer throughout the development phase. This principle also extends to cost and schedules, especially in the case of new systems where development or operations may lead to unexpected changes.

Mission engineers develop models to forecast system capabilities, refining these models as testing and operational experience accumulate. System models gain fidelity as the design advances and the interactions between the constituent systems in the SoS mature. Mission engineers must manage the maturity of individual system models and the federation of verified and validated system models representing the SoS. Failure to maintain the systems basis can result in the loss of understanding regarding why certain procedures or specifications were defined. This becomes particularly problematic for aging or obsolete systems, especially as they approach the generational gap.

5.6.2 Practical Principles for Application

The application principles discussed in this section provide the foundational truths on applying the MEI discipline across organizational boundaries, systems development, and information flows to produce effective and safe SoS for end users.

Principle 1: MEI considers the budget and schedule constraints when integrating the systems and the disciplines

This application aligns with ***Postulate 5***. Budget and schedule serve as constraints for both the integration of systems and the integration of disciplines involved in developing or operating the systems. Note that budgets represent the allocated amount for executing the systems' development or operation, not the actual cost. The primary focus of MEI is to ensure that the cost remains within the budget for individual systems and across the SoS. Additionally, MEI plays a crucial role in recommending adjustments when the solution space defined by budget and schedule does not align with the intended applications.

Principle 2: Complex systems build complex systems

This principle is fundamental to MEI execution. The mission engineer must navigate both the complexity of the system (organization) responsible for developing the systems and the intricate systems and resulting SoS themselves. This dual focus serves as the foundation of the overarching MEI approach. It encompasses both the mission context and systems integrating physics, as well as the organization structure and information flow. ***Postulates 4*** and ***5*** also encapsulate this duality, acknowledging the mission engineer's responsibility for both integrating the systems discipline functions (as per ***Postulate 2***) and managing the development organization disciplines (as defined in ***Postulate 3***).

Principle 3: The focus of MEI during the development phase is to progressively gain a deeper understanding of the interactions, sensitivities, and behaviors of the systems within the resulting SoS

This principle aligns with the application of ***Postulate 7***. The upfront work should not constrain SE and SoS engineering, nor should it diminish as the mission engineer advances through system development. Instead, the knowledge captured, maintained, and enhanced by MEI deepens as interdisciplinary organizations complete their development work and system functions are integrated. This continuous understanding enables the necessary SE decisions to produce an effective system.

MEI's focus lies in understanding the interactions of constituent, heterogeneous systems, many of which may not become apparent until systems integration (e.g., physical integration, logical integration). Current SE tools often lack the depth needed to fully comprehend system interactions, leading to a continuous reduction in system uncertainties and identification of system sensitivities. The mission engineer is tasked with understanding the system's behavior, including emergent behaviors, well before the operational phase. As development progresses, the mission engineer seeks the optimal balance of performance, cost, schedule, and risk across the SoS to provide mission capabilities.

Principle 4: MEI approach translates operational concepts into SoS requirements and into the individual system requirements

This principle aligns with the application of ***Postulate 6***, emphasizing a deeper understanding of system interactions, sensitivities, and behaviors. Requirements and models should accurately reflect this understanding of the system under design. The precision and completeness of system requirements and models must be in line with the depth of understanding achieved. A system that is not well understood can result in poorly stated requirements, gaps in requirements, and inaccurate system models and representations. An objective of MEI is to comprehend the SoS, which subsequently leads to the proper specification of requirements and accurate representation of systems in the system models. This ensures that the requirements and models align with the actual behavior and characteristics of the system.

Requirements represent specific, mutually agreed-upon preferences established by the developing organization. These preferences are individual attributes for systems, requiring the organization to consolidate these individual preferences and reach a consensus on specific values (e.g., performance, cost, schedule). These agreed-upon preferences, along with some agreement on the uncertainty in their measure, constitute the system requirements. These requirements are specific to the system being developed, and those essential for the successful completion of the resulting SoS should be carefully defined as part of MEI. These requirements are necessary for the successful integration of the different disciplines involved in developing or operating the systems. Requirements and design are progressively defined as development advances. Mission requirements are established early in the understanding of systems as part of the mission context. The remaining technical requirements are derived based on SoS design decisions, including design trade-offs at the SoS level that evolve throughout the development phase.

As systems progress through development, the mission engineer gains a deeper understanding of the organizational relationships necessary for system development success. New relationships may be defined, and the magnitude of these relationships may change as the design matures. Organizational groups that do not share information in early development may become instrumental in sharing information later in the development process. Similarly, organizational groups critical at the concept development phase may become less crucial to information flow as development matures. This dynamic understanding of organizational relationships is a key aspect of MEI.

5.7 Need for MEI Approach

MEI is a mission-based approach to understanding operational requirements and the design of the required SoS for specific missions. This approach is essential for aligning development efforts with specific operational objectives and ensuring the effectiveness of operations. MEI places missions at the forefront, acknowledging the dynamic and evolving nature of the tasks. By focusing on specific mission requirements, the design process becomes more targeted, enabling the development of capabilities that directly contribute to achieving mission success. An example is military operations, which often involve the coordination of diverse systems across different domains and a mission-based approach to recognize the interconnectedness of these systems.

MEI promotes a holistic understanding of the operational environment, encouraging a comprehensive analysis of potential threats, challenges, and the need for collaborative functioning among systems. Moreover, a mission-based design ensures that resources are optimized across a mission or capability portfolio, directing investments strategically toward capabilities that align with operational needs. This approach also emphasizes adaptability and flexibility, allowing use of modular and interoperable systems that can respond effectively to changing mission requirements. Ultimately, a mission-based approach is outcome-oriented, focusing on the end results of operations and mitigating risks associated with mission failure. It provides a structured and purposeful framework for developing SoS capabilities that are mission-driven, adaptable, and aligned with the overall strategic goals of the organization.

The transition from platform-centric acquisition to mission-based acquisition signifies a fundamental shift in the way capabilities are developed and procured. For example, the military departed from the traditional focus on acquiring specific platforms as the primary building blocks of military power. Adopting a mission-based approach placed a central emphasis on meeting specific operational objectives.

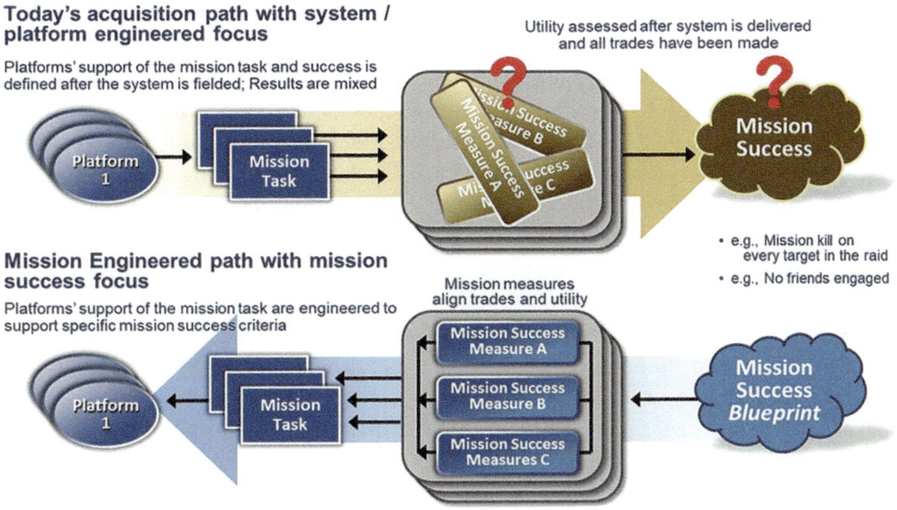

Fig. 5.6 Mission engineering and integration aligns mission success measures

Figure 5.6 depicts this critical paradigm shift recognizing the dynamic and interconnected nature of modern warfare, where success is contingent on the effective coordination of various systems across different domains.

By prioritizing mission requirements over standalone platforms, this MEI approach promotes adaptability and flexibility in responding to evolving threats. It encourages a holistic view of capability development, considering the integration and collaboration of diverse systems to achieve overarching mission goals. Mission-based acquisition optimizes resource allocation by directing investments toward capabilities that align precisely with operational needs. The approach also fosters a risk-aware mindset, enabling a more precise identification and mitigation of risks associated with mission failure. Overall, the move from platform-centric to mission-based acquisition reflects a strategic orientation toward a more agile, outcome-oriented, and interconnected organization.

5.7.1 Developing Mission Success Measures

Developing mission success measures involves establishing clear and actionable standards to evaluate mission success, mission essential task effectiveness, and system performance. First, the mission engineer must define the mission objectives by clarifying specific, measurable, achievable, relevant, and time-bound (SMART) goals, ensuring alignment with stakeholder expectations. The next step is to determine Measures of Success (MoS) to gauge mission success from the stakeholders' perspective. These measures reflect mission objectives and include indicators such as the number of people safely evacuated in disaster relief efforts or the increase in target sales following a marketing campaign. The mission engineer then formulates Measures of Effectiveness (MoE) to assess whether the mission meets its intended objectives, achieves the required proficiency level for MoE threshold, and accomplishes overall impact for each of the mission essential tasks. These tasks include forming the mission thread and focusing on outcome-based indicators such as a reduction in threat incidents for security operations, or improvements in patient health outcomes for healthcare missions. At the bottom of the hierarchy, the Measures of Performance (MoP) are established to evaluate the efficiency and effectiveness of the systems mapped to the mission essential tasks to assess the actions taken to meet the desired functions during mission execution. Examples of these process-oriented, quantifiable measures might include the number of units produced per hour in manufacturing processes or the average response time to customer inquiries in a service operation.

To ensure the criteria are actionable, baseline data is collected to establish current performance levels, and specific targets and thresholds are set to define acceptable performance ranges. The mission engineer identifies data collection methods and tools and implements a regular review process to compare mission performance and outcomes against the established criteria. Continuous improvement is facilitated through a feedback loop, where strategies and operations can be adjusted based on findings, and stakeholders are kept informed with regular updates. Finally, clear documentation of processes and thorough training for team members are essential to ensure everyone understands the mission success criteria and their roles in achieving them.

Goals and objectives are fundamentally shaped by our beliefs and values, serving as the foundation when establishing mission success criteria and measures. These core beliefs and values guide what we consider important, influencing the definition of mission objectives and success parameters. Beliefs are things we consider to be true, shaping our understanding of the world. Values, on the other hand, act as a moral compass, guiding us on what is right and wrong. These values establish our standards for what we are willing and unwilling to do, essentially defining what we deem important. Importantly, values do not directly drive our actions; rather, they inform us about the right and wrong actions to take, and therefore have a major influence on the definition of mission success and the corresponding objectives. To illustrate this point, here is a healthcare example:

> In a healthcare organization, values like compassion, integrity, and patient-centered care do not directly dictate the day-to-day actions of doctors and nurses. However, these values inform healthcare professionals about the right and wrong ways to treat patients, guiding their decisions and behavior. For instance, a doctor may face a situation where following standard protocol conflicts with a patient's specific needs. Compassion and integrity help the doctor prioritize personalized care, thus influencing how the organization defines mission success—ensuring patient well-being and trust. The corresponding objectives, such as reducing patient wait times or improving communication, are shaped by these values.

5.7.2 System-of-Systems Analyses and Design

SoS analyses and design constitute a comprehensive approach to addressing the complexities inherent in modern interconnected environments. This complex examination and development effort considers the collection of independent, heterogenous systems that, when integrated, form a more extensive and complex system capable of achieving overarching objectives. The MEI approach recognizes that the interactions and interdependencies among these systems are central determinants of overall SoS effectiveness and the resulting capability. The design phase focuses on harmonizing the interoperable SoS capability so that it optimally leverages the capabilities of each constituent part. Considerations include modularity, adaptability, and scalability to accommodate evolving mission requirements. SoS design encompasses not only the technical aspects but also the organizational, operational, and social dimensions. The goal is to achieve synergy among diverse systems, thereby ensuring a cohesive and effective response to multifaceted challenges. Successful SoS analyses and design require a holistic understanding of the mission context, an awareness of the interconnections among systems, and a proactive approach to addressing emergent properties and complexities that may arise during the system lifecycle development process. Ultimately, SoS analyses and design are integral components of a strategic approach to building interconnected systems that can operate collaboratively to meet the demands of contemporary and future challenges.

Examples of SoS span various domains, including military capabilities, transportation systems, healthcare systems, and communication networks to name a few. In the military context, an SoS could be composed of interconnected systems such as aircraft, ground vehicles, and communication networks working together to achieve strategic objectives. For military applications, the effectiveness of an SoS can be measured in terms of its ability to deliver a physical force that surpasses the structural load capacity of the opposing system or SoS. This represents a thermodynamic relationship where the focus is on the energy transfer and structural considerations. This thermodynamic relationship becomes apparent in military scenarios when analyzing the impact of a single system on another system. This occurs when the force exerted by a military unit exceeds the structural capacity of an opponent's defense system. When dealing with multiple systems within an SoS, military operators must consider the distribution of capabilities corresponding to the distribution of the opposing SoS.

Incorporating engineering statistics becomes crucial in this context to represent the distributions and uncertainties in the parameters of the SoS. This statistical approach allows for a more nuanced understanding of how variations in the capabilities of individual systems within the SoS can impact the overall effectiveness and performance against an opposing force. It considers the variability in capabilities and takes into account factors such as reliability, efficiency, and effectiveness. Thus, the statistical approach provides a more comprehensive assessment of the SoS in realistic and uncertain operational conditions. An influence diagram describes

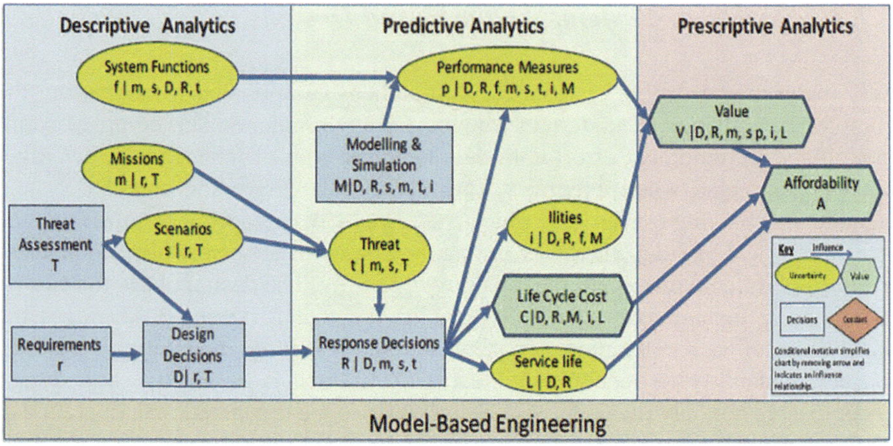

Fig. 5.7 Framework for integrated analysis of alternatives

the integrated trade-off analytics framework in Fig. 5.7 [SPECKING18a, SPECKING18b].

5.7.3 *Flexibility in Design*

Flexibility in design is a critical attribute that enables systems to adapt to evolving requirements, uncertainties, and constraints without incurring excessive costs or delays. Set-Based Design (SBD) inherently fosters this flexibility by maintaining multiple design alternatives in parallel, allowing for continuous refinement and informed decision-making. The SBD process, as illustrated in Fig. 5.8 [BERNSTEIN98], emphasizes the iterative refinement of design options, fostering a dynamic and informed progression toward a well-informed and optimized solution.

Unlike Point-Based Design (PBD), which commits to a single solution early in the process and refines along a fixed trajectory, SBD provides the freedom to explore and compare various possibilities before selecting the most optimal path forward. This approach is particularly valuable in complex and dynamic environments where mission requirements, technological capabilities, and stakeholder needs are subject to change. By deferring key decisions until sufficient data is available, SBD reduces the risk of costly rework and ensures that design choices remain aligned with the broader mission and operational context. The iterative nature of SBD allows for adjustments as new insights emerge, making it a powerful tool for managing uncertainty and mitigating risks in system development.

Additionally, SBD aligns with the principles of lean thinking by optimizing resource allocation and minimizing waste. Instead of prematurely discarding less favorable options, SBD preserves design alternatives until a more informed stage of the development process is reached. This measured approach not only enhances

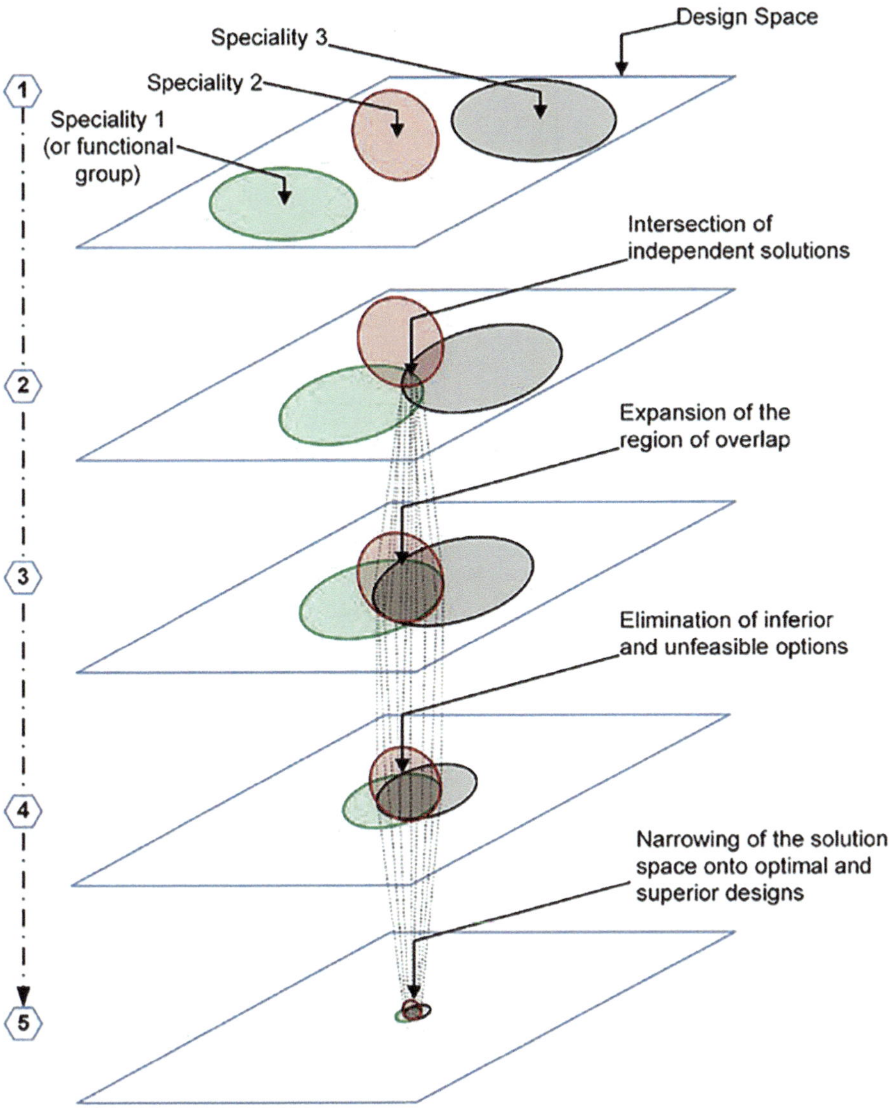

Fig. 5.8 Set-based design approach. (Adapted from [BERNSTEIN98])

adaptability but also encourages collaboration and creative problem-solving within design teams. By fostering an open-minded and systematic evaluation of trade-offs, dependencies, and constraints, SBD leads to the development of more robust, resilient, and high-performing systems. This difference between PBD and SBD is illustrated in Fig. 5.9 with PBD represented in (a) and SBD in (b).

SBD offers several advantages over PBD as identified in Fig. 5.9, particularly in managing the cost of changes, reducing technical debt/risk, and enhancing

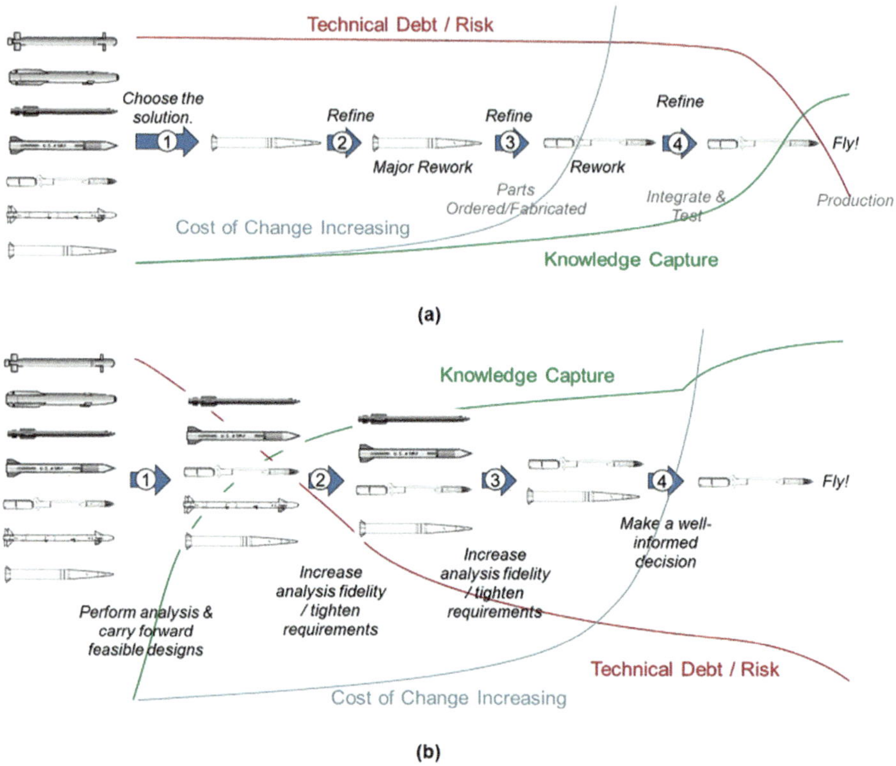

Fig. 5.9 Advantages of Point-Based Design (PBD) (**a**) to Set-Based Design (SBD) (**b**)

knowledge capture. One of the key benefits of SBD is its ability to mitigate the high costs associated with late-stage design changes. By maintaining multiple design alternatives in parallel and gradually narrowing down options based on emerging data, SBD avoids the costly rework often seen in PBD, where early design commitments can lead to expensive modifications if requirements evolve. Secondly, SBD reduces technical debt and design risk by ensuring that decisions are made with a more comprehensive understanding of trade-offs and constraints. Instead of locking into a single solution prematurely with the potential to lead into suboptimal performance or unforeseen integration challenges, SBD enables engineers to systematically refine solutions, minimizing the likelihood of failure or last-minute redesigns. Furthermore, SBD enhances knowledge capture by fostering an iterative and collaborative approach, where lessons learned from exploring multiple alternatives are documented and leveraged to inform future projects. This contrasts with PBD, where discarded design paths often lead to a loss of valuable insights. By promoting a more informed, adaptive, and risk-conscious approach, SBD ultimately leads to more robust, cost-effective, and mission-aligned solutions.

Engineers operating within the SBD framework are tasked with enhancing the precision and detail of their design options as the design timeline unfolds. This approach aims to narrow down the set of available options based on accumulating information rather than arbitrary decisions. The systematic decision-making process is characterized by a continual influx of knowledge and increasing detail. This enhances the likelihood of achieving convergence toward the final product. In an era of rapid technological advancements and complex mission environments, the ability to remain flexible throughout the design process is essential. By integrating SBD principles, mission engineers and designers can proactively navigate challenges, drive innovation, and develop solutions that are both effective and future-ready.

5.7.4 Multidisciplinary Analysis and Optimization (MDAO)

Multidisciplinary Analysis and Optimization (MDAO) is a computational approach that plays a vital role in the design and optimization of complex systems involving multiple disciplines. In MDAO analyses, engineers and researchers use integrated computational models to simultaneously evaluate and optimize various aspects of a system across different engineering disciplines. This approach is particularly valuable when dealing with intricate systems, such as aircraft, spacecraft, or military platforms, where the interactions among multiple disciplines significantly impact overall performance. In the MEI approach, MDAO serves as a powerful means to model and analyze the integrated operation that results from the collaboration of individual systems, forming a new or extended capability within the SoS. The strength of MDAO lies in its ability to balance and optimize the shared variables that represent the intricate interactions among systems within the SoS. By employing MDAO, engineers can explore a wide range of design alternatives and consider the implications of each system's performance on the overall SoS capability. This approach facilitates informed decision-making, allowing for the identification of optimal solutions that align with mission objectives. The iterative nature of MDAO permits adjustments and refinements throughout the design process which is important to make sure the integrated SoS operation is well-balanced and effective. MDAO brings together experts from various engineering domains, including aerodynamics, structures, propulsion, and controls, who work collaboratively to balance competing objectives such as reducing weight, improving efficiency, and meeting mission requirements. Through iterative computational simulations, MDAO enables engineers to explore the design space and identify solutions that satisfy multiple, often competing, criteria.

The engineer using the MDAO process typically develops integrated models, establishes design variables, and applies optimization algorithms to find the most efficient and effective design solutions. Today's engineers developing complex systems and SoS need to use this approach in order to understand the interdependencies across different disciplines, ensuring that changes in one aspect of the system are appropriately reflected in others. Conventional SE processes often rely on hierarchical decomposition to organize design and development tasks, resulting in a linear and simplified structure with straightforward relationships. However, this approach encounters challenges when applied to complex SoS. In such intricate environments, the interactions among systems become overwhelming, and the hierarchical decomposition may not capture the full complexity. Examining system couplings addresses this complexity by providing the medium through which the interactions of system functions and their relationships with the environment can be understood. Unlike the linear structure of hierarchical decomposition, system couplings allow for a more nuanced and interconnected representation of system elements. These couplings manifest in various forms, encompassing both physical and logical connections between different components of a system. Despite their significance, these couplings are not always immediately apparent and can give rise to emergent behaviors. Emergent behavior refers to phenomena that arise from the interactions of system elements and which lead to outcomes that may not be predictable based solely on the behavior of individual components. In the context of complex SoS, emergent behavior can modify existing system couplings, introduce new connections or alter the responses of known couplings. This dynamic nature adds layers of complexity and highlights the importance of understanding not just the individual systems. But also, their interdependencies and emergent behaviors within the broader SoS framework.

MDAO research places significant emphasis on developing frameworks that facilitate system optimization by appropriately modeling inherent couplings in physics and analysis. Computer simulations play a crucial role in capturing system interactions, enabling a comprehensive system analysis integrated into the larger optimization process. The engineer employs constraints to embody design requirements while carefully selecting the objective function to best represent preferences for the system design. Given the intrinsically interdependent nature of physics models, achieving convergence in the system analysis requires an iterative process. This iterative approach is essential for assessing the impacts of design variables on individual systems and the overall SoS. MDAO achieves convergence by initiating a system with a set of design variables and iteratively progressing through the coupled analysis until convergence is reached. This advanced computational approach improves design efficiency by simultaneously integrating multiple engineering disciplines. By doing so, it promotes innovation and enables engineers to identify optimal solutions that balance complex and often conflicting design objectives.

Integrating MDAO and SBD creates a powerful synergy that enhances decision-making, fosters innovation, and improves efficiency in complex system development. MDAO's computational capabilities enable the rapid evaluation of multiple design alternatives, providing high-fidelity simulations and optimization techniques that support SBD's emphasis on maintaining a broad solution space. This integration enhances flexibility by allowing engineers to explore diverse options while systematically narrowing choices based on evolving constraints, stakeholder input, and mission requirements. MDAO automates evaluations and facilitates real-time adjustments, reducing the risk of costly rework and ensuring design decisions are based on robust, data-driven insights. By capturing and analyzing knowledge throughout the design process, this combined approach mitigates technical risk, improves adaptability to changing requirements, and streamlines system development. Ultimately, leveraging MDAO and SBD together leads to more resilient, optimized, and high-performing system architectures.

With a solid foundation in **postulates and principles for MEI**, we now shift our focus to how these guiding concepts apply in real-world scenarios. Successful MEI requires more than just technical soundness; it demands a deep understanding of operational context, constraints, and real-world mission demands. The design of SoS solutions must account for factors such as mission objectives, environmental conditions, interoperability challenges, and evolving threats to ensure systems can operate effectively in dynamic and often unpredictable settings. In the next chapter, we examine how these **operational context and design** factors influence design decisions, shaping MEI strategies that are not only theoretically sound but also practical, adaptable, and mission-ready. Through this lens, we explore how engineering principles are applied to enhance operational effectiveness, ensuring that integrated systems can perform reliably and efficiently in their intended environments.

References

1. Ashby, W. R. (1958). General systems theory as a new discipline. *General Systems, 3*, 1–6.
2. Bernstein, J. (1998). *Design methods in the aerospace industry: Looking for evidence of set-based practices*. Master thesis, Massachusetts Institute of Technology.
3. Von Bertalanffy, L. (1968). *General system theory: Foundations, development, applications*. George Braziller.
4. Forrester, J. W. (1968). *Principles of systems*. Productivity Press.
5. Kurtz, C. F., & Snowden, D. J. (2003). The new dynamics of strategy: Sense-making in a complex and complicated world. *IBM Systems Journal, 42*, 462–483.
6. Rosen, R. (1971). Dynamical system theory in biology. In *Stability theory and its applications* (Vol. 1). Wiley.
7. Specking, E., Parnell, G., Pohl, E., & Buchanan, R. (2018a). A foundation for system set-based design trade-off analytics. In *2018 ASEM International Annual Conference*. American Society for Engineering Management.
8. Specking, E., Parnell, G., Pohl, E., & Buchanan, R. (2018b). Early design space exploration with model-based system engineering and set-based design. *Systems, 6*(4), 45.

Chapter 6
Operational Context and Design

6.1 Operational Planning Development Philosophy

Employing the instruments of national power, Diplomatic—Information—Military—Economic, requires a sound strategy to synchronize and integrate the individual elements with the goal of achieving national, operational, and multinational objectives. Another dimension of this strategy encompasses the art and science of exploring alternative futures to establish the framework for defining the desired end state or condition (ends). The strategy involves:

- Communicating this information across all levels of a force or competitive landscape across any industrial sector for sustained efforts
- Evaluating potential courses of action (ways)
- Identifying the necessary authorities and allocating resources (means such as time, personnel, equipment, and funding) to effectively achieve the planned objective while actively mitigating associated risks

Framing the operational environment involves defining the context within which an organization or entity operates [ADRP12]. This encompasses various factors such as the current state, objectives, challenges, stakeholders, and resources. The mission engineer must understand the operational environment, which in turn requires a comprehensive analysis of both internal and external elements that impact the organization's activities. This analysis serves to clarify the organization's current state, identify its desired future end state, and outline the necessary and essential steps to bridge the gap between the two. First, the mission engineer must thoroughly assess the current state of the operational environment to baseline the conditions and determine the problem statement or operational need. This involves examining the organization's internal dynamics, including its structure, processes, capabilities, strengths, weaknesses, and existing challenges. Additionally, external factors such as market conditions, competitive landscape, regulatory environment, technological

J. D. Moreland, Jr., *Mission Engineering and Integration*,
https://doi.org/10.1007/978-3-032-32132-9_6

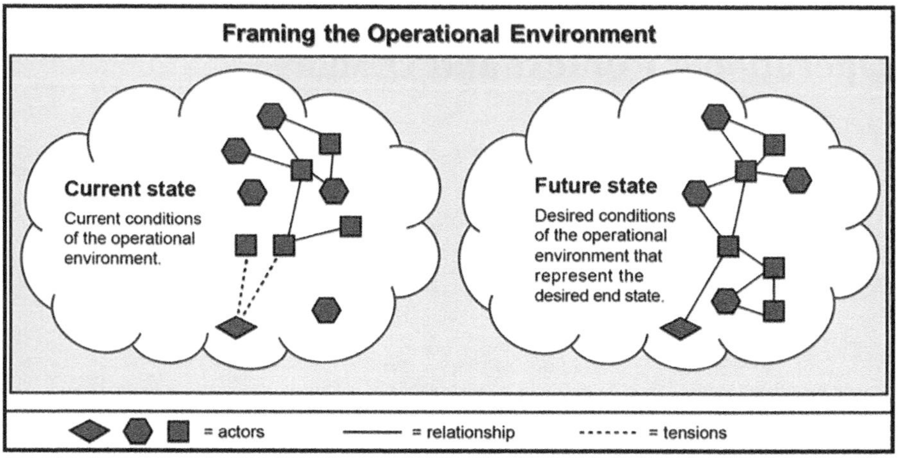

Fig. 6.1 Determining the future desire end state

trends, and socio-political factors must be analyzed to gain a holistic understanding of the environment in which the organization operates. Once the current state is established, the desired future end state or objectives must be clearly defined. By articulating the desired future end state, the organization can establish a clear direction and purpose for its activities, see Fig. 6.1.

Next comes framing the operational environment. Framing involves identifying the key factors and variables that influence the organization's ability to achieve its objectives. This includes understanding the needs and expectations of stakeholders, assessing potential risks and opportunities, and evaluating the resources available to the organization. Finally, based on the analysis of the current state and desired future end state, the mission engineer can develop strategic initiatives and action plans to bridge the gap between the two. These initiatives should outline the steps, timelines, responsibilities, and resources required to achieve the organization's objectives. By framing the operational environment in this way, organizations can effectively navigate complexities, make informed decisions, and drive progress toward their desired outcomes.

Effective and efficient operations must be synchronized across these operational planning activities with a clear, concise message of intent from leadership to flow down through all levels of the force or executing entities. Joint operations depend on the coordinated employment of capabilities and activities across the Joint Force to accomplish common objectives [JP1]. In addition, Joint operations require the integration and synchronization of capabilities, activities, and functions in time and space to accomplish operational objectives [JP3]. Coordinating activities and efforts is crucial to avoid duplication, reduce conflicts, and ensure a seamless integration of tasks. Effective communication channels and coordination mechanisms enhance the efficiency of operations spanning movement, engagement of predetermined essential activities, and supporting activities. This intricate process, which entails coordination, collaboration, and rehearsal, is pivotal for achieving precise mission success.

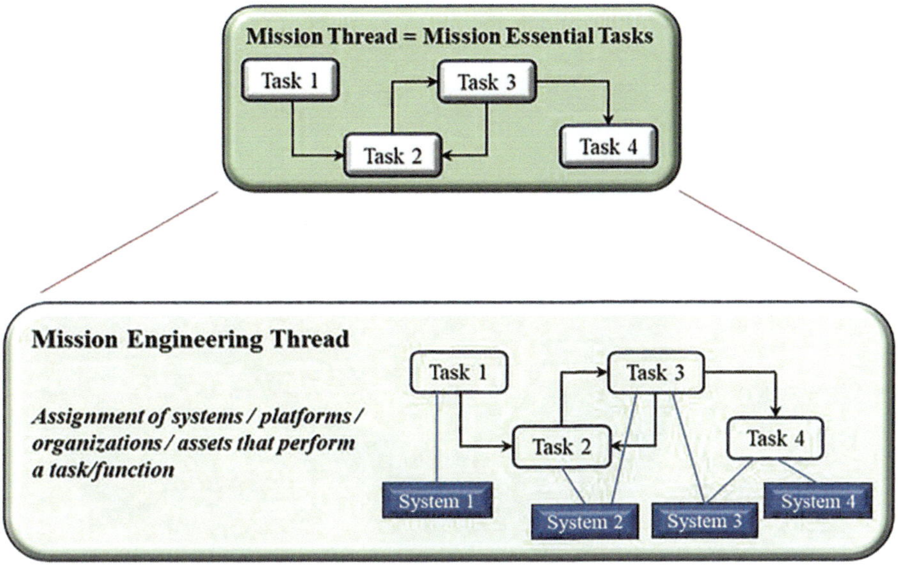

Fig. 6.2 Forming mission engineering threads for mission threads

This surgical precision is necessary to maximize effectiveness with close attention paid to the temporal aspect of operations for timing and sequencing of events and the spatial dimensions to understand the physical layout of the environment.

As mentioned previously, operational planning is an intentional process that outlines the execution of strategic guidance by establishing the connection between how (the ways) capabilities are employed (the means) within the temporal and spatial dimensions of the operational context to accomplish objectives (the ends) [JP5]. This approach is necessary to increase the probability of achieving mission success and maintaining an acceptable level of risk. Although the intention is to conduct events with precision, it is crucial to quantify the uncertainties associated with each individual action and their cumulative effects across the entire system-of-systems (SoS). This is essential for providing decision-makers with a clear understanding of the level of risk involved in the operational event. Without this comprehensive analysis, the decision-maker may lack crucial insights into potential vulnerabilities, leading to an incomplete assessment of the risks that could impact the success or safety of the operation. The ultimate goal is to translate strategic objectives into operational activities and mission essential tasks. Further, the aim is to optimize the application of the mission engineering thread/web formed from SoS capabilities to reach mission success based on mission success criteria. Figure 6.2 illustrates a pivotal aspect of Mission Engineering and Integration (MEI) by mapping systems to the mission thread forming the mission engineering thread, which consists of a sequence or network of mission essential tasks that are intricately woven to achieve specific mission objectives leading to mission success. These intricately interwoven tasks often overlap in execution, as they require handoffs and

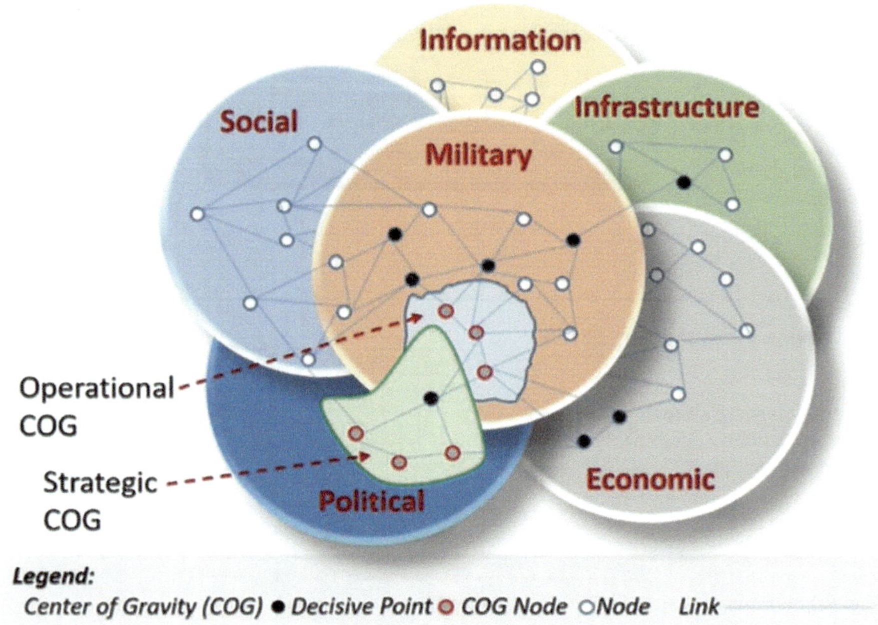

Fig. 6.3 Centers of gravity within elements of national power

synchronization between functions. Furthermore, there is constant iteration and adaptation throughout the process due to the potential rapid changes in conditions when operating in a dynamic environment with many moving parts, necessitating real-time adjustments to maintain mission alignment and effectiveness.

A focus on military operations using the military instrument of national power requires a good understanding of the linkage and effects to the other instruments of national power as shown in Fig. 6.3. National security goals and objectives are usually focused on predetermined centers of gravity as these are considered strengths with regard to achieving national security objectives. As with operational planning, Mission Engineering and Integration (MEI) MEI starts with the desired end state of mission success in mind, providing a unified purpose across an entire force to surgically execute mission essential tasks. This provides efficient, effective allocation of resources. The targeted audience for this sort of operational planning resides at the operational and tactical levels of execution; however, many stakeholders are involved as a "Whole of Government" approach is considered across the United States. Government title codes identify designated positions and roles within federal, state, and local governments, providing a standardized means to distinguish organizational responsibilities, decision-making authority, and command or coordination relationships. This "Whole of Government" approach is a comprehensive strategy that brings together various governmental agencies, departments, and sectors to work collaboratively toward shared objectives, such as national security, public policy, or crisis management. By integrating the diverse capabilities, operating philosophies, resources, and expertise of these entities, this approach seeks to enhance coordination, reduce redundancies, and ensure a unified response to complex, multidimensional challenges. It emphasizes interagency cooperation, cross-sector partnerships, and the alignment of policies and operations to achieve more effective and efficient outcomes.

This operational planning process is dynamic due to changing conditions, and therefore requires a recursive, threat assessment-informed process which is not linear. The original problem statement used to define the operational objectives may need to be revisited on a regular basis to incorporate issues discovered later and adjustments made to plans to stay on the right vector for mission success. Operational planning consists of four related functions to include operational art, operational design, operational approach, and operational need. Typically sequential, these four functions are frequently carried out simultaneously to enhance discussions among all pertinent stakeholders and expedite the overall planning process, as discussed below.

Operational Art

Operational art is the development, coordination, and implementation of ends, ways, and means to execute policy and promote national interests. It refers to the application of creative and critical thinking to the design and conduct of military campaigns and major operations. A critical part of this function is the evaluation of the adversarial actors and the operational environment to understand the operational context. The purpose of this function is to bridge the gap between overarching strategic goals and tactical execution within the operational environment while also providing a coherent, enduring plot to navigate from the present state to the future desired state. A desired outcome of this function is for all relevant stakeholders to visualize and understand their support in achieving strategic success.

> *From a problem-solving perspective, Operational Art will make it possible to take an unstructured problem and give it sufficient structure to ensure that further planning can lead to useful action. Operational Art will remain essential when recognized as a methodology that enables the effective planning and execution of all operations. At its best, Operational Art can play a pivotal role in military success when skilled leaders apply it in its full dimensions—functioning as true artists to give expression to a nation's strategic vision.* [OLSEN11]

Operational Design

Operational design is a methodology that applies critical and creative thinking to comprehend, visualize, and articulate complex, ill-defined problems while formulating strategies to address these problems. US General James N. Mattis stated that "design does not replace planning, but planning is incomplete without design" [MATTIS09]. Operational design involves a systematic, analytic framework to conduct analysis on the operational environment viewed as a complex interactive system. For example, the framework elements for military operations and strategic planning are political, military, economic, social, information, and infrastructure (PMESII). Operational design focuses on shaping the conditions necessary for mission success and adapting to the changing circumstances and serves as the foundation for operational planning. As part of this function, when sharing results, mission engineers must consider how to translate inconsistent and confusing military jargon and concepts to very diverse stakeholders in a universally understood language to inform interagency partners and other functional instruments of national power. Operational design is intertwined with operational planning in a manner to help fill in knowledge gaps and provide guidance to aid in the planning for the complexities of the operational environment. In support of MEI, operational design develops courses of action and concepts of operation with the highest probability of reaching mission success.

Design must be complemented with engineering, which eventually produces a detailed plan of action. As the mission engineer understands the conditions in the operational environment, the boundary conditions and performance parameters can be determined for the SoS providing the capability. The defined capability can be translated down to the individual, heterogeneous systems that form the SoS capability. To provide a formal link across the levels of operations as well as across the engineering of each constituent system within the SoS, this understanding provides the necessary information to establish the measures of success at the mission level down through the system level. Determining the vulnerabilities in the "as-is" SoS architecture serves as a link between design and engineering; it makes design logic operational. It also provides avenues to collect feedback and identify the mission success criteria that eventually return as input to the design assessment and reframing process. Problem framing involves determining the root cause of a problem so that the mission engineer does not chase symptoms without resolution of the problem but can identify actual operational gaps.

Operational planning consists of both the cognitive application of design to develop broad concepts of engagement based on threat assessments and the subsequent translation of those concepts into a complete and practical engineering design. This is a continuous, iterative and often cyclic process involving multiple stakeholder perspectives to capture the dynamic operational conditions. The transition between design and detailed planning is aligned by executing the threat evaluation and weapon assignment process for military operations to match threat vulnerabilities against friendly capabilities. To achieve mission success, the mission engineer must revisit the operational problem statement and be willing to reframe it if required based on accuracy of the desired end state or desired conditions. These elements of the MEI approach can change as the operational environment evolves during mission execution. Necessary updates may be required in the operational design as the starting conditions may change or no longer be possible. Table 6.1 captures some of the key cognitive processes for design and engineering when characterizing the operational environment.

The primary goals of the operational design process are to:

- Understand ill-structured problems
- Anticipate change, create opportunities

Table 6.1 Cognitive processes for design and engineering

Design	Engineering
Framing the Problem	Solving the Problem
Innovative Thinking	Creating Coherent Design
Dealing with Limits of Existing Knowledge	Working within Defined Parameters
Assumptions and Methods	Established Procedures and Laws
Conceptual	Physical and Detailed
Gains Insight and Understanding	Develops Products and Capabilities
Set Paradigms and Principles	Accept and Apply Paradigms and Principles

- Recognize interdependencies between the elements
- Manage transitions for integration and interoperability

Operational environments are characterized by complexity due to the many moving parts, both behavioral and functional. By analyzing both the context and operational variables of a given problem set together, the mission engineer can determine the essential inputs to frame ill-structured problems with sufficient success. Understanding complexity is integral to comprehending an operational environment and devising mission plans with effective mission essential tasks to effect desired changes in the environment. Effective design should offer a model for innovative and adaptive thinking to successfully understand a problem space; however, the risk and uncertainty need to be identified and quantified as part of the assessment for the decision-maker to make an informed decision without perfect knowledge. The design process was developed to put every actor on a vector of solving the "right problem" without creating collateral problems. The goal is to reach a state of shared understanding of the operational context based on factual data from authoritative sources of truth. The solution to operational problems needs to be driven by mission success criteria that will produce the conditions defining the desired end state for mission success. The MEI approach depicts the high-level operational plan using an effects web framework. This framework denotes lines of effort in mission engineering threads which provides a graphic link among mission essential tasks, objectives, threat evaluation and weapon assignment, SoS actors, conditions, and the desired end state.

Operational Approach

An operational approach outlines the sequence of actions, allocation of resources, and coordination of forces to achieve operational and strategic objectives. It considers factors such as maneuver, operational tempo, and synchronization of different elements in the operational environment. Engaging a cognitive approach, all pertinent stakeholders

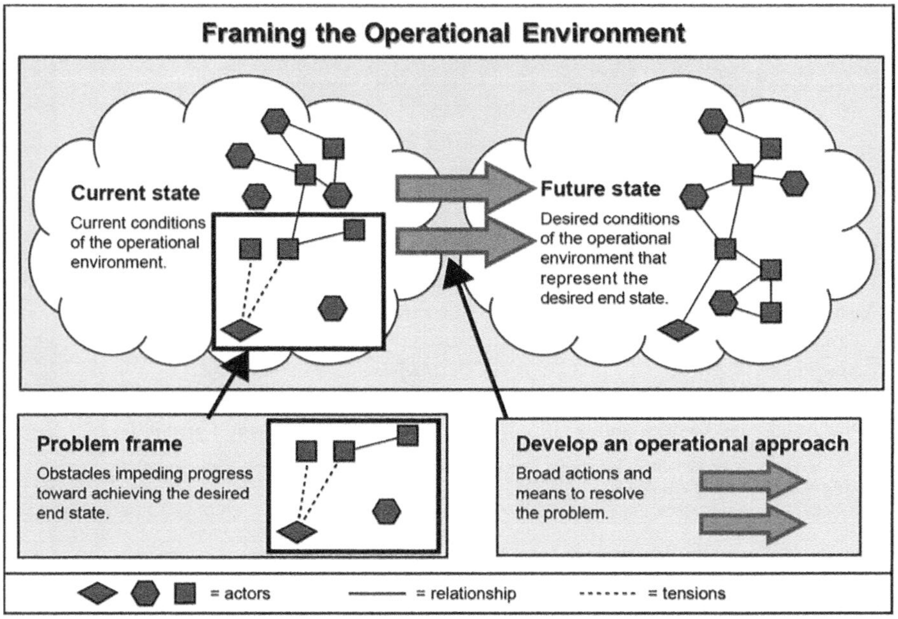

Fig. 6.4 Identifying operational need (problem frame)

contribute their skills, knowledge, experience, creativity, and judgment to formulate strategies and operations for the organization and deployment of forces or actors. This process integrates ends, ways, and means while continuously assessing risks. While planning often follows a scientific methodology, planning for conflict and challenges is grounded in principles derived from operational art and operational design.

Operational Need

The operational need function analyzes the "as-is" SoS architecture against a specific threat within the operational environment to determine capability gaps. These identified gaps in existing capabilities significantly impede achieving the defined mission objectives and ultimate mission success. These needs identified in this operational planning function are based on the objectives of the operation and the conditions of the operational environment. The problem frame in Fig. 6.4 shows the operational need for this situation.

Combining these operational planning functions in a structured, but iterative process allows the relevant stakeholders to examine the adaptation to the dynamics of the mission and operational environment, while concurrently carrying out operations in alignment with the current approach and adjusting plans as necessary. Figure 6.5 provides a diagram capturing the relationship between these four operational planning functions as driven by strategic objectives and guidance.

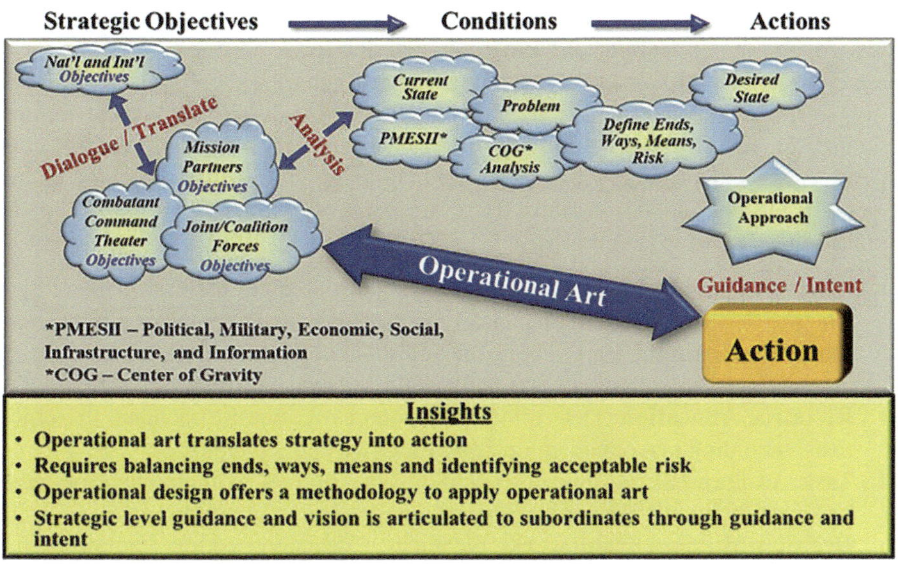

Fig. 6.5 Relationship between Operational Art and Operational Design

6.2 Threat Evaluation and Competitive Landscape

In defense operations, threat evaluation and weapon assignment (TEWA) involves assessing potential threats in real time and assigning appropriate defensive resources, such as weapons or countermeasures, to neutralize those threats efficiently. This dynamic process ensures that critical assets are protected while optimizing resource allocation under high-stakes conditions.

In the context of non-defense industrial sectors, a comparable term might be competitive threat analysis and resource allocation (CTARA). This concept involves evaluating competitive threats—e.g. market entrants, technological disruptions, or shifts in consumer behavior—and assigning strategic resources, such as marketing budgets, R&D investment, or operational adjustments, to address those threats effectively. Similar to TEWA, CTARA requires a balance of real-time decision-making, prioritization, and strategic execution to maintain a competitive edge in rapidly changing environments. While TEWA is primarily used in military contexts for planning and executing missions, similar principles can be applied to setting objectives for non-military mission execution across industrial sectors. In non-military contexts, the focus shifts from neutralizing threats and deploying weapons to achieving specific goals and objectives. This involves identifying and analyzing potential market threats, such as new competitors, disruptive technologies, changing consumer preferences, or regulatory shifts. The goal is to assess the potential impact on market position and guide strategic actions, such as innovation, diversification, or market repositioning.

The link between these two competitive landscape processes lies in their shared objective of understanding and mitigating external risks to achieve mission or strategic goals. Both processes rely on data analysis, scenario modeling, and prioritization to make informed decisions under uncertain conditions. While the military focuses on security and operational integrity, businesses emphasize market competitiveness and profitability, yet the methodologies and decision-making frameworks often overlap significantly. Here are some ways objectives can be set for non-military mission execution:

- **Risk Assessment**: Identify potential risks and threats to the mission's success, whether they are related to financial, operational, environmental, or other factors.
- **Objective Setting**: Clearly define the goals and objectives of the mission, ensuring they are specific, measurable, achievable, relevant, and time-bound (SMART).
- **Resource Allocation**: Determine the resources required to achieve the objectives, including personnel, equipment, finances, and technology.
- **Task Assignment**: Assign tasks and responsibilities to individuals or teams based on their skills, expertise, and availability.
- **Decision-Making Framework**: Establish a decision-making framework that outlines how decisions will be made, who will be involved, and how information will be communicated.
- **Monitoring and Evaluation**: Implement mechanisms for monitoring progress toward objectives and evaluating the effectiveness of strategies and actions.
- **Adaptation and Flexibility**: Remain flexible and adaptable to changing circumstances, adjusting objectives and strategies as needed to overcome obstacles and achieve success.

For example, in disaster response or emergency management situations, TEWA principles can be adapted to assess various threats or hazards and assign resources (such as personnel, equipment, and supplies) to mitigate those threats and accomplish mission objectives. This might involve evaluating risks posed by natural disasters, coordinating rescue efforts, and allocating resources to areas in need of assistance. Similarly, in business or project management settings, managers can utilize TEWA concepts to evaluate potential risks, allocate resources effectively, and assign tasks to achieve project objectives. This might involve assessing market risks, optimizing resource allocation, and assigning roles and responsibilities to team members. Overall, while the specific applications may differ, the fundamental principles of assessing threats or risks, allocating resources, and assigning tasks to achieve objectives can be adapted from military contexts to various non-military missions.

The US grapples with a fragile global order strained by increasing competition among major powers, heightened transnational challenges, and numerous regional conflicts with significant consequences. Concurrently, advancements in technology, vulnerabilities in public health, and environmental shifts are occurring more frequently, often with global ramifications that are difficult to predict. These turbulent periods will shape the future world order, influenced by persuasive arguments on governance, societal organization, and effective systems for economic growth. The actions of both state and non-state actors in addressing transnational issues and regional crises will also play a crucial role. Amid these challenges, opportunities for collective action, involving various actors, will arise.

Given the evolving global order, these regional and localized conflicts pose significant concerns. Instabilities spanning from East Asia to Africa to the Western Hemisphere, worsened by global issues, have the potential to spill over into various territorial domains, affecting the United States, its allies, partners, and the global community. In addition, economic strain is exacerbating global instability as numerous states grapple with mounting debt, economic repercussions stemming from the conflict in Ukraine, and escalating losses from extreme environmental events. Many countries face persistent domestic food price inflation and precarious food security vulnerable to economic and geopolitical shocks. Simultaneously, the world confronts a myriad of common challenges and competition necessitating collective global solutions. This competition exploits technological advancements, including artificial intelligence (AI), biotechnology, and emerging quantum developments to shape global narratives and influence geopolitical dynamics. Advances in AI and biotechnology are rapidly progressing, with the convergence of various scientific fields likely to yield significant breakthroughs. Consider the escalating impacts of climate change, which are disproportionately affecting populations in low- and middle-income countries and increasing their vulnerability to extreme weather events, food and water insecurity, and humanitarian crises. The impacts also include migration and the heightening risk of future pandemics as pathogens adapt to changing environments.

It is imperative for the United States to remain vigilant in assessing and monitoring both direct and indirect threats to United States and allied interests. This entails conducting thorough diagnostics to identify the most significant immediate- and long-term threats. It demands analysis of the evolving global order and other macrotrends that may shape these threats' direction and impact. However, such assessments must acknowledge the inherent uncertainty and risk involved, as predictions are often made with incomplete or imperfect information about the threat landscape. The capacity to analyze and anticipate potential future scenarios and their key characteristics is essential for developing effective capabilities to address emerging challenges.

6.2.1 TEWA Process

Both TEWA and CTARA processes require identifying potential risks, prioritizing them based on severity, and responding with a combination of mitigation and proactive measures. CTARA for businesses mirrors TEWA by using systematic data analysis, risk prioritization, and dynamic response planning to maintain an edge in a competitive environment. As used in the military, TEWA is a two-phased complex dynamic system that operates within an ever-changing and intricate environment. The initial phase involves evaluating and ranking targets based on their threat level to a defended asset. Subsequently, in the second phase, reactive assignment of weapon systems to these threats occurs. These activities align with the broader joint targeting process, which connects targeting priorities and desired effects with available capabilities to support operational objectives [JP3-60]]. When the number of attacking threats (n) is minimal, the execution of TEWA is simplified requiring trivial solutions. For instance, when n equals 1, threat neutralization is straightforward with attention on this single threat without the additional complexity of prioritization. However, in a scenario with multiple targets ($n > 1$), the complexity of TEWA intensifies with increasing n. This complexity arises because there are more viable options to consider for threat ranking, resource allocation prioritization, and neutralization. Effective threat ranking is crucial as it determines which threats pose the highest danger and should be neutralized first. Threats can be prioritized based on their expected degree of harm to a high value unit, often referred to as the "Threat Index." Let T_i represent the number of threats, where $1 \leq i \leq n$. The outcome of the evaluation phase results in a ranked list, denoted as $T = \{T_1, T_2, T_3, \ldots, T_n\}$, such that the Threat Index follows the order: $I_1 > I_2 > I_3 > \ldots > I_n$.

Two primary models are commonly employed for Threat Evaluation (TE): the capability-based TE model and the intent-based TE model. The capability-based TE model ranks targets based on their capability index, which assesses a target's potential to cause damage to a defended asset. On the other hand, the intent-based TE model evaluates a target's intent, representing the target's willingness to inflict harm on a defended asset, to calculate its threat index. Various parameters are utilized to estimate both intent and capability of a threat. For instance, parameters such as target type, speed, direction, heading, weapon envelope, and lethality are considered for calculating the capability index. Similarly, factors like velocity, altitude, heading (bearing and course), speed, detection of emissions from the fire control radar, and flight pattern to include unique maneuvers are considered for intent index calculation. Each model, capability, and intent holds its own significance in TE. However, an optimal TE model can be achieved by integrating capability and intent parameters, resulting in an opportunity-based TE model. This combined approach enables a more comprehensive assessment of threats and enhances the effectiveness of TE processes. After calculating the parameters for each identified threat, the next step is to determine the final Threat Index for each threat. Research suggests that Bayesian networks and fuzzy logic rules are viable methods for addressing this challenge. Johansson and Falkman, in their study [JOHANSSON08], compared these two approaches and highlighted their respective advantages and disadvantages. Bayesian networks offer a strong mathematical foundation within probability theory, making them adept at handling missing evidence. They tend to compute threat values with clear separation but require more development time compared to fuzzy inference systems. Fuzzy logic is a mathematical framework that handles reasoning with uncertainty by using degrees of truth rather than binary true/false values, while fuzzy inference is the process of applying fuzzy logic to map inputs to outputs by using rules and membership functions to make decisions in uncertain or imprecise environments. As compared to fuzzy inference, fuzzy logic systems may struggle to handle uncertain evidence explicitly and are generally quicker to implement than Bayesian networks. According to Johansson and Falkman, an optimal solution can be achieved by combining elements of both approaches. By leveraging the strengths of Bayesian networks'robust mathematical framework and the efficiency of fuzzy logic systems, a hybrid approach can provide a balanced and effective solution for calculating the final Threat Index.

Once threats have been ranked, they must be neutralized by assigning each threat to the most suitable weapon system. This assignment process is known as the weapon assignment (WA) phase. The output of the TE phase influences the selection of weapons for specific targets in the WA phase. However, relying solely on this data is inadequate for making informed decisions. The WA phase is complex and demands careful consideration of additional parameters associated with both the defended asset and the weapon systems. Weapon assignment involves selecting the most appropriate weapon system to neutralize each threat in a near-optimal manner. Extensive literature expounds on the weapon assignment problem; however, TEWA, being a real-time activity requiring a real-time system, is subject to uncertainties and a stringent set of external constraints. In the dynamic and ever-changing operational environment, assignment problems may need to be readdressed, further complicating the design and scheduling implementation processes. Typically, a system comprises multiple weapon systems of various types, each with its own parameters such as lethality index, priority, rate of fire, field of fire, and elevation angle. Based on these parametric values, weapon systems can be effectively assigned to neutralize each threat. In essence, WA can be viewed as a specialized scheduling problem, demanding careful consideration of numerous factors to achieve optimal threat neutralization in real-time scenarios.

6.2.2 Data Fusion for TEWA

Data fusion plays an integral role in the TEWA processes. The challenge of drawing meaningful conclusions from multiple pieces of data is inherent in all biological cognitive processes and virtually every automated approach to information utilization. Data fusion is defined as a process dealing with association, correlation, and combination of data and information from single or multiple sources. This data fusion process thereby achieves refined position and identity estimates, and complete and timely assessments of situations and threats, and their significance. The process is characterized by continuous refinement of its estimates and assessments, and evaluation of the need for additional sources, or modification of the process itself which improves results. To be more concise, data fusion is the process of combining data to refine state estimates and predictions given the wide range of similar data fusion problems encountered in engineering, analysis, and cognitive situations across both military and civilian domains. For TEWA, an effective data fusion capability provides decision-makers with a richer, timelier, and more accurate understanding of the operational environment. Decision-makers can then make better-informed decisions, allocate resources more effectively, and respond more efficiently to threats and challenges based on the following benefits:

- **Enhanced Situational Awareness**: Data fusion aggregates information from diverse sources such as sensors, intelligence reports, and surveillance systems. This consolidated view provides decision-makers with a more comprehensive understanding of the operational environment, facilitating more accurate threat evaluation or risk assessments.
- **Improved Decision-Making**: By integrating data from multiple sources, data fusion enhances the quality of information available for decision-making activities. Decision-makers can make more informed choices about threat prioritization, resource allocation, and mission planning, leading to better outcomes.
- **Real-Time Intelligence**: Data fusion enables real-time integration and analysis of information, allowing decision-makers to respond promptly to emerging threats or changing circumstances. This agility is critical in dynamic operational environments and supports effective threat evaluation and risk assessment processes.
- **Optimized Resource Allocation**: Through data fusion, decision-makers can identify patterns, trends, and correlations in the data, enabling more efficient resource allocation for mission execution. By understanding the capabilities and limitations of available resources, decision-makers can deploy resources effectively to address identified threats or risk situations.
- **Adaptive Response**: Data fusion facilitates the detection of anomalies or deviations from expected patterns in the data. This early detection enables decision-makers to adapt their strategies and tactics in threat evaluations or risk assessments to mitigate emerging threats or exploit new opportunities.

6.2.3 TEWA and Command and Control

In early military operations, before the nineteenth century, Command and Control (C2) typically fell under the purview of a single commander. This commander would undertake all aspects of planning, direction, and monitoring, often with the assistance of limited aides and messengers. As military operations progressed into the twentieth century, the approach to C2 evolved significantly, leading to the development of the concept of a C2 system. This C2 system encompasses the facilities, equipment, communications, procedures, and personnel essential to a commander for planning, directing, and controlling operations. This system description highlights the comprehensive nature of modern C2 systems, which involve not only the commander but also a network of resources and infrastructure necessary for effective planning, execution, and oversight of military operations. Some forms of C2 focus on procedural or technical aspects, such as the management of air traffic and airspace. Other forms include coordinating supporting arms or Fire Control (FC) of weapon systems. On the other hand, there are broader C2 functions that encompass

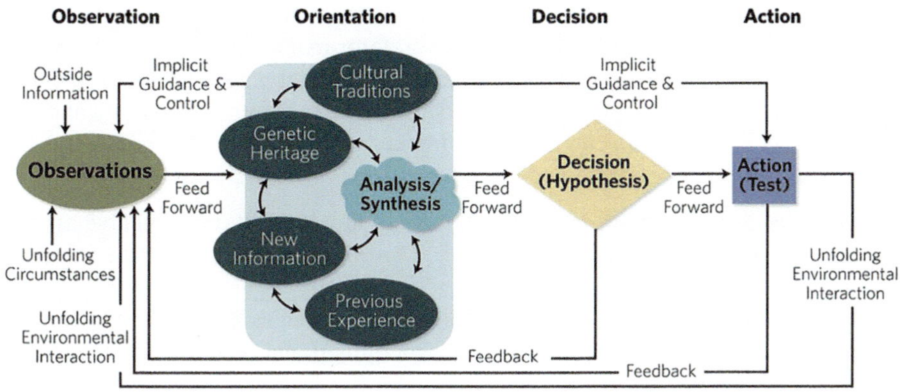

Fig. 6.6 John Boyd's OODA Loop

the overall conduct of military actions. These actions involve the formulation of concepts, deployment of forces, allocation of resources, and supervision of forces. The scope of operation of a TEWA system spans both procedural/technical C2 functions and broader military action coordination. Therefore, the design engineer must have a comprehensive understanding of the operational processes of C2 before designing an efficient TEWA system. This understanding ensures that the TEWA system effectively integrates with existing C2 structures and processes to support decision-making and operational effectiveness.

Stepping back from this intricate C2 process is the fundamental and widely recognized model known as Boyd's Observe-Orient-Decide-Act (OODA) loop [BOYD87], illustrated in Fig. 6.6. Colonel John Boyd, a military strategist, devised the OODA loop to be a continuous and iterative process that individuals or organizations use to adapt to changing situations and gain a competitive advantage. The OODA loop is widely acknowledged as a decision cycle model applicable to all C2 systems, whether they pertain to friendly forces or adversaries. It's important to note that while there are several other recognized decision process models, they share similar functional elements. Some of these models are Plan–Detect–Control–Engage–Assess (PDCEA), Sense–Process–Compare–Decide–Act, Monitor–Assess–Plan–Execute (MAPE), Input–Process–Output, and Find–Fix–Track–Target–Engage–Assess (F2T2EA).

Regarding MEI and the SoS required to execute the mission, the OODA loop applies to the defined stages in the following way:

Observe: During the observation phase, the mission engineer gathers relevant information that is pertinent to the decision being made. This information can be collected from either internal or external sources. Internal sources typically involve feedback loops within the decision-making entity itself. This could include data generated from previous actions, reports from personnel within the organization, or information derived from internal systems and databases. External sources, on the other hand, typically involve sensors or information sources located outside the decision-making entity. This could include data from surveillance systems, reconnaissance missions, intelligence reports, or information obtained from external agencies or partners. By gathering information from both internal (system monitoring and external sources during the observation phase (external environment observation)), decision-makers can gain a comprehensive understanding of the situation and make informed decisions accordingly.

- **Internal System Monitoring**: Continuously monitor the status and behavior of the constituent systems within the SoS. This includes collecting data on performance, health, and interactions.
- **External Environment Observation**: Observe changes in the external environment that may impact the SoS, such as emerging threats, technological advancements, or changes in user requirements.

Orient: Boyd emphasized this phase of the decision-making process as the phase where most of the cognitive effort is concentrated. Two sub-phases: destruction and creation comprise this phase.

During the destruction phase, a decision-making entity breaks down or decomposes a problem until the sub-problems closely resemble situations for which the decision-maker has a plan. This process involves dissecting the problem into more manageable components. The decision-maker's familiarity with these sub-problems is acquired through education, training, experience, and instructions. Solutions to these sub-problems are often viewed as compilations of doctrine-based contingency plans.

In the creation phase, the decision-maker matches the current situation to one they have encountered, considered, or been instructed on previously. They then apply the planned solution to the sub-problem. In essence, problems are matched with their corresponding contingency plans, which are subsequently combined or adapted into an overarching plan of action.

Overall, the orient phase involves the cognitive effort of breaking down complex problems, applying learned solutions, and synthesizing contingency plans into a cohesive strategy of movements required based on gathered situational awareness and contextual analysis to understand the situation at hand in the operational environment.

- **Situational Awareness**: Analyze the information gathered during the observation stage to develop a comprehensive understanding of the current state of the SoS and its surrounding environment.
- **Contextual Analysis**: Understand the relationships and dependencies between different systems within the SoS. Consider factors such as interoperability, data sharing, and communication protocols.

Decide: If a decision-maker formulates a single feasible plan during the orientation phase, the subsequent decision phase is straightforward whether to execute the plan. However, if multiple overall plans or combinations of contingency plans exist, the mission engineer must select the most suitable course of action. This decision-making process often involves weighing the risks or costs associated with each plan against their potential benefits. When a single superior choice emerges, the decision can be made quickly and confidently. Conversely, if several good feasible plans are available, the decision-making process may take longer as the decision-maker evaluates the various options to make an informed decision. Likewise, when faced with many poor choices, the decision-making process may also take longer. In such cases, the focus shifts from selecting the plan with the highest probability of success to choosing the plan with the lowest risk of failure. Ultimately, the decision-making process is influenced by the balance between the potential benefits and risks associated with each available plan.

- **Decision-Making**: Based on the observed information and orientation, make informed decisions about potential actions to be taken. This involves assessing risks, identifying opportunities, and considering the overall mission or objectives of the SoS.
- **Prioritization**: Determine the priority of actions and allocate resources accordingly. Consider the potential impact of decisions on the entire SoS.

Act: The action node symbolizes the implementation of a selected course of action or plan. These actions can encompass a range of activities, including physical attacks or movements, the issuance of orders, or directing attention toward sensors to enhance observation in subsequent cycles of the process. In addition, the feedback mechanism within the OODA loop is critical to provide situational awareness during operations to recognize and react to changing dynamic conditions in the operational environment.

- **Implementation**: Execute the decisions made in the previous stage. Implement changes or actions to address identified issues, improve performance, or respond to evolving requirements.
- **Feedback Loop**: Monitor the outcomes of the actions taken and gather feedback. This feedback is then fed back into the observation stage to start the OODA loop anew.

Note that the OODA loop is not a linear process; rather, it is an iterative and continuous loop. The goal is to cycle through the stages rapidly, allowing the SoS to adapt and respond effectively to dynamic and complex environments. In the context of SoS engineering and MEI, the OODA loop can be a valuable framework for managing the complexity and uncertainty associated with the interactions between different systems. It supports a proactive and agile approach to decision-making, enabling the SoS to navigate changing conditions and optimize performance.

The operational environment of the future will allow for more autonomous decisions by moving away from embedding decision systems in platforms to the use of microservices and applications which reside in a secured combat cloud environment on the tactical edge. For example, TEWA traditionally executed in the combat system would now be executed using microservices in the cloud and, therefore, would eliminate the extra step of going to a platform. This combat cloud stands as a potentially decisive medium for enhancing multi-domain C2 with a beneficial reduction in the OODA decision loop. If anything, the combat cloud represents an opportunity to break away from the overindulgence in permissions by adding information to connectedness. Thus, the operator on the battlefield has access to the same exact information as the leader in the operations center. The leader once again can trust his or her subordinates to execute their orders appropriately and with much greater speed. This capability links operations moving at the speed of light with operations moving at the speed of sound. The combat cloud represents a profound shift from viewing platforms as simply consumers of information to that of sensors and from a series of individually networked platforms to a broader SoS integrated across all domains.

6.3 Mission Command and Control

This section builds on some earlier C2 discussions under the generic OODA Loop framework in Sect. 6.2.3. In contemporary military and civil operating environments, operations face increasing interconnectedness, interdependence, and challenges. The rapid technological advancement and widespread intellectual property proliferation have diminished comparative advantages over adversaries in capabilities and operational approaches. Operations in contested environments against these adversaries necessitate a different approach than those in permissive environments of the past. To address this futuristic operational environment requires a structured MEI approach focusing on how the force observes, orients, decides, and acts to achieve mission success. Two key imperatives driving this approach are:

- Ensuring resiliency in C2 systems and processes for enhanced effectiveness.
- Facilitating effective integration, interoperability, and synchronization of actions across all action domains.

The tenet of centralized control and decentralized execution allows for the synergy between an individual capable of balancing tactical needs with strategic requirements and tactical operators fostering initiative and flexibility. This tenet provides a common framework for developing new operating concepts, organizational approaches, and material solutions to address future operating environment challenges. The main challenge is ensuring the resilience of networks, datalinks, and communications under varying accessibility conditions. During these moments, distributed control (rather than centralized control) ensures continued effective communication across a distributed force.

Distributed control involves delegating or assuming control activities through orders or protocols to synchronize operations, maintain initiative, and achieve mission objectives. It occurs explicitly or implicitly at the operational and tactical levels of mission execution, based on delegated or assumed control authority. Distributed control is the process of transitioning control authority from one entity to another for short or long periods, without delegating command authorities or responsibilities across leadership levels. Trust is key in this process, and trusted autonomy allows individuals to act with initiative, supported by decision-making stakeholders.

Individuals with trusted autonomy are more likely to seize initiative implicitly, contributing to mission success. Maintaining initiative is essential in dynamic operational missions, requiring clear communication of mission intent.

For example, in a fast-paced tech startup developing a new application, maintaining initiative is essential for navigating the rapidly changing market; the project manager must clearly communicate the mission intent to the development team, ensuring everyone understands the project's goals and priorities. This communication structure enables the team to swiftly adapt to user feedback and emerging trends while driving the project forward.

Whereas centralized control is not universally applicable and must be calibrated to operational needs, resilient C2 in challenged access environments requires a combination of centralized command, distributed control, and decentralized execution. The reliance on unconstrained network-enabled operational environments has been challenged by sophisticated adversaries capable of denying and disrupting existing C2 architectures. Distributed control's strength lies in its ability to link and synchronize otherwise disconnected units, increasing the overall resilience of mission capabilities.

While no solution can create a perfect operational environment, new technical solutions, such as the combat cloud concept, contribute to effective and efficient execution. The combat cloud federates all systems that gather, process, store, and disseminate information, allowing independent operation without a specified ranking hierarchy. In the context of MEI, it allows for the simultaneous operation of various systems, contributing to a more dynamic and adaptive mission execution. The combat cloud's decentralized structure aligns with the principles of distributed control, enabling greater flexibility, initiative, and resilience in responding to evolving mission requirements. This architecture reduces the vulnerability of a central repository that emphasizes many independent feeds of information. Challenges include processing speeds, data latency, and susceptibility to corrupt or malicious data. Secure and reliable connections between nodes, along with advanced algorithms for data fusion, updating, and cropping, are essential to ensure the accuracy and reliability of data in the combat cloud.

6.3.1 Universal Mission Command

Mission command emphasizes the importance of understanding context and exercising disciplined initiative within the framework of the decision-maker's intent to navigate transitions between offensive, defensive, and stability operations, as well as between centralized and decentralized operations. It underscores how leaders must possess the capability to grasp, conceptualize, make decisions, give guidance, lead, and evaluate the situation effectively by fostering a shared understanding of the operational context. This approach recognizes the dynamic nature of operations and the need for adaptable leadership that can respond to changing circumstances while remaining aligned with overarching strategic objectives. By empowering

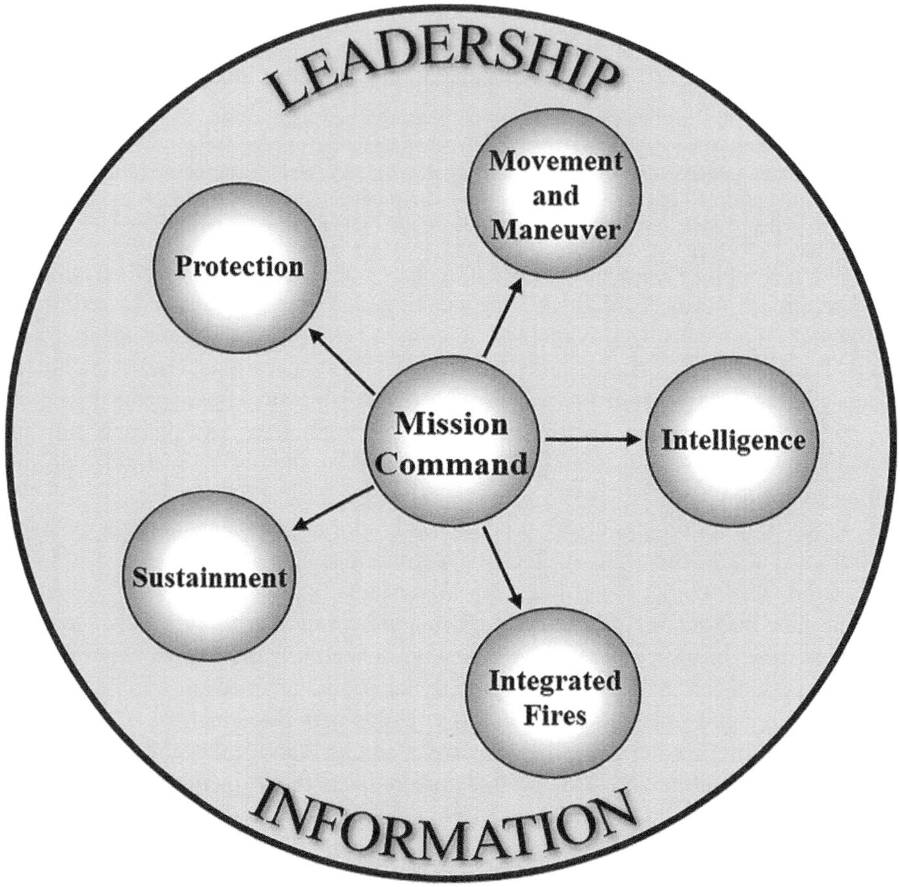

Fig. 6.7 Mission command as a warfighting function

subordinate leaders and fostering a shared understanding of the mission's purpose and goals, mission command enables decentralized execution while maintaining unity of effort and purpose across the organization to effectively address complex challenges and achieve mission success.

As discussed earlier in this chapter, operational design serves as a methodology that applies critical and creative thinking to comprehend, visualize, and articulate complex, ill-structured problems while promoting innovative approaches. It forms the foundation for the decision-maker's responsibility in guiding advanced initiatives throughout the operations process. As depicted in Fig. 6.7, representing mission command as an operational function provides a more accurate portrayal of mission execution. C2 is closely associated with the technical aspects of the network, sometimes neglecting the human dimension. Moreover, the term C2 is insufficient to depict the decision-maker's role in today's operational environment from a holistic perspective.

The universal mission command perspective drives an agile C2 environment. It has emerged as the optimal response to the complexities inherent in modern conflicts, characterized by unpredictable change, and interconnected systems. This approach calls for a departure from traditional command hierarchies toward flatter C2 networks, accompanied by greater delegation of authority guided by command intent.

For instance, in a humanitarian crisis response, a universal mission command perspective allows various organizations—such as government agencies, non-government organizations, and local authorities—to work collaboratively in an agile C2 environment. By establishing shared objectives and flexible communication channels, these entities can quickly adapt to the evolving situation, such as sudden weather changes or shifts in population needs. This approach ensures that resources are allocated efficiently and effectively to meet the urgent demands of the affected communities.

Three crucial aspects define agile C2: the extensive and integrated use of information networks, enhanced collaboration, and the implementation of mission command. Transitioning to agile C2 requires a cultural shift toward information-centric capability user requirements, enabling information to flow horizontally rather than in vertical stovepipes. While practical constraints like security and data volume remain, information processes must evolve toward dynamic lateral networks shaped by user needs. Automation plays a significant role in data analysis and distribution, allowing users to subscribe to relevant information services. While traditional C2 hierarchies may still have a role in solving certain problems, future operational art requires the ability to choose the appropriate C2 model based on the conflict's requirements.

6.3.2 *Levels of Operational Dependencies*

The three levels of operations commonly used are **strategic**, **operational**, and **tactical** [CJCSM23]. Each level is deeply interconnected with the adjoining levels, with each level influencing and depending on the others. Together, they form a hierarchical framework that links high-level national objectives to specific actions on the ground, ensuring that tactical decisions align with broader strategic goals. Understanding the interactions and dependencies between these levels is crucial for effective mission planning, execution, and adaptation.

Strategic Level: At the strategic level, the focus is on defining overarching objectives, resource allocation, and policy direction to achieve national or organizational goals. This level provides the vision and intent that drive operational planning and tactical execution. For example, a strategic goal might involve securing critical infrastructure in a region or ensuring supply-chain resilience across a theater of operations.

Operational Level: The operational level serves as the bridge between strategy and tactics, translating high-level objectives into coordinated campaigns and missions. Planners at this level allocate resources, define objectives for specific areas of responsibility, and synchronize efforts across multiple tactical units. Dependencies at this level include ensuring that tactical units receive the resources and guidance needed to execute their missions effectively while remaining adaptable to evolving strategic priorities.

Tactical Level: The tactical level focuses on the execution of specific tasks and missions, often under immediate constraints such as time, geography, and adversary actions. Tactical decisions and actions directly impact the success of operational objectives, which in turn contribute to strategic goals. Dependencies at this level include clear communication and alignment with operational priorities, as well as feedback mechanisms to inform higher levels of potential challenges or opportunities.

Scenarios and vignettes play a critical role in understanding the relationships and dependencies between these levels of operation. By simulating real-world or hypothetical situations, they provide mission planners with a framework to analyze how decisions and actions at one level influence and interact with those at others.

Scenarios: Scenarios are broad, overarching representations of potential operational environments, often incorporating variables such as geopolitical context, adversary capabilities, and environmental factors. They help define the strategic objectives and operational constraints within which missions must be planned. For example, a scenario involving humanitarian assistance in a disaster-stricken region might outline strategic goals such as ensuring regional stability and operational objectives like delivering aid efficiently.

Vignettes: Vignettes are more focused, detailed representations of specific events or mission threads within a scenario. They provide a micro-level view of tactical actions, interactions, and outcomes, enabling planners to test and refine operational plans. For instance, a vignette might describe the deployment of an unmanned air vehicle to deliver medical supplies to a remote village, highlighting tactical challenges such as terrain, weather, or adversary activity.

The effective interaction of scenarios and vignettes plays a critical role in bridging the gaps between the strategic, operational, and tactical levels of operations, ensuring cohesive planning and execution as depicted in Fig. 6.8. Scenarios provide a bridge to and from the strategic context by defining high-level objectives, constraints, and potential challenges, such as the decision to deploy forces to a specific region based on geopolitical goals. These strategic decisions cascade down to the operational level, where planners develop campaigns and missions that align with the overarching objectives, and further to the tactical level, where specific actions like securing supply routes or engaging with local populations are executed. Conversely, tactical outcomes significantly influence higher levels of planning, as successes or failures provide vital feedback that can reshape operational strategies and even strategic priorities. For instance, delays or challenges encountered during tactical operations may prompt operational adjustments, such as reallocating resources or extending timelines, to ensure alignment with the broader mission.

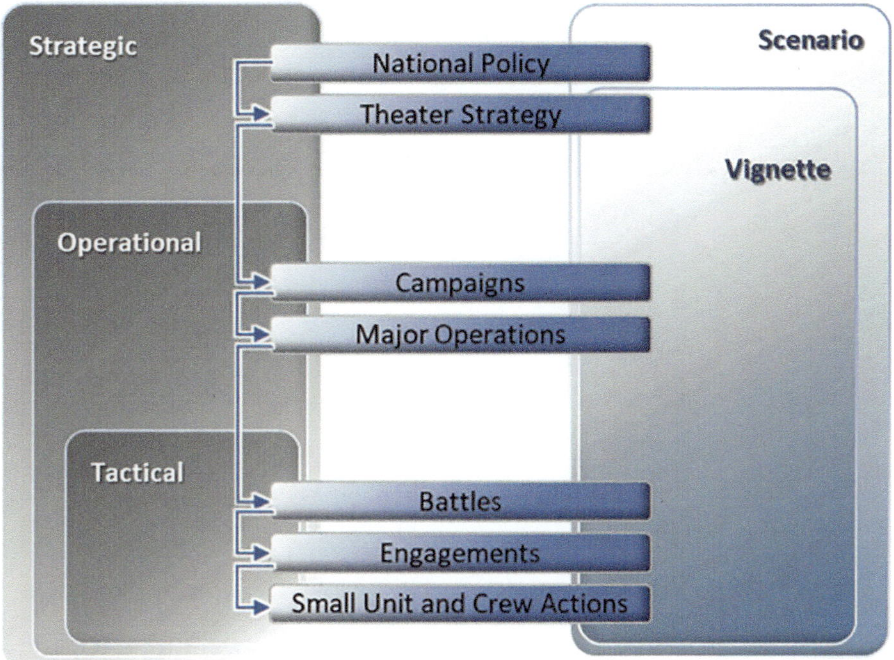

Fig. 6.8 Generalized hierarchy and overlapping relationships of levels of operations

Vignettes, as detailed representations of specific mission threads and a bottoms-up perspective, are particularly valuable for highlighting misalignments across levels, such as tactical units being under-resourced to fulfill operational objectives or operational plans failing to support strategic goals effectively. By integrating scenarios and vignettes into MEI, planners can identify these gaps and dependencies, enabling the design of flexible, resilient, and interconnected systems and operations. This iterative approach ensures that actions at all levels are harmonized, enhancing the overall effectiveness of mission execution and achieving strategic objectives with precision and reliability.

6.3.3 Multi-Domain Operations

Multi-Domain Operations (MDO) serves as a contemporary operational execution strategy designed to acknowledge and leverage the interconnected nature of all operational domains, including land, air, maritime, space, and cyber. The essence of MDO lies in the integration and synchronization of operations across these domains to attain a comprehensive advantage over adversaries. This strategy recognizes that success in one domain often hinges on actions and capabilities in others. This is exemplified by the coordination between aerial operations and cyber capabilities to

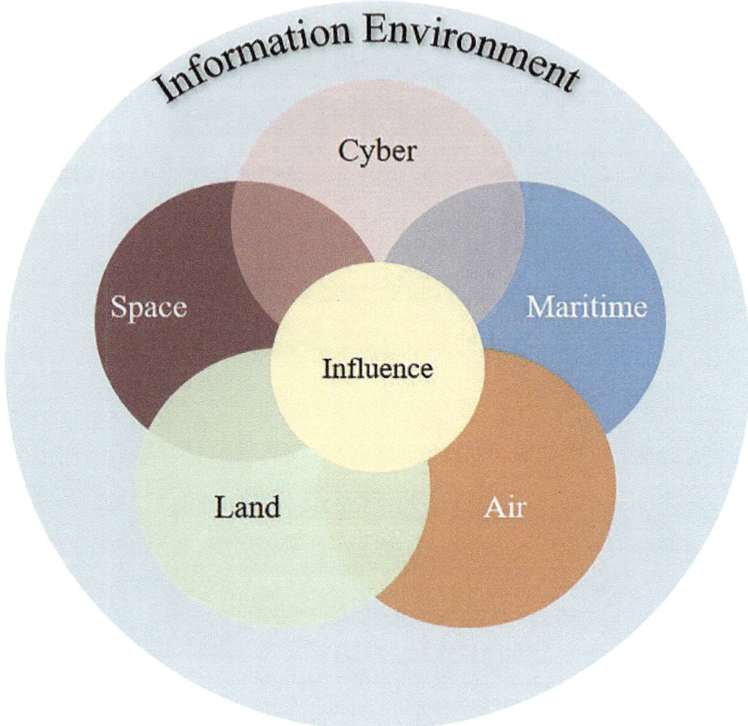

Fig. 6.9 Information environment and five operating domains achieve influence

achieve air superiority. MDO underscores the imperative for a joint and integrated approach, where different military services collaborate seamlessly to achieve strategic objectives. Advanced technologies, data analytics, and rapid decision-making processes are harnessed to enhance the agility and effectiveness of operations under MDO. Moreover, the interconnectedness of domains underscores the need for a holistic understanding of the operational context, encompassing both physical and non-physical elements such as cybersecurity, electronic warfare, and information operations. Furthermore, MDO emphasizes joint interoperability with partner nations and allied forces, as well as the integration of capabilities from various military services. By operating across multiple domains simultaneously, MDO aims to create dilemmas for adversaries, presenting them with challenges that transcend traditional military boundaries.

At the core of enhancing joint action lie three central ideas: exploiting information, achieving greater integration as a force, and fostering adaptability to changing circumstances. These principles form the framework for orchestrating activities that contribute to a full-spectrum approach in defense operations. Achieving influence necessitates a clear focus on target audiences and desired effects. Figure 6.9 illustrates integrating and synchronizing of both kinetic and non-kinetic activities across physical and virtual domains. This holistic approach ensures that defense operations remain effective and responsive to evolving threats and challenges in an increasingly complex and interconnected global landscape.

Adapting to changing complex environments, rather than attempting to exert control over them, emerges as a fundamental principle for the future force or actor. The foremost quality of this MDO force is its capacity for adaptation, coupled with the agility to execute such adaptation seamlessly. Both attributes hinge on the effectiveness of the assessment process at strategic, operational, and tactical levels. Adaptability extends beyond merely responding to external conditions or adversaries; it encompasses the ability to recognize and respond to internal stimuli for change while actively seeking opportunities for advantage. However, in some instances, adaptation alone may prove insufficient, necessitating innovation to maintain superiority over potential adversaries. Innovation challenges establish routines and systems that often require substantial changes in doctrine, structure, and technology.

Deterrence and freedom of action in contested domains require a resilient future force at both individual and organizational levels. This resilience is not about rigidly maintaining form but about enduring shocks, adapting, and prevailing. Creative thinking, robustness, redundancy, and fallback modes will ensure our freedom of action, acknowledging that it is impossible to predict, plan, or prepare for every contingency. Material losses, system degradation, casualties, and surprises will occur, so we must be able to rapidly replace and expand both capacity and capability. Operational resilience will demand more than just physical protection; legality, legitimacy, and public support will become increasingly crucial. Additionally, hybrid attacks by adversaries targeting the nation's will and cohesion, possibly by hitting soft targets at home, are likely to rise. This requires a broader, more coherent view of protection—a whole of society resilience and a full-spectrum approach.

6.4 Information Processing in Intelligence Operations

In the context of MEI and the global information environment, a significant operational challenge arises in efficiently handling vast amounts of data and information and presenting outcomes promptly and effectively. The risk of processing capabilities being overwhelmed by the sheer volume of available information looms. An intelligence unit must possess the capacity to gather, process, and distribute a diverse array of data and information derived from an extensive range of intelligence-collection methods and socio-political sources.

6.4.1 Information Processing for Intelligence Operations

Intelligence processing stands as a critical component of C2, given its pivotal role in achieving the most accurate situational awareness of the operational environment before making decisions and undertaking activities. This requirement remains

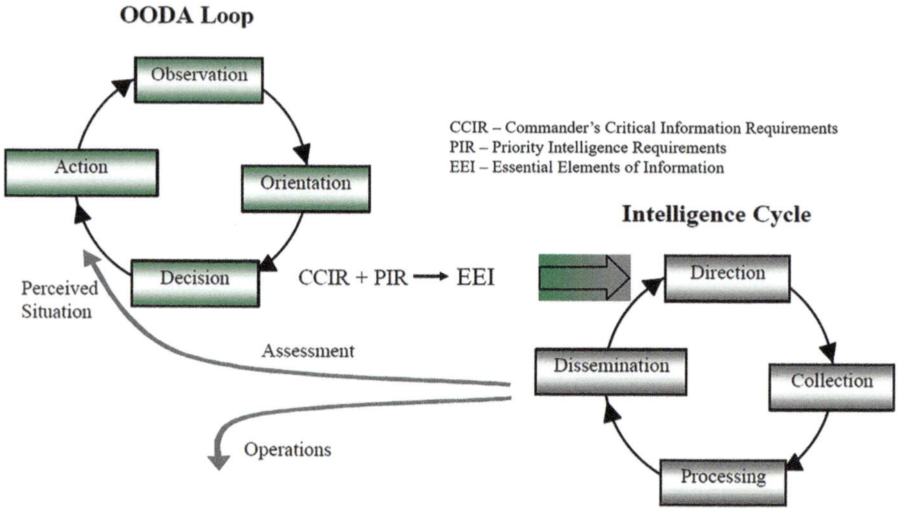

Fig. 6.10 Intelligence cycle interface with OODA loop

fundamental, irrespective of the ever evolving and diverse nature of potential conflicts. Intelligence cells process and assess incoming information to meet the needs of various users and furnish timely and reliable depictions of the operational space. This involves ascertaining the precise location, strength, and activities of all actors, while inferring their likely intentions. A broad array of information from sensors and human sources must be gathered, filtered, processed, and disseminated. This process follows a structured and systematic series of operations known as the intelligence cycle, comprising four key phases:

- *Direction*: Identifying intelligence needs, strategizing collection efforts, issuing directives to collection agencies, and ensuring ongoing monitoring of their effectiveness.
- *Collection*: Utilizing sources from collection agencies and transferring acquired information to the pertinent processing unit for intelligence production.
- *Processing*: Compiling, assessing, analyzing, integrating, and interpreting information and/or other intelligence.
- *Dissemination*: Delivering intelligence promptly in an appropriate format and through suitable channels to intended recipients.

These four distinct operations ultimately result in disseminating the finalized intelligence product. The process is cyclic, as intelligence necessitates continual reassessing and updating to remain pertinent to users' requirements. While information flows, is processed and disseminated as intelligence, the operations overlap, occurring concurrently and continuously rather than sequentially. The depiction of the intelligence function in Fig. 6.10 illustrates the integration of the OODA loop decision process with the direction phase of the intelligence cycle.

6.4.2 Intelligence Products Supporting Operations

Intelligence does not have its own agenda, rather it provides critical information essential for intelligence consumers to execute their missions. It functions as a service to all stakeholders, from individual end users to high-level decision-makers who request products tailored to their operational constraints. Consequently, a single piece of data can contribute to multiple pieces of information, each serving a distinct operational purpose. Data inherently forms part of all necessary common products, often depicted as reference images infused with operational context. This contextual specificity identifies the features and conditions of the specific environment of the operation and underscores the essence of delivered intelligence products: they are raw data transformed into information, infused with the operational nuances of their recipients.

This dynamic extends to the processing of data across varying levels of conflict intensity. On the one hand, we can rely on stable, well-established foundations for understanding and processing data in high-intensity conflicts. High-intensity conflicts are characterized by rapid changes in tactical situations, significant levels of violence, the involvement of conventional military forces, complex multidimensional engagements, extensive humanitarian impacts, and the use of advanced technology in operations. From another perspective, we encounter unpredictable scenarios in which threats remain undisclosed, and adversaries are deeply embedded and sustained by the local population and environment. For example, in the context of some humanitarian crises, aid organizations encounter unpredictable scenarios where adversarial forces operate covertly within civilian areas, using the local population as a shield and gaining support through the provision of services, complicating efforts to deliver humanitarian assistance and putting both aid workers and civilians at risk. Hence, it is imperative that the description and comprehension of system requirements and metrics for automated intelligence processing support closely mirror the environmental and informational context experienced by end users within the operational environment. These factors will serve as crucial decision-making prerequisites that cannot be overlooked. Relying on outdated paradigms, without considering the realities of urban complex operations spanning from peaceful discussions to escalated actions, will render it impossible to develop intelligence systems capable of delivering high accuracy performances.

A comprehensive functional analysis of both operational and processing aspects is an essential prerequisite for understanding the intricate and sophisticated nature of information, as well as the various processing levels necessary for transforming data into intelligence. Figure 6.11 depicts this transformation as a "Knowledge Pyramid." The heightened complexity arises from the operational context, where an ever-increasing significance of implicit contextual information is not explicitly provided.

Facts without problem context and a guiding question lack inherent meaning. For example, an intelligence officer is inevitably confronted with a deluge of information, including partial, false, unreliable, irrelevant, and redundant data, which must

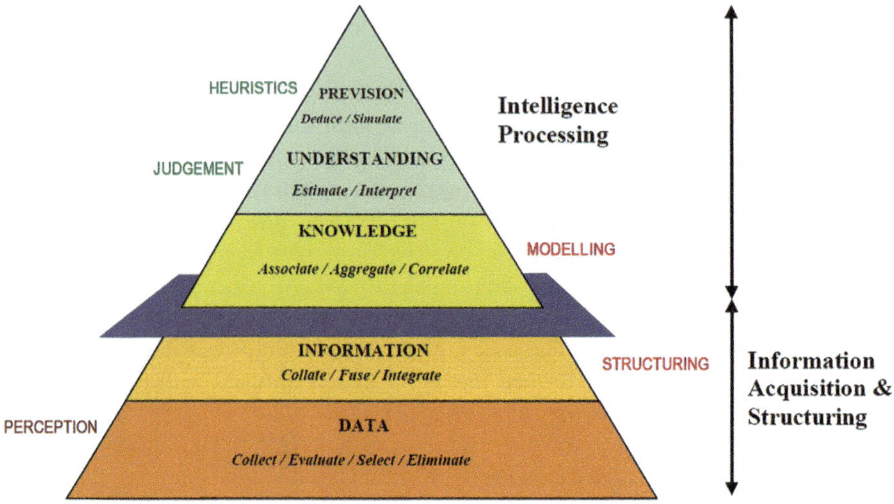

Fig. 6.11 The "Knowledge Pyramid"

be sifted through. The human brain instinctively filters and incorporates pertinent information from its internal mental understanding and model of the situation, injected with insights gleaned from past experiences with similar issues. This process enables the correlation and aggregation of data to construct a coherent understanding of the situation within the brain, thereby resulting in deciphering the significance of the input data pieces.

6.4.3 Developing Realistic Operational Scenarios

To analyze the methods of intelligence processing effectively, the mission engineer must work within a scenario that encompasses all pertinent aspects of real-life processing. This entails incorporating information regarding the operational context, as well as expert knowledge about operational behavior and activities, which provides an understanding of operational dynamics. Using both scenarios and vignettes to execute the MEI approach offers a comprehensive approach to understanding and preparing for complex operational missions. Scenarios provide a broad overview of the operational environment, incorporating various factors such as terrain, weather, adversarial capabilities, and friendly forces. They offer a holistic view of the mission landscape, allowing for a thorough analysis of potential challenges and opportunities. In contrast, vignettes offer focused snapshots of specific situations within the larger scenario. These vignettes highlight critical moments or dilemmas that may arise during the mission, providing a detailed examination of specific aspects of the operation. Together, scenarios and vignettes provide a multidimensional representation of the mission environment, enabling operators to develop critical

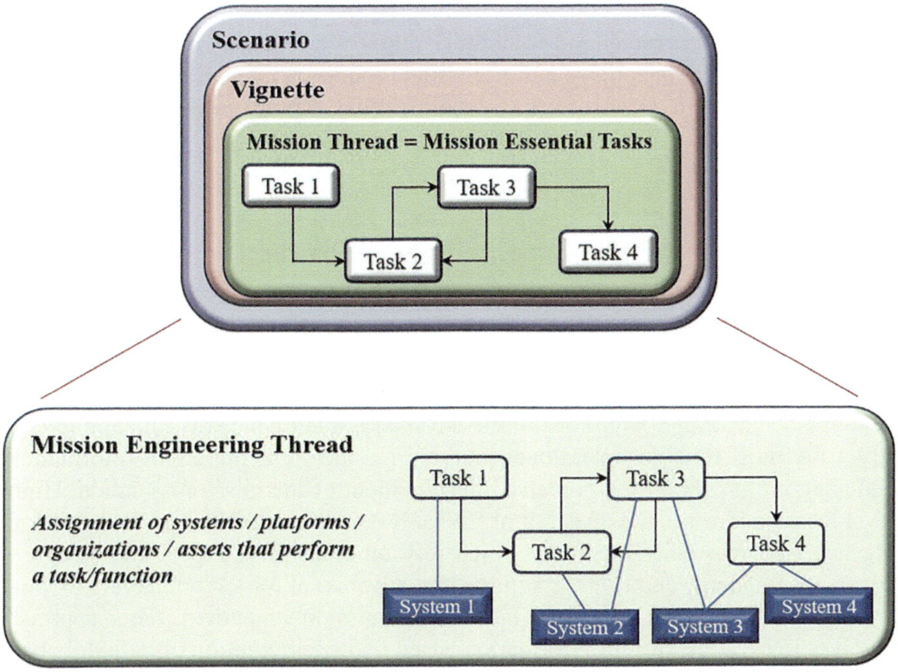

Fig. 6.12 Scenario and vignette usage

thinking skills, decision-making abilities, and situational awareness. They serve as valuable training and preparation tools, helping individuals simulate and practice responses to a range of potential scenarios. Additionally, scenarios and vignettes serve as decision-support tools for key decision-makers and planners. They facilitate evaluation of different courses of action and assessment of risks and uncertainties. By fostering communication, collaboration, flexibility, and adaptability, scenarios and vignettes contribute to the MEI approach as illustrated in Fig. 6.12.

A scenario designed for executing a maneuver differs from one intended for elaborating a process. In the former, the objective is to produce a collection of operational facts detailing the specifics of the maneuver. Conversely, in the latter scenario, the aim is to furnish a comprehensive set of information that establishes a coherent background, in addition to the factual elements present in a pure maneuver scenario. This entails providing non-factual data, such as immaterial fields, environmental conditions, and factors related to psychology and sociology to capture operational behaviors. By incorporating these additional elements, the scenario becomes more nuanced and realistic, offering a deeper understanding of the context. A scenario lacks effectiveness if it fails to incorporate certain aspects of real-life situations, including irrelevant or misleading information, which can intentionally disrupt or skew intelligence deductions. It is important to recognize that not all parameters can be feasibly included in a scenario. Moreover, scenarios should strike a balance, avoiding excessive complexity and breadth, as this can lead to challenges in maintaining detail and coherence. Managing the level of detail and coherence within a scenario is crucial and sometimes requires a focused vignette. These factors can expand exponentially and necessitate the implementation of a coherence management system to ensure the scenario remains practical and useful for its intended purpose.

6.5 Modeling the Human Intelligence Process

In heuristic intelligence processing, the focus shifts from numerical data to symbolic information, such as documents containing text and images. The primary objective is not to identify and comprehend individual objects, but rather to comprehend the behavior and intentions of the adversary. Intelligence officers aim to construct the most reliable operational picture possible from imperfect information, deducing its impact on the perceived development of the mission situation. High-level fusion addresses the question of "What does it mean?" This involves employing heuristic reasoning methods, which rely on end users' expert knowledge of adversary standards and rules, assuming that adversarial forces will adhere to these regulations and constraints to achieve their mission objectives. This approach involves integrating current information based on the assumption of default behavior. This processing method is grounded in behavior modeling, such as doctrinal templating or profiling, and employs a qualitative approach using descriptive knowledge representation. Complex elements of the operational environment are generally depicted using text and graphics. These descriptions encompass their relationships, dynamic behaviors, mission intentions and concepts, ethnic or religious positions, moral beliefs, and political involvements, as available.

For high-level fusion systems supporting intelligence, two distinct types of models are required:

1. **Behavior Models**: Describe the tactics employed by potential adversary factions and all necessary preconditions for their aggressive activities. The primary objective of high-level fusion in this case is to detect indicators of a pattern of activity as defined by the behavior model. This approach is commonly utilized in both low- and high-intensity conflicts.
2. **Normality Models**: Outline the common and unsuspicious behavior exhibited by defined subgroups of the population. The goal of high-level fusion is to identify deviations from the established pattern of "normal" behavior. This method is often applied when the aim is to define indicators of suspicious activities and identify potential threats.

By using these two types of models, high-level fusion systems can effectively analyze and interpret intelligence data to detect patterns of activity and deviations from normal behavior, thus enhancing the ability to identify and mitigate potential threats and security risks.

Based on this discussion, the following requirements for developing effective automated high-level fusion systems can be derived:

1. **Definition of High-Level Fusion Problem**: It is essential to define the high-level information fusion problem at a sufficient level of abstraction and develop appropriate task models to guide system development.
2. **Development of Suitable Fusion Methods**: Suitable fusion methods need to be developed to effectively integrate and analyze information from various sources.

In addition to these requirements, a high-level fusion system must automatically access the semantic content of current input and background information. Given that high-level fusion often relies on document input, automated extraction of content from individual documents is crucial. The semantic meaning of documents must be accessible for subsequent automated context-based information analysis and integration. To achieve this, methods and procedures for automated text extraction and analysis are necessary. Tools equipped with efficient linguistic methods and adequate ontological models for each application domain are needed to understand the semantic meaning of document input and manage the information workload in network-centric intelligence and other C2 activities. While this semantic text analysis is not the information fusion problem itself, it is closely connected to the modeling problem. Therefore, it is a major part of the broader challenge of developing high-level fusion systems, particularly due to its ontological approach to semantic text analysis.

Information processing is conceptualized as a context-dependent and template-based heuristic reasoning process, where the real situation is characterized by the assumption that forces operate in a structured and reasonable manner. This representation relies on rules, doctrines, or modus operandi as templates for default reasoning. The analysis and fusion of incomplete and imperfect information is thus allowed in automated systems. These models are developed through a thorough analysis of real processes in intelligence and C2 activities. While describing the operational environment using default behavior models serves as a solid foundation for applying area knowledge, the relevant information context is dynamic rather than fixed. Particularly in new missions with limited background intelligence, the initially available knowledge may prove insufficient and require modifications to existing models or the creation of new ones. Any concept not covered by an ontology cannot be accounted for during semantic analysis, thereby leading to blind spots in fusion systems. To address this learning problem, an interactive high-level fusion system may offer a solution. Such a system would provide formally correct fusion hypotheses using all available information but would defer knowledge maintenance and final decisions to human operators. By supporting operators in adapting to new and evolving situations, interactive systems offer adaptability and flexibility in addressing complex intelligence and C2 challenges.

Given the descriptive and qualitative nature of behavior models represented by non-numeric methods and the challenge of deducing from incomplete information, it is likely that mission engineers will employ qualitative, non-numeric methods to match imperfect information with these models. However, use of high-level fusion could also derive benefits from quantitative mathematical methods such as fuzzy logic, evidential reasoning, and probability. This potential benefit hinges on achieving a suitable abstraction of the problem, one that enables the formulation of detailed problems that can reasonably be addressed using quantitative approaches. By integrating both qualitative and quantitative methods, high-level fusion systems can effectively navigate the complexities of incomplete information and make informed decisions in uncertain environments.

6.5.1 Representation of Knowledge

During the initial phase of intelligence processing, known as collation, the intelligence system collects various types of input documents, ranging from structured to unstructured formats, including texts, pictures, and sounds. To facilitate further processing tasks (such as classification, correlation, aggregation, and interpretation), the stored raw data must undergo conversion into a structured format that can be interpreted by machines. This conversion process involves processing, transforming, and representing the raw input data as both numerical and symbolic information. A crucial consideration at this stage is establishing a formal representation framework capable of structurally representing diverse real-world information types. This formal representation should encompass entities, concepts, relations, intentions, rules, and doctrines, while also accommodating uncertainties and conflicts that may arise within the data. By adopting such a structured representation, the intelligence system can effectively manage and interpret the complexities inherent in the collected information, enabling more robust analysis and decision-making processes downstream.

In addition to receiving incoming intelligence information, the knowledge base must also be supplemented with background knowledge, which serves as the conceptual model of the mission space. This ontological information forms the foundation for the development of inference engines and fusion functions, enabling the interpretation of available information to identify significant facts, generate hypotheses, and infer potential future activities. An intelligence knowledge base must manage two distinct types of knowledge: the conceptual model of the mission space, which primarily deals with abstract classes of objects, and the collected intelligence information, which predominantly involves real instances of these objects. To effectively manage such a knowledge base, specialized knowledge representation techniques commonly studied in the field of AI are employed. Every intelligent entity engaged in reasoning requires a structured knowledge representation, as reasoning occurs internally while the entities it addresses exist externally. Establishing a common data dictionary may be necessary to define critical domain-specific terminology and ensure the maintenance of operational context in interpretations. An essential consideration in knowledge representation is the degree of fidelity between the surrogate representation and the real object it represents. It is important to understand which attributes of the original object are captured and made explicit, and which are omitted. However, achieving perfect fidelity is generally unattainable, both in practice and in principle. In principle, perfect fidelity is impossible because the only completely accurate representation of an object is the object itself, rendering all other representations inherently inaccurate to some extent. A representation of this effect is the following:

> Imagine trying to capture the essence of a sunset through a photograph. While the image may beautifully portray the colors and shapes, it can never fully encapsulate the experience of witnessing the sunset in person, with its shifting light, the surrounding atmosphere, and the emotions it evokes. Every representation falls short of complete fidelity.

Various methodologies exist for knowledge representation and reasoning, including First Order Logic, Fuzzy Logic, Production Systems, Neural Networks, Petri-Nets, and Semantic Networks. These methodologies vary in their structure and organization, with some being highly structured and others less so. Structured methodologies offer significant advantages over non-structured models. They allow for the representation of knowledge bases in a modular fashion, facilitating easy understanding and concurrent, efficient access. This structured approach enhances the manageability and accessibility of the knowledge base, contributing to more effective decision-making and problem-solving processes. When processing text, the brain does not register documents as composed of hundreds or thousands of lines. Neurobiological experiments suggest that the brain associates concepts, which are essentially sentences made up of words. These sentences share words and collectively form a network. Our brain organizes information based on this networked structure. While we may have learned each sentence by reading a document, such as a newspaper, our brain doesn't register the sequences of sentences found on each page of the newspaper. However, we typically store information in document form on our computers. This disparity in how information is represented in brains versus computers significantly hinders the optimization of man-machine cooperation in intelligence. Alternatively, we can opt to use computers to electronically simulate how brains represent information, namely as a network of sentences that share words. These representation techniques have been extensively explored in the field of AI under the designation of "semantic networks" since the early 1950s.

An additional semantic technology technique is the combination of Resource Description Framework (RDF) triples with Large Language Models (LLMs) to increase the efficiency and effectiveness of automating the extraction of data from structured and unstructured data formats within a domain [KOSKI21]. These integral components of the semantic web facilitate the organization and representation of knowledge in a machine-readable format. RDF triples provide a structured format for representing data by defining relationships between entities through subject-predicate-object assertions. The subject represents the resource being described, the predicate denotes the relationship between the subject and object, and the object identifies the value or resource associated with the subject through the predicate. This structure, as depicted in Fig. 6.13, illustrates how the three components of subject, predicate, and object interact to represent a relationship within the RDF

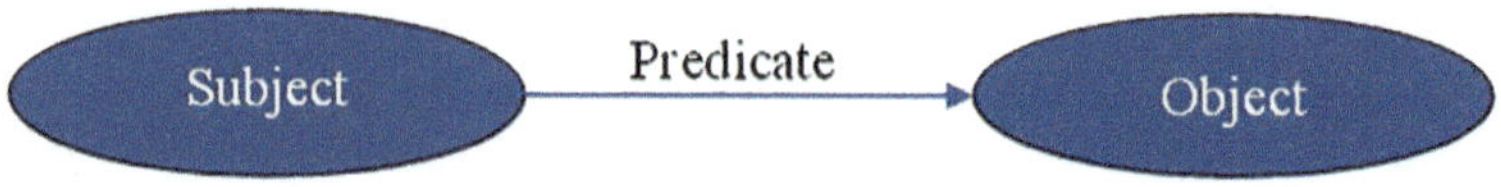

Fig. 6.13 RDF triple relationship

data model. On the other hand, LLMs possess advanced natural language processing capabilities, enabling them to understand and generate human-like text based on vast amounts of data. By integrating RDF triples with LLMs, organizations can enhance the automation of data extraction from both structured and unstructured data formats. LLMs can parse unstructured text data, extracting relevant information and converting it into RDF triples. This process enables the conversion of textual data into a structured format that can be easily processed and analyzed within a semantic framework. Additionally, LLMs can also assist in generating textual descriptions or explanations based on RDF triples. This capability enhances the interpretability and usability of semantic data, making it more accessible to human users.

In an effective knowledge representation system, addressing uncertainty and temporality is crucial. For semantic networks, incorporating uncertainty and time information can be achieved by integrating such data into nouns and relations. For example, the existence of a noun can be constrained within a specified time range, such as birth and death dates of a person. Similarly, the duration of an event can be defined by its start and end times, as in the statement "The meeting starts at 3 pm EST and ends at 4 pm EST." Additionally, the strength of a relation can be quantified by attaching a probability value, indicating the likelihood or confidence level associated with the relationship. By incorporating uncertainty and time constraints into semantic networks, knowledge representation becomes more robust and nuanced, allowing for a more accurate and comprehensive understanding of the underlying data. This approach enhances the system's ability to handle complex real-world scenarios where uncertainty and temporality are inherent aspects of the information being represented.

There are several ways to represent knowledge as discussed. Selecting a knowledge representation method and applying it to a domain poses its challenges. However, another critical hurdle lies in populating the knowledge base with essential background information. Without this foundational data representing the conceptual model of the mission space, no fusion process can effectively extract significant facts or predict future activities from incoming intelligence information. Hence, acquiring authoritative sources of truth and expert knowledge to enrich the knowledge bases becomes paramount. This task extends beyond knowledge representation. The procurement of accurate, comprehensive domain expertise is necessary to inform the knowledge base. Collaborating with domain experts and sourcing credible information are vital steps in ensuring the knowledge base reflects the complex realities of the operational environment.

6.5.2 Management of Intelligence Predictions

Uncertainty in observations and limited data coverage can lead to a wide range of potential estimates for a given situation, making it challenging to draw accurate conclusions or make informed decisions based on the available information. Uncertainty in observations and limited data coverage can result in a wide range of potential estimates for a given situation. In intelligence processing, automated systems must be able to manage multiple hypotheses and predictive states of capability, especially in the early stages of an operation. Formalizing prediction management can significantly enhance intelligence processing and subsequently aid in operation management by improving efficiency, objectivity, and communication with the goal of enhancing the effectiveness of operations. While formal guidelines may seem to constrain flexibility, they offer several advantages:

- Serve as the foundation for automation, enabling the testing of numerous predictive states that would be impractical for manual processing by human operators.
- Allow for the rapid, simultaneous treatment of many hypotheses, free from subjective biases or blind spots in human consciousness.
- Facilitate the efficient transfer and accumulation of experience, surpassing the limitations of human knowledge management alone.
- Promote communication and enhance process transparency, enabling others to comprehend and monitor activities effectively.

The primary objective of an intelligence service should not focus solely on estimating the most probable state or situation based on available data. Instead, effective situational awareness requires identifying all potential explanations of the information that hold significance within the current operational context. Therefore, assessing the potential costs and risks associated with decisions and subsequent actions should be central to situational awareness. Each individual observation, when considered in isolation, may carry uncertainty and seem insignificant for assessing potential threats. These observations could include various data points such as car routes, correlations between movements of vehicles and individuals, activities at suspected storage locations, communications, and observations of suspicious individuals. However, when these disparate pieces of information are aggregated and analyzed holistically, patterns may emerge that align with known threat scenarios or indicate potential risks. These patterns, often represented as templates for information matching, can then prompt further investigations. Automated formal procedures play a crucial role in identifying and highlighting patterns, allowing human operators to interpret data more readily.

6.5.3 *Intelligence Processing Functional Modeling*

Figure 6.9, presented earlier in this chapter, offered a high-level depiction of the intelligence cycle divided into distinct functions. The processing task is the primary capability for producing end user products based on intelligence requirements. The processing phase involves converting the collected information, as directed by the end user, into intelligence products. This phase comprises a structured series of actions, which, although typically sequential, may also occur concurrently. It encompasses the production of intelligence through collation, evaluation, analysis, integration, and interpretation of information and/or other intelligence. Figure 6.14 provides a simplified representation of the intelligence processing phase.

The processing phase in the intelligence cycle involves several key components:

- **Collation**: Gathering and organizing the collected information into a coherent format for analysis.
- **Evaluation**: Assessing the reliability, relevance, and significance of the collected data to determine its value for intelligence purposes.
- **Analysis**: Examining the information in-depth to identify patterns, trends, and insights that can inform decision-making.
- **Integration**: Incorporating relevant data from various sources and disciplines to create a comprehensive understanding of the intelligence picture.
- **Interpretation**: Making sense of the analyzed information by deriving meaning, drawing conclusions, and generating actionable intelligence products.

These components work together synergistically to transform raw data into actionable intelligence that supports decision-makers in achieving their objectives for mission success.

During the collation process, incoming information is initially logged and stored in the appropriate repository. Subsequently, the information received is decomposed into individual items, which are then categorized and cross-referenced with previously received information items. In this step, the engineer conducts a preliminary semantic analysis to associate each new report with an appropriate category of intelligence. The specific categories used depend on the type of operation conducted and

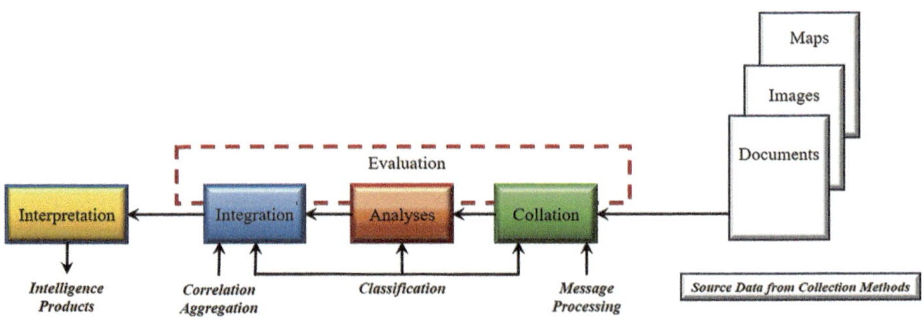

Fig. 6.14 Representation of processing phase of intelligence cycle

are derived from the end user's intelligence requirements. These categories define the main aspects of the situation and the information context necessary to provide a comprehensive description reflecting the capabilities and intentions of the adversary, facilitating assessment. However, it is important to note that the association of a single report with a category is primarily guided by experience and correlation, with no further interpretation performed at this stage. For reports in free-text format, preprocessing is necessary to extract and exploit their semantic content effectively.

Classification is the process of establishing the identity of an item by determining its membership in a specific class or type within a class. Essentially, classification involves matching an item with predefined categories and answering the question, "Who or what is it?" The quality of classification results is heavily influenced by the richness and appropriateness of the domain representation model, which includes a comprehensive set of attributes for observed or deduced objects, events, etc. While determining the category in the collation phase, high-level classes such as people, places, events, organizations, and equipment are used to categorize incoming information. This initial categorization can be seen as a raw classification process and serves as a basis for automation. Following classification, the correlation process combines data or information from similar sources about the same item, activity, or occurrence. Correlation involves associating or linking different information items (classification matches) that share common attribute values or meet specific criteria. Correlation focuses on identifying connections between different objects based on shared attributes or criteria.

In the final stage of the processing phase, the received information items are aggregated or integrated. This process involves fusing different information items, which are elements of the problem domain, to deduce new and more valuable information items. During aggregation, the input information items that are fused typically conform to the pattern of an object belonging to a more complex domain class. This conformity is based on basic intelligence, including background knowledge, doctrines, and related requests for information (RFI). Through aggregation, disparate pieces of information are combined to generate insights that may not have been apparent from individual data points alone.

This section devoted to intelligence processing functional modeling underscores the inadequacy of traditional information processing methods and explains why the adoption of new approaches tailored to dynamic operational environments and unstructured data are necessary. Semantic networks, characterized by highly structured data representation, show promise in generating machine-readable intelligence data closely aligned with operational information processing needs. Moreover, the ability to deploy such systems without extensive technical assistance offers a significant advantage in rapidly evolving operational situations. Consequently, it becomes apparent that certain aspects of intelligence processing, particularly in collation and analysis, can be largely automated with minimal human intervention, provided all relevant environmental information is filtered through operational requirements. Furthermore, operational personnel can effectively manage support systems utilizing semantic networks and default reasoning, both of which closely mirror human cognitive processes.

6.6 Governance Structure

In general, governance is defined such that it emphasizes the collection of rules, practices, and processes by which an organization is directed and controlled. This framework is essential for ensuring that organizations operate efficiently and effectively, aligning their actions with both strategic goals and ethical standards. In the context of SoS, governance has similar significance with an extension into the management of complex interdependencies between various constituent systems, fostering collaboration, and aligning individual system objectives with the broader goals of the SoS. Adhering to governance principles helps organizations make well-informed decisions that not only ensure operational effectiveness but also contribute to the sustainability and long-term value of the SoS as a whole. Good governance within the SoS can improve decision-making at all levels, from individual system owners to the overarching network of connected systems. In turn, this leads to the creation of more resilient, adaptable, and innovative systems, ensuring that they can evolve in line with changing needs and priorities while maintaining alignment with strategic objectives. Thus, effective governance becomes a foundational element in driving long-term value creation and ensuring the continued success and relevance of the SoS.

A robust governance structure is vital for overseeing the health of an SoS supporting critical national security missions. These missions, marked by their complexity and sensitivity, rely on the integration of interdependent elements, including technological systems and human resources, to achieve success. Effective governance provides a framework for oversight, decision-making, and alignment with strategic objectives essential for ensuring the integrity of the system. By clearly defining roles, responsibilities, and accountabilities for stakeholders, the governance structure enables everyone involved to understand their contribution to the mission's success. Transparent decision-making processes support informed choices about portfolio selection, prioritization, and resource allocation—critical factors in maintaining stakeholder confidence and advancing security objectives. Additionally, risk management strategies integrated within the governance framework mitigate potential threats, while mechanisms for communication, reporting, and regulatory compliance enhance transparency and accountability. A well-defined governance structure, therefore, is essential for managing the SoS effectively, ensuring optimal resource use, robust risk management, and the achievement of mission objectives.

Mission-oriented policy builds on this foundation, emphasizing coordinated efforts to achieve specific, ambitious, and time-bound policy goals. Achieving such missions requires the establishment of robust governance structures and coordination mechanisms. These ensure that policymaking institutions—often spanning multiple industrial sectors, domains, and levels of government—align their actions effectively to meet defined targets. Furthermore, missions demand active engagement with diverse stakeholders beyond government, including organizations with differing perspectives and limited histories of collaboration. This necessitates effective horizontal and vertical coordination, along with proactive multi-level governance to harmonize efforts. However, governance and structural frameworks for missions are often treated as secondary considerations, constrained by existing institutional setups. This oversight presents a significant challenge to realizing the transformative potential of missions aimed at driving systemic change.

Within each program office, the technical authority structure plays a crucial role in guiding the design, implementation, and integration of systems within the SoS. Comprising subject-matter experts and technical leaders, this structure provides specialized expertise that enhances governance and portfolio management. By incorporating the leadership of individual systems and their technical authority structures, the governance framework achieves multiple benefits. First, it promotes clear communication and accountability among the portfolio management team, system leaders, and technical authorities, fostering collaboration and alignment toward shared objectives. Second, it integrates technical expertise and domain knowledge into decision-making processes, improving strategic planning and resource allocation. Third, it identifies and mitigates technical risks and challenges early in the portfolio lifecycle, reducing the likelihood of delays or setbacks during implementation. Leveraging the expertise and leadership embedded within each system and technical domain optimizes the SoS's performance and resilience, ensuring it meets critical mission requirements with rigorous verification and validation of capabilities.

6.6.1 Working Across Governance Structures

Inspired by the principles in "Team of Teams" by General Stanley McChrystal [MCCHRYSTAL15], the governance structure for effective information flow and rapid decision-making integrates adaptability, shared consciousness, and decentralized decision-making. This approach aligns strategic objectives with tactical execution while maintaining flexibility to address dynamic challenges. The structure operates through three interconnected layers—strategic, operational, and tactical—linked by robust communication mechanisms as discussed in Chap. 6. At the strategic level, the focus is on defining the organization's vision, allocating resources, and setting priorities. A senior leadership team serves as the decision-making hub, supported by an information analysis unit synthesizing data to guide strategy. Regular briefings align teams on key objectives, while a clear articulation of strategic intent ensures operational and tactical units understand the broader context. This level emphasizes long-term objectives, risk management, and organizational purpose.

The operational layer translates strategic intent into actionable plans, coordinating efforts across teams. Cross-functional groups collaborate to address interdependencies and refine approaches, supported by real-time insights from a shared operations dashboard. Regular synchronization meetings allow teams to adapt to evolving circumstances, ensuring alignment and resource optimization. Bridging the gap between strategy and execution, this layer focuses on the connective layer of planning and cross-departmental collaboration. At the tactical level, teams are empowered to make real-time decisions within the framework of strategic intent.

Tactical leaders leverage continuous feedback loops to address immediate challenges while keeping higher level personnel informed of critical insights. Daily stand-up meetings or huddles that focus on short-term objectives and adaptability enable teams to respond effectively to emerging situations.

To foster shared consciousness across all levels, the governance structure incorporates transparent information-sharing practices and advanced communication tools. Daily synchronization calls and real-time digital platforms are vital to maintain situational awareness for all stakeholders, breaking down silos, and cultivating a unified organizational mindset. Transparency, combined with decentralized authority and leadership development, builds trust and empowers teams to make autonomous decisions. By embedding adaptability and networked communication into its core, this governance structure equips organizations to operate with agility and resilience critical for success in complex environments.

6.6.2 Characteristics of Effective Governance Structures

Effective governance in an SoS requires a comprehensive approach that balances contributing factors to ensure seamless integration and operation of complex, interconnected systems. Good SoS governance is characterized by several essential traits that address the unique challenges of such environments. These characteristics include the need for collaboration and authority to facilitate decision-making and coordination, alongside motivation and accountability to drive system performance. Governance must also accommodate multiple models to allow flexibility in addressing diverse system needs, while anticipating evolution to adapt to changing circumstances. With highly fluid processes, SoS governance supports continuous adaptation, whereas minimal centrality decentralizes control, promoting agility and responsiveness. Together, these features form the foundation of effective SoS governance.

6.6.2.1 Collaboration and Authority

In a traditional, stand-alone system, program managers from both the acquirer and contractor possess clear control and authority within their respective organizations, enabling effective enforcement of governance. Even in multi-organizational environments, the governance structure is typically hierarchical, with contractual relationships that outline well-defined lines of authority. However, the dynamics of an SoS are fundamentally different due to their inherent complexity. Unlike traditional systems, no single organization holds ownership (and therefore no exclusive authority) over the entire SoS capability. This distributed ownership introduces significant challenges in establishing control across a network of independent systems, especially in the absence of a centralized governing body. Even when a central authority has some oversight over the constituent systems, the stakeholders of those systems generally remain primarily loyal to their individual interests. As a result, it becomes difficult to impose top-down control. In such an environment, governance must shift toward a more collaborative, community-driven approach, where shared policies are developed and adhered to collectively. Collaborative governance promotes deliberative, multi-stakeholder engagement in rule-setting, fostering dynamic and adaptable processes that can evolve over time. Rather than relying on a rigid, hierarchical model, SoS governance thrives on peer-to-peer relationships, where independent authorities cooperate without the need for formal, centralized oversight. This approach requires a shift from a "chain of command" to a "web of shared interest," encouraging cooperation and flexibility even in the absence of formal agreements.

6.6.2.2 Motivation and Accountability

In most hierarchical organizations, accountability is traditionally enforced through imposed standards and coercion. While these approaches may be effective initially, particularly in less dynamic environments, they become challenging to sustain in complex, highly dynamic contexts like an SoS. The key challenge is designing a governance structure that can balance decision-making and actions at both the SoS level as well as at the individual system or component owner level. A well-designed structure allows individual system or component owners to collaborate by building consensus, which in turn makes the governance process more inclusive and effective. A critical element of this governance is the creation of policies that incentivize individuals and organizations to adopt the SoS perspective. Without including appropriate incentives, existing reward structures and performance measures may unintentionally discourage behaviors that are vital for SoS capability development and overall mission success. For example, program managers are typically incentivized based on the performance of individual systems or components, with an emphasis on meeting budget and schedule requirements. However, in the context of SoS governance, incentives may need to be redefined to include broader measures such as service-level agreements or performance metrics that promote system-wide cooperation and reduce interoperation failures. These incentives, paired with visible performance metrics, help shift organizations toward a more collaborative approach.

Enforcing such policies in a distributed SoS environment is challenging, particularly because no single entity holds enforcement power. However, one effective strategy is to make behaviors transparent and visible to all participants. By creating and publicizing performance metrics, organizations can foster positive behaviors and discourage actions that undermine the SoS objectives. In this way, the self-interest of participants is leveraged to ensure that actions align with the larger goals of the SoS.

6.6.2.3 Multiple Models

SoS governance policies are likely to vary depending on the roles of participants, but variations may also arise based on factors such as the timing of governance and the relationships between systems or components within the SoS. One key distinction is the shift from governance primarily focused on design-time, where policies may specify standards like "Use standard X" or "Document design by standard Y," to governance that spans deployment and runtime. This shift highlights the need for different governance approaches across various stages of the system's lifecycle.

Varying types and levels of governance are particularly crucial when considering the relationship between individual systems or components and the broader SoS. A relevant example is the automotive industry, where a company like Mack Trucks sources parts from a range of suppliers. Mack may produce some parts in-house, have other parts made to Mack's specifications, or adopt parts from suppliers who adhere to their own standards (e.g., Mack uses Bendix stability control systems, also used by other manufacturers like International Truck and Engine and Volvo). Enforcing a single, uniform governance model across all sources of parts in this scenario is not feasible. Instead, a more flexible approach, such as a matrix view of SoS governance, could be adopted. This matrix might classify parts by their source (internal, externally supplied but controlled, or purely external) and by the phase of the part's lifecycle (development, deployment, and runtime).

However, this classification approach may not be effective for highly dynamic and interconnected systems or components where real-time information sharing and collaboration are necessary to create a comprehensive, accurate picture of the operational environment. In such environments, a rigid, one-size-fits-all governance model is less appropriate. Governance needs to be adaptable to accommodate the needs of various systems or components within the SoS. This is where the concept of "neighborhoods" comes into play, which is defined as the grouping of systems or components that collaborate closely on a particular task or mission thread. These neighborhoods and the interactions between them may require distinct governance models tailored to specific mission requirements such as mission survivability or the trustworthiness of information and its sources. Because different neighborhoods within the SoS may have different operational priorities and risk profiles, their governance structures must be flexible to meet both the evolving needs of the broader SoS and the specific functionality of the constituent systems. Thus, SoS governance in complex, dynamic environments requires a nuanced and adaptable approach that can accommodate the diverse roles of systems or components, their varying lifecycle stages, and the unique needs of each neighborhood. This approach ensures that governance supports the SoS's goals of collaboration, mission effectiveness, and resilience while also accounting for the programmatic success of acquiring individual systems or components.

6.6.2.4 Expectation and Evolution

Multiple forms of SoS evolution must be considered including the evolution of the individual systems or components and the evolution of the SoS itself. As previously discussed in Chap. 5 postulates and principles, a key characteristic of an SoS is that its constituent systems evolve at different rates and in an uncoordinated manner. While an organization might attempt to impose synchronization, such as through policies, this authority often extends only to the boundaries of the organization's control. Establishing control over the entire SoS is rare, and when it is attempted, it is unlikely that synchronization policies can be fully enforced across all programs acquiring the individual systems or components. Given this lack of centralized control over the independent evolution of the constituent systems or components, governance must focus on reducing the harmful effects of uncontrolled changes. To achieve this desired outcome requires policies that guide and regulate how portfolio systems or components within mission capabilities evolve. For instance, in a case where no consideration is given to the SoS as a whole during the early development of a legacy system replacement, the replacement developers may simply build new systems as instructed while being unaware that they were replacing most of the interfaces of the legacy system. Without proper guidance or requirements for integrating the new system within the SoS, the entire SoS capability can fail irrespective of the success of the individual system. Even if the legacy system's maintainers take steps to ensure the replacement does not disrupt the overall SoS, no formal requirement or guarantee is in place for this level of coordination.

At a minimum, governance for the evolution of individual systems or components within the SoS must include the following rules and guidelines:

- Inform other constituent systems or components (when known) of changes to interfaces and functionality.
- Coordinate schedules between constituent systems or components to allow for simultaneous updates when backward compatibility cannot be maintained.
- Maintain multiple versions of the system or component when schedules cannot be synchronized to provide flexibility.
- Develop systems or components to minimize the impact of changes in other constituent systems or components using modular open system architecture principles.
- Minimize disruptions to interfaces when making changes to individual systems or components.

The second form of evolution involves the SoS itself. While this evolution may be directed, it often occurs organically when new constituent systems or components are added to the SoS. If systems or components are integrated without careful consideration, unanticipated interactions between systems or components may lead to undesirable behaviors (i.e., emergent behaviors). For example, the accumulation of systems or components in an SoS once led to an incident where a sheriff's department inadvertently gained access to medical records from a local hospital. Such unintended interactions highlight the need for thoughtful governance when adding new systems or components to advance capabilities. It is difficult to specify exactly which policies are required or how they should be enforced, particularly because governance is distributed among various system or component owners. In any case, some form of modeling is essential to understand the potential effects of adding new systems or components to the SoS. Through careful modeling and assessment, the stakeholders can evaluate the impact of new systems or components and mitigate the risks of unintended consequences.

6.6.2.5 Highly Fluid Processes

As future SoS are expected to adapt rapidly to different contexts and requirements by remaining highly dynamic, their governance processes must be equally flexible. The ability to respond quickly to change will be essential, and planning for rapid evolution in SoS governance will be crucial to their success. One key aspect of flexible governance is developing strategies that allow rapid policy adjustments. For example, governance frameworks might include mechanisms for temporarily relaxing security policies to address an urgent goal, with the ability to re-tighten those policies once the need has been addressed. Similarly, flexible enforcement processes will be required to quickly verify and deploy updated systems or components, especially in crisis situations.

Multi-level governance models as discussed earlier in this section facilitate the rapid adaptation of functional capabilities and governance policies at different levels of an SoS. These models allow for quick responses to changing conditions at one level while maintaining more controlled evolution at other levels. For example, neighborhood governance within an SoS might involve local processes that support rapid consensus-building and immediate implementation of changes. A neighborhood of closely related systems, for instance, may be the first to detect a problem with a current component or process and must act quickly to address it. In such cases, neighborhood governance policies might allow for negotiation of SoS governance adjustments specific to that neighborhood. In more fluid neighborhoods, the governance policies might be adjusted more dynamically to reflect the changing needs. At the same time, it is essential that stability is maintained beyond the immediate neighborhood. While rapid, localized governance adjustments are critical, broader portions of the SoS must maintain more stable governance. Some governance policies are necessary to support a wider consensus and reflect the input of a larger, more diverse set of "neighbors" involved in determining appropriate responses. For example, a global decision such as the migration between internet protocol versions (IPvs), e.g., from IPv4 to IPv6, would not be made hastily in response to a sudden shift or urgent condition but would require a more stable and deliberate aspect of SoS governance. However, within specific neighborhoods, local systems might be allowed to transition to IPv6 if it addresses local conditions, provided they maintain compatibility with more distant neighbors by supporting agreed-upon interfaces (e.g., IPv4).

6.6.2.6 Minimal Centrality

While we have primarily advocated for a decentralized approach to governance in SoS, centrality is likely to occur in two key scenarios: (1) when there is a dominant system or organization within the SoS, and (2) within the SoS infrastructure itself. In the first case, the presence of a dominant system or organization effectively eliminates the need for decentralized governance. For example, a company that has significant control over a local market might change its business practices and require its suppliers to adhere to these new practices under the threat of losing business. Despite the involvement of many systems, the influence of the dominant company means the SoS operates more like a single, controlled system, governed by the company's policies and decisions. The second case involves the SoS infrastructure, where some level of central governance is essential to prevent chaos and ensure interoperability. Without central governance in the infrastructure, systems within the SoS are unable to communicate with one another effectively, resembling a "Tower of Babel." In this scenario, the infrastructure provider plays a critical role in governance by:

- Setting rules for system integration, such as defining the protocols and standards required for communication between systems.
- Creating metadata repositories or system registries, where the capabilities and services of different systems are documented and discoverable. These registries define the boundaries and scope of the SoS.

It is possible to have multiple infrastructure providers within an SoS. A historical example occurred in the early days of the Internet, when different regions and organizations operated their own localized networks, such as Advanced Research Projects Agency Network in the United States. Each network provider enforced its own policies, such as naming conventions, which were not compatible with one another. Over time, advancements in technology introduced gateways that allowed messages to be routed between networks, even when naming structures differed. As networking technology evolved and the individual policies were tested and refined, these infrastructure providers began to coordinate and collaborate, eventually adopting a federated governance model which enforced a set of common policies.

With a clear understanding of the **operational context and design** for MEI, we now turn to the architectural tools and frameworks that provide structure and analytical rigor for complex systems. Architecture frameworks, such as the DoD Architecture Framework (DoDAF), Unified Architecture Framework (UAF), and NATO Architecture Framework (NAF), offer standardized methodologies for visualizing, organizing, and managing the intricate relationships and interdependencies within an SoS. These frameworks help define system capabilities, identify integration challenges, and ensure alignment with mission objectives by offering multiple viewpoints. In the next chapter, we will explore how **architectural design and frameworks** support MEI by enabling better decision-making, enhancing interoperability, improving lifecycle management, and ensuring that each constituent system functions cohesively within the broader mission landscape.

References

1. Army Doctrine Reference Publication (ADRP) 5-0. (2012). *The operations process*. U.S. Army.
2. Boyd, J. (1987). *Destruction and creation*. US Army Command and General Staff College.
3. Chairman Joint Chiefs of Staff Manual (CJCSM) 3500.04G Series. (2023). Universal joint task manual.
4. Johansson, F., & Falkman, G. (2008). A comparison between two approaches to threat evaluation in an air defense scenario. In *Modeling decisions for artificial intelligence* (pp. 110–121) ISBN:978-3-540-88268-8.
5. Joint Publication 1-0. (2023). *Volume 1: Joint Warfighting*. U.S. Department of Defense.
6. Joint Publication 3-0. (2017). *Joint operations*. U.S. Department of Defense.
7. Joint Publication 5–0. (2011). *Joint planning*. U.S. Department of Defense.
8. Joint Publication 3-60. (2013). *Joint targeting*. United States Department of Defense.
9. Koski, S. R., & Moreland, J. D., Jr. (2021). Performing information extraction and ontology development for mission engineering applications. *Journal of Integrated Design and Process Science, 25*, 25–37. https://doi.org/10.3233/JID-210012
10. Mattis, J. N. (2009). *Vision for a joint approach to operational design*. U.S. Joint Forces Command.
11. McChrystal, S., Collins, T., Silverman, D., & Fussell, C. (2015). *Team of Teams: New rules of engagement for a complex world*. Portfolio.
12. Olsen, J. A., & Van Creveld, M. (Eds.). (2011). *The evolution of operational art: From Napoleon to the present*. Oxford Academic.

Chapter 7
Mission Architectures and Engineering Frameworks

7.1 Architecture Design

ISO/IEC/IEEE 42010 describes architecture as: [ISO22]

> The fundamental concepts or properties of a system in its environment embodied in its elements, relationships, and in the principles of its design and evolution.

This excerpt defines a "system" broadly, encompassing various entities that can be analyzed or managed using a systemic approach. Specifically, it includes:

- **Product**: Any tangible item produced to meet a need or want.
- **Service**: Activities or benefits provided to fulfill a need or want.
- **Information System**: A combination of technology, people, and processes managing and processing information.
- **System-of-Systems (SoS)**: A complex configuration of multiple interacting systems.
- **Enterprise**: An organization or business entity structured to achieve specific objectives.

This inclusive definition emphasizes that a systemic approach is applicable to a wide range of entities, regardless of their nature or complexity. However, a description of architecture can be started before any identification of systems. This is the case when the description begins with a pure operational description or a set of operational capabilities that explain what the user needs. By focusing initially on operational aspects, the architecture can be aligned with the specific mission essential tasks, functions, and services required to meet operational objectives and mission success, independent of the systems that will eventually be identified or developed to support these operations. This path aligns with the Mission Engineering and Integration (MEI) approach and ensures that the architecture is driven by user requirements from the outset.

J. D. Moreland, Jr., *Mission Engineering and Integration*,
https://doi.org/10.1007/978-3-032-32132-9_7

An architecture achieves benefits such as customer collaboration, system development, and technology transfer. Through customer collaboration, it validates the needs and requirements of users, ensuring that their expectations are met. In terms of system development, it provides a concise and comprehensive description of the system's architecture to developers and maintenance engineers, facilitating a clear understanding and implementation. Architecting complex SoS requires consideration of the systems as a whole, including interactions among its elements, competing objectives and constraints, and the tradeoffs necessary to develop an effective solution [RECHTIN91]. Additionally, an architecture supports technology transfer by enabling the reuse of architectural styles and patterns, promoting efficiency and consistency across projects. Common themes across architecture definitions highlight these key aspects:

- A complex entity, such as a system, SoS, or enterprise, is scoped and partitioned into components with specified roles, activities, and functions.
- Interfaces and rules are established between the internal components of the complex entity.
- Interfaces and rules are established between the internal components and those external to the complex entity.
- Components, activities, rules, and interfaces must evolve over time.

7.2 Reference Architectures

Reference architecture provides a predefined template or blueprint of a generic solution for a particular domain or industry. It offers standardized guidelines, principles, and best practices that organizations can use as a starting point when designing specific architectures [PATTON23]. It focuses on specific domains, technologies, or industries and provides a reusable model or framework that can be adapted and customized to fit the needs of individual organizations. The main purpose of the reference architecture is to promote consistency, interoperability, and efficiency by providing proven solutions and patterns that can be applied to similar problems across different projects, programs, or organizations. It typically includes architectural patterns, standards, specifications, and recommended technologies relevant to the domain it addresses. As captured in its name, it serves as a reference point to guide architects and developers in designing and implementing systems that align with industry standards and best practices.

A reference architecture serves as an authoritative source of information for a specific subject area, guiding and constraining architectures and solutions through common information, guidance, reusable components, and direction. By presenting standardized information and components, reference architectures create patterns for addressing similar problems across multiple domains, thereby driving efficiency and interoperability. These architectures operate at any level and originate from both internal and external sources to the primary stakeholders. Mission threads describe the end-to-end set of activities and systems required to execute a specific mission, representing a horizontal segment of an architecture. Each mission thread has an owner who collaborates with architects to define a capability by integrating tasks, conditions, standards, and methods to achieve a desired effect. These threads are informed by doctrine, training, readiness manuals, and other authoritative capability-related data sources. Mission threads help reduce interoperability risks by highlighting critical interfaces when solutions are applied in real-world contexts. This ensures that all components work seamlessly together to achieve mission objectives efficiently and effectively.

Information Support Plans (ISPs) are technical documents essential for programs in providing program managers with a means to identify and resolve potential information support implementation issues and risks. Development of ISPs ensure programs can be operationally employed to meet current and future mission requirements. As authoritative documents, ISPs directly inform the program's Test and Evaluation Master Plan, establishing threshold and objective operational parameters and supporting the validation of a program's eligibility for interoperability certification. They support building authoritative architecture products by identifying information technology and information needs, dependencies, and interfaces for all acquisition categories. This is achieved by focusing on interoperability, information and system supportability, and information sufficiency concerns. ISPs assess a program's progress toward integrated operations, describe cybersecurity compliance, and address supportability topics such as intelligence, bandwidth, and spectrum. They include an operational employment concept, system interface descriptions, required information exchanges, and information technology information supporting requirements. In addition, ISPs encompass the associated integrated architectures, potential issues or risks, and proposed solutions or risk mitigation plans. Figure 7.1 shows an example of typical ISP data.

7.3 Authoritative Source of Data and Information Access

Sharing and using project-sensitive data requires guidance by a set of governing rules and principles to ensure that the right type, amount, and format of data is exchanged properly. Careful attention needs to be taken on requirements to disclose a minimum set of critical data elements to achieve integration and interoperability which in some cases provides competitive advantage. In addition, this exposure may potentially violate intellectual property rights and put key competition-sensitive discriminators at risk for some parties as well as data-loss prevention through cyber means. At the same time, the right amount of data must be shared between connected models to maintain the accuracy of the technical baseline to represent the authoritative source of truth such that decision-makers are able to make better decisions faster and more efficiently. Each stakeholder must have access to the appropriate data whenever needed and be able to communicate and collaborate on the needed information within the context of a configuration-controlled baseline for constituent

INFORMATION EXCHANGE RATE #	EVENT/ ACTION	INFORMATION CHARACTERIZATION	SENDING NODE	RECEIVING NODE	CRITICAL	FORMAT	TIMELINESS	DATA CLASSIFICATION
1	Receive GPS update	Time and position data	GPS Satellite	System A	Y	Data	1 sec	U
2A	Control aircraft: Attack enemy maritime targets (CV)	System A track data, engagement directive, air control data	System A	System B	Y	Data	JDN<16s	S
2B	Control Aircraft: Attack enemy maritime targets (DCA) (CV)	Status, engagement reports	System B	System A	Y	Data	JDN<16s	S
3A	Control aircraft: Attack enemy air targets (DCA) (CV)	System A track data, engagement directives, time	System A	System B	Y	Data	JDN<16s	S
3B	Control Aircraft: Attack enemy air targets (DCA) (CV)	Status, engagement reports	System B	System A	Y	Data	JDN<16s	S
4A	Engage enemy air targets based on remote sensor data	Track data, engagement data and orders to controlled aircraft	System A	System B	Y	Data	JCTN<1.5s	S
4B	Engage enemy air targets based on remote sensor data	Track data, engagement data and orders to controlled aircraft	System B	System A	Y	Data	JCTN<1.5s	S

Fig. 7.1 Information support plan example

systems within the mission capability system-of-systems (SoS). Achieving the right flow of critical data is key to fully realize the benefits of model-based systems engineering (MBSE) within the digital engineering ecosystem.

The ultimate goal is to achieve complete and seamless sharing of the right data at the right time in the right format that arrives at the right place for all relevant stakeholders involved. An integrated digital environment on a connected infrastructure needs all stakeholders to have access to the necessary data and tools in the format or view appropriate to their discipline, and all stakeholders to be able to collaborate across organizational boundaries.

Right information. Government and industry stakeholders need access to selected elements of the relevant models, government or industry furnished, that provide Key System Attributes (KSAs) and system specifications from developers, operations concepts, architecture models, and mission performance data. The right information must be fully transparent into all aspects of the development models that relate to their discipline.

Right time. Timeliness of access depends on the program phase within which the stakeholder work begins. For example, some participants will be involved throughout, and others will access the models as needed. It is also critical to have real-time access to this data to avoid receiving outdated or stale data that is no longer considered the authoritative data of truth.

Right format. Ultimately, government and industry standards, tools, and processes will mature to the point that every stakeholder is able to access the digital engineering model directly through a dashboard or similar medium. Another necessary step involves the development and implementation of open standards and interfaces to promote this vision. For the immediate future, however, data must be provided to some stakeholders in custom forms or views, including documents and file transfers for special tool interfaces.

Right Place. Delivering the data to the right location reinforces the concept of sharing critical data elements by verifying actual receipt by the appropriate stakeholders.

The basic requirement of a successful data-driven or "transparent" operations are the ability to collect data and to store data centrally. But that data is largely useless if it does not enable improvement, corrective action, and the pursuit of operational excellence. This can only be done if the data is properly mined and appropriately displayed to the right people at the right time so that the data can be acted upon appropriately. The phrase, "I can't see the forest for the trees," is often used when data is delivered to a single source, in a poorly organized format. The ability to process and act upon data starts with the ability to visualize that data in a way that can be analyzed and interpreted. This does not mean delivering all the data to one person with a large volume of performance indicators or measurements but instead delivering the data needed to do the best job. For a line operator, the data required may be minimal and may reflect simple elements like system performance, up-time, or shortages. An engineer might need data on the performance of a particular product in a test to analyze the reasons for failures. And the production planner will need data that allows better planning and consideration of "what-if" scenarios around increased volumes or supply-chain disruptions.

All stakeholders with a responsibility in fielding these SoS capabilities have a vested interest in timely precise data for fielded capabilities. Data that provides the operator with an advanced capability will contribute to the data that allows the Chief Operating Officers in individual organizations to make the right capital-equipment investment or outsourcing decisions. A digital ecosystem provides transparency across the stakeholder enterprise to provide a clear and undeniable value of data sharing for critical elements required for making information visible and useful to the enterprise. The data visibility required to achieve a transparent factory is related directly to the activities and roles of those who require information in their day-to-day jobs. The data needs are either real-time or historic with delivery mechanisms spanning several formats such as dashboards, analytics, reports and, most recently, mobile applications for those needing access to data on the move.

Real-time information is most commonly delivered in what is currently called dashboards. A process engineer or production-line operator will often use these data to maintain and improve the performance of the line. The data must be instantaneous because any delay will likely cause or extend down-time and subsequent impact further along the line. This data delay also causes quality challenges and reduces yield; latent data used for designs or faulty production assemblies continue to increase while an error is detected and corrective action taken. These data are best delivered in a simple graphical format with strong visual and even audio alerts delivered when a potential problem is likely to occur. The need for instant data here is obvious: the faster the alert, the faster the reaction and eventual solution.

Historical data is needed for production control, engineering, quality, test, and to allow management to monitor, analyze, and adjust production based on current and predicted status. This is where traceability enters the equation. Traceability, like visibility, is an essential cornerstone for any program emphasizing operational excellence that seeks to add value, reduce costs, and mitigate risk in fielding a mission capability. The historical data's delivery method and presentation style can take many forms. Among these are reports and detailed analytics as the basics of any transparent system, and details informing stakeholder decisions about product selection, development, operation, and maintenance solutions.

7.3.1 Data Dictionary

A data dictionary is a collection of metadata such as object name, data type, size, classification, and relationships with other data assets. Think of it as a list along with a description of tables, fields, and columns assembled to help stakeholders understand data assets. A data dictionary is used by data administrators, analysts, and engineers to understand and trust data assets. The semantic understanding of critical data elements about systems helps to create authentic, transparent, and consistent data throughout the organization. This data dictionary is developed from the following common components:

- A listing of ***data objects*** (names and definitions)
- Detailed ***properties*** *of* ***data elements*** (data type, size, nullability, optionality, indexes)
- ***Entity-relationship*** (ER) and other ***system-level diagrams***
- ***Reference data*** (classification and descriptive domains)
- ***Missing data*** and quality-indicator codes
- ***Business rules***, such as for validation of a schema or data quality
- ***Data source*** (data warehouse, data lakes, databases, applications)
- ***Date*** and ***time*** when the property was created or changed
- ***Descriptive statistics*** that go beyond missing values, such as min-max values and histogram distribution
- ***Owners*** and ***editors*** of data sets that contain these variables

A data dictionary documents data assets with relevant context, making it easy for stakeholders to use, analyze, and discuss data. Key benefits of building and implementing a data dictionary are listed below:

- ***Detect anomalies quickly and avoid data inconsistencies.*** Modern data platforms automatically generate data quality metrics and statistics so that data quality is understood. Since the data dictionary displays descriptive statistics (minimum, maximum, count, frequency, mean, and median), spotting anomalies in data is simplified, which helps to avoid inaccuracies or inconsistencies.

- ***Acquire more trustworthy data that is easy to analyze.*** A data dictionary offers context by documenting metadata as well as data sources/origins, owners, creation dates, which helps to validate data. This emphasizes the importance of reliability and accuracy of information used for decision-making events. Additionally, modern data platforms such as a data catalog also allow for the visualization of overall data flow, making it easier to interpret the impact on upstream or downstream applications.
- ***Establish consistent processes to collect, document, and use data.*** If constructed correctly, a data dictionary can help establish certain ground rules for collecting, documenting, and using data. This, in turn, simplifies regulatory compliance across all stakeholders to identify misuses and fix bad data practices.
- ***Build transparency and enable self-serve within data teams.*** The data dictionary equips all stakeholders with a common repository for data definitions, standards, metrics, and more. Therefore, stakeholders better understand the meaning of any element within a data set without having to consult an expert. This reduces dependencies, helps stakeholders use the data in the same way, and makes onboarding of new personnel easier.

7.3.2 Data Security

In addressing challenges with data security to include intellectual property and data rights, a common data framework is required like Zachman's Framework (shown in Fig. 7.3 and discussed in Sect. 7.4.3) to determine the right questions to be addressed. The critical questions in Zachman's Framework are located across the column headings and are known as the five Ws and H (how) approach to data problem solving. Looking through the lens of intellectual property and data rights leads to the following questions:

- What specifically do those digital artifacts contain?
- Where will the information be accessed and stored?
- Who can see and modify the digital artifacts?
- When and how frequently will the information exchange(s) happen during the system lifecycle?
- Why is the data being exchanged necessary for various stakeholders with access?
- How will the information be used by key decision-makers at various stages of acquisitions?

Existing intellectual property and data rights policies and standards need to be modified and new ones developed to meet the needs of a digital ecosystem. Information management requirements are most effective when the guidelines below are considered:

- Recommend customer access to in-process baselines in contractor infrastructure during competitive phases to limit risk of information spillage
- Provide information access based solely on acquisition decision-making needs
- Revise existing data item descriptions using government-industry collaboration forums to support transition to the digital ecosystem
- Use an agreed to structured process to determine key information types and protection requirements by program phase and product type to include cybersecurity standards
- Create a common information access management ontology usable by current and future information technology for data loss prevention, and network access control applications to manage access

7.4 Levels of Architecture

Figure 7.2 shows how this analytical tool considers the critical translations from enterprise to individual systems.

7.4.1 Enterprise Architectures

Enterprise architecture is a comprehensive framework used to align an organization's business and operational processes, information technology infrastructure, information flows, and technology with its strategic goals and objectives. It encompasses the entire enterprise, including its business and operational units, processes, information systems, and technology infrastructure. The primary goal of an enterprise architecture is to ensure that all components of the enterprise work cohesively to achieve objectives efficiently and effectively in leading to mission success. From an MEI perspective, enterprise architecture includes various domains such as mission architecture, SoS architecture, and system architecture. The enterprise architecture provides a holistic view of the organization and guides decision-making related to information technology investments, business and operational transformation, and operational efficiency.

7.4.2 Mission Architectures

Enterprise architecture is the interrelation and integration of a mission (business) architecture and a solution architecture and is fundamental to building a roadmap describing the current state ("as-is") and future desired state ("to-be") from both an operational and technical perspective, along with the sequencing plan to achieve the future state. The mission architecture models the enterprise's mission, needs,

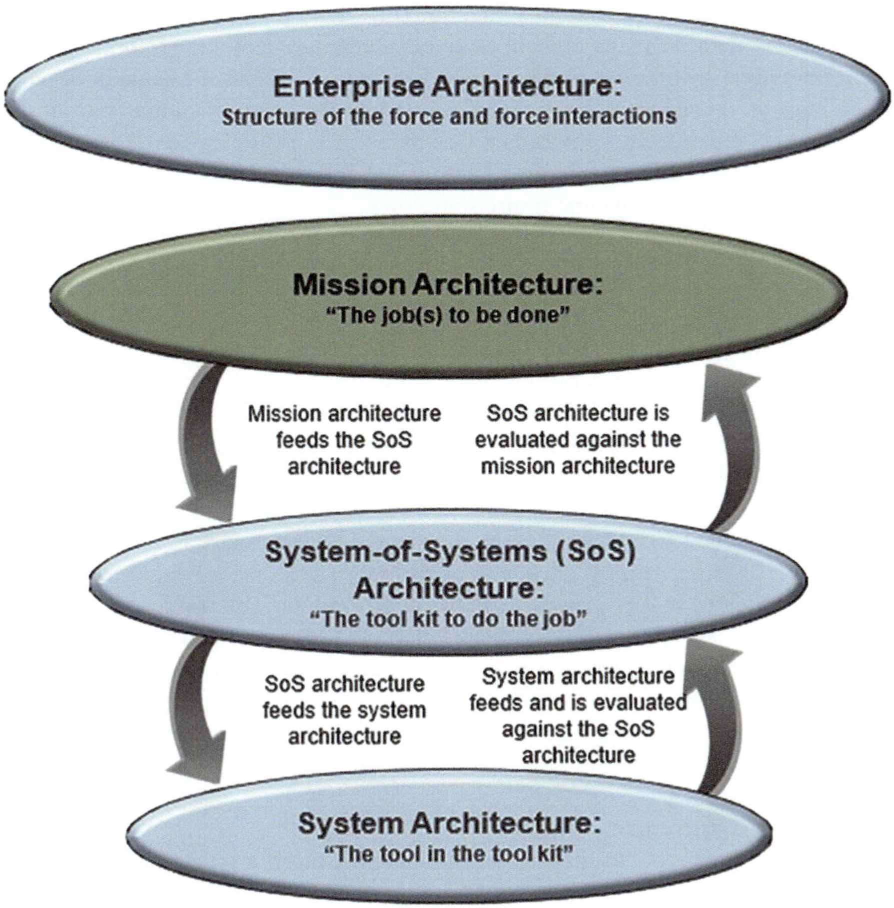

Fig. 7.2 Levels of architectural products

strategies, goals, business rules, business processes, information flows, and supporting operational structure. The solution architecture models the technical aspects of a system, or SoS encompassing elements such as data, applications, and enabling technologies. This includes defining how data is stored, processed, and accessed within the system or SoS, specifying the various applications and their functionalities, and determining the technologies and platforms that support operations. It provides a blueprint for designing and implementing the technical components of the system or SoS, ensuring that they align with the overall objectives and requirements. It addresses issues such as scalability, performance, security, and interoperability. This guides the development team in making informed decisions about technology selection, integration approaches, and system implementation. By modeling the technically focused architectural aspects of the system, the solution architecture enables stakeholders to understand how different components interact and collaborate to fulfill the system's functionality. This understanding is crucial to ensure the system meets its objectives and can adapt to evolving requirements and environments over time. A commitment to enterprise architecture entails developing a mission architecture to guide and align with the solution architecture. It also involves recognizing the importance of modeling both the current state and the desired future state of the enterprise, alongside crafting a transition plan outlining the steps needed to achieve the future state.

- **"As-Is" Architecture**
  Analyzing the "as-is" mission architecture against adversarial threats serves as a valuable mechanism to identify and understand operational gaps within the current mission framework. This strategic assessment involves evaluating the existing architecture's vulnerabilities, weaknesses, and potential points of failure in the context of potential adversarial actions or threats. By conducting a threat-centric analysis, mission engineers can pinpoint areas where the mission architecture may be susceptible to disruptions, security breaches, or other adversarial activities. This process facilitates a comprehensive understanding of the mission's resilience and its ability to withstand external pressures. It allows mission engineers to assess the effectiveness of current security measures, identify potential blind spots, and prioritize areas for improvement. The insights gained from this analysis enable the development of targeted strategies and enhancements to fortify the mission architecture against adversarial threats. In essence, scrutinizing the as-is mission architecture in light of adversarial threats provides a proactive approach to risk management and ensures that mission-critical operations can continue with minimal disruption, even in the face of potential challenges posed by adversaries.
- **"To-Be" Architecture**
  The development of "to-be" mission architecture offers a systematic approach to determining future solutions and devising strategies for their seamless integration into the existing SoS. This forward-looking analysis involves envisioning the desired state of the mission architecture, considering technological advancements, operational requirements, and evolving mission objectives. By assessing the to-be architecture, mission engineers can identify opportunities for innovation, optimization, and the incorporation of emerging technologies. It provides a mechanism for envisioning how new solutions, capabilities, or systems can be integrated cohesively into the existing SoS framework. This proactive approach allows mission engineers to plan for the evolution of the mission architecture, so that it is adaptable and scalable to best meet changing demands and objectives. The as-is assessment helps to identify potential challenges, dependencies, and integration points and provides a foundation for developing a roadmap for implementing future solutions in a phased and strategic construct. This progressive perspective enables mission engineers to align the mission architecture with long-term goals, enhance overall performance, and maintain relevance in dynamic operational environments.

Mission architectures are a foundational framework employed in the discipline of MEI to strategically plan, organize, and execute intricate missions. Serving as a comprehensive blueprint, the mission architecture delineates key elements, relationships, and processes crucial for achieving mission objectives. At its core, this structured architecture using a defined and accepted architectural framework plays a pivotal role in guiding mission design; it ensures coherence and promotes the effective collaboration of various mission components. Using a systematic approach to define mission objectives, mission architectures identify essential mission elements, establish inter-component relationships, assess risks, allocate resources judiciously, and conduct scenario analyses. The architecture serves as a decision-support tool throughout the mission lifecycle, providing insights for informed decision-making.

Moreover, the development of architecture products emphasizes integration and interoperability, addresses lifecycle considerations, and ensures the products are designed to adapt to evolving technological, environmental, and operational factors. Collaboratively developed by mission architects, engineers, and stakeholders, the architecture is a dynamic tool that evolves in tandem with the mission's goals, promoting resilience and efficiency across the entire mission spectrum, from planning and development to operations. The key elements involved in building effective architecture products are identified in Table 7.1.

Mission architecture encompasses several key elements essential for the success and effectiveness of a mission. First, clearly defined mission objectives and goals lay the groundwork, outlining the purpose and desired outcomes. An operational concept provides a detailed plan of execution, specifying the roles and interactions of various components, systems, and stakeholders. The identification and specification of the SoS and system components form another crucial element. This involves detailing the hardware, software, personnel, and other entities contributing to mission success. Seamless interconnectivity and integration of these systems are vital, requiring well-defined communication protocols, data exchange mechanisms, and interfaces to ensure cohesive functioning. A clear timeline with milestones and performance measures facilitates progress tracking and keeps the mission on schedule. The adaptability and flexibility of the architecture allow for adjustments to changing requirements, technology, or external factors. In addition, in the digital age, cybersecurity measures are paramount to safeguard systems and data, thereby enhancing overall mission resilience.

7.4.3 Architecture Frameworks

Architecture frameworks play a crucial role in MEI by providing structured approaches and methodologies for designing, analyzing, and managing complex systems and their interactions. Several widely used architecture frameworks contribute to providing effective and efficient MEI analyses. Among these frameworks is the Zachman Framework™, a schema formed at the intersection of two historical classifications that has been used for hundreds of years. The first classification of

Table 7.1 Key elements of a mission architecture

Element	What it Provides
Mission Definition / Objective	**Focus on the Job**
Threat Definition	**Focus on the Why**
Graphical Overview (i.e., OV-1)	**Vision**
Functional Flow	**How the Job is Done**
Timelines	**Urgencies and Synchronization**
Nodes and Interactions	**Focus on the Who**
Function Node Task Table	**Mapping of the Who to the What**
Operation Resource Flow	**Focus on the Interactions**
Attributes and Measures	**How Well the Job is Done**

Zachman presents the fundamentals of communication, which are encapsulated in what Zachman calls "the primitive interrogatives": What, How, When, Who, Where, and Why. These questions, when answered collectively, provide a comprehensive and detailed description of complex ideas. The second classification stems from the concept of reification, the process of transforming an abstract idea into a concrete instantiation. In the Zachman Framework™, this process is articulated through six stages: Identification, Definition, Representation, Specification, Configuration, and Instantiation.

> John Zachman defines an enterprise architecture as that set of descriptive representations (i.e., models) that are relevant for describing an enterprise such that it can be produced to management's requirements (quality) and maintained over the period of its useful life (change). [ZACHMAN96]

The Zachman Framework™ significantly enhances architecture frameworks by providing a structured, systematic approach to organizing and understanding complex systems. By employing a two-dimensional classification system, it ensures comprehensive coverage of all aspects of an enterprise architecture. The columns represent fundamental questions (What, How, When, Who, Where, and Why), while the rows depict levels of reification (Identification, Definition, Representation, Specification, Configuration, and Instantiation) as illustrated in Fig. 7.3 [ZACHMAN24]. This detailed structure facilitates better communication among stakeholders by clearly and consistently categorizing information and architectural artifacts. Moreover, the framework guides the development process by identifying gaps and overlaps, ensuring all necessary components and relationships are addressed. It aligns technology initiatives with operational objectives by considering both perspectives. This provides a foundational structure that other architecture

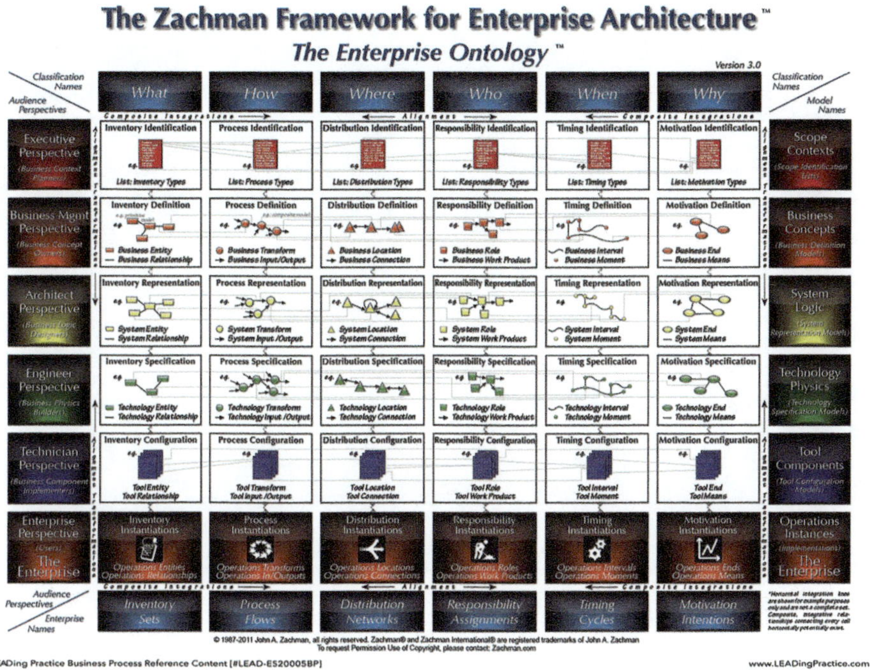

Fig. 7.3 Zachman framework for enterprise architecture

frameworks can adapt and extend. Additionally, by modeling the current state, desired future state, and transition plan of the enterprise, the Zachman Framework™ supports the evolution and transformation of architectures, ensuring their adaptability to changing operational needs and technological advancements.

The major architecture frameworks include:

Department of Defense Architecture Framework (DoDAF)

Overview: DoDAF is a framework specifically designed for the US Department of Defense. It provides standardized approaches for visualizing, understanding, and managing the defense enterprise architecture.

Focus: It emphasizes the importance of using an authoritative source of data to provide the foundation for many viewpoints that capture different aspects of the architecture, such as operational, system, and technical views.

The DoDAF has served as the traditional architecture framework in the US DoD for numerous years. The views offered by this framework have grown over the years to include the set of views shown in Fig. 7.4 [DODAF24].

In the initial phase of functional analysis using the DoDAF, the emphasis is placed on the Operational Viewpoint. This first order of analysis aims to provide a comprehensive understanding of the system's functionality within the operational context of the DoD as illustrated in Fig. 7.5. The analysis involves identifying and

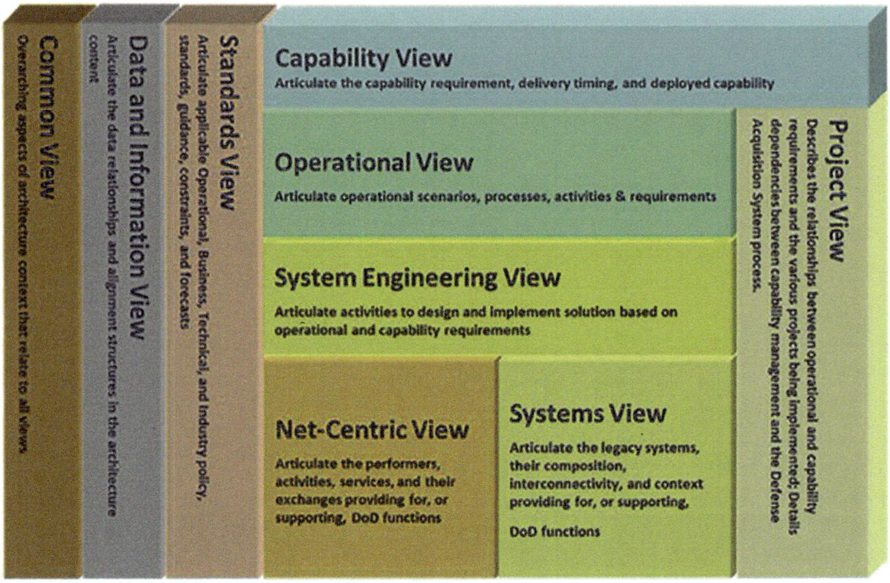

Fig. 7.4 DoD architecture framework views

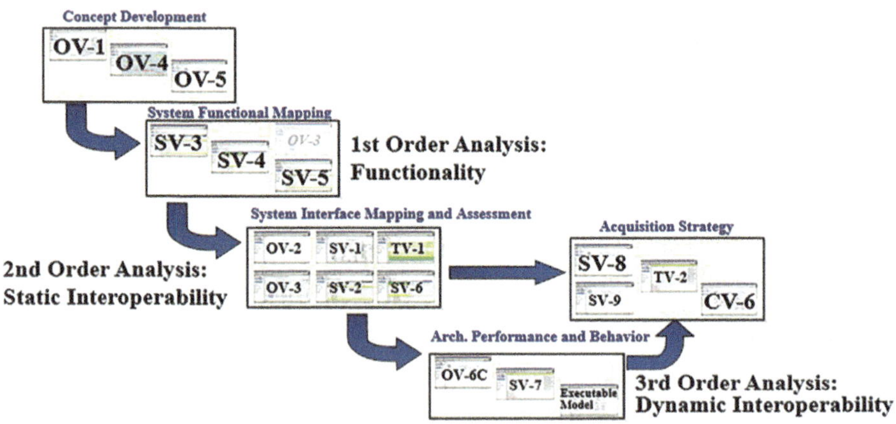

Fig. 7.5 DODAF views for MEI depth of analysis

defining key operational activities aligned with the overarching mission and objectives. Operational nodes, representing locations or entities where these activities occur, are specified along with their interrelationships. Operational scenarios are developed to illustrate the system's role in different situations, emphasizing its contribution to mission tasks. MoE and MoP metrics associated with each operational activity are identified to establish criteria for evaluating the system's impact on mission success.

In the second order of analysis within the DoDAF, the focus shifts to static interoperability, specifically addressing how the system interacts with other systems and components in a stable or non-dynamic state. The key viewpoint for this analysis is the System Interface Description or SV-1. This phase involves identifying and characterizing the various systems and entities with which the analyzed system needs to interoperate as part of the static relationships and interfaces between the subject system and external entities. This includes specifying data exchanges, communication protocols, and other pertinent interface details. The goal is to ensure that the system's components can seamlessly exchange information and collaborate with other systems, fostering cohesion within the broader mission architecture.

Lastly, the third order of analysis within the DoDAF is the extended focus to dynamic interoperability. This emphasizes how the system engages with other entities and adapts to changing conditions in operational scenarios. This phase centers around the Operational Viewpoint, particularly the OV-6c (Operational Event-Trace Description) model. The analysis involves examining the dynamic interactions, events, and functional exchanges between the subject system and external entities during operational activities. By leveraging the OV-6c model, analysts can depict event traces, illustrating the chronological sequence of system behaviors and their impact on operational events.

Subsequent orders of analysis can further explore aspects such as system architecture, resource utilization, and capability evolution. Each order of analysis builds upon the previous, creating a layered understanding of the system's role, integration, and effectiveness within the broader DoD force architecture. This flow-down process involves a structured approach to translating operational requirements into detailed system specifications using architecture products. These products ensure a seamless alignment between high-level mission objectives and the specific functionalities and capabilities of individual systems within a comprehensive architectural framework.

The Open Group Architecture Framework

Overview: ***The Open Group Architecture Framework (TOGAF)*** is an enterprise architecture methodology and framework that offers a systematic approach to designing, planning, implementing, and governing an enterprise's information technology architecture. This structure is illustrated in Fig. 7.6 [OMG24].

Focus: While not exclusively designed for MEI, TOGAF's comprehensive structure can be adapted to address mission-specific concerns within an enterprise context.

The NATO Architecture Framework

Overview: The NATO Architecture Framework (NAF) provides a standardized approach to enterprise architecture, enabling NATO-member nations and organizations to design, document, and analyze architectures related to defense and military operations. Its views are structured in accordance with the general configuration shown in Fig. 7.7 [NATO24].

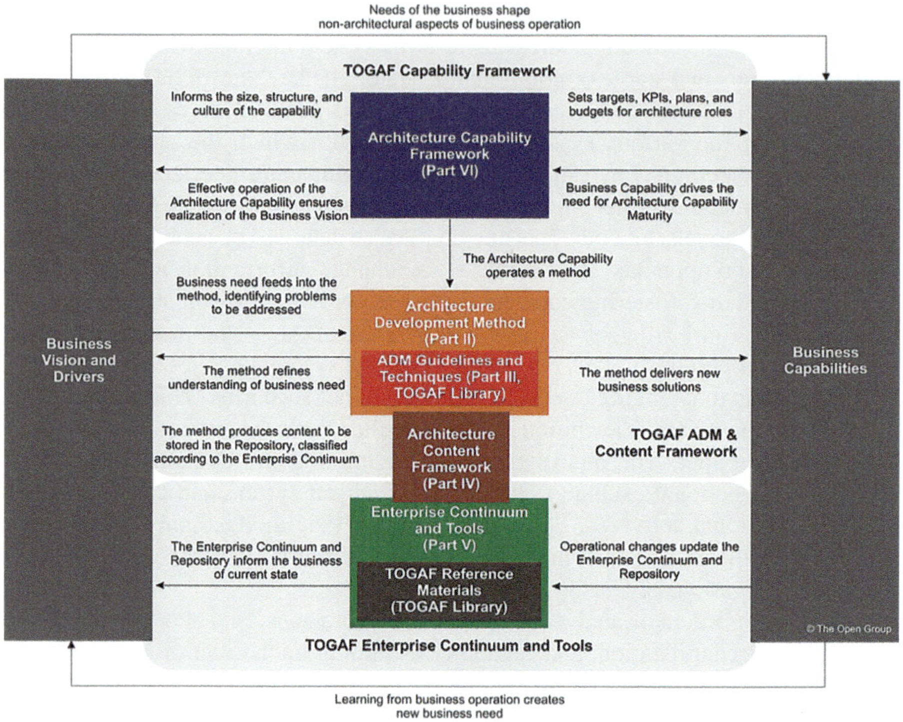

Fig. 7.6 The OMG architecture framework structure

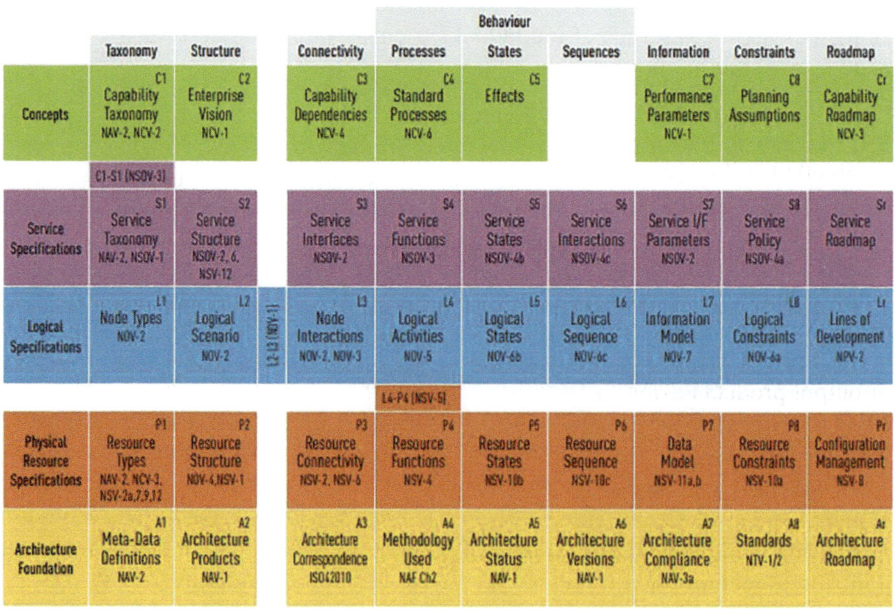

Fig. 7.7 NATO architecture framework views

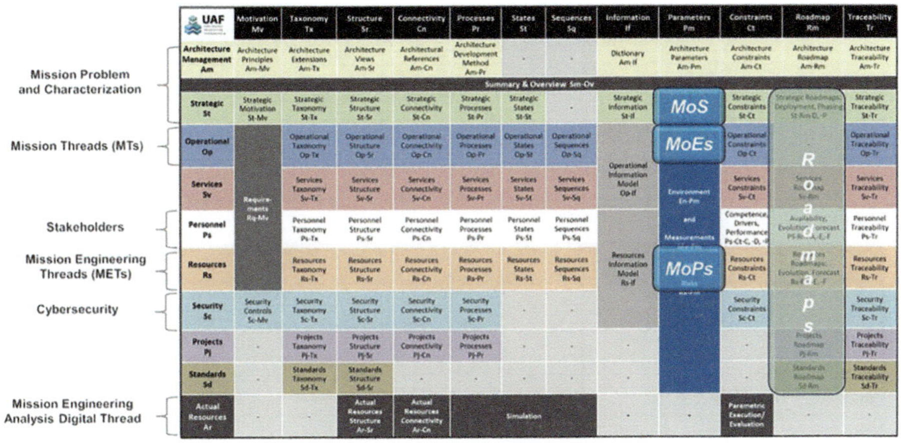

Fig. 7.8 OMG unified architecture framework

Focus: NAF is a framework developed by the North Atlantic Treaty Organization (NATO) to guide and support the development of architectures within the NATO community.

The OMG Unified Architecture Framework

Overview: The Object Management Group (OMG) Unified Architecture Framework (UAF) is an architecture framework developed by OMG to provide a standardized approach for modeling and designing complex systems and enterprise architectures. It places specific focus on MEI and provides general views according to the structure in Fig. 7.8 [OMG/UAF24].

Focus: UAF is designed to be a comprehensive framework that supports multiple architecture domains by being domain-agnostic, including system, software, and enterprise architectures. This framework emphasizes the integration of different architectural viewpoints to address various stakeholders' concerns. Based on industry standards, the UAF is compatible with widely adopted modeling standards and practices and aligns with OMG standards, such as UML, to ensure consistency and interoperability.

This framework provides structures, methodologies, and visualizations that help mission engineers systematically address the complexities of designing and managing systems and SoS to achieve mission success. The choice of a framework depends on the specific requirements, context, and objectives of the MEI efforts. Examples of output products using the UAF are shown in Fig. 7.9 for the historical events of Desert Storm. In this case, these products play a critical role in structuring and analyzing complex military operations. The ability to construct mission threads, sequence diagrams, and an order of battle is essential for ensuring that operational planning and execution are based on legitimate, data-driven inputs. In Desert Storm, integrating these architectural outputs enabled a more structured representation of joint force operations, illustrating the coordination of assets, decision points, and

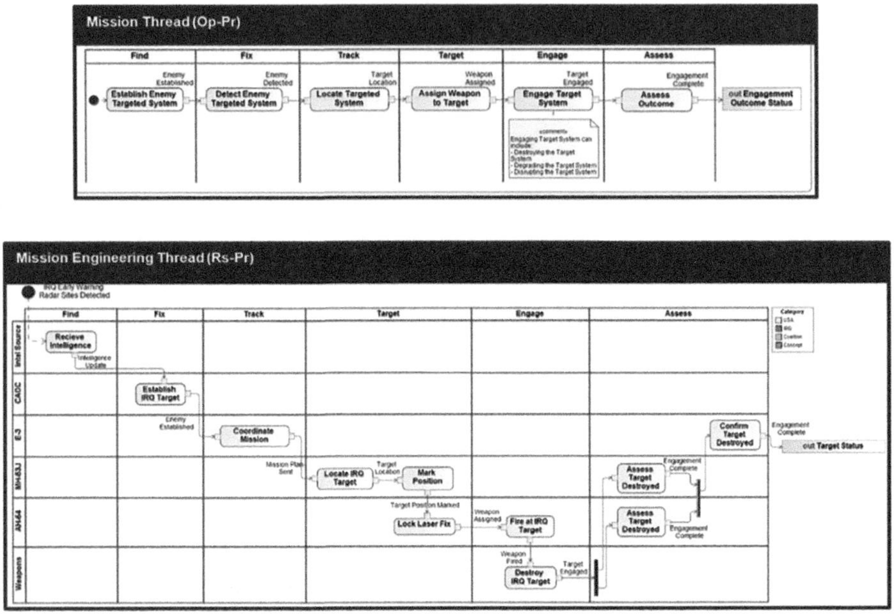

Fig. 7.9 UAF desert storm products

engagements across multiple domains. Mission threads captured the end-to-end operational flow, while sequence diagrams mapped interactions between key entities, ensuring synchronized actions. The order of battle provided a detailed depiction of friendly and adversary force composition, capabilities, and disposition. Together, these elements contributed to a comprehensive mission engineering thread, aligning operational objectives with system capabilities and constraints. This structured approach is critical to enhancing situational awareness, decision-making, and adaptability in real-world missions and reinforces the necessity of robust architectural outputs in modern MEI.

Building on the architectural strength of the UAF mission engineering threads, Fig. 7.10 illustrates how baseline and alternative mission engineering threads can be compared and contrasted to optimize operational planning. In the baseline scenario, multiple intelligence sources are leveraged to identify key adversarial targets, setting the stage for the execution of a mission. In contrast, the alternative mission thread introduces an added layer of redundancy by incorporating information from another intelligence source. This alternative approach provides an additional communication path between critical locations and assets. This additional fail-safe mechanism ensures continuity in the event of connectivity loss, enhancing the reliability and resilience of mission execution. Comparing these mission engineering threads helps the engineer select the most effective operational plan and enhances flexibility and robustness in mission execution.

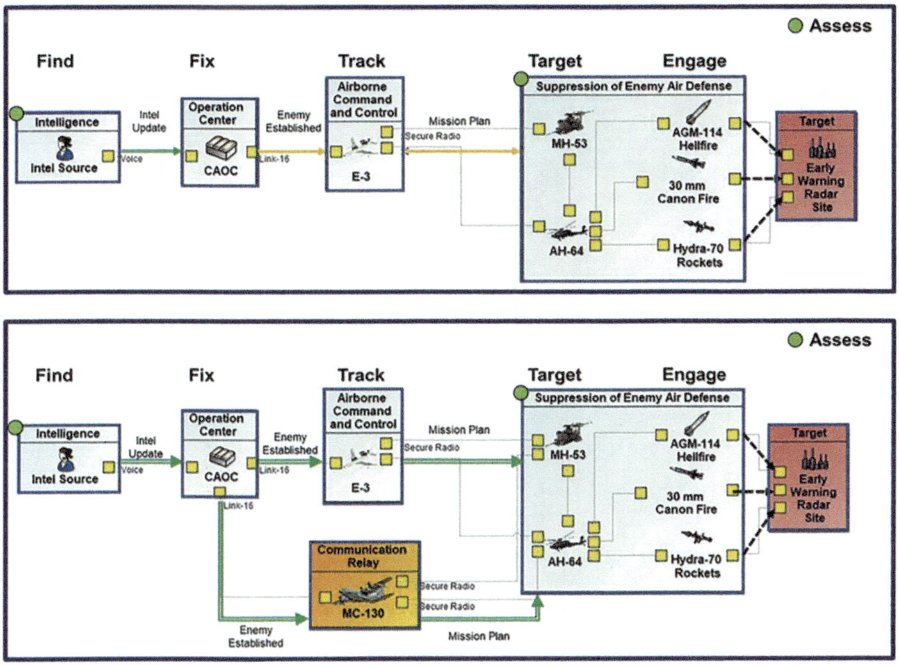

Fig. 7.10 Using UAF products for mission options

7.5 Integrated Capability Framework

The Integrated Capability Framework (ICF) establishes standards for data elements, relationships, products, and implementation guidance essential for MEI analysis across all operational stakeholders. This includes determining operational needs, conducting experiments, defining requirements, designing systems, testing, training, and operational deployment. Its objective is to facilitate the seamless integration of platforms and systems into mission architectures and serve as a communication conduit among the relevant stakeholders. The ICF provides structured guidance for organizing complex mission and system information, ensuring a standardized framework that accommodates the insertion and evaluation of new capabilities. Figure 7.11 shows the enterprise architecture structure of the ICF.

Creating the ICF involved a systematic process centered on understanding user needs and addressing organizational constraints. The first step was identifying the diverse group of expected users, including both creators of architectures and users of ICF-compliant architecture products.

Some common themes that emerged in the development of the ICF were the following:

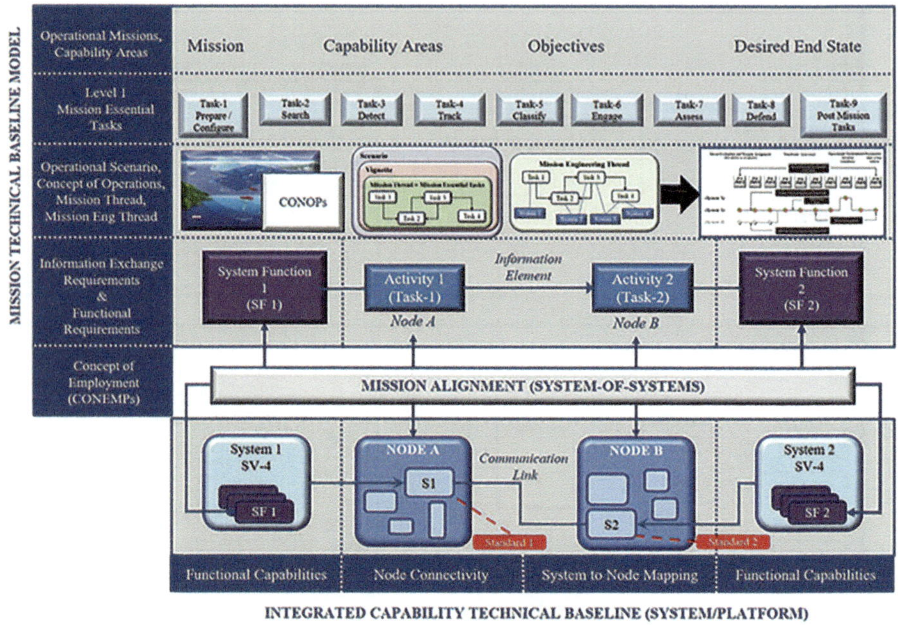

Fig. 7.11 System to mission integrated capability framework

- Effective architectures needed consistent taxonomies.
- Architecture data must be from authoritative sources.
- Architectures must be based on operational mission execution as defined by the end user.
- Mission and system architecture products must provide a mechanism to map systems to mission essential tasks and corresponding functions.

The ICF described in this section includes the following elements:

The ICF Metadata Model

Shown in Fig. 7.12, the ICF metadata model establishes the data structure, taxonomy, and relationships necessary for precise and comprehensive mission definition. It ensures consistency across developed products and enhances communication and collaboration using a standardized taxonomy derived from authoritative sources.

A foundational element of working across many mission-based products is a structured metadata structure to relate and store crucial information for easy retrieval in a tool-agnostic way. This framework develops contextual relationships across an enterprise, which allows for an effective working relationship and better understanding of roles and responsibilities when working between technical disciplines and tactical tribes. This metadata structure needs to support the alignment of crucial technical baseline elements and products to define the mission requirements and engineering development details across programmatic boundaries using the appropriate architecture products with an emphasis on effective mission success. The goal

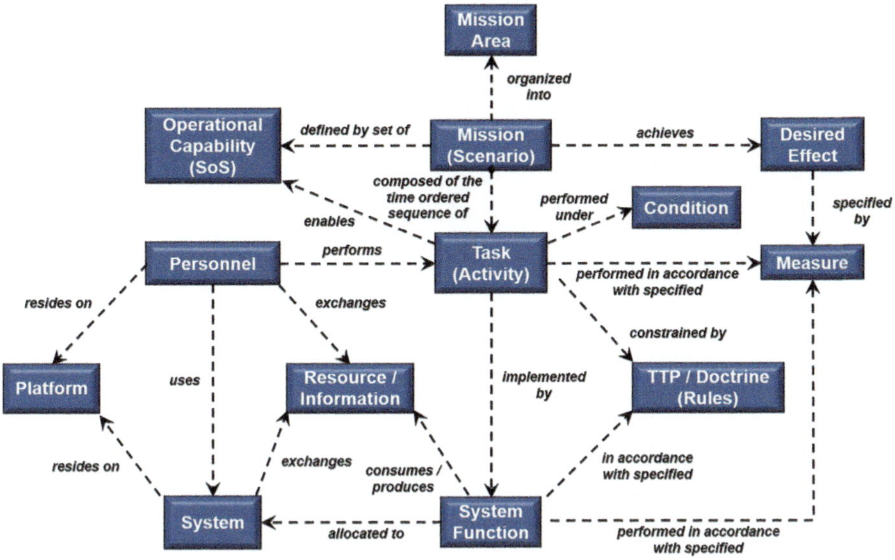

Fig. 7.12 ICF metadata model

is to define the required SoS linkages to provide an integrated warfighting capability and provide a technical reference document to drive synchronization.

The ICF Mission Model

The mission model delineates essential operational capabilities and leverages relationships defined in the ICF data model to specify functional, interface, and performance requirements for systems and platforms. The model is articulated through two primary components: Mission Technical Baselines (MTBs) and Integrated Capability Technical Baselines (ICTBs). While these artifacts can be documented on paper, their preferred format involves using models and an SoS modeling language such as SySML. Like the ICF data model, establishing a common taxonomy, incorporating authoritative data sources, and developing a unified methodology for architecture development is crucial for the ICF mission model. A good start toward resolving language and definition inconsistencies is to use the DoDAF Meta-Model (DM2) to identify the minimum set of data and relationships necessary for defining a mission-based architecture. This common taxonomy ensures shared understanding across stakeholders and facilitates seamless integration of architectures. In addition, authoritative sources need to be identified for each data type so that definitions and terms are consistent with operational practices and policies, such as using established universal joint tasks lists for operational tasks if they exist.

This framework defines a minimum set of products and views organized into MTBs and System/Platform Technical Baselines (S/PTBs) with each derived from the authoritative data model. MTBs capture the mission details and include stakeholders' objectives, desired effects, operational constraints, rules of engagement

(RoE), performance measures for success criteria, failure conditions, environmental factors, and threat descriptions. S/PTBs outline the systems and platforms involved, their functions, and interoperability requirements. Specificity in defining systems, platforms, and functionalities results in a comprehensive analysis of mission capability and interoperability. The final step involves integrating MTBs and S/PTBs to create the Integrated Capability Technical Baseline (ICTB). The ICTB segment of the mission model captures the following elements:

- tactical situation (TACSIT) or operational situation (OPSIT)
- mission threads with associated tactics, techniques, procedures (TTPs)
- mission engineering threads
- performance metrics
- information exchange requirements
- necessary system functionalities

In addition, this unified baseline provides detailed linkages between mission requirements, system capabilities, and operational tasks. It enables in-depth analysis of communication pathways, system performance, and mission readiness, which facilitates coordination among operational, training, testing, and acquisition communities. By linking operational tasks to system and operator performance, the ICTB supports gap analysis, prioritization of needs, and development of training curricula and testing scenarios. The S/PTBs in the ICTB delineate the capabilities of systems and platforms. These technical baselines specify functional capabilities, performance metrics, mappings of systems to nodes, networks, communication paths between systems and nodes, and applicable standards. The level of detail and accuracy in these technical baselines varies based on the specific requirements of users utilizing ICF artifacts. The system/mission alignment module utilizes relationships defined in the ICF data model to connect S/PTBs with mission models captured in the MTBs, fully describing one or more missions. This process generates a cohesive set of products that furnish the necessary context for

- defining missions
- developing mission success criteria
- analyzing capabilities
- evaluating mission capabilities
- identifying mission requirements

Using the ICF facilitates consistent and comprehensive definition of stakeholder needs leading to clarity from initial concept through testing and training phases for any new or upgraded system. By employing the same mission definitions from the outset, training and testing efforts better align with developmental and operational requirements, thereby validating systems and preparing operators effectively for planned missions. When user-defined operational requirements are used as captured through ICF mission models, the process of defining and designing system and platform requirements is greatly enhanced. Validated and comprehensive mission context is achieved, and planned operations crucial for system development can be defined. Moreover, using this approach establishes clear interoperability requirements, minimizing redundancy among existing capabilities. By integrating operational and system/platform metrics tied to mission objectives, the ICF facilitates analysis of MEI challenges and identifies operational gaps. Through this systematic approach, the engineer is able to track and resolve issues using an SoS framework, promoting efficient closure and continuous improvement of operational capabilities with a move away from the suboptimization of individual programs and systems. The desired outcome is the delivery of robust, mission-aligned architectures capable of supporting operational and strategic goals.

7.5.1 Governance Structure for ICF

The ICF is designed to support mission-level use cases across various stakeholders and functions, including acquisition, engineering, training, testing, and operations; however, it works across many levels of the enterprise, so a governance structure is necessary. While the specific implementation and involved stakeholders may differ depending on the use case, the following roles are generally essential: mission-level stakeholders and authorities required to develop and validate ICF mission model products, and system/program-level stakeholders who characterize current and planned system and platform capabilities.

Mission Model Authority
Mission model products are designed to comprehensively capture the required end-to-end operational capabilities and guide the derivation of related system and platform interfaces. Consequently, their development and validation require the involvement of multiple stakeholders. The ICF should establish the mission model authority, an authorized board led by representatives across the stakeholder community. This authority is responsible for developing and validating mission model products for each mission area. Additionally, the mission model authority collaborates with relevant program authorities to ensure the chosen level of alignment and granularity isolates necessary system capabilities. The exact composition of the mission model authority varies depending on the use case and the scope of the mission under evaluation.

Program Authority
The program authority is responsible for several key tasks. These include characterizing current and planned system and platform capabilities and capturing and validating these capabilities in the system/platform baseline products. Additionally, the program authority defines system capabilities in terms of system functions, ensuring traceability to system documentation. It also provides system-to-node mapping that details current and planned system installations. Finally, the program authority is tasked with validating the alignment of the program's capabilities with the associated mission models, confirming consistency and coherence is maintained between the systems and their intended operational use.

Technical Authority

The technical authority serves as the independent agent responsible for making decisions and providing recommendations on technical issues across various programs. This role includes ensuring technical feasibility during acquisition reviews, guaranteeing the interoperability of technical solutions across programs, and advising programmatic decision-makers. The technical authority reviews program architectures and technical requirements to identify and resolve conflicting requirements, clarify ambiguities, ensure consistency, and verify that the program architecture will meet system requirements.

At its core, the ICF provides the essential knowledge needed to integrate effective material solutions into mission architectures while minimizing the risk of interoperability issues. This ICF governance structure breaks down and manages the desired end-to-end operational capabilities into system and platform interfaces, behaviors (such as functions and timing), and environmental requirements across all levels of the enterprise. For transparency and understanding, current and planned system capabilities must be captured in a standardized format and taxonomy, allowing them to be linked to mission models. Once the SoS is comprehensively defined and traceable to the end-to-end mission requirements, it can be analyzed to estimate performance improvements and identify the necessary system and platform interface as well as the behavior requirements needed to achieve the required mission capabilities.

7.5.2 Mission Models

ICF mission model data elements and products capture operational requirements and identify the necessary capabilities at the mission task and interface levels. This detailed decomposition enables the discovery and definition of the required system and platform interfaces and behaviors needed to support end-to-end mission capabilities. These data elements will then be organized into ICF MTB views to facilitate mission analysis workflows and use cases.

The MEI approach aims to establish a valid multi-stakeholder authoritative mission architecture. Currently, many mission capability-based assessments define their mission requirements using unique methods and sources. Operational subject-matter experts decompose mission objectives into mission threads and operational views without an enterprise capability to share lessons learned or reuse data across assessments. Consequently, organizations often conduct multiple assessments on the same campaign, timeframe, and mission area simultaneously without cross-pollination. This lack of alignment in key mission assumptions leads to siloed acquisition recommendations that may be duplicative or non-interoperable.

At the top level, the ICF model is organized around critical mission areas, capability areas, operational objectives, and desired effects leading to the development of their associated mission essential tasks. This high-level mission decomposition establishes a common operational framework for MEI activities. Beyond this, ICF

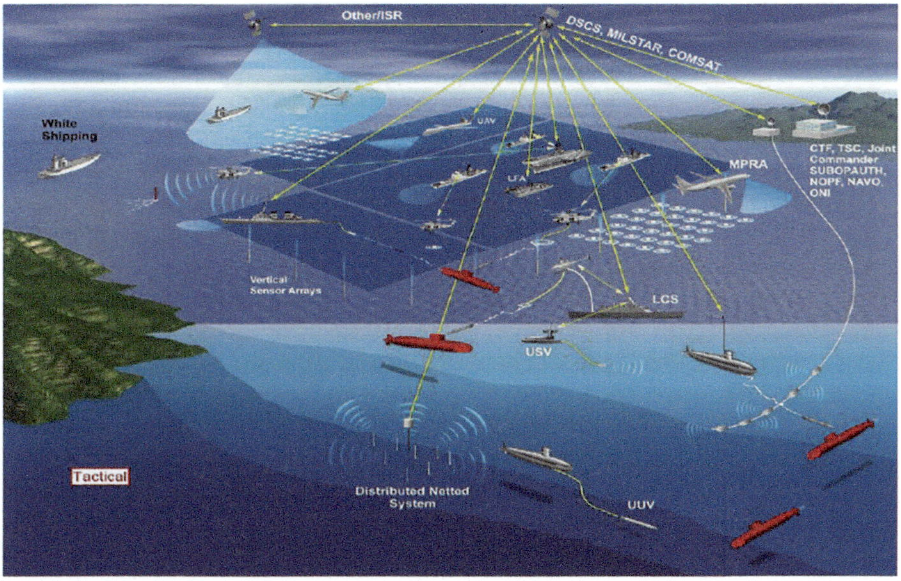

Fig. 7.13 Example operational view (OV-1)

users need flexibility to represent their specific operational contexts and perspectives. The ICF mission model guides technical implementation and enables the operational authority to represent specific assumptions and mission contexts at the level of detail required to achieve consistent operational assumptions across mission-level assessments.

The next layer of the ICF mission model produces a description of the operational scenario, detailed in textual Concept of Operations (CONOPs) and a high-level operational concept (OV-1) graphic. The CONOPs is a verbal or graphic statement of an end user's assumptions or intent regarding an operation or series of operations. Its intent is to provide an overall picture of the operation for added clarity of purpose. The CONOPs aims to address the following areas at a minimum:

- the problem being addressed (with threats defined by authoritative intelligence sources)
- the mission
- the stakeholders' intent (desired effect)
- an operational overview (including conditions and TTP/doctrine)
- tasks to be executed
- the roles and responsibilities of involved organizations

The OV-1 is a pictorial representation of the CONOPs, capturing high-level operations, organizations, and the geographical distribution of the scenario in a graphical format to facilitate human communication and detailed discussion of assumptions. Figure 7.13 shows an example of an OV-1.

Building on this top level of the ICF, the mission-model products then focus on the required capabilities. This product documents the capabilities necessary to achieve the desired mission effects, including their interrelationships, the organizations or performers responsible for providing these capabilities, and the tasks that facilitate their implementation. Next, the ICF mission-model product deals with the representation of the command structure or relationships among human roles, organizations, or types of organizations crucial to the architecture. It elucidates the diverse relationships existing within and between organizations, both internally and externally. These products lead to the development of the mission essential tasks or mission thread, which includes information about the rules model and the event trace description. This step captures operational doctrine and tactics, techniques, and procedures relevant to the specific scenario as well as setting parameters for the operational scenario outlined in the CONOPs/OV-1. This is achieved by describing a sequence of operational activities (tasks) and interactions involving associated organizational roles and responsibilities. With the mission essential tasks and systems/platforms identified, it becomes critical to identify the information exchange requirements as part of the mission model. Within DoDAF, information exchange requirements are documented in the operational resource flow matrix. These information exchange requirements are presented in tabular format and capture the mission thread, along with associated information elements. Each information exchange within the mission thread is detailed through one or more information exchange requirement, as defined by the information elements.

Information elements outline the information flow between tasks at a level of detail necessary for guiding, rather than limiting, solution architectures. Establishing a common ontology for information elements is crucial for effectively managing interfaces across system and platform boundaries. The final ICF mission model product is the functional requirements product. Additional data definitions in other products may be developed based partly on the previous mission-model products discussed and the data specified in this section. For instance, a capability phasing plan is essential for monitoring mission gap closure over time and for analyzing potential mission-level issues in the future. Other potentially necessary products include an operational resource flow description, organizational relationships identification, and operational activity model.

7.5.3 System/Platform Models

To achieve mission success in the MEI approach, the mission engineer must identify, engineer, track, and deliver system interfaces and behaviors for interoperability from a mission-based enterprise perspective. The ICF serves as this perspective, enabling systems and platforms to align themselves and facilitate the definition and evaluation of innovative capabilities that enhance mission performance. The ICF system/platform model products provide standardized templates and implementation guidance to ensure consistent representations of system and platform capabilities. These baseline products are designed to seamlessly integrate with mission model products so that required interfaces and behaviors for both existing systems and newly needed capabilities can be defined. ICF system/platform model data elements will be compiled into ICF ICTB views to support mission analysis workflows and various use cases. Time is an important factor in all ICF ICTB products. Operational missions rely on systems that span various stages of their lifecycle and maturity levels. The ICF model products specify when specific system functions will be available. Knowing this availability facilitates enterprise alignment of system development schedules and ensures comprehensive delivery of capabilities to the end user.

The ICF system/platform model product starts with functional capabilities representations to include systems functionality descriptions and system measures. These products document the capabilities delivered by a specific system, detailing its functions and the performance criteria associated with each function. As explained in the mission model section, system functions establish the connection between system and platform models and the mission model. In addition to detailing the functions delivered by systems, the duplicative or overlapping functional allocations need to be identified. This analysis is essential to ensure that all data and decisions within a platform and across the stakeholders originate from a single authoritative source. When two systems within the enterprise perform identical functions, there is a substantial risk of generating conflicting data or decisions, which can potentially affect mission capability. This issue underscores some of the root causes of modern integration and interoperability challenges.

Next, the ICF system/platform model provides the system-to-node mapping, which refines system functional assignments based on node suitability and identifies the overall system communication requirements. This mapping is documented in a system interface description. This description outlines the platforms (nodes) where a specific system is currently deployed, as well as planned installations over time. It also delineates the logical communication needs between systems, focusing on end-to-end requirements rather than the physical communication details. Detailed information regarding current and future system installations and their interfaces is typically found in system fielding plans and requirements documents. This data is best maintained and tracked by authoritative source of truth databases. To define future system capabilities and assess potential upgrades, the system interface description is necessary for outlining the platforms and systems with which the system must interact during planned missions and within specified timeframes. This architectural view identifies necessary system communications based on the functions performed by each system and the specific inputs and outputs required for each function.

The node-connectivity can now be documented as part of the ICF system/platform model and captured in the system resource flow description and systems resource flow. These products comprehensively capture the nodes (systems, platforms, and personnel) and their physical connectivity, communication paths, and protocols. Moreover, the node-connectivity product enables analysis of existing and potential future system interfaces that align with the required end-to-end communication requirements for mission execution. This analysis includes assessing interface performance requirements and identifying potential issues, such as bandwidth limitations and communication speed. Additionally, this analysis verifies that system function input/output (I/O) requirements can be satisfied through physical communication paths. It ensures alignment between system functionalities and the capabilities of physical communication interfaces.

The ultimate product to capture all pertinent standards applicable to the system in the system/platform model is the system/interface standards. This data is needed to specify the standards that any new or upgraded system must adhere to and must be meticulously detailed for each system and all planned installations within the operational capability. These standards and their implementations play a pivotal role in identifying interoperability risks associated with system interfaces. They ensure the system interfaces align with established standards, thereby mitigating potential compatibility issues across systems and platforms.

7.5.4 *System/Mission Alignment Models*

The ICF aims to facilitate the evolution of mission capabilities over time, adapting to changes in threat assessments, readiness evaluations, and defense priorities. As technology advances and new doctrines and training methods are adopted, certain aspects of the baseline architecture (as-is) may become obsolete, even as the overall mission remains relatively stable. The system/mission alignment models are designed to facilitate this ongoing evolution of mission readiness. Figure 7.14 illustrates the

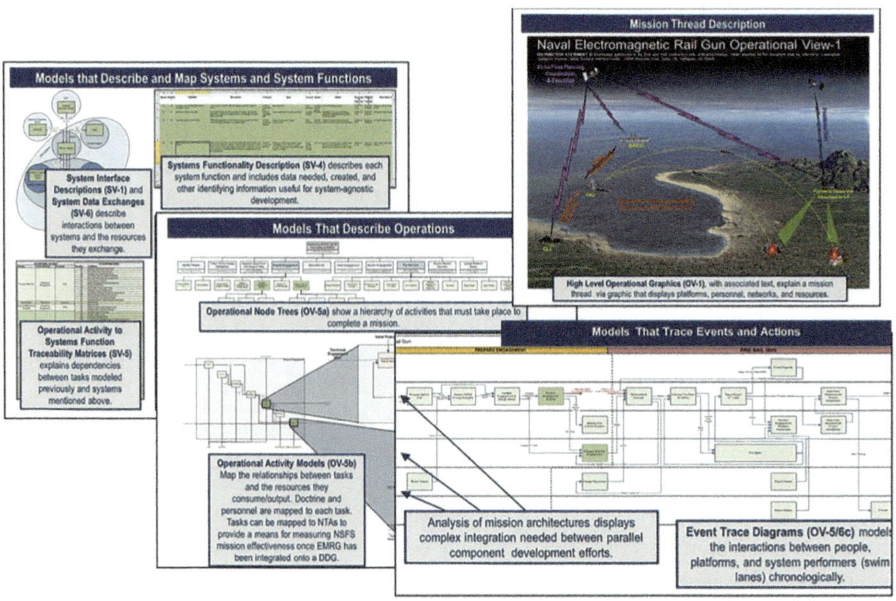

Fig. 7.14 Mission to system model representation

integration of mission model and system/platform model products discussed earlier, along with the system/mission alignment model. This alignment model ensures that systems and platforms are aligned with missions, identifying both existing and emerging interfaces and behaviors. The outcome is an SoS architecture that connects upward to the mission model and downward to individual system/platform models. ICF system/mission alignment data elements will be compiled into ICF ICTB views to support mission analysis workflows and various use cases.

The process of mapping systems and platforms to the mission thread pinpoints the interfaces necessary to support the mission model Information Exchange Requirements. This approach yields a clear set of interfaces and behaviors that can be traced back directly to operational requirements. Additionally, these products include information elements linked with each interface, along with alternative communication pathways. This comprehensive dataset forms the basis for assessing interoperability risks and exploring alternative interface designs that align with operational capabilities. Once established, the SoS supporting a specific mission model can undergo evaluation for performance and interoperability issues using the defined ICF relationships. The next section demonstrates how the mission engineer derives the performance metrics at both the individual system and SoS levels. This derivation starts from the mission-level requirements to facilitate evaluations of mission performance. Furthermore, this section on mission success criteria explores how standards and implementations specified at the system/platform model layer can be utilized to conduct interoperability assessments at the SoS level. By leveraging these standards, it becomes possible to assess whether the interfaces and interactions between systems within the SoS meet the required levels of interoperability, ensuring seamless integration and operation during mission execution.

7.5.5 Decomposing Mission Success Criteria

At the core of defining the mission success criteria, there are three fundamental levels of measures: mission, task, and system, illustrated in Fig. 7.15. Mission-level measures are determined by the intended outcomes of the mission in question. The mission owner specifies the outcomes and articulates the outcomes, which are broken down into specific and measurable components. The mission objective is translated into measurable criteria that define success, aligning with the overarching goals of the operation. At this mission level, the mission success criteria are the Measures of Success (MoS). Beyond outlining the overarching desired effect, it is crucial to define specific measures that align with anticipated conditions and threats during mission execution.

As outlined in the mission models section, the subsequent phase following the definition of the overall mission involves specifying the necessary tasks to accomplish the mission's objectives and establishing the interrelationships among these

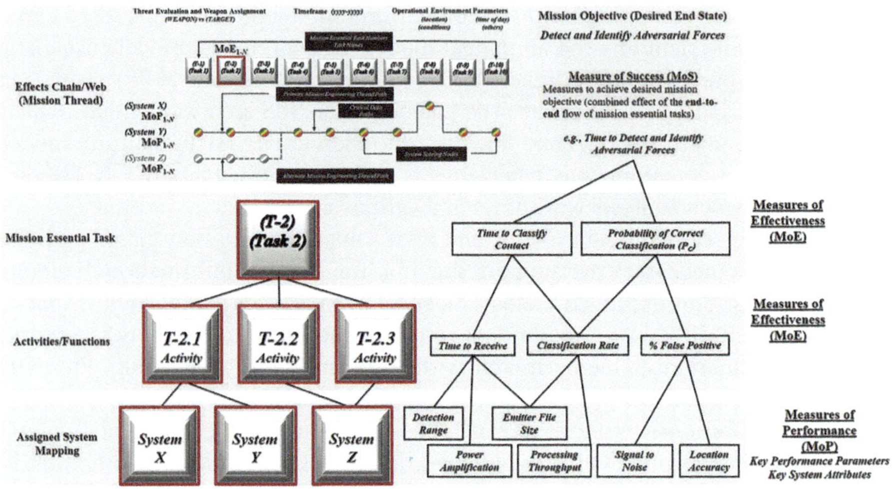

Fig. 7.15 Decomposition of mission success criteria

tasks. This process encompasses detailing the conditions and specific metrics associated with each task. These metrics, known as task measures or Measure of Effectiveness (MoE), serve as the performance indicators required to fulfill the overarching desired effect of the mission. The specific metrics for each task should be derived directly from the metrics established for the desired effect or the MoS. Note that the measures may not cover the correct detection and classification of every threat; this may seem contradictory to the goal of protecting the force from all threats. However, these measures can still be effective and align with mission objectives if adequate platforms are assigned to the same task with similar measures, and if the mission incorporates strategies to address missed threats through countermeasures. The overall mission objectives should be outlined comprehensively, examining the entirety of the mission's operational context and requirements.

The last tier of measures is the system/platform measures. Similar to mission measures, system/platform measures hinge on the particular mission and the sequence of system functions used to accomplish the mission essential tasks. The mission success criteria for systems are commonly known as Measures of Performance (MoP) and are represented by key performance parameters (KPP) and key system attributes (KSA). After breaking down the mission model into systems, their functions, and the connections between these functions (system threads), specific performance measures for system functions can be outlined. The specified system measures can additionally establish the parameters for design trade-offs.

The intricate connections link system/platform baselines to mission models for identifying necessary system interfaces that support the mission. Once identified, these interfaces can be engineered for interoperability through Interface Control Documents (ICDs) managed by the program owners of the interfacing systems. An ICD is an approved information assurance (IA) document detailing both physical and logical interfaces for a system. It specifies the parameters and standards required for each interface, encompassing IA aspects. These documents explicitly outline the standards and technical implementations necessary to ensure interoperability across all operational conditions defined at the mission model level, aligning with the systems' standards profiles. Connecting the standards views of two interfacing systems enables SoS interoperability assessments, ensuring uniform implementation across physical, data, application, and security layers. Besides verifying that standards are aligned the mission engineer must be sure that technical implementations are consistent. For instance, while two systems might share a message data standard with identical format and field names, discrepancies like different date formats could lead to interoperability issues. The systematic application of ICF across component systems within an SoS facilitates such comparative analyses.

7.5.6 ICF Applications

The ICF as described in this section has defined technical baseline concepts to methodically document and deconstruct mission requirements into functions, interfaces, performance criteria, and subsequent acquisition steps necessary to fulfill the mission. These technical baselines comprise a Mission Technical Baseline (MTB), along with three tiers of Integrated Capability Technical Baselines (ICTB-1, ICTB-2, and ICTB-3). Collectively, these baselines encapsulate the acquisition community's most comprehensive understanding of the mission requirements and the systematic arrangement of systems deployed to fulfill those requirements. Essentially, these technical baselines formalize the current program of record strategy for meeting military needs.

The MTB provides a detailed depiction of a specific military engagement, including its location, targeted objectives, desired outcomes, and the threats involved. This baseline is rigorously controlled across stakeholders to ensure consistency. Each MTB articulates the stakeholders' intent, enabling a thorough evaluation of each scenario's capability to achieve the intended outcomes. Collectively, the set of MTBs within a specific operational domain comprehensively delineates the desired effects, the designated targets, and the environmental conditions which include physical elements and potential threats that must be navigated and resolved.

Mission Technical Baseline (MTB)

- **Timeframe:** Define specific time of mission assessment to understand existing capabilities at that time.
- **Threat/Competition:** Understand the threat or competition to baseline how good SoS capability needs to be to accomplish mission success.
- **Location:** Define specifics about location of operations to include operational environment and conditions.

- **Stakeholders Intent/Mission Objectives:** Set mission objectives based on stakeholders' goals and values to understand desired end state for mission success which defines MoS.
- **Risk:** Provide a risk assessment to understand severity and likelihood that an event may occur during operational execution.
- **Measures of Success/Desired Effects:** Define MoS, MoE, and MoP leading to a desired effect or desired end state.

The ICTB is developed as a multi-level or multi-tier product, roughly in line with a multi-tier concept:

Integrated Capability Technical Baseline Level-1 (ICTB-1) describes a Concept of Employment (CONEMP) aimed at achieving specific effects as detailed in an MTB. It encompasses various approaches for engaging a specific target set, presenting multiple options for consideration. ICTB-1 also delineates how the CONEMP, typically involving kinetic missions but not exclusively, integrates with supporting non-kinetic operations that may precede or run concurrently with the mission. This baseline is the basis for discussions regarding the integration of CONEMP within broader campaign plans. This technical baseline maps to the as-is architecture as the preliminary solution architecture by considering the working arrangements between the system under design with other new systems as well as existing fielded systems.

ICTB Level 1—*Employment*

- **Concept of Employment**
- **Mission Measures of Effectiveness**

Integrated Capability Technical Baseline Level-2 (ICTB-2) aligns effectiveness criteria outlined in ICTB-1 CONEMP with role-specific performance requirements assigned to individual platforms. A mission area authority oversees configuration control for each ICTB-2. This technical baseline maps to the "as-programmed" architecture to consider specific performance requirements of the system under design with the other constituent systems as part of the solution architecture and represents the next level of maturity in the solution architecture.

ICTB Level 2—*Functional*

- **Operational Analysis (gaps)**
- **Platform Performance in Effects Chain/Web**
- **Technology Improvements**
- **Common SoS Tactical Enablers**
- **Interoperability Functions**

Integrated Capability Technical Baseline Level-3 (ICTB-3) represents the transition from mission area management oversight to program management control. Within the ICTB-3, comprehensive documentation outlines the governance and administration of the SoS. This technical baseline meticulously tracks requirements across all participating programs, ensuring that any changes to requirements within one program are effectively communicated and integrated across the entire system.

This technical baseline maps to the to-be architecture representing the final design of the system under design for the solution architecture and the mature state of the SoS capability.

ICTB Level 3—*Allocated*

- **Allocated SoS Performance:** Cumulative performance assessment of SoS capability after comparing MoP to MoE for entire mission engineering thread.
- **Detailed Interoperability Descriptions:** Assessment of both interface functionality between systems as well as delivery mechanism to include both syntactic and semantic interoperability.
- **Detailed Tactical Enablers Specifications and Standards:** Defined list of specifications and standards for the design and integration of the individual systems to form the SoS capability.

7.6 Foundational Architecture Tools

In the development of mission architectures for MEI, the complexity of SoS requires use of tools and methodologies to effectively model, analyze, and address interactions and interfaces between individual systems. Although not by far an exhaustive list, Functional Flow Block Diagrams (FFBDs) and N^2 matrices are two effective tools to capture system functionality and interdependencies, which enable engineers to visualize, design, and evaluate mission-critical operations.

- **Functional Flow Block Diagrams** are graphical representations of the sequential, time-based flow of functions in a system. They illustrate the system's logical behavior, identifying the sequence of operations and how functions are linked to achieve mission objectives. FFBDs are hierarchical in nature, starting from high-level mission objectives and progressively decomposing into detailed sub-functions.

  These FFBDs are essential tools in MEI, offering a structured way to represent the sequential flow of functions within a complex SoS. Each system function is represented by a block; arrows between blocks depict the order of execution, clearly illustrating how one function transitions to another across the SoS. Decision points introduce conditional paths, showing alternative flows based on specific conditions or outcomes. FFBDs also support hierarchical decomposition, enabling high-level diagrams to be broken down into detailed functions from SoS to system to components for in-depth analysis. These features provide mission engineers with a visual roadmap of operations, helping them understand the sequence of actions required to meet mission objectives. FFBDs are particularly valuable for identifying gaps or redundancies in processes, and verifying that all functional steps are accounted for and optimized. Additionally, they support testing and validation by serving as a foundation to verify that system behavior aligns with mission requirements. For instance, in a mission involving the deployment of an unmanned aerial vehicle, an FFBD can

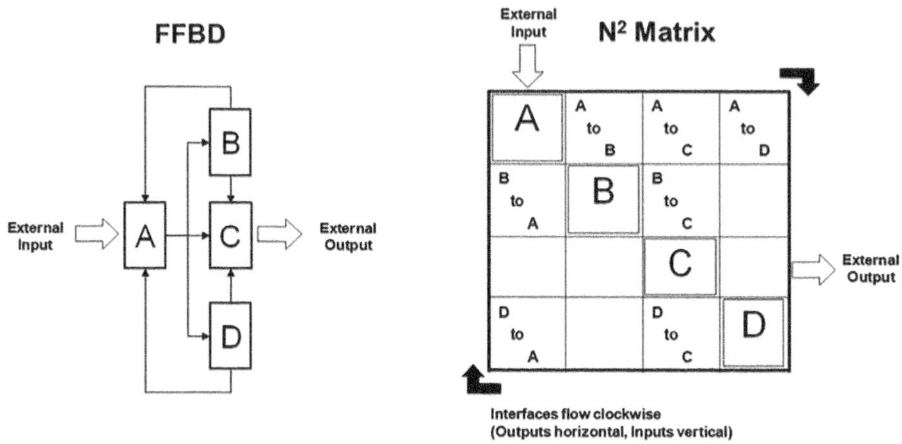

Fig. 7.16 Translation of FFBD to N^2 matrix

outline the sequence of operations from pre-mission checks to flight initiation, target acquisition, data collection, and return to base. With this clear visualization, mission planners can effectively design and analyze the end-to-end operation. Figure 7.16 illustrates the use of a generic four system FFBD translation into a representative N^2 matrix.

- **N^2 Matrix** is a tool used to model and analyze the interactions and interfaces between components or systems within a larger SoS framework. It is a square matrix that displays the relationships between elements, with the same list of elements along both the rows and columns. The diagonal cells represent the system components and therefore determine the size of the matrix, while the off-diagonal cells indicate the interactions or interfaces between these components. N^2 matrices are powerful tools in MEI, designed to model and analyze the interactions and interfaces between systems in an SoS and system components at the lower levels of functional decomposition. Each row and column of the matrix corresponds to a specific system or component, while the off-diagonal cells define the type and direction of interactions, such as data exchange, internet protocols, communication paths, control signals, or power flow. Annotations or arrows in these cells indicate whether the interactions are unidirectional or bidirectional. N^2 matrices are highly scalable, making them suitable for analyzing systems ranging from small systems to complex SoS. Their key benefits include providing a clear visualization of interfaces, helping engineers understand how systems interact, and identifying critical interfaces that require careful management. Additionally, N^2 matrices reveal dependencies between systems, ensuring that changes to one component do not disrupt others. This in turn supports risk mitigation by mapping potential points of failure. For example, in a missile defense mission, an N^2 matrix illustrates the interactions between radar systems, command and control units, communication links, and interceptor platforms. Off-diagonal cells would then capture details such as radar detection data sent to the command center or launch commands issued to interceptors, enabling engineers to ensure seamless and reliable integration.

The combined use of FFBDs and N^2 matrices provides a comprehensive approach to MEI by addressing both functionality and system interactions. FFBDs focus on sequencing and functionality, helping engineers visualize the logical progression of operations and ensuring each function aligns with mission objectives. In contrast, N^2 matrices emphasize interfaces and dependencies, highlighting how systems communicate and collaborate to ensure seamless integration and prevent bottlenecks. Together, these tools allow mission engineers to model both the flow of operations and the interactions between systems, creating a holistic view of the SoS. For instance, in a humanitarian aid mission involving unmanned aerial vehicles, logistics systems, and ground teams, FFBDs can define the sequence of operations, such as unmanned aerial vehicle reconnaissance, data analysis, and deployment of aid. Simultaneously, an N^2 matrix can map data exchanges and communications among these systems, ensuring smooth coordination and effective information flow. This integrated approach is critical for managing complex, multidisciplinary missions where even minor misalignments in functionality or interfaces can jeopardize mission success. By leveraging the strengths of both FFBDs and N^2 matrices, mission engineers can design, integrate, and validate SoS with precision and reliability.

7.7 Future Direction of Architecture Tools

The future of reference architectures and frameworks will be influenced by technological progress and changing business needs, with several key trends expected:

- **Universal Interoperability and Open Standards:** As digital ecosystems become more interconnected, developing universal interoperability driven by open standards will be essential. These standards will ensure seamless data exchange and integration across diverse systems and industries.
- **Adoption and Integration of Emerging Technologies:** Expectations of future frameworks are the probable integration of AI solutions, blockchain, digital twins, and Internet of Things, leading to more adaptable and intelligent models. This integration will improve decision-making, security, and transparency within digital ecosystems.
- **Modularity and Customization:** To tackle scalability, composability, and integration challenges, a shift toward more modular and customizable frameworks is anticipated. This approach will enable selective adoption and easier scalability and will align more effectively with the unique needs of individual organizations.
- **Governance and Compliance:** As regulatory complexities increase, the next generation of frameworks will incorporate advanced governance and compliance features. These improvements aim to streamline adherence to regulations and mitigate the risks associated with non-compliance.

- **Evolutionary Agile Development:** Embracing an agile approach to continually develop these frameworks will help maintain their relevance and effectiveness as industry landscapes and technological innovations evolve.

By tackling these challenges with targeted strategies and embracing these emerging trends, organizations will be better able to fully leverage reference architectures and frameworks, maintain competitiveness, and expand adaptability in a rapidly changing business and operational environment.

Having established a foundation in **architectural design and frameworks**, we now move from structure to execution by applying these tools within the **Mission Engineering and Integration (MEI) approach**. Chapter 8 serves as a pivotal point in understanding MEI by focusing on how to systematically assess "as-is" SoS configurations to identify operational gaps and needs. From there, we transition into developing "to-be" configurations to evaluate the impact of potential solutions and determine the best way to integrate them into a complete, workable package. By leveraging the MEI approach, the engineer can effectively design, analyze, and field operational capabilities that enhance functionality, effectiveness, and mission success.

References

1. Department of Defense Architecture Framework (DoDAF). (2024). Chief Information Officer Website. Retrieved from https://dodcio.defense.gov/Library/DoD-Architecture-Framework/dodaf20_viewpoints/
2. ISO/IEC/IEEE 42010:2022. *Software, systems and enterprise - Architecture description.*
3. North Atlantic Treaty Organization (NATO) Architecture Framework (NAF). (2024). North Atlantic Treaty Organization Website. Retrieved from https://www.nato.int/nato_static_fl2014/assets/pdf/2021/1/pdf/NAFv4_2020.09.pdf
4. The Open Group Architecture Framework (TOGAF). (2024). The Open Management Group (OMG). Retrieved from https://pubs.opengroup.org/architecture/togaf9-doc/arch/welcome.html
5. Unified Architecture Framework (UAF). (2024). The Open Management Group. Retrieved from https://www.omg.org/spec/UAF/1.2/DMM/PDF
6. Patton, J., & Moreland, J., Jr. (2023). Chapter 32. Reference architecture. In D. Verma (Ed.), *Systems engineering for the digital age book.*
7. Rechtin, E. (1991). *Systems architecting: Creating & building complex systems.* Prentice-Hall.
8. Zachman, J. (1996). *Enterprise architecture: The issue of the century*
9. Zachman, J. (2024). *The Zachman framework for enterprise architecture.* Zachman International Website. Retrieved from https://zachman-feac.com/zachman/about-the-zachman-framework

Chapter 8
Mission Engineering and Integration Approach

8.1 Understanding the Problem

To provide definitive focus for the execution of the Mission Engineering and Integration (MEI) approach, the development of a clear, concise, and precise problem statement is essential. At the foundational state, the problem statement answers the question "What is the nature of our problem?" The problem statement should provide factual information and contain root cause of the situation while setting the fundamental contextual hypothesis in complex and ambiguous environments. Moderate differences in the hypothesis can drive substantial differences in the resultant operational plans. Revisiting problem statements may be required for refinement or validation as operations unfold and the operational environment becomes better understood or changes.

It is important to distinguish between symptoms and root causes when defining problems. Before applying the MEI approach, all stakeholders should ask, "What problem are we really being asked to solve?" The root cause is not always articulated in top level guidance. Identifying root causes should begin the stakeholder discussions at the top level and then drive down to the operational levels. Understanding the problem enables the identification and definition of the desired goals and objectives as part of determining mission success. If the objectives and desired end state are clearly understood, a Measure of Success (MoS) and mission essential tasks can be articulated to address the solution of the defined problem rather than addressing only symptoms of the problem.

Mission capabilities are composed of system-of-systems (SoS) with a diverse array of connected elements creating a density of linkages and complexity. Because of this architecture characteristic, a small fluctuation in one constituent system may cause extreme emergent behaviors across the SoS exhibiting unpredictability. However, complex is different than being complicated and requires a different approach when trying to understand performance and behaviors. Complicated

J. D. Moreland, Jr., *Mission Engineering and Integration*,
https://doi.org/10.1007/978-3-032-32132-9_8

systems usually have many parts with defined linkages, both physical connection points from system to system as well as simplified interfaces for information sharing across system boundaries. In simple machines, the turning of a single part causes another joined part to turn, demonstrating the dependency between the individual parts to accomplish tasks.

The design of complicated systems like an automobile braking system may be confusing but can be decomposed into its individual pieces with deterministic relationships. This structured breakdown allows for a higher probability of predicting the performance when parts are operated or altered to gain critical insights on sensitivity parameters. On the other hand, complexity is identified by a drastic increase in the interactions and interdependencies between the parts. This unpredictable growth in interdependencies is driven by the exponential growth of probable paths due to high-density interactions which may alter predictions. Due to dynamics in inputs, conditions, constraints, and bounds on mission success criteria, a combinatorial explosion is likely as both the means and ends in obtaining results grow rapidly with many probable paths. Combinatorial explosion refers to the rapid increase in the number of possible combinations as the size of a problem or the number of elements involved grows. In the context of MEI, the relationship between combinatorial explosion and mission complexity is significant. The challenges associated with combinatorial explosion can affect various aspects of mission planning, design, and execution. Addressing the challenges associated with combinatorial explosion in MEI often involves leveraging optimization algorithms, artificial intelligence, and advanced planning techniques. It also emphasizes the importance of interdisciplinary collaboration among mission planners, system designers, and algorithm developers to find solutions that balance mission complexity with feasibility and efficiency. In addition to this complexity, SoS exhibit nonlinear change and behavior, and therefore are unpredictable and defy our understandings on composability and scalability.

8.2 Historical Perspective and Evolution of MEI

As of the writing of this book, Mission Engineering (ME) can refer to different concepts depending on the context. In the context of Aerospace or Systems Engineering (SE), ME generally involves the process of designing and optimizing complex systems to meet specific mission requirements. It can apply to a wide range of fields, including space exploration, defense, and other critical missions. The origins of ME at NASA can be traced back to the early days of the space exploration program, particularly during the Mercury (1958–1963), Gemini (1961 and 1966), and Apollo (1961 and 1972) missions.

During the mid-twentieth century, as nations began launching satellites and manned missions, the need for systematic planning and engineering for mission success became evident. The Apollo program, which landed the first humans on the Moon, is a landmark example of ME. Engineers and scientists had to design and coordinate complex systems, including spacecraft, launch vehicles, and life-support systems, to achieve the ambitious goal of lunar exploration. The development and operation of the Space Shuttle marked another phase in ME based on the design for reuse, adding new layers of complexity to the engineering challenges. In addition, mission planning had to account for the intricate logistics of transporting payloads, conducting experiments, and facilitating human spaceflight.

Over time, ME has become closely associated with SE principles. This involves an interdisciplinary approach that considers the entire system's lifecycle, from concept to retirement. Systems engineers work to optimize the performance, reliability, and cost-effectiveness of systems designed for specific missions. This emergence to modern applications continues to evolve in response to advances in technology and the expanding scope of space exploration, defense missions, and other complex endeavors. This includes unmanned missions, planetary exploration, satellite deployments, and more. In fact, the defense applications involve the execution of multi-missions simultaneously, which has added another layer of complexity to the single mission cited above at NASA.

MEI originated in 2010 within the Navy's Systems Commands (SYSCOM), the Integration and Interoperability Initiative, in response to the Chief of Naval Operations' concern over the effective execution of mission engineering threads across fleet forces [LEADINGEDGE14]]. Addressing this concern required a better understanding of interactions and dependencies among systems and SoS required to successfully execute operational missions. During this time, Dr. James D. Moreland, Jr. was asked to lead a multi-SYSCOM team of experts to assess the root cause and determine technical fixes to identify operational gaps. Starting in 2014, Dr. Moreland as the Executive Director—Naval Warfare under Office of the Secretary of Defense Acquisition, Technology, Logistics (OSD (AT&L)) developed the MEI discipline that links engineering rigor to operational context necessary to identify requirements and material solutions in alignment with the mission success criteria for multi-domain, multi-mission, and across joint/coalition forces. In 2020, OSD Research and Engineering (R&E), published the Mission Engineering Guide (*MEG*) *v1.0* and defined MEI as the "deliberate planning, analyzing, organizing, and integrating of current and emerging operational and system capabilities to achieve desired warfighting mission effects [MEG1.0]."

Maturing the Mission Engineering Guide further, OSD R&E released the *MEG v2.0* in December 2023. Also in 2023, the Defense Acquisition University (DAU) in their forthcoming fundamental MEI credential, recognized MEI as "a systematic, quantifiable, and timely analysis of operational planning concepts and emerging threats to evolve end-to-end, SoS architectures and capability attributes to define requirements. It establishes data driven technical architecture baselines and builds upon a SE framework requiring authoritative data and a common framework."

While the discipline was originally developed within the Department of the Navy as part of the Integration and Interoperability Initiative in 2010 and then as MEI in AT&L in 2014, it now resides within A&S and R&E. The application of this mission-based capability discipline is intended to integrate the operational communities, academic institutions, and industry. Full stakeholder integration ensures new systems and Concept of Employment (i.e., technologies, and Tactics, Techniques, and Procedures) are engineered and assessed in alignment with accurate operational needs and meet performance requirements in isolation as well as part of the complex SoS required for desired operational capabilities. Therefore, MEI uses systems and SoS in an operational mission context to ensure the right system is built correctly and guides capability maturation tied to operational mission needs. Given the complexity of current and future weapons, the engineering community must ensure the mission and SoS levels are thoroughly documented to drive appropriate design from concept through sustainment.

Effective use of MEI requires skilled practitioners and conscientious leaders to ensure the MEI team has access to critical data. Skillful application of MEI, in collaboration with other relevant tools and processes, is critical to addressing the National Security Strategy and executing mission success for national security. Title X of the U.S. Code gives service commanders authority over decisions related to military acquisitions and performance requirements for their individual branches (e.g., Army, Navy, and Air Force). This means each branch has control over the specific tools, technologies, and standards it needs to achieve its mission. However, today's military operations require a joint force—all branches working together seamlessly across land, maritime, air, space, and cyber operational domains. For this level of coordination, the equipment and systems each branch uses must be fully compatible and interoperable with those of the other branches, regardless of individual acquisition decisions. At the writing of this book, the MEI approach is employed on a regular basis across the DoD as well as interests from multiple allied partners to help ensure acquisition and training decisions are appropriately data-informed and support interoperability and effective mission execution for combined force operations.

8.3 Definition and Scope

Definition

MEI is an interdisciplinary approach that involves the systematic and holistic design, analysis, planning, and execution of complex missions. These missions can span various domains, including aerospace, defense, space exploration, healthcare, business, and more. The primary goal of MEI is to ensure the success of a mission by considering a wide range of factors, including mission objectives, system architecture, interoperability, risk management, resource optimization, and adaptability to dynamic environments. The original definition of MEI within the Department of the Navy (DoN) in 2010 and during the initial introduction of the approach to the DOD in 2014 was as follows:

> The planning, analyzing, organizing, and integrating current and emerging operational concepts for the purpose of evolving the end-to-end operational architecture and capability attributes, across the DOTMPLF-P spectrum, including anticipated BLUFOR and OPFOR

> behaviors, that are needed to inform the communities of interest involved in fulfilling mission needs statements. [MORELAND19]

The corresponding methodology used to execute this definition:

- Warfighting Capability Baseline (WCB):
  - Assessment agenda: Select and prioritize existing effects/kill chains experiencing problems within the "as-is" SoS configuration. This step identifies specific weapon-target pairings composed of specific weapons and their intended targets to form the basis for kill chain assessments.
  - Kill chain assessment: Construct the kill chains within Tactical Situations (TACSITs) to capture the sequence of mission essential tasks and decision nodes required to execute a specific mission. These kill chains are linked to TACSITs, which define the operational context, including threats, platforms, environmental conditions, and communication pathways. Performing kill chain assessments using a root cause failure process involves analyzing weak or broken links within the chain to identify underlying issues that hinder mission success. This method traces failures back to their root causes, such as system interoperability issues, gaps in communication links, or limitations in platform capabilities, enabling targeted solutions to improve overall kill chain performance.
  - Developing scoring criteria: Scoring criteria are established for each mission essential task, linked to critical measures of mission success. A color-coded system (GREEN, YELLOW, RED, WHITE) is used to evaluate performance against these criteria.
  - Scoring kill chains: Principal investigators evaluate test data to assign scores to individual tasks within a kill chain. Weak or broken links in the chain highlight areas requiring improvement.
- Capability Solution Management (CSM):
  - DOTMLPF-P Solution Development: Systematic evaluation of operational gaps to identify comprehensive solutions across multiple domains and disciplines. Each element of the DOTMLPF-P (Doctrine, Organization, Training, Materiel, Leadership, Personnel, Facilities, and Policy) framework is analyzed to address capability deficiencies, ensuring that proposed solutions align with end-user needs and operational priorities. This process considers trade-offs and synchronization across the spectrum, resulting in integrated recommendations that enhance mission effectiveness while optimizing resources and readiness.
  - Integrated Capability Package (ICP): Develop ICP with a complete set of solution recommendations to close the associated operational gaps identified by the WCB.
- Office of the Chief of Naval Operations (OPNAV)/Assistant Secretary of the Navy for Research, Development, and Acquisition (ASN(RDA)) Resourcing Decisions:

- Prioritize and endorse gap closure solutions developed in CSM phase and documented in the ICPs for submission to OPNAV to incorporate ICP prioritization into Navy budget development plans.
- Monitor status of ICPs for inclusion in Navy budget and resulting funding resources to incorporate operational fixes.

In October 2023, this definition evolved into the following text with the release of the *MEG v2.0* by the Office of the Secretary of Defense Requirements and Engineering (OSD R&E) replacing *MEG v1.0* from November 2020:

> An interdisciplinary process encompassing the entire technical effort to analyze, design, and integrate current and emerging operational needs and capabilities to achieve desired mission outcomes. [MEG2.0]

The main difference between *MEG v1.0* and *MEG v2.0* is the expanded scope and methodology in the matured *MEG v2.0*. *MEG v2.0* introduces a more robust, iterative approach to MEI as compared to *MEG v1.0*, emphasizing the integration of digital engineering tools, reusable data models, and adaptive frameworks to address emerging threats. Additionally, *MEG v2.0* places greater focus on collaboration across operational, acquisition, and engineering communities to enhance mission-driven capability development and SoS analysis. From the beginning, the vision and execution of MEI have focused on a structured methodology for decomposing missions into their constituent parts to evaluate and improve the performance of systems and SoS within complex operational contexts. This interdisciplinary approach integrates engineering, operational analysis, and stakeholder collaboration to ensure that technology, capabilities, and systems are tailored to meet specific mission outcomes.

Similar to the original MEI definition, the *MEG v2.0* definition is focused on mission capability in a threat-informed context to evaluate mission effectiveness of the proposed solution sets, advise on development of requirements, and inform technology investment decisions. This is accomplished by identifying enhanced capabilities, technologies, system interdependencies, and architectures to determine and close operational mission gaps. The new process in *MEG v2.0* is an iterative, data-driven approach to decompose and characterize the mission and problem set, evaluate a potential mix of solutions, identify measurable trade-offs, and then inform the implementation of the most promising solutions. The process has five broad phases: (1) identifying mission and purpose; (2) characterizing the mission and operational parameters; (3) building mission architectures; (4) evaluating solution trades and mission efficacy; and (5) informing the integrated mission solution. Figure 8.1 provides a high-level mapping of the *MEG v2.0* phases on the left to the *MEI Guidebook* 10-step process on the right. The 10-step process will be covered in detail later in this chapter.

Scope

The scope of MEI is broad and encompasses a wide range of activities and considerations related to the design, analysis, planning, and execution of complex missions across operational domains and industrial sectors. In addition, MEI reflects an

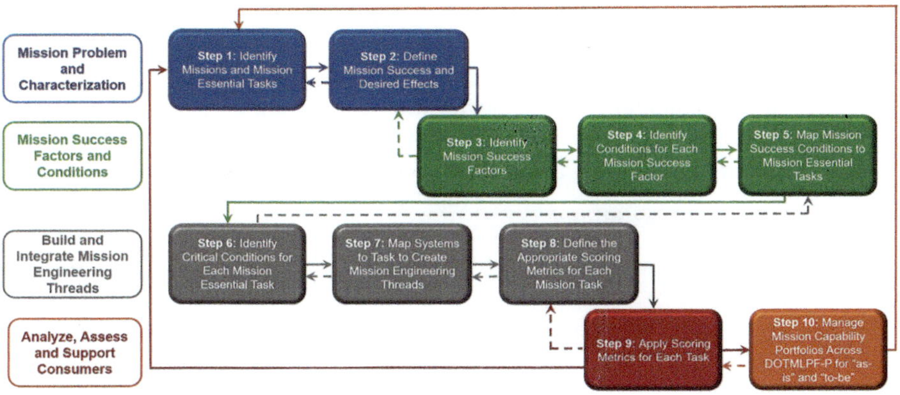

Fig. 8.1 Mapping of MEI guidebook to *MEG v2.0*

interdisciplinary nature, drawing upon principles from SE, project management, information technology, and other relevant fields. It emphasizes a holistic approach to mission success, considering not only the technical aspects but also the human, organizational, and environmental factors that can influence mission outcomes. Key aspects that define the scope of MEI:

- Systematic Mission Analysis:
  - Understanding and defining mission objectives, goals, and requirements
  - Conducting a thorough analysis of the mission context, including environmental factors and constraints
- System Architecture and Integration:
  - Designing the architecture of systems involved in the mission
  - Ensuring effective integration of systems and SoS for seamless operation capability
- Interoperability:
  - Establishing interoperability between different systems, technologies, and entities
  - Implementing standards and protocols to enable communication and collaboration
- Risk Management:
  - Identifying, assessing, and managing risks associated with the mission
  - Developing strategies to mitigate potential challenges and uncertainties
- Data Management and Analysis:
  - Managing data generated during the mission, including collection, storage, and analysis

  - Utilizing data for decision-making and improving mission outcomes

- Mission Planning and Execution:
  - Developing comprehensive plans for mission execution
  - Implementing strategies to achieve mission objectives while considering adaptability to changing conditions
- Adaptability and Flexibility:
  - Designing systems and plans that can adapt to unforeseen changes and challenges
  - Incorporating flexibility to respond to dynamic mission-operational environments
- Collaboration and Stakeholder Engagement:
  - Facilitating effective communication and collaboration among diverse stakeholders
  - Engaging with teams, departments, and external partners to ensure a coordinated approach
- Ethical and Legal Considerations:
  - Addressing ethical considerations in mission planning and execution
  - Ensuring compliance with legal and regulatory frameworks relevant to the mission
- Emerging Technologies and Trends:
  - Staying informed about and integrating emerging technologies that can enhance mission capabilities
  - Considering trends and advancements that may impact the future of MEI
- Continuous Improvement and Lessons Learned:
  - Evaluating mission outcomes and performance for continuous improvement
  - Documenting and applying lessons learned from past missions to enhance future endeavors

Currently, there are several limitations associated with executing MEI in terms of workforce education and developing and using advanced tools. One significant challenge lies in the availability of skilled personnel with the requisite expertise in MEI methodologies and techniques. The shortage of professionals adequately trained in MEI principles hampers the effective implementation of MEI. Furthermore, the complexity of MEI requires continuous education and training to keep pace with evolving methodologies and technologies. In addition, the development and adoption of advanced tools, methodologies, and technologies for MEI are impeded by various factors, including limited funding, interoperability issues among existing systems, and the need for specialized training to effectively utilize these tools. This includes a lack of standardized frameworks and guidelines for integrating advanced tools into MEI practices, resulting in inconsistency and inefficiency in their implementation. Addressing these limitations requires concerted efforts to invest in workforce development, provide comprehensive training programs, and promote collaboration among stakeholders to overcome barriers adopting advanced tools and technologies in MEI practices. By addressing these challenges, the DoD enhances its capacity to leverage MEI for more effective mission planning, execution, and integration across the enterprise.

Due to current limitations, the major focus areas for the application of MEI are devoted to the integration, coordination, and alignment among various systems, platforms, and capabilities deemed essential for critical national security missions. MEI efforts prioritize means for addressing challenges related to interoperability, SoS integration, capability deficiencies, and resource optimization. By identifying and rectifying these critical areas, MEI enhances the overall effectiveness and efficiency of operations while ensuring that resources are allocated and utilized efficiently to meet mission objectives. This targeted approach allows MEI efforts to address key challenges while fostering synergy and cohesion across the diverse spectrum of missions. Some characteristics driving the use of MEI involve:

- Complex, joint, and high-priority missions and problem sets
- Extensive collaborative and dynamic engineering analysis requirements
- High level of risk or uncertainty in solution decisions
- Emphasis on optimization or more than one DOTMLPF-P area
- Continuing evolution of systems, networks, and capabilities
- Simultaneous M&S of a wide range of variables to understand the behavior of systems operating within a broader system
- Rigorous interface design and management of the capability across multiple platforms or locations, which might involve combinations of military and industry standards at multiple layers

8.4 Foundational Concepts

Understanding and managing the interdependence of systems are crucial in MEI. It allows for efficient collaboration and coordination, ensuring that each system contributes effectively to the overall success of the mission. While the systems operate independently in their specialized functions, their interdependence enhances the overall capabilities and performance of the entire mission. The challenge lies in designing and orchestrating these systems to work cohesively toward common mission goals. While the systems are functionally dependent on each other for specific tasks, they retain independence in their core functions. Some foundational concepts from systems theory that have played a crucial role in the definition and formation of the MEI discipline to optimize effectiveness, enhance resilience, and foster innovation are:

- ***System Darkness:*** This principle states that complete knowledge about a complex system is not achievable. The system can be defined only through its representation (model), which is always incomplete; therefore, no team member has a complete picture of everything associated with the system. If the team does not plan and decide together, there is a high probability that the design will be incomplete since individual team members have an incomplete mental model of the entire system.
- ***Complementarity:*** Truths about a system can be captured only through multiple perspectives (models) as different perspectives reveal truths about the system which are neither entirely independent nor entirely compatible. Another dynamic of the concept of complementarity is the realization that different perspectives of a single system can appear to have different or even contradictory properties. In the quantum theory domain, complementarity is an essential principle as it relates to the Heisenberg Uncertainty principles [HEISENBERG27]. This relates to the inability to predict both the position and velocity of a particle simultaneously.
- ***Self-Organization/Governance:*** The principle of self-organization is a concept derived from complex systems theory, where systems can spontaneously organize and adapt without external control to achieve specific goals or respond to changes in their environment. Self-organization allows mission architectures to dynamically reconfigure in response to changing conditions, threats, or opportunities which yields adaptive improvement of mission resilience.
  The application of the self-organization principle in MEI aligns with the goal of creating adaptive, resilient, and efficient systems capable of operating in complex and dynamic environments. Self-organization refers to the spontaneous emergence of order, patterns, or structures in a system without external guidance or central control. Instead, these systems exhibit the capacity to organize themselves based on local interactions and feedback loops among their components.
- ***Minimum Critical Specification:*** In the context of MEI, defining the minimum critical specifications involves identifying the mission essential tasks, requirements, and parameters for mission success. These specifications form the baseline criteria that must be met to ensure the mission achieves its objectives. By systematically defining and prioritizing these minimum critical specifications, MEI creates a solid foundation for the design, development, and execution of successful missions as well as the mission engineering threads/webs. The specifications serve as criteria for evaluating the mission's performance and ensuring its overall success through the development of the Measures of Success (MoS), Measures of Effectiveness (MoE), and Measures of Performance (MoP).
- ***Synergism:*** Synergism refers to the interaction or cooperation of two or more systems to produce a combined effect that is greater than the sum of their individual effects. In other words, it is the idea that the collaboration of different elements results in an outcome that is more powerful, efficient, or effective than what each element could achieve on its own. This cooperative interaction is enabled through the design and implementation of open interfaces.

- ***Open Interface:*** An open interface concept pertains to the ability to design and construct systems and SoS with permeable boundaries. Opposite to closed systems, natural open systems have permeable boundaries enabling them to exchange energy and information with other systems and with the surrounding environment to allow for learning, adaptation, and evolution. In engineered SoS, interfaces are governed through adherence to open standards and specifications.
- ***Conservation:*** Conservation enables the SoS to conserve resources by optimizing performance and eliminating waste. The idea is to ensure that the overall SoS maintains its functionality, effectiveness, and resilience over time by optimally utilizing resources. Conservation can be applied to various aspects of an SoS, including resources, energy, information, and interoperability. Unnecessary duplication of resources leads to waste reduction and hinders efficiency.
- ***Modularity:*** Modularity focuses on the development of modular open systems and SoS architectures allowing plug and play, mix and match, ability to evolve, and be compatible. Modularity is a design technique that involves breaking down a system or SoS into smaller, independent, and interchangeable modules or components. These modules encapsulate specific functionalities, and their interfaces allow them to interact with each other in a standardized and well-defined manner.
- ***Emergence:*** The concept of emergence refers to the phenomenon where complex and often unexpected patterns, behaviors, or properties arise from the interactions of systems within an SoS. In emergent systems, the collective behavior of individual systems gives rise to properties or behaviors that cannot be easily predicted or directly derived from the properties of the individual systems.
- ***Symbiosis:*** Mutualism is a type of symbiotic relationship between two different systems in which both systems benefit from the association. This form of interaction is characterized by cooperation and interdependence, where each system provides some form of advantage to the other. Successful development and sustainment of an SoS are largely dependent on symbiotic collaboration among the stakeholders and systems comprising it.
- ***Re-Configurability:*** Re-configurability refers to the ability of a system, device, or component to be easily modified, adjusted, or adapted to changes in its environment, requirements, or operational conditions. A re-configurable system is designed to allow for dynamic changes in its configuration, structure, or behavior without significant downtime or manual intervention.

  The ability to modify the architecture based on dynamic operational conditions and environmental changes is necessary to provide relevant capability without external intervention. An engineered SoS adaptation or reconfiguration is largely dependent on its open architecture design, open standards and specifications adherence to define its open interfaces, and adequacy of its technical baseline and configuration control documents.
- ***Modular Open Systems Architecture/Weapon Systems Open Architecture (MOSA/WOSA):*** In the context of MEI, MOSA plays a crucial role in ensuring interoperability, flexibility, and scalability of the constituent systems within the larger framework. MOSA is a design approach that emphasizes the use of standardized, modular components and interfaces, allowing for greater integration

and interchangeability of systems. The MOSA structure enables the seamless integration of heterogeneous systems from various domains and vendors. By adhering to open standards and interfaces, MOSA facilitates the integration of new capabilities and technologies into the SoS architecture without requiring extensive redesign or reconfiguration of existing systems. This design pathway promotes the development of reusable, interoperable components that can be easily integrated into different systems within the SoS. Furthermore, MOSA supports the evolution and adaptation of SoS architectures over time by allowing for the incremental upgrade and expansion of SoS capabilities without necessitating costly and disruptive overhauls of the entire architecture when mission requirements evolve or new technologies emerge. This same technique is important to the design of futuristic weapon systems which is referred to as WOSA.

- ***Composability:*** Composability refers to the ability of individual systems or components to be combined or arranged in various ways to create larger, more complex systems or solutions. Composable systems or SoS are designed in a modular fashion, allowing their individual pieces to work together seamlessly, and enabling flexibility, scalability, and adaptability in the overall structure. This concept is of vital importance in distributed SoS and networks of such complex architectures. The basic challenge confronting us is to be able to develop, configure, and operate systems and networks of systems with high levels of trustworthiness, with respect to critical requirements for security, reliability, fault tolerance, survivability, performance, and other behavioral criteria without too seriously sacrificing the desired functionality.

The following examples provide insight on these foundational concepts that drive collaboration, innovation, and value creation across a wide range of industrial sectors. Utilizing these concepts enables organizations to leverage the strengths of diverse stakeholders and technologies to achieve common goals and deliver superior products and services discussed below:

- **Integrated communication systems:** Consider a communication system where multiple systems handle different communication channels (e.g., voice, data, and video). While each system operates independently in managing its specific type of communication, they are functionally dependent on an overarching control system that coordinates and manages the overall communication strategy.
- **Manufacturing processes:** In a manufacturing environment, various production systems may operate independently in terms of specific tasks (e.g., machining and assembly). However, these systems could be functionally dependent on a centralized quality-control system that monitors and ensures product quality across the entire manufacturing process.
- **Transportation networks:** Think of a transportation network with independent modes of transportation (e.g., roads, railways, and airways). While each mode operates independently, their schedules and coordination may be functionally dependent on a centralized traffic management system to optimize overall efficiency.

8.5 Relationship to Systems and SoS Engineering

MEI plays a crucial role within the broader framework of systems and SoS engineering, contributing specialized knowledge and practices tailored to the unique challenges and objectives of specific missions. Here are key aspects of the role of MEI in systems and SoS engineering:

- **Mission Definition, Objectives, and Alignment**
  - **Role of MEI:** Mission engineers work on defining the specific objectives, goals, and requirements of a mission as well as ensure that the objectives and requirements of the individual systems within the SoS align with the overarching mission goals. They take into account the mission's purpose, constraints, and environmental factors that shape the design and operation of the system.
  - **Impact on systems and SoS engineering:** The mission definition provided by MEI serves as a foundation for systems and SoS engineers to develop a system architecture that aligns with the mission's objectives and requirements. This is accomplished by aligning the individual systems toward achieving common mission objectives. It provides the mission context necessary for effective integration.
- **System Architecture and Design**
  - **Role of MEI:** Mission engineers contribute to the design of the system architecture, considering the unique demands of the mission. They determine the overall structure of the system, including subsystems and their interactions as well as define the interactions and dependencies among individual systems to ensure they work collaboratively in achieving mission goals.
  - **Impact on systems and SoS engineering:** Systems engineers build upon the MEI architecture, refining it to ensure that it meets broader systems and SoS engineering principles, such as modularity, scalability, and integration.
- **Interconnected and Integration of Systems**
  - **Role of MEI:** Mission engineers address the integration of systems within the context of the specific mission, ensuring that they work together cohesively to achieve mission objectives.
  - **Impact on systems and SoS engineering:** Systems engineers build on MEI system integration efforts, applying broader systems and SoS engineering principles to ensure seamless communication, coordination, and interoperability among systems. This is critical for defining the interconnectedness of systems and understanding how they contribute collectively to the mission's success.

- **Trade-off Analysis**
  - **Role of MEI:** Mission engineers analyze trade-offs between competing objectives, such as mission capability, resource utilization, and mission duration at the SoS level versus suboptimizing the individual, constituent systems.
  - **Impact on systems and SoS engineering:** System and SoS engineers conduct trade-off analyses to optimize system parameters to accomplish mission success, considering factors like cost, performance, schedule, and technical feasibility.
- **Mission-specific requirements**
  - **Role of MEI:** Mission engineers define requirements that are specific to the mission's objectives, taking into consideration the environmental conditions, performance criteria, and mission constraints.
  - **Impact on systems and SoS engineering:** System engineers incorporate mission-specific requirements into the overall requirements framework, ensuring that the system meets the specialized criteria set by the MEI team.
- **Risk Analysis and Mitigation**
  - **Role of MEI:** Mission engineers identify and analyze risks associated with mission execution, considering uncertainties, technical challenges, and external factors.
  - **Impact on Systems and SoS Engineering:** Systems engineers use the risk analysis provided by MEI to implement risk mitigation strategies and develop resilient systems that can adapt to unforeseen challenges.
- **Optimization of Mission Performance**
  - **Role of MEI:** Mission engineers focus on optimizing the performance of the overall mission, considering the interactions and contributions of individual systems to achieve mission objectives efficiently.
  - **Impact on Systems and SoS Engineering:** Systems engineers leverage the MEI performance metrics, further refining and optimizing system components to meet broader mission goals, including reliability, efficiency, and maintainability. It is critical to develop and maintain a linkage between the mission success measures of MoS at the mission level, MoE at the mission essential tasks level, and MoP at the system level.
- **Human-in-the-Loop Considerations**
  - **Role of MEI:** Mission engineers recognize the role of human operators and decision-makers in mission execution, especially in dynamic or unforeseen situations. This work also includes the determination as to whether the human should be on-the-loop or completely off-the-loop.
  - **Impact on systems and SoS engineering:** Systems engineers incorporate human factors considerations into system design, ensuring that interfaces,

controls, and decision-support mechanisms are designed to facilitate effective collaboration between humans and the systems.

- **Mission Lifecycle Considerations**
  - **Role of MEI:** Mission engineers consider the entire lifecycle of the mission, from planning and design to execution, operation, and eventual retirement or decommissioning.
  - **Impact on systems engineering:** Systems engineers adopt a lifecycle approach, building on the MEI framework to ensure that the system is designed, developed, and managed such that it aligns with the full mission lifecycle.
- **Mission Execution, Monitoring, and Adaptive Coordination**
  - **Role of MEI:** Mission engineers are actively involved in overseeing and monitoring the execution of the mission, making real-time decisions as needed for real-time adjustments based on the mission's evolving requirements.
  - **Impact on systems and SoS engineering:** Systems engineers design monitoring and control systems that align with MEI requirements, allowing for effective oversight and adaptability during mission execution.
- **Collaborative Decision-Making**
  - **Role of MEI:** Mission engineers contribute to the development of decision-making processes that align with the mission's unique challenges.
  - **Impact on Systems and SoS Engineering:** Systems engineers extend collaborative decision-making approaches to encompass the collective decision-making requirements of the interconnected subsystems within the SoS.
- **Post-Mission Analysis and Lessons Learned**
  - **Role of MEI:** Mission engineers conduct post-mission analyses to gather data, assess mission performance, and extract lessons learned for future improvements.
  - **Impact on systems and SoS engineering:** Systems engineers incorporate lessons learned from MEI into continuous improvement processes, refining system designs and engineering practices for future missions.

In essence, MEI provides a mission-centric perspective and serves as a crucial bridge between individual systems and SoS within the broader framework of SE. Collaboration between the two disciplines is crucial for developing effective and efficient systems that meet the unique demands of specific missions. The specialized insights and requirements generated by MEI inform and guide the broader SE process. As depicted in Fig. 8.2, MEI resides at the top of the SE hierarchy, encompassing the overarching process of defining and aligning system objectives, functions, and requirements with the broader strategic goals and mission objectives.

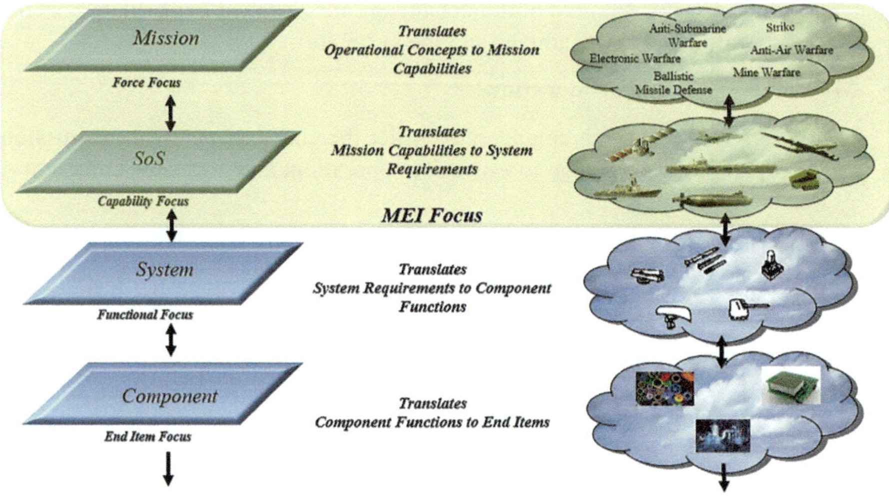

Fig. 8.2 MEI focus within systems engineering hierarchy

8.6 Relationship to DoD Acquisition

The relationship between MEI and DoD acquisition is integral to the efficient development and acquisition of military capabilities. The MEI approach serves as the foundational process that aligns operational needs with technological solutions. By thoroughly analyzing mission requirements, risks, and constraints within a comprehensive framework, MEI provides a clear roadmap for the acquisition process. This includes defining the scope of mission objectives, understanding the operational environment, and identifying critical technical attributes. The insights derived from MEI directly inform the acquisition strategy, guiding decisions related to technology selection, procurement, and system development. Figure 8.3 depicts this relationship in a decision tree structure to provide insight on the interdependencies across the SoS for a particular mission area.

In turn, the DoD acquisition process operationalizes the MEI insights into tangible capabilities. This involves the identification of suitable technologies, the formulation of acquisition plans, and the execution of procurement strategies. The structured framework provided by MEI aids in setting realistic expectations for program outcomes, ensuring that acquisition efforts align closely with mission requirements. Moreover, the relationship extends to the continuous feedback loop between MEI and acquisition. This allows adjustments and refinements as technological advancements and operational needs evolve over time. The continuous feedback is represented using the multiple, overlapping SE V-Models in Fig. 8.4 to synchronize across the SoS for a specific mission capability. Overlapping SE V-Models facilitate iterative feedback loops, enabling engineers to assess trade-offs, manage risks, and

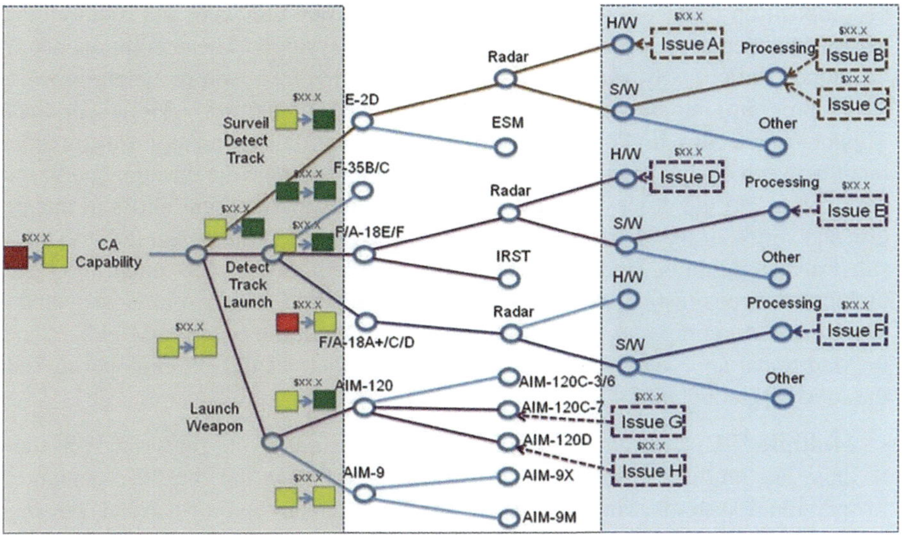

Fig. 8.3 Decision tree structure for acquisition investments

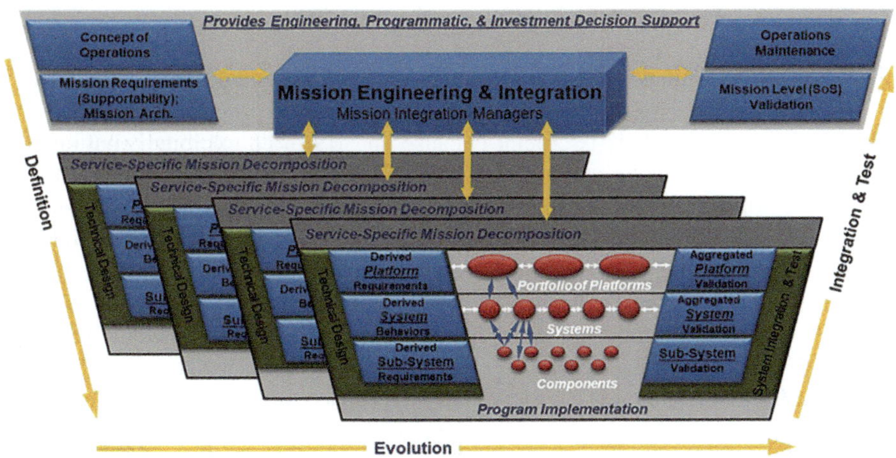

Fig. 8.4 Synchronizing SoS capabilities across acquisition lifecycles

ensure that individual system capabilities align with SoS-level requirements, ultimately supporting the seamless execution of complex, multi-domain missions.

In SE, the traditional SE V-Model is a popular framework used to represent the development lifecycle of a system. It emphasizes verification and validation, where the left side of the SE V-Model represents the development of system requirements translated from the decomposition of mission to the identification of the required SoS capability and the design of the proposed solution. This process continues with the critical phase along the bottom of the SE V-Model transitioning from decomposition to design to integration where abstract concepts and requirements evolve into concrete, actionable deliverables and systems. These components are developed individually and often in parallel and therefore require synchronization across programs and between systems using an overlapping SoS V-Model approach. This bottom section highlights the iterative nature of development, ensuring alignment between design and implementation and minimizing risks through continuous verification. As the process progresses to the right side, the focus shifts to integration and verification, where the components are assembled incrementally into larger subsystems, systems, and eventually the complete SoS. Each integration step includes rigorous testing to ensure that the system and SoS perform as expected at every level. As compared to the traditional SE V-Model, the SoS V-Model extends the traditional approach to address the unique challenges of SoS engineering with the following adaptations:

- **Multiple V-Cycles**: Instead of one SE V-Model and cycle for a single system, there are multiple, overlapping SE V-cycles representing the development of individual systems within the SoS. These systems evolve in parallel, but their interdependencies need to be carefully managed.
- **Interoperability and integration**: At higher levels of the SoS V-Model, the focus shifts from integrating various components within a single system to ensuring entire systems can work together cohesively. This requires robust interfaces and protocols.
- **Emergent properties**: The right side of the SoS V-Model emphasizes validating not only the systems themselves but also the emergent properties that arise from their interactions. This is crucial because an SoS often exhibits behaviors that cannot be predicted by examining individual systems alone.
- **Continuous iteration**: Unlike traditional systems where the SE V-Model typically represents a more static lifecycle, the SoS V-Model is often more iterative and ongoing. The SoS continuously evolves; engineering must accommodate constant changes and updates across the constituent systems.
- **Governance and coordination**: The SoS often involves multiple stakeholders and organizations, each with its own goals and processes. Therefore, the SoS V-Model incorporates higher levels of governance and coordination to manage these relationships and ensure alignment across the system.

Key Considerations for SoS V-Model:

- **Scalability**: Handling multiple systems with complex interdependencies.
- **Resilience**: Ensuring the SoS is robust against failures in individual systems.
- **Interoperability**: Developing interfaces and protocols that allow disparate systems to communicate effectively.
- **Evolution**: Allowing for continuous updates without disrupting the entire SoS.

As an example, the synergy between MEI and DoD acquisition is crucial for mitigating risks, optimizing resource allocation, and enhancing the overall effectiveness of military capabilities. By integrating processes, the DoD fosters a more agile and adaptive approach to acquisitions, aligning technology solutions with mission objectives to meet the dynamic challenges faced by joint/coalition forces. This collaborative relationship ensures that the DoD acquires systems and technologies not only meet immediate needs but also possess the flexibility and scalability required for evolving operational landscapes based on the planned engagement strategy. In the context of the Minimum Viable Product (MVP) concept, emphasis on agile and adaptive approaches to fielding operational capabilities at rapid speeds is crucial. The MVP concept focuses on the simplest version of a product that delivers enough value to users to test assumptions, gather feedback, and validate the concept. Priority is placed on rapid development and deployment with minimal effort, focusing on the core functionality necessary to meet immediate objectives. Its iterative nature allows developers to refine and expand the product based on real-world feedback, optimizing resources and reducing the risk of developing features that do not align with user needs. A combined focus on the MVP concept and MEI approach increases the probability of fielding working capabilities at desired speeds with a focus on delivering a functional solution that meets the core operational requirements of a mission and does not require reworking.

8.7 Considering Potential Future States for National Security

Whether in business or defense, strategic planning begins with identifying and defining objectives and the mission success criteria to drive the evaluation for pursuing the objectives. Determining the environment or operational context for this pursuit requires the reconciliation of the many competing assessments of the anticipated outlines of potential futures. The goal is to develop a baseline consensus of the probable future states or at least some common characteristics across multiple alternative futures. Figure 8.5 [MORELAND09] depicts a general framework of considering multiple probable futures with the intent of determining common characteristics across potential outcomes defining desired end states.

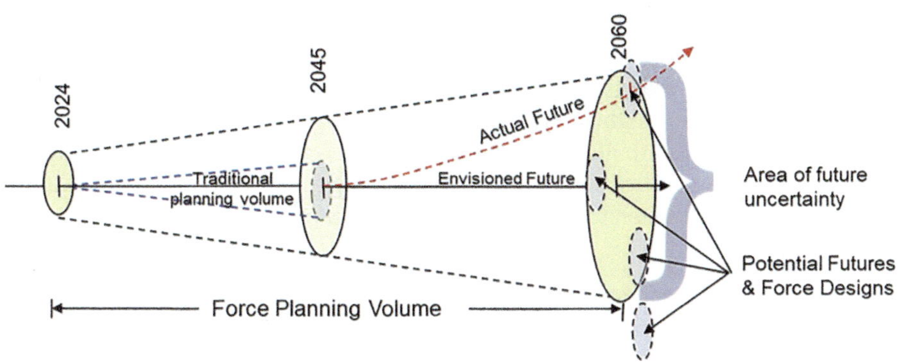

Fig. 8.5 Future force planning framework

Trying to reach a consensus view of the future has conceptual and practical limitations. These limitations begin with the difficulty in comparing a mixture of assessments that use different techniques. These techniques vary from estimates, forecasts, and scenarios. Estimates utilize assessments based on current conditions to identify possible future events. The priority in this technique is accuracy, which requires a relatively short time horizon. On the other hand, forecasts represent longer-range assessments with primary reliance on trends-based analyses. Finally, scenarios are affiliated with a range of forecasts, which tend to be elaborately developed depictions of alternate futures based on plausible changes in current trends.

As the pace of change accelerates across all facets of life in the United States, the military enterprise faces increasingly intricate economic, demographic, social, technological, and security challenges. The nation is presented with a crucial choice: either proactively shape the future or risk being shaped by it according to someone else's agenda. To effectively address these challenges, our defense capabilities must undergo transformation through a synergistic evolution of technological concepts and organizational relationships, with cultural change being the primary focus (e.g., a shift in attitudes, beliefs, and values). Establishing trust and cooperation cannot be achieved overnight; it requires a gradual process integrated into the organizational culture over time. This cultural transformation is imperative due to the profound shifts occurring in the strategic landscape:

- Transition from the industrial age to the information age
- Changes in the threat environment
- Diminishing barriers to competition across various defense domains, including the convergence of commercial technologies and business models

Predicting the future state with absolute accuracy is impossible; nonetheless, some businesses adopt a long-term perspective by creating a series of alternative futures, known as scenarios, to gain insights. This planning technique is employed by commercial organizations to inform critical decisions regarding expensive, long-term capital investments. One alternative future prediction example for long-term capital investments is in renewable energy infrastructure, where investors must consider long-term changes in energy demand, climate policy, and technology advancements.

For instance, a company investing in large-scale solar farms must predict not only the future costs of solar technology but also anticipate regulatory shifts, public policy trends toward carbon reduction, and changes in energy storage technology. One scenario might project strong policy support and rapid technological improvements, resulting in high returns from renewable investments. An alternative scenario could assume slower policy changes or technological advancements, leading to extended payback periods and more cautious returns.

By evaluating multiple future scenarios like these, investors can make more informed decisions, managing the risks and potential rewards of long-term capital investments in a shifting energy landscape. These decisions must be accurate to ensure a successful return-on-investment and sustain the viability of the organization. The alternative to this approach would be to make uninformed guesses about the future direction and rely on reactive measures, which may prove challenging to manage effectively. The challenge is to deploy military force that safeguards United States vital interests while simultaneously fostering greater collective security, stability, and trust. This necessitates a holistic approach that embraces innovation, adaptability, and collaboration to navigate the complexities of the modern era effectively.

The absence of insights into potential future scenarios inevitably results in speculation regarding defense requirements, the incorporation of current concepts into future force structures, and the trajectory of technological advancements. The primary objective of this strategic analysis framework is to explore potential future world environments, formulate a force structure compatible with those environments, and then assess the framework's performance and implications for acquisition planning. This process begins by creating multiple force designs tailored to various alternative futures. Key characteristics that contribute to a resilient force are identified. These characteristics then inform the design of the future force. By adopting this "reverse engineering" approach, envisioning the future ensures that operational concepts are grounded in a solid intellectual framework. Furthermore, this methodology helps delineate the spectrum of potential force characteristics and determine the capabilities needed to address a diverse array of threats in future environments.

Change must be driven by making critical risk judgments during the initial decision-making process instead of relying on the current stove-piped approach, which only seeks to mitigate risk based on a narrow view of needs. Furthermore, an uncertain future necessitates increased flexibility in designing and building future systems. Merely adjusting current capabilities to achieve a new static balance and mix will ultimately fall short of meeting enterprise needs due to imbalances in response strategies to an ever-changing security environment. Instead, capabilities must be defined to promptly execute the strategy, continually assess the magnitude and type of demand signal for these capabilities and develop the appropriate system and technology mix based on real-time demands. Incorporating these key elements establishes a strong link between strategy and system capabilities with inherent agility. In this new approach, the ability to swiftly transform the configuration of the SoS based on environmental triggers through adaptive and innovative techniques is essential. For example, currently, the DoD faces a disconnect in demand signals, where the strategy calls for capabilities that the force does not provide, and force generators produce force structures that the strategy no longer requires. Closing this gap requires a proactive and responsive approach to align force capabilities with evolving strategic needs. Figure 8.6 [MORELAND09] illustrates the required relationship between military strategies and demand signals when attempting to build an agile force structure with the ability to effectively and efficiently flex at numerous security environments.

Four overarching characteristics described below are expected to drive and advance the futuristic operational technologies.

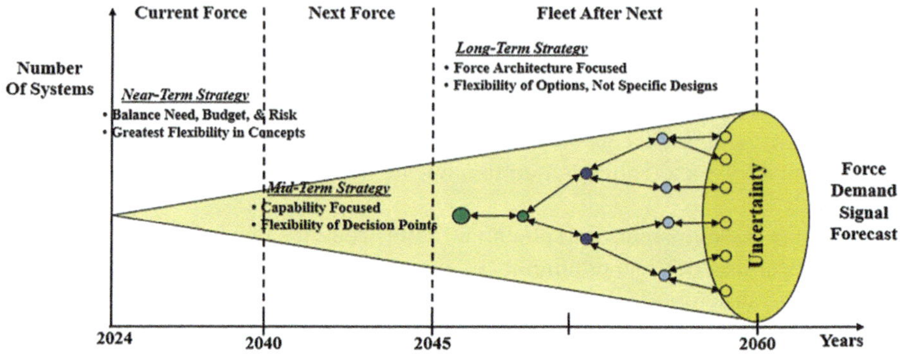

Fig. 8.6 Capability strategies drive force design

- **Intelligent:** Utilize integrated Artificial Intelligence (AI) systems, knowledge-focused analytic capabilities, and symbiotic AI-human intelligence to create disruptive applications spanning various technological domains
- **Interconnected:** Leverage the interconnected network of virtual and physical realms, which includes networks of sensors, organizations, individuals, and autonomous agents. This network is interconnected using innovative encryption methods and distributed ledger technologies
- **Distributed:** Utilize decentralized and widespread large-scale sensing, storage, and computation capabilities to achieve novel disruptive military effects
- **Digital:** Blend human, physical, and informational domains digitally to facilitate innovative disruptive effects
- **Modular:** The common open-modular approach is aimed at advancing the fielding of heterogeneous systems and components by fostering interoperability, enhancing commonality across platforms, promoting the reuse of hardware and software components, and facilitating access to cutting-edge technologies by multiple vendors. This approach emphasizes the development and adoption of open standards and modular architectures, thereby enabling different systems and components to seamlessly work together, regardless of their origin or specific technology stack. By adhering to common standards and modular design principles, mission essential systems become more flexible, adaptable, and cost-effective, while also fostering innovation and competition among vendors. Ultimately, the common open-modular approach seeks to improve the efficiency, effectiveness, and agility of operator capabilities, ensuring that defense forces can adapt swiftly to evolving threats and operational requirements.
- **Adaptive and capable:** The future force must possess adaptability and capability to effectively respond to a wide array of dynamic and uncertain threats. To achieve this, the organization must shift toward a Complex Adaptive System (CAS) model. CAS is characterized by its interdependent elements and decentralized control. It exhibits coherent yet non-deterministic behavior, aligning closely with the demands of modern warfare. Building an agile force structure requires elements to operate in a representative CAS, allowing for flexible

responses to evolving situations. Merely networking legacy systems and platforms will not suffice; instead, the future force must embody the principles of CAS. Thus, the development process must leverage current knowledge of CAS dynamics to ensure readiness for complex operational environments.

- **Responsive, persistent, and sustainable:** Future forces must address a diverse range of threat scenarios, spanning strategic and operational levels, each presenting unique complexities and scales. Depending on the nature of the threat or risk, force structure requirements may vary, ranging from swift responses with compact units to enduring engagements necessitating larger deployments. Regardless of the specific scenario, the force must exhibit persistence and self-sustainability, ensuring its ability to maintain operations over extended periods without external support. This adaptability and resilience are essential characteristics for effectively navigating the dynamic and multifaceted landscape of modern challenges.
- **Joint, Interagency, Multinational (All Synchronized):** Future forces must seamlessly integrate a wide spectrum of joint, interagency, national, and international resources, ensuring their smooth incorporation into a unified and cohesive force structure. This integration extends beyond military entities to encompass the various organizations associated with the instruments of national power, including diplomatic, informational, military, and economic elements. Understanding the interdependencies among these components is crucial for synchronizing and maximizing operational effectiveness. Striking the right balance between utilizing both soft and hard national powers of influence is essential to leveraging the "Whole -of-Government" approach effectively [PETERS15]. In the United States, the Whole-of-Government approach is facilitated through various legal authorities and frameworks that empower different agencies and departments to work across legal responsibilities. These "Title" authorities, among others, provide the legal framework for implementing the Whole-of-Government approach, which emphasizes collaboration and coordination among all elements of national power to address complex challenges and achieve common objectives. By diversifying the range of motivational instruments employed and introducing unpredictability into adversary perceptions, this approach mitigates the risk of overburdening any single power while enhancing the likelihood of mission success.

Technologies possessing these characteristics are poised to enhance operational and organizational effectiveness in several ways. First, they will contribute to the development of a knowledge and decision advantage, leading to more informed and strategic decision-making processes. Second, these technologies will leverage emergent trusted data sources, ensuring the reliability and accuracy of information used for decision-making and operational activities. Third, they will enhance the effectiveness of meshing capabilities across all operational domains and instruments of power, thereby enabling seamless communication and coordination between different elements of operational missions. Last, these technologies will facilitate adaptation to a future security environment characterized by the widespread availability of inexpensive, distributed technologies on a global scale. The synergistic integration

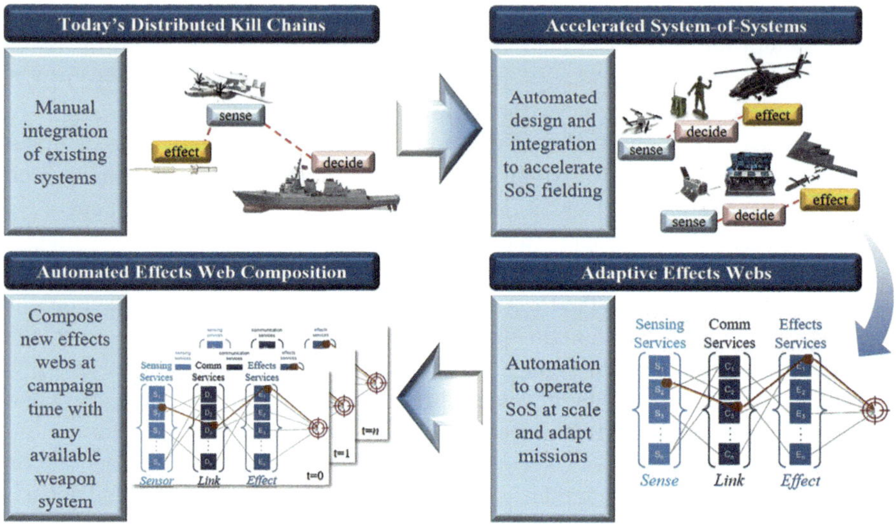

Fig. 8.7 Evolution of MEI capability states

of autonomy, big data, and AI, facilitated by intelligent, widely distributed, and cost-effective sensors and alongside autonomous physical or virtual entities, holds the promise of providing a significant strategic and operational decision-making advantage for mission execution. This combination harnesses new technologies and methodologies to enhance the operator's ability to collect, process, and analyze vast amounts of data in real time. The result is better informed and timely decision-making at both strategic and operational levels.

In defense, the evolution of MEI capability states progresses through four key stages, progressing through increasing levels of autonomy and adaptability in operational systems. Moving from left to right across the top of Fig. 8.7, *Today's Distributed Kill Chains* enable rapid sensor-to-shooter linkages across multiple domains, leveraging resilient networks to deliver precise effects. As capabilities advance, they transition into *Accelerated SoS*, where platforms and sensors collaborate more dynamically, reducing latency and increasing decision-making speed. Continuing clockwise around Fig. 8.7, the next stage, *Adaptive Effects Webs*, integrates AI-driven automation, enabling real-time orchestration of sensors, shooters, and C2 elements to generate mission-tailored effects. Finally, the most advanced state, *Automated Effects Web Composition*, introduces fully autonomous, AI-enabled mission execution, where machine-driven decision-making continuously adapts to the battlespace, dynamically composing and reconfiguring kill chains to maintain operational advantage in contested environments. This progression enhances warfighter effectiveness by increasing speed, adaptability, and resilience in future conflicts.

8.8 Performing Mission Decomposition

Mission decomposition involves breaking down a complex mission into manageable activities while considering various mission success measures and operational conditions such as MoS, MoEs, MoPs, and mission execution attributes. At the outset, the mission engineer defines the mission's overarching objective, providing a clear direction for decomposition. Each mission essential task is identified and associated with specific attributes, including specificity, measurability, achievability, relevance, and time-bound nature. These attributes ensure that each component contributes directly to the mission's success and can be effectively evaluated. Furthermore, the mission engineer assures that the mission success criteria at each level of mission execution (MoS, MoEs, and MoPs) are established for each mission activity and system assignment to assess its impact on achieving the mission's objectives. MOEs provide a quantitative or qualitative means of evaluating the success of completing the individual mission essential tasks and their contribution to the broader mission goal. Additionally, MoPs are defined to assess the efficiency and effectiveness of the individual systems assigned to each of the mission essential tasks. Most important is linking the measures of success across all levels of mission execution to confirm alignment and coherence throughout the mission. This linkage ensures that the success of individual systems contributes to the overall success of the mission. By establishing clear connections between MoS, MoEs, and MoPs, mission engineers can effectively monitor progress, identify areas for improvement, and optimize resource allocation to maximize mission success. This integrated approach to mission decomposition enables organizations to achieve their strategic objectives efficiently and effectively in dynamic and challenging environments.

Moving the focus from a "systems only" approach to one that deliberately addresses tasks, and the mission involves a fundamental shift in perspective toward mission-centered planning and execution. Instead of solely concentrating on the development and optimization of individual systems, this approach emphasizes the integration of systems within the context of specific tasks and overarching mission objectives. By adopting this approach, organizations can ensure that the selection, design, and deployment of systems are aligned with the requirements of the targeted tasks and the broader mission. This involves considering not only the technical capabilities of systems but also their suitability, interoperability, and effectiveness in achieving desired outcomes. Moreover, by prioritizing task and mission requirements, decision-makers make better informed choices regarding system acquisition, deployment, and utilization. This may involve trade-offs between different systems or configurations to optimize overall mission effectiveness and efficiency. This shift toward a task and mission-focused approach promotes a more holistic and strategic perspective on system development and deployment. Investments in technology and resources are thus directed toward achieving tangible mission objectives and delivering value to stakeholders.

Describe the mission in terms of objectives and desired effects(outcomes).

Identify the Mission Statement(s), Objectives, and Desired Effects from authoritative sources:

- Joint Capabilities Integration Development System (JCIDS) documents (Integrated Capability Document (ICD), Capability Development Document (CDD), and Capability Production Document (CPD))
- Analysis of Alternatives
- Joint Mission Thread (if available)
- Joint/Service Doctrine/Concept of Operations (CONOPS)
- Subject Matter Expert input

Fig. 8.8 Mission description details

Decompose the mission into relevant tasks.

Mission(s) decomposed into tasks (activities) and sub-tasks with key nodes identified:

- Functional performers/roles identified that perform the mission
- Nodes identified as the "means" to perform tasks (e.g. from DoD Architecture Framework (DODAF) products, joint mission threads, joint publications, Concept of Operations)
- Universal Joint Tasks Lists (UJTLs) and Service task lists may be included

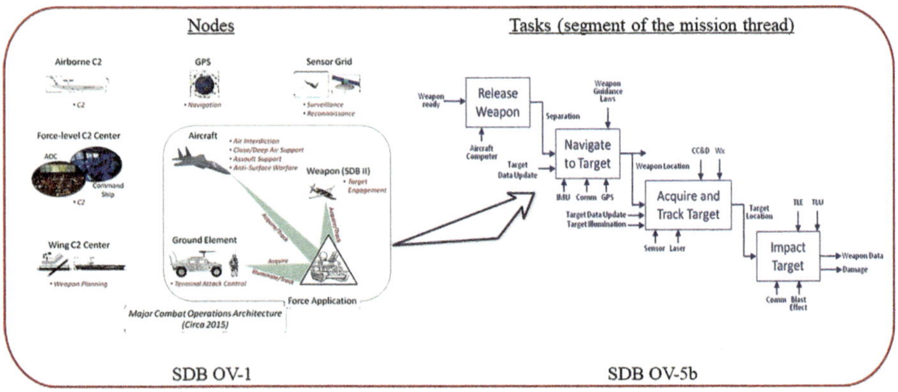

Fig. 8.9 Tasks description and details

Mission decomposition is achieved by following these general steps:

- **Describe mission:** Describe the mission in terms of objectives and desired effects (outcomes), see Fig. 8.8.
- **Describe task:** Decompose the mission into relevant tasks, see Fig. 8.9.
- **Develop mission and tasks attributes:** Identify the attributes (characteristics) of desired effects and tasks, see Fig. 8.10.
- **Determine mission and tasks measures:** Ensure there are separate measures for the military effect (mission accomplishment), task performance, and system function, see Fig. 8.11.

Identify the attributes (characteristics) of desired effects and tasks.

- Important and relevant characteristics of Desired Effects and Tasks are identified
- JCIDS prioritized list of capability attributes for enabling Joint Capability Areas: Battlespace Awareness, C2, Net-centric, Logistics)
- Dimensions of performance attributes (time, space, quality, action, etc) are directly linked to task and sub-task descriptions

	Task/Sub-Task Attributes			
Operational Task/Sub-Tasks	Accurate	Timely	Networked	Lethality
A5. Engage Mobile Target				
A51. Release Weapon		X	X	
A52. Navigate to Target	X		X	
A53. Acquire and Track	X		X	
A54. Impact Target	X			X
A6. Assess Effectiveness	X	X		

Table showing SDB task/sub-tasks vs attributes

Fig. 8.10 Mission and tasks attributes with example

Ensure there are separate measures for the military effect (mission accomplishment), task performance, and system function.

- Mission Measures
  - Should assess an attribute of a desired effect
  - Consists of a scale and a description
- Task Measures
  - At least one measure for each task-attribute pairing (more may be required)
  - In addition to JCIDS, may come from the joint/service task lists (UJTL, etc.)

Objectives: 1) Achieve battlefield effects against Mobile Targets 2) Minimize Undesired Effects

	Measures				
Attributes	Percent of actions where CAS was employed and effectively reduced risk to attack	Time to effectively conduct CAS against mobile targets	Percent of CAS missions where fratricide (including bodily harm) occurred as a direct result of CAS	Percent of CAS missions where collateral damage from CAS was acceptable	Percent of systems integration that are successful
Precision	X		X	X	
Lethality	X				
Timeliness					

	Task/Sub-Task Measures								
Attributes	Impact distance to center of target	Probability of Single Shot (Pssk)	Pct post release comm acknowledge	Pct successful target updates & retargeting	Pct successful weapon location updates	Pct successful weapon aborts	Avg delta in actual target location & weapon data target location	Pct accurate Bomb Hit Indication reports	Avg time to correctly assess BDA
A52. Accuracy							X		
A53. Accuracy							X		
A54. Accuracy	X						X		
A61. Accuracy								X	
A61. Timeliness									X

Fig. 8.11 Mission and tasks measures with example

Ensure system attributes are traceable to task and mission.

- The JCIDS ICD has a Capability Gap Table that connects capabilities (task-based) with attributes and metrics (standards for assessment)
- Look for a connection between the measured system attributes (Key Performance Parameters (KPPs), Key System Attributes (KSA), etc.) and the capability gap as expressed in the ICD

Priority	Tier 1 & 2 JCA	Description	Measure	Minimum Value	KPP, KSA, other attributes
1	e.g. Force Application Engagement	Capability 1			
		--Attribute 1	Description	Value	6.1, 6.2
		--Attribute *n*	Description	Value	6.2, 6.3, *n*
2	e.g. Force Application Engagement	Capability 2			
		--Attribute 1	Description	Value	6.1, 6.2. 6.3
		--Attribute *n*	Description	Value	6.4, 6.5, *n*

ICD Capability Gap table with system attributes traced to task.

Fig. 8.12 Traceability from mission to task to system

- **Show traceability from mission to task to system:** Ensure system attributes are traceable to task and mission, see Fig. 8.12.

Mission decomposition offers several key benefits in mission planning and execution as listed below:

- Provides a disciplined, repeatable, and sufficient process for developing mission, task, and system measures
- Enables objective understanding of a system's contribution to the SoS, task performance, and mission effectiveness
- Provides traceability to end-user requirements
- Enables validation of capability gap closure
- Moves the focus from "system only" to task and mission
- Helps develop requirements and design tests using design of experiments in accordance with the intent of the mission while enhancing mission-based test design

8.9 Determining Mission Essential Tasks

For military missions, the OODA Loop (Sect. 6.2.3) has set the foundation for proceeding through military operations and mission execution and therefore serves as a framework for the development of the sequential mission essential tasks.

8.9.1 Leveraging Critical Path Methods

In MEI, the Critical Path Method (CPM) is utilized to identify and prioritize mission essential tasks. This step-by-step resource allocation technique provides a tool to help decompose the mission into its mission essential tasks, mapping them in a flow diagram to capture dependencies. By calculating the mission duration based on estimated timeframes for each task, CPM highlights tasks that are essential with defined execution times. The critical path, as determined by CPM, represents the lengthiest sequence of tasks in the mission plan that must be completed on time for the mission to achieve its desired end state. This method is crucial for ensuring that mission essential tasks are identified, sequenced, and executed efficiently to enhance the overall success of the MEI process.

The adoption of CPM for project scheduling originated during the late 1950s with two distinct projects conducted simultaneously. The U.S. Navy's Fleet Ballistic Missile (Polaris) Program faced schedule delays, prompting the need for a resolution. The proposed solution involved breaking down the project into numerous tasks, representing each task with an arrow, establishing proper sequencing of the arrows, estimating task durations, and calculating both the project duration and the level of schedule criticality for each task. Concurrently, the American chemical company DuPont was encountering delays in its plant turnarounds, a project aimed at reorganizing production facilities for various products. They also sought assistance, and the suggested approach mirrored that of the Polaris Program.

The creators of the Polaris Program's strategy labeled their planning solution as the Project Evaluation and Review Technique (PERT), whereas the DuPont methodology was termed CPM. While these approaches share similarities, they employ distinct techniques for estimating task duration. PERT incorporates three different time estimations for each task duration, calculating the probability of project completion at any given time. In contrast, CPM utilizes a single time duration for each task and focuses not only on project completion time but also on analyzing additional costs that would arise if the project duration was reduced. While sharing a common purpose, PERT primarily diverges from CPM by assuming that task completion times are uncertain and independent of each other. PERT models the uncertainty of task completions based on assumptions about the distribution and likelihood of the time required for each specific task. This technique demands additional input from the operator regarding these distributions, which may not be intuitive or readily available. However, when executed properly, PERT can unveil interdependencies or issues that would not be discovered with a purely deterministic approach. Military operations are marked by substantial uncertainty which makes assumptions about the distributions or variance of random quantities challenging. This is especially true for those operations related to actions or responses to adversarial forces. Justifiably, the stochastic nature of PERT renders it more complex and challenging to implement effectively as a user-friendly application compared with CPM. This complexity has led to PERT being more commonly used in research and development projects, whereas CPM finds widespread application in various commercial enterprises, including construction and industrial production.

8.9.2 Identifying Mission Essential Tasks

Mission essential tasks are considered necessary, indispensable, or critical execution activities toward reaching mission success. The combination of these tasks is referred to as the mission thread. Each mission essential task encompasses particular conditions and standards that define a successful task contribution to achieving mission objectives. Evaluating mission essential tasks forms the basis for capabilities-based reporting and provides a standardized mechanism to understand the operational capabilities and limitations in performing designated functions within a specific capability across military services and DOD agencies.

Mission essential tasks are developed based on the Universal Joint Task List (UJTL). Force units at each organizational level have a mission to carry out. This mission is their fundamental purpose or "core" mission, designated operational plans, named operations, or a combination of these elements. Core tasks are the functions that the organization is structured to accomplish. Designated operations are responses to unforeseen crises that a unit is presently engaged in or has been activated to execute. Numerically designated plans are currently limited to high-level plans outlined in the United States Secretary of Defense's Contingency Planning Guidance (CPG) and the Chairman of the Joint Chiefs of Staff's (CJCS) Joint Strategic Capabilities Plan (JSCP).

In generating mission essential tasks, a starting point from a generic perspective is to structure thinking around the sequence of "plan, detect, control, engage, assess," or "find, fix, track, target, engage, assess," to frame the development of mission essential tasks in the context of military operations. A brief description of each mission essential tasks using this generic framework follows as an outline to plan, execute, and evaluate operations to accomplishment mission objectives:

- **Plan (Mission Planning)**
  - Develop detailed mission plans outlining objectives, strategies, and task assignments
  - Consider factors such as intelligence, logistics, coordination, and risk assessment in the planning process
- **Detect (Situational Awareness)**
  - Establish and maintain situational awareness through surveillance, reconnaissance, and intelligence gathering
  - Utilize technology, human intelligence, and other resources to detect potential threats or critical information
- **Control (Command and Control)**
  - Implement command and control measures to ensure effective coordination and communication
  - Establish protocols for decision-making, information flow, and coordination among units and command levels

- **Engage (Operational Execution)**
  - Execute planned actions based on the information gathered during detection
  - Deploy forces, initiate maneuvers, and engage with identified threats or targets according to the established plan
- **Assess (Operational Evaluation)**
  - Evaluate the outcomes of the engagement phase and overall mission execution as part of battle damage assessments using battle damage indicators
  - Assess the impact on the enemy, mission success, and the effectiveness of the plan
  - Gather feedback and lessons learned to inform future planning and enhance operational capabilities

The mapping from "Plan, Detect, Control, Engage, Assess" (PDCEA) to "Find, Fix, Track, Target, Engage, Assess" (F2T2EA) can be summarized as follows:

- **Plan (P) to Find (F):** The planning phase in PDCEA involves determining the strategy and objectives for identifying and engaging targets. In F2T2EA, the "Find" step corresponds to locating and identifying potential threats or targets based on the established plan.
- **Detect (D) to Fix (F):** Detection in PDCEA involves using sensors or intelligence gathering methods to identify potential targets. In F2T2EA, the "Fix" step involves pinning down the location of the target once it has been detected.
- **Control (C) to Track (T):** Control in PDCEA refers to maintaining situational awareness and coordinating actions to engage targets effectively. In F2T2EA, the "Track" step involves continuously monitoring the movement and behavior of the target to ensure accurate engagement.
- **Engage (E) to Target (T) and Engage (E):** Engagement in PDCEA involves taking offensive action against identified targets. In F2T2EA, the "Target" and "Engage" steps involve selecting the target for engagement and neutralizing it using appropriate tactics or weapons.
- **Assess (A) to Assess (A):** Assessment in PDCEA involves evaluating the outcomes of engagement and determining the effectiveness of actions taken. In F2T2EA, the "Assess" step similarly involves evaluating the results of engagement to determine success and adjust future actions as necessary.

Overall, while the terminology may vary slightly between the two frameworks, the core processes align in terms of identifying, engaging, and assessing targets. However, in some cases where mission essential tasks do not exist or need major adjustments, the CPM introduced in the Leveraging Critical Path Methodology, Sect. 8.9.1, can be used to develop these critical activities for operational missions. There are six steps in applying CPM for this purpose:

- **Specify each activity**—Identify every mission-related activity, focusing on higher level tasks to avoid excessive complexity in critical path analysis.

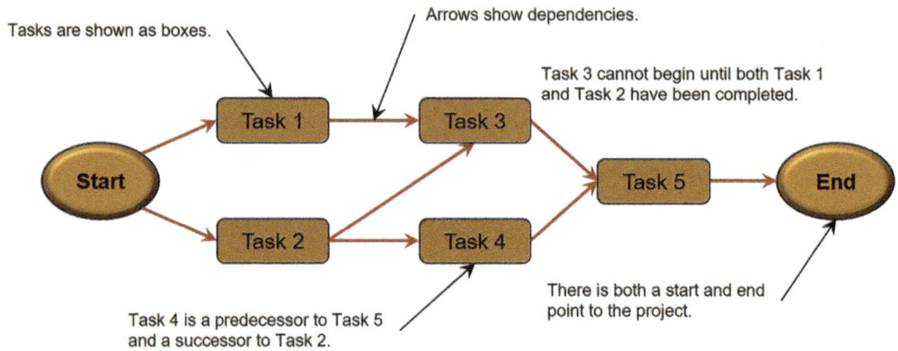

Fig. 8.13 CPM network diagrams

- **Establish dependencies (activity sequence)**—Define dependencies by identifying activities that rely on the completion of others. Answer questions like which tasks precede, occur simultaneously, or follow each identified activity from the previous **Specify each activity** step.
- **Draw the network or activity diagram**—Create a visual representation of the activity order based on dependencies using the CPM analysis chart, known as the network or activity diagram. See Fig. 8.13
- **Estimate activity completion time—**Utilize experience or the expertise of an operational planner to estimate the time required for each activity. Clearly identify the units used for estimating total activity time.
- **Identify the critical path**—Identify the critical path either by visually identifying the longest path on the network diagram or by using the forward pass/backward pass technique to determine the earliest and latest start and finish times for each activity. If multiple critical paths exist, network sensitivity may affect the project schedule. A sensitive project schedule is more likely to change during execution.
  - **Forward pass:** Determine the earliest start and finish times for each activity by considering the finish date of the previous activity. Continue until the last activity is reached.
  - **Backward pass:** Identify the latest start and finish dates by working backward from the last activity to the first in the mission-plan timeline.
- **Update the critical path diagram to show progress**—As the project advances, measure actual activity completion times. Update the network diagram with this information for a more accurate depiction of mission execution completion status to reach mission essential, task-acceptable standards under relevant conditions. Regular updates support recalculations and provide insights into whether the mission is on track or falling behind schedule.

Efficient algorithms developed for certain network and graph models offer a more effective solution compared to traditional linear optimization techniques when solving the complexity of multiple operational paths. Notably, CPM analysis of networks in project management, initially applied to large-scale complex engineering scheduling problems, proves applicable to military operational planning. CPM analysis reveals the intricacies and interrelationships among project or mission activities, facilitating in-depth analysis, easy modifications, and flexible evaluations to inform decision-making. The resulting critical path in the CPM network represents a chain or web of activities that, if delayed or extended in duration, would prolong the overall project or mission completion time. Activities on the critical path are deemed essential, allowing no room for deviations in time.

8.10 Developing Mission Engineering Threads/Webs

The MEI approach builds around a horizontal integration framework to address the interaction of platforms, sensors, systems, and weapons that form the mission engineering threads/webs within the foundation of the mission thread (mission essential tasks). The intent of MEI is to:

- Assess mission areas end-to-end, across system and platform boundaries, to identify and close gaps in mission critical capabilities
- Provide a common framework to compare like functionalities and capabilities, while expanding the trade space from the system and platform level to the mission level
- Improve warfighting capability by identifying gaps between systems and platforms while considering mission wholeness by providing gap closure recommendations across the range of DOTMLPF-P. Once identified, decision-makers can consider cost-effective solutions to the gaps to include materiel and non-materiel options, ensuring efficient system integration and effective force interoperability

8.10.1 Common Framework

Mission engineering threads/webs represent a series of integrated end-to-end mission essential tasks with multiple paths linked by interoperable interfaces. These interfaces are required to employ a specific blue force weapon against a specific red force target within a given tactical situation to achieve a desired effect. Figure 8.14 shows a representative effects web framework.

The effects web framework represents a mission capability through the execution of the identified mission thread. As such, the effects web framework provides a mechanism to map systems, platforms, sensors, and weapons to mission essential tasks to produce the resulting mission engineering thread in the context of a specific tactical situation. The mission essential tasks form the columns of this product to represent the mission thread functions performed for the specific mission under

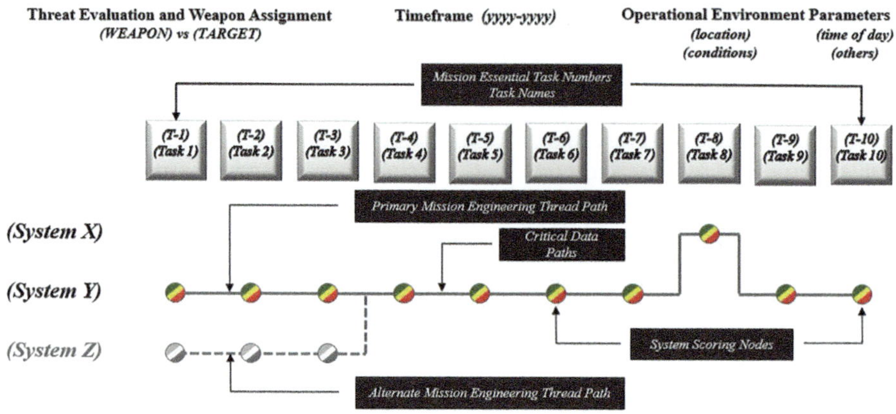

Fig. 8.14 Effects web framework using common taxonomy

assessment. The rows provide the platforms, sensors, systems, and weapons that represent the blue force employment against a specific threat environment and form an SoS construct. The assessment of these fielded/fielding systems forms the foundation for informing efforts to improve the seams between the existing ("as-is" architecture) and future systems ("to-be" architecture).

8.10.2 *Mission Engineering and Integration Approach Structure*

The MEI approach is structured as a 10-step process starting with the prioritization of operational mission areas and ending with the continuous management of end-to-end mission engineering threads to maintain the execution health of specific mission capabilities. The implementation of this process comes with challenges associated with governance structure, data availability and collection, stakeholder coordination across the DoD, multiple system lifecycles due to maturation (legacy to new), and workforce/tool development to name a few. However, the effects web framework provides a mechanism to translate what the DoD plans to procure to the resulting capability.

The 10-step process for MEI follows:

STEPS

1. Identify missions and mission essential tasks.
2. Define mission success and desired effects.
3. Identify mission success factors.
4. Identify conditions for each mission success factor.
5. Map mission success conditions to mission essential tasks (mission thread).

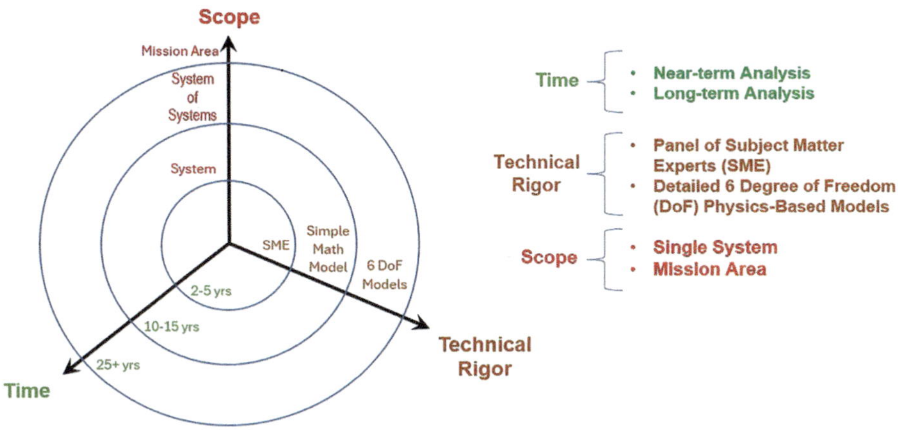

Fig. 8.15 Mission engineering and integration scalability

6. Identify critical conditions for each mission essential task.
7. Map systems to mission essential task of the mission thread to form the mission engineering thread/web.
8. Define appropriate scoring criteria for each mission essential task.
9. Apply the scoring metrics for success criteria to mission engineering thread/web for assessment.
10. Manage the mission engineering thread/web for assigned mission areas as part of capability portfolio management.

Later parts of this section use a sample mission as an example to characterize the 10-steps in the MEI approach. Users must tailor the steps to the specific mission under evaluation to the time available to conduct an analysis and to the information available at each step along the way. Figure 8.15 illustrates the multidimensional complexities associated with performing MEI analyses.

These multiple levels of consideration in performing MEI analyses enhance the mission analysis process by providing a comprehensive view of the technical aspects, allowing for informed decision-making and planning that consider the breadth and depth of technical tasks, timelines, and the required technical standards. Figure 8.16 provides some additional details for each dimension of this three-dimensional framework.

Technical rigor scoring evaluations involve assessing the precision, reliability, and adherence to technical standards for individual systems assigned to mission essential tasks, see Fig. 8.17. By incorporating technical rigor scoring evaluations, relevant stakeholders can verify tasks meet defined technical standards, promoting high-quality outcomes and reliability in their operations. This approach helps maintain consistency, transparency, and continuous improvement in technical practices.

The scope of analyses for MEI is broad and multifaceted, encompassing various dimensions. Figure 8.18 identifies the main areas when considering the complexities of mission execution. These drivers are dynamic and adaptive, focusing on the

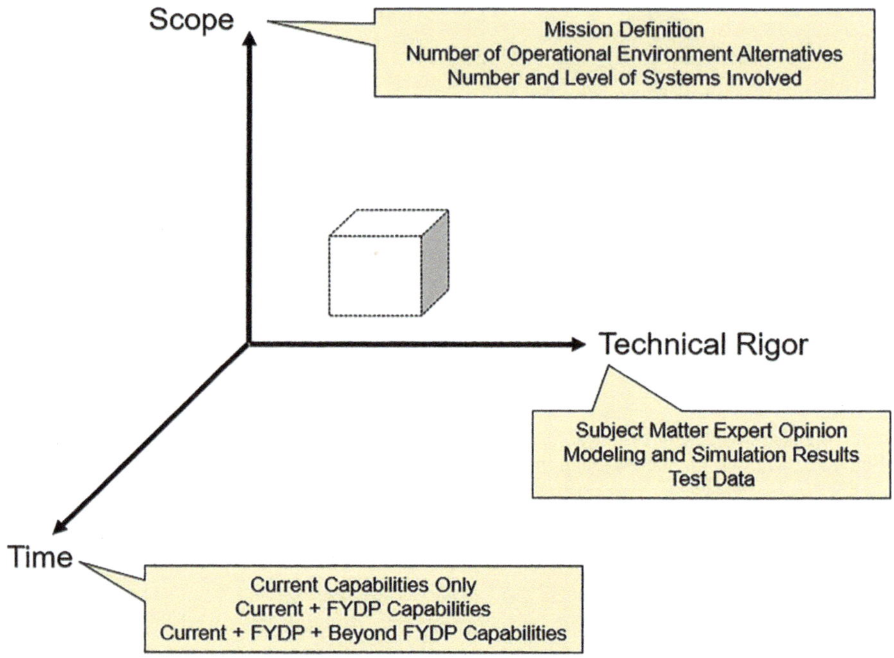

Fig. 8.16 Detailing the three-dimensional space for MEI analyses

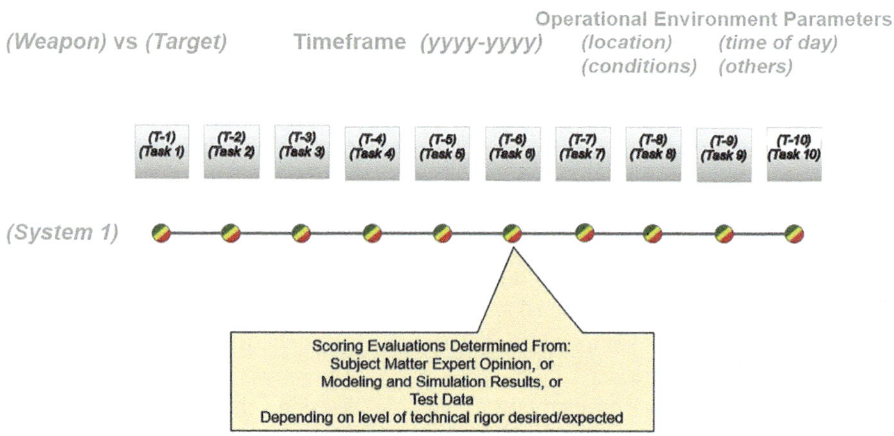

Fig. 8.17 Technical rigor considerations

specific context, objectives, and challenges of each mission. To reach reasonable and accurate solutions, the comprehensive and integrated MEI approach is necessary to ensure that all aspects are thoroughly evaluated for successful mission outcomes.

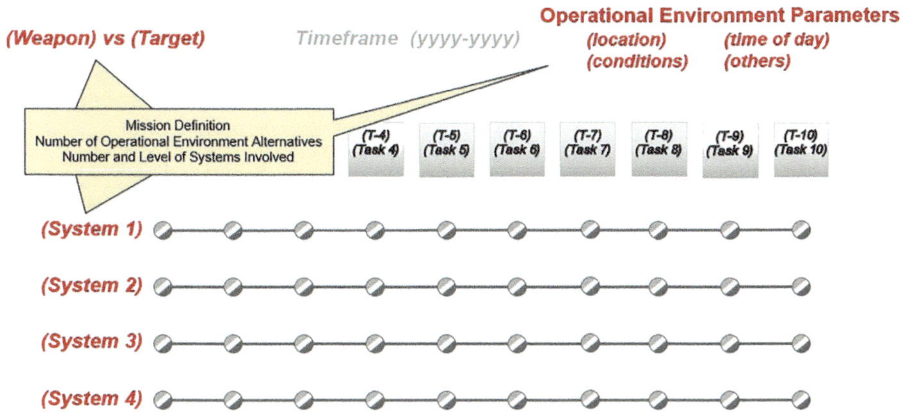

Fig. 8.18 Scope of mission considerations

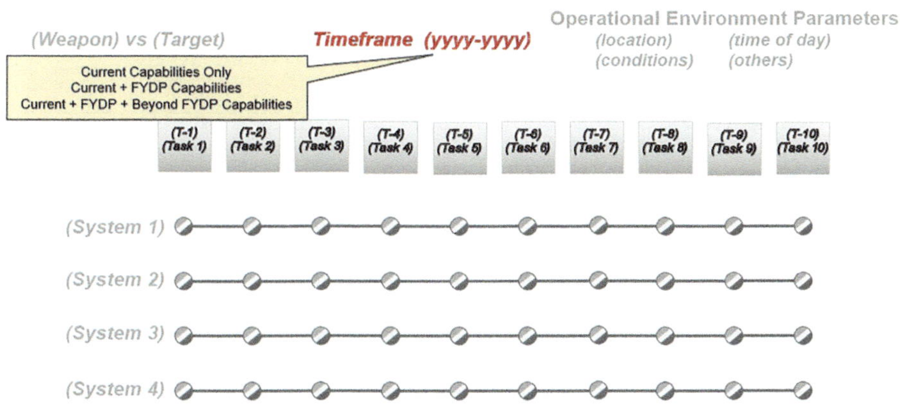

Fig. 8.19 Timeframe of analyses considerations

The timeframe takes on heightened significance when executing MEI analyses, particularly in response to rapid technological advances witnessed from both adversarial (red) and friendly (blue) forces. Figure 8.19 illustrates where this dimension of time is captured in the effects web framework. The swift evolution of technology necessitates a continuous and vigilant assessment within well-defined timeframes. Adversarial forces constantly seek to exploit the latest technological developments for strategic advantage, making timely analysis imperative for understanding and countering their adaptations. Conversely, friendly forces must rapidly integrate innovative technologies into mission strategies to maintain a competitive edge. The timeframe becomes a critical factor in aligning mission planning with the cycles of capability development and in enabling agile responses to emerging threats. It helps minimize the lag between technology introduction and integration into mission systems and SoS while maximizing the assurance that blue forces can effectively

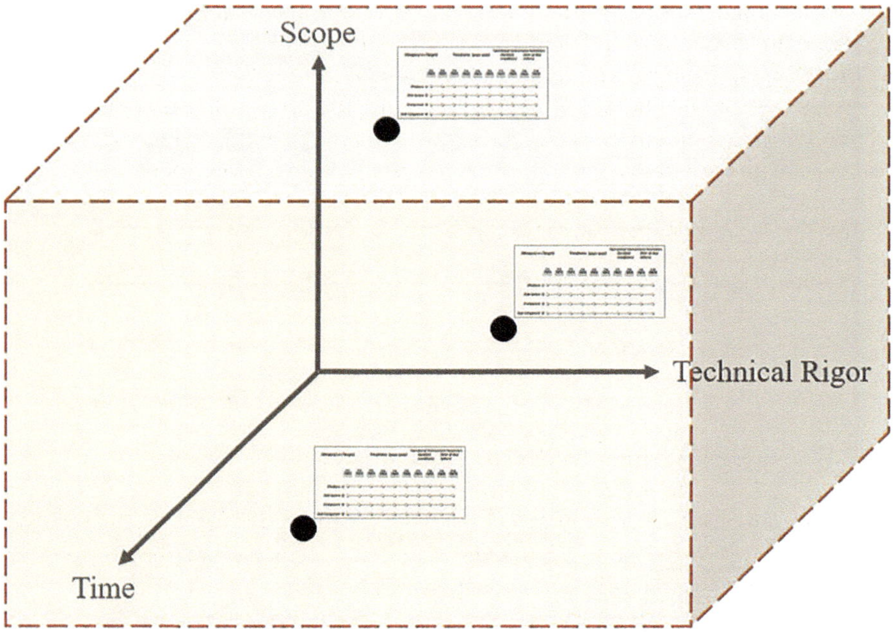

Fig. 8.20 MEI analyses flexibility supports the entire 3D trade space

harness the latest advancements. Moreover, the timeframe is essential for identifying and evaluating emerging threats, particularly in the realm of cybersecurity where the sophistication of attacks escalates. In essence, a well-structured timeframe facilitates agile decision-making, resource allocation, and continuous monitoring, and thereby ensures the MEI analyses remain dynamic and responsive to the ever-evolving technological landscape.

Figure 8.20 shows the potential outcomes when performing MEI analyses in the three-dimensional framework discussed. Flexibility associated with incorporating the distinct boundaries of time, scope, and technical rigor is essential for adapting to the dynamic nature of contemporary missions. In this multi-dimensional approach, time delineates the temporal constraints within which mission objectives must be achieved and new systems delivered. Planners are prompted to synchronize activities, set deadlines, and respond agilely to changes. The scope dimension defines the breadth and depth of mission essential tasks and the size of the SoS required for capability. Understanding the scope promotes a comprehensive understanding of the mission's complexity. From a scope perspective, flexibility enables the ability to adjust objectives, tasks, and resource allocations as the operational environment evolves. The technical rigor dimension establishes the precision and adherence to standards required for mission success based on defined MoEs and MoPs. A flexible approach to technical rigor allows the mission engineer to incorporate cutting-edge technologies while simultaneously maintaining a balance between innovation and reliability. Overall, this three-dimensional framework offers stakeholders the flexibility to navigate the intricate interplay between time, scope, and technical rigor, fostering adaptability, resilience, and optimized mission outcomes in the face of evolving challenges.

In addition, a focus on both business and operations from strategies, processes, policies, product, and service structures to regulations and so forth requires other dimensions of complexity for consideration as described below:

- ***Dynamic complexity*** describes matters in every aspect. When a subject requires extensive explanation, it is complex in terms of depth. When a subject has different dimensions, it demonstrates complexity in terms of breadth. Moreover, the rules change continuously, which makes it hard to keep up with reality.
- ***Social complexity*** occurs when there are many interactions within a fast-changing environment. For instance, operations usually involve multiple stakeholders, each having different and sometimes conflicting requirements. Often the mission engineer experiences difficulty dealing with all these different requirements because each requires dedicated interaction and negotiation skills.
- ***Emerging complexity*** is characterized by disruptive change and the uncertainty that these disruptions create. New entrants and existing competitors able to deal with the risk of uncertainty will outperform organizations or units that cannot.

With experience, teams will learn to combine certain steps, to iterate around a set of steps to refine the products, and to identify the key aspects of the mission needing further study. Below are several factors inherent in the MEI approach that influences how to apply the scoring criteria and interpret the results.

- The basis for MEI analysis is the TACSIT. Users must tailor the definition of measures and scoring criteria to specific threats.
- Because overall platform scores reflect a roll-up of subsystem performance as well as tactics, doctrine, etc., a score that reflects no capability does not necessarily indicate that the platform is at fault. The fault may lie in the doctrine or tactics in use. The users must determine and document the specific causes of mission failure.
- Deficiencies in one mission essential task do not carry forward to succeeding mission essential tasks. The assessment of a specific mission essential task assumes that all preceding mission essential tasks are fully effective, regardless of the actual assessment of those preceding mission essential tasks.
- The MEI approach focuses heavily on the examination of test data, with modeling and simulation used only as a fill for areas where no test data exists. To score a complete mission engineering thread/web, data will likely piece together from separate events occurring under different sets of conditions. Consequently, when interpreting results and drawing conclusions, users must consider the availability, applicability, and consistency of test data during the scoring process.

8.10.3 Conducting Mission Engineering and Integration Approach

The overall goal of the MEI approach is to assess the current health of the mission engineering thread/web within a specific mission area. Users accomplish this goal by determining how well particular platforms and systems support the mission essential tasks that make up the mission engineering thread/web using the effects web framework. This section outlines the 10 steps that define the assessment process using a specific scenario of an air warfare mission:

STEP

1. **Identify missions and mission essential tasks.** The first step in the MEI approach involves the identification of the operational functions and mission essential tasks necessary to execute critical mission areas. With the emphasis on operational relevance, the user must prioritize mission areas and determine the DoD plan for fighting these missions. Figure 8.21 provides an example of mission essential tasks for the air warfare mission.

2. **Define mission success and desired effect.** The second step in the MEI approach involves defining mission success and desired effects. Users must tie the measures to the scoring criteria to assess the health of the mission engineering thread/web to mission success. Users should consider fratricide (or minimizing fratricide) when characterizing mission success. Table 8.1 defines mission success and desired effects for the air warfare mission example.

3. **Identify mission success factors.** The third step of the approach identifies factors that have a major impact on the ability to achieve mission success while under control of friendly forces. Examples of factors that would not be under control of friendly forces include threat type, raid size, and spacing, as well as weather and other environmental conditions. Table 8.2 lists and describes the mission success factors. Using the effects web framework, users must consider these factors during the assessment process.

4. **Identify conditions for each mission success factor.** In the fourth step, users will define the conditions for each mission success factor. The conditions expand the definition of the mission success factors into more specific statements or

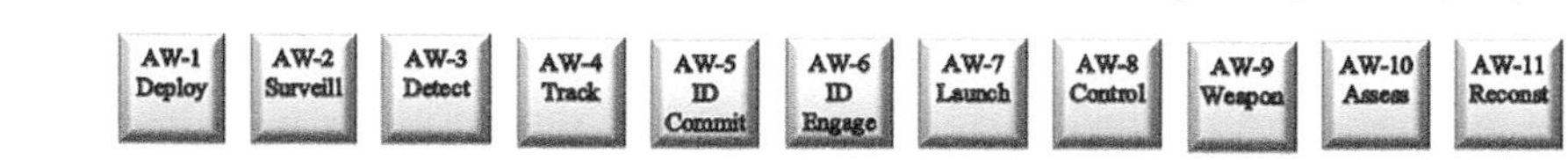

Fig. 8.21 Mission essential tasks (mission thread)

Table 8.1 Definition of mission success and desired effects

Mission success defined	Desired effects
All hostile force threats defeated with no friendly force losses or collateral damage.	• Detect and identify hostile forces • Neutralize detected threats • Prevent damage to battle group assets

Table 8.2 Mission success factors

Mission success factor	Description
Battle management	The ability to achieve and maintain situational awareness in the operational area to effectively combat threats to the force
Battle space	Consists of the amount of decision time afforded to the shooter and the time the shooter has to engage a track (i.e., the engagement window)
Communications link effectiveness	The ability to achieve required connectivity (range, performance against threats), timely (latency), bandwidth (capacity), and interoperability
Track management	The ability to maintain unique tracks on threats with correct identification (ID) for a sufficient duration and of sufficient quality to support successful weapon engagement
Engagement decision	Determines which tracks to engage and which weapon to use. It culminates with the issuing of an engagement order (EO) in accordance with the ID policy and rules of engagement (ROE)
Engagement rate	Determines the total number of tracks that can be engaged and the number of missiles that can be scheduled over a given time
Engagement effectiveness	Refers to the ability to achieve a mission kill on the assigned target. It includes the contribution of engagement support as well as the missile's capability to guide to and destroy the assigned threat
Re-engagement decision	Determines the need to re-engage a track
Readiness	Refers to the availability for the duration of the operation of all the elements in and supporting an effects web

questions that describe mission success. Table 8.3 identifies some representative conditions for the mission success factors in the air warfare mission example.

5. **Map Mission Success Conditions to Mission Essential Tasks (Mission Thread).** The fifth step relates the conditions for mission success to each mission essential task. This relationship should be the result of some logical action, for example: "To what extent does Condition A support the accomplishment of Mission Task B?" Table 8.4 is an example of relating mission success conditions to mission essential tasks.

6. **Identify critical conditions for each mission essential task.** Using the results from steps 3 to 5, users identify the critical conditions that define mission success for each mission thread task. Table 8.5 shows the example critical conditions for each mission task.

Table 8.3 Conditions for mission success

Mission success factor	Conditions
Battle management (BM)	BM1) Does the platform provide data that enhances situational awareness in support of battle management execution? BM2) Does the platform sufficiently enhance situational awareness when underlying intelligence data is missing or incomplete? BM3) Can platform capability be reconstituted within required timeframes for re-deployment/re-engagement?
Battle space (B)	B1) Does the platform detect targets of interest at a range to support or exceed the desired weapons employment or enhance decision time? B2) Does the platform detect targets of interest with sufficient accuracy to cue TRACK sensor? B3) Does the platform detect targets of interest when there are missing or incomplete intelligence data about targets of interest? B4) Does the platform meet/exceed firing doctrine requirements for the shooter?
Communications link effectiveness (CL)	CL1) Can required communication links achieve required connectivity in terms of necessary range against expected threats? CL2) Are required communication links timely (i.e., data are received without too much latency)? CL3) Do required communication links have necessary bandwidth (capacity) to support the kill/effects chain purpose? CL4) Are required communication links interoperable to allow for the exchange of data needed to perform the kill/effects chain mission? CL5) Are platforms interoperable above the basic communication link level (e.g., Layer 1, 2, 3 of the 7 Layer Protocol stack)?
Engagement decision (ED)	ED1) Is a correct decision (engage RED and do not engage BLUE) made on each track consistent with rules of engagement? ED2) For each track to be engaged, is an engagement order issued? ED3) Is an appropriate/preferred weapon ordered for the engagement?
Engagement rate (ER)	ER1) Can every track with an engagement order be scheduled for engagement? ER2) Can every missile in the salvo be scheduled for launch? ER3) How well is the weapon integrated into the combat system/launch platform (fully integrated, semi-integrated, and not integrated)?
Engagement effectiveness (EE)	EE1) Are all missiles scheduled for an engagement successfully launched? EE2) Is adequate in-flight support provided to all launched missiles? EE3) Does the asset/weapon have sufficient kinematics to defeat the threat? EE4) Does the missile guide to and destroy the assigned target?
Track management (TM)	TM1) Can the platform provide a weapon quality track? TM2) Is the required range to each target based on weapon kinematic range available? TM3) Does the platform track targets sufficiently when it has incomplete mission data? TM4) Is the potential target classified correctly to commit resources to that area? TM5) Are all BLUE and RED forces correctly identified?

(continued)

Table 8.3 (continued)

Mission success factor	Conditions
Re-engagement decision (R)	R1) Is the outcome of each engagement correctly assessed?
Readiness (RR)	RR1) Is the platform available/or can it be deployed or made ready within required timelines? RR2) How long can the elements stay on before requiring maintenance? RR3) How long does it take to resupply/reconstitute the capability? RR4) How long does it take to update mission systems to resolve combat ID ambiguities? RR5) How much degradation occurs because of incomplete or missing intelligence and mission data for what we know to be current data gaps?

7. **Map Systems to Mission Essential Task of the Mission Thread to Form the Mission Engineering Thread/Web.** Determines whether joint/coalition military service systems perform the functionality required of the mission essential tasks and make assignments to form the mission engineering thread/web. The user identifies alternate paths or mission engineering threads to accomplish the mission essential tasks using multiple systems for added resiliency to form the mission engineering web. Figure 8.22 is an example of a system to mission essential task mapping.

8. **Define Appropriate Scoring Criteria for Each Mission Essential Task.**

   (a) Users define scoring criteria to assess how well a platform or system supports one of the mission essential tasks that comprise the mission engineering thread/web using the effects web framework. In this example, three criteria define the scoring for each task: Full Capability to support (GREEN), Limited Capability (YELLOW), and No Capability (RED). The scoring criteria must relate to the tasks' critical conditions identified in Step 6 and must be appropriate to the platforms, systems, sensors, and weapons being assessed.

   (b) Assessment scoring criteria are necessary to evaluate the ability of the individual systems engaged in the mission engineering thread/web to meet mission success criteria and therefore achieve the desired effect of the mission. Each mission essential task shown below includes a general description of the task and examples of critical conditions with corresponding scoring criteria:

   Deploy Scoring (DEP).

   - DEP-1 covers availability of the platform to deploy.
   - DEP-2. Platform is available/or can be deployed or made ready within required timelines. (scoring criteria—GREEN: yes; RED: no).

   Surveillance Scoring (SUR). SUR provides non-weapon quality tracks and data to support early warning, situational awareness, ID, and cues to

Table 8.4 Mission success conditions correlated to mission thread tasks

Mission success factors	Conditions	Mission tasks										
		Deploy	Surveil	Detect	Track	Identify: commit	Identify: engage	Launch	Control	Weapon	Assess	Reconstitute
Battle management (BM)	BM1) Does the platform provide data that enhances situational awareness in support of battle management execution?											
	BM2) Does the platform sufficiently enhance situational awareness when underlying intelligence data is missing or incomplete?		H									
	BM3) Can platform capability be reconstituted within required timeframes for re-deployment/ re-engagement?											H

Battle space (B)	B1) Does the platform detect targets of interest at a range to support or exceed the desired weapons employment or enhance decision time?			H			M					
	B2) Does the platform detect targets of interest with sufficient accuracy to cue TRACK sensor?											
	B3) Does the platform detect targets of interest when intelligence data about targets of interest are missing or incomplete?											
	B4) Does the platform meet/exceed firing doctrine requirements for the shooter?							M				

(continued)

Table 8.4 (continued)

Mission success factors	Conditions	Mission tasks										
		Deploy	Surveil	Detect	Track	Identify: commit	Identify: engage	Launch	Control	Weapon	Assess	Reconstitute
Communication links effectiveness (CL)	CL1) Do required communication links achieve required connectivity in terms of necessary range against expected threats?											
	CL2) Are required communication links timely (i.e., data is received without too much latency)?											
	CL3) Do required communication links have necessary bandwidth (capacity) to support the kill/effects chain purpose?											
	CL4) Are required communication links interoperable to allow for the exchange of data needed to perform the kill/effects chain mission?											
	CL5) Are platforms interoperable above the basic communication link level (e.g., Layer 1, 2, 3 of the 7 Layer Protocol stack)?											

Readiness (RR)	RR1) Is the platform available/or can it be deployed or made ready within required timelines?	H										
	RR2) How long can the elements stay on before requiring maintenance?											
	RR3) How long does it take to resupply/reconstitute the capability?											
	RR4) How long does it take to update mission systems to resolve combat ID ambiguities?											
	RR5) How much degradation occurs due to incomplete or missing intelligence and mission data for what we know to be current data gaps?											

(continued)

Table 8.4 (continued)

Mission success factors	Conditions	Mission tasks										
		Deploy	Surveil	Detect	Track	Identify: commit	Identify: engage	Launch	Control	Weapon	Assess	Reconstitute
Track management (TM)	TM1) Can the platform provide a weapon quality track?				H							
	TM2) Is the required range to each target in a group provided?				M							
	TM3) Does the platform track targets sufficiently when mission data re less than complete?				M							
	TM4) Is the potential target classified sufficiently to commit resources to that area?					H						
	TM5) Are all BLUE and RED forces correctly identified?						M					
Engagement decision (ED)	ED1) Is a correct decision (engage RED and do not engage BLUE) made on each track consistent with rules of engagement?						H					
	ED2) Is an engagement order issued for each track to be engaged?						M					
	ED3) Is an appropriate/ preferred weapon ordered for the engagement?						M					

Engagement rate (ER)	ER1) Can every track with an engagement order be scheduled for engagement?							H				
	ER2) Can every missile in the salvo be scheduled for launch?							H				
	ER3) How well is the weapon integrated into the combat system/launch platform (fully integrated, semi-integrated, not integrated)?							M				
Engagement effectiveness (EE)	EE1) Are all missiles scheduled for an engagement successfully launched?							M				
	EE2) Is adequate in-flight support provided to all launched missiles?								H			
	EE3) Does the asset/ weapon have sufficient kinematics to defeat the threat?									H		
	EE4) Does the missile guide to and kill the assigned target?									H		
Re-engagement decision (R)	R1) Is the outcome of each engagement correctly assessed?										H	

H—High, M—Medium, L—Low, [blank]—None

Table 8.5 Critical conditions for mission thread tasks

Mission task	Critical conditions
Deploy	RR1) Is the platform available/or can it be deployed or made ready within required timelines?
Surveillance	BM2) Does the platform sufficiently enhance situational awareness when underlying intelligence data is missing or incomplete?
Detect	B1) Does the platform detect targets of interest at a range to support a desired weapons employment or enhance decision time? B2) Does the platform detect targets of interest with sufficient accuracy to cue TRACK sensor? B3) Does the platform detect targets of interest when intelligence data about targets of interest are missing or incomplete?
Track	TM1) Can the platform provide a weapon quality track? TM2) Is the required range to each target in a group provided? TM3) Does the platform track targets sufficiently when mission data are less than complete?
Identification: Commit	TM4) Is the potential target classified sufficiently well to commit resources to that area?
Identification: Engage	TM5) Are all Blue and Red forces correctly identified? ED1) Is a correct decision (engage Red and do not engage Blue) made on each track? ED2) Is an engagement order issued for each engagement track? ED3) Is an appropriate/preferred weapon ordered for the engagement? B1) Does the platform detect targets of interest at a range to support or exceed the desired weapons employment or enhance decision time?
Launch	ER1) Can every track with an engagement order be scheduled for engagement? ER2) Can every missile in the requested salvo size be scheduled for launch? ER3) How well is the weapon integrated into the combat system/launch platform? EE1) Are all missiles scheduled for an engagement successfully launched? B4) Does the platform meet/exceed firing doctrine requirements for the shooter?
Control	EE2) Is adequate in-flight support provided to all launched missiles?
Weapon	EE3) Does the asset/weapon have sufficient kinematics to defeat a non-cooperative threat? EE4) Does the missile guide to and kill the assigned target?
Assess	R1) Is the outcome of each engagement correctly assessed?
Reconstitute	BM3) Can the platform capability be reconstituted within required timeframes for re-deployment/re-engagement?

weapon quality sensors. Scoring reflects enhancing the strike group's situational awareness. Examples of critical measures are:

- SUR-1. Platform covers surveillance area as required by TTP
- SUR-2. Data availability, integration, and correlation capability (scoring criteria—GREEN: sufficient; YELLOW: degraded; RED: unable to support mission in TACSIT environment).

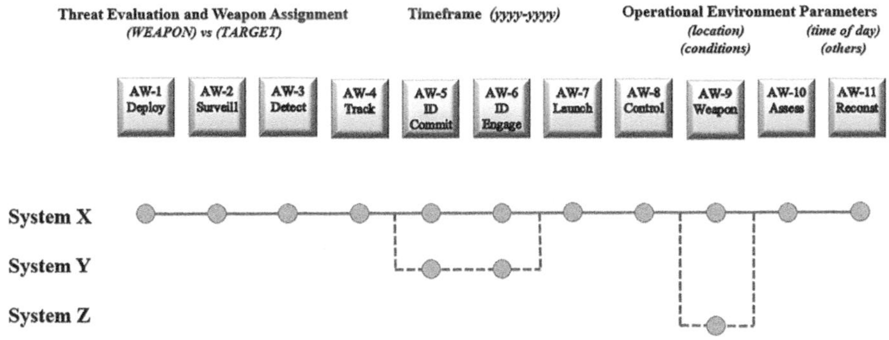

Fig. 8.22 Systems mapped to mission essential tasks (mission thread)

Detect Scoring (DET). DET is the perception of an object of possible military interest, but unconfirmed by recognition. The detecting sensor typically must cue a weapons-quality tracking sensor. As a result, detect scoring is based on the degree to which cues from the platform enhance the shooter's battle space. Examples of critical measures are:

- DET-1. Does the platform detect targets of interest at a range to support TTP requirements (desired weapons employment, decision time, or standoff range)? (scoring criteria—GREEN: yes; RED: no)
- DET-2. Does the platform detect targets of interest with sufficient accuracy to cue the track sensor? (scoring criteria—GREEN: yes; RED: no)
- DET-3. Does the platform detect targets of interest when it has incomplete mission data? (scoring criteria—GREEN: yes; RED: no)

Track Scoring (TRK). TRK maintains weapon quality tracks with sufficient accuracy to support engagement and the desired damage criteria. This means that the track is continuous, unique to each threat, and has sufficient accuracy to support missile uplink and illuminator pointing throughout the engagement (not required for active missiles). Examples of critical measures are:

- TRK-1. Can the platform provide a weapon quality track? (scoring criteria—GREEN: yes; RED: no)
- TRK-2. Is the required range for each target in an available weapon group based on kinematic range? (scoring criteria—GREEN: >x; YELLOW: >y; RED: <y)
- TRK-3. Does the platform track targets sufficiently when mission data are less than complete?
  (scoring criteria—GREEN: yes; RED: no)

Identification—Commit Scoring (IDCOM). IDCOM is the ability to classify tracks sufficiently to warrant the allocation and subsequent commitment of manned and unmanned assets to provide a positive identification of the tracked object and potentially engage it. For this iteration of the MEI, TACSITs with cruise missile targets will not consider IDCOM in the scoring of its platforms (insufficient reaction time to reposition assets). Examples of critical measures are:

- IDCOM-1. Can the platform classify the target as an adversarial threat at a range appropriate to making an allocation decision?

(scoring criteria—GREEN: classification is made before the target enters classification, identification, and engagement area; YELLOW: classification (threat) or procedural identification (unknown based on target profile, lack of friendly identification friend or foe, etc.) is made before the target enters the vital area; RED: classification/procedural identification is not made before the target enters the vital area).

- IDCOM-2. Track identification correlation capability.

(scoring criteria—GREEN: platform can make a timely identification correlation in the TACSIT track density environment to support commit before the target enters classification, identification, and engagement area; YELLOW: platform can make an identification correlation in the TACSIT track density environment to support commit before target enters vital area; RED: the platform is unable to correlate the identification of the target effectively to support the commit decision in the TACSIT track density environment).

Identification—Engage Scoring (IDENG). The IDENG score is a measure of a platform/sensor's ability to identify the target sufficiently well to permit appropriate action and ultimately to release the weapon. The implication of this criteria is an "approval to engage" that is generated based on the data and existing ROEs. Examples of critical measures are:

- IDENG-1. Is the engage identification criteria (defined by TTP/ROE/ Special Instructions) achieved in time to meet the TTP engagement range guidelines?

(scoring criteria—GREEN: type identification (minimum attributes required to optimize specific weapon employment and lethality per TTP) achieved prior to reaching the desired TTP engagement range; YELLOW: the type identification or procedural identification is achieved prior to reaching the minimum TTP engagement range; RED: engage identification criteria are not achieved prior to reaching the minimum TTP engagement range)

- IDENG-2. Track ID correlation capability.

(scoring criteria—GREEN: the platform can make a timely identification correlation in TACSIT track density environment to support engagement before the target enters the desired TTP engagement range; YELLOW: the platform can make an identification correlation in the TACSIT track density environment to support engagement prior to reaching the minimum TTP engagement range; RED: the platform is unable to correlate the identification of target effectively to support the engagement prior to reaching the minimum TTP engagement range).

Launch Scoring (LNCH). LNCH is the execution of a sequence of events required to schedule, initialize, and fire a weapon. Weapon Launch is the ability to launch successfully without degrading the desired firing doctrine and associated depth of fire. Examples of critical LNCH measures are:

- LNCH-1. Can every weapon in the requested salvo size be scheduled for launch? (scoring criteria—GREEN: yes; RED: no).
- LNCH-2. Does the launch platform survive the enemy weapons employment given an appropriate post-launch maneuver determined by TTP? (scoring criteria—GREEN: yes; RED: no).

Control Scoring (CTL). CTL scoring represents the post-launch interactions between the launch platform and the weapon. The control function assesses the accuracy and latency of the weapons data link. Examples of critical measures are:

- CTL-1. Capable of providing all data to guide the missiles to the required area to utilize full capability?

(scoring criteria—GREEN: yes; RED: no).

- CTL-2. Capable of controlling number of missiles required by TTP? (scoring criteria—GREEN: yes; YELLOW: can control a reduced number of missiles; RED: no)

Weapon Kinematics (WPNKIN) and Lethality Scoring. WPNKIN is the ability to reach the lethal radius for a given target with a certain state of end-game conditions (i.e., airspeed, G-force (G), terminal angle). Kinematics is applicable primarily when a target is non-cooperative, for example, a manned aircraft and can maneuver to deny engagement success.

- At a range defined by the concept of operation or TTPs, does the missile guide to the assigned target given a certain post-launch maneuver?

  WPNKIN-1. (Beyond Visual Range: > x miles):
  (scoring criteria—GREEN: x Mach (M) or x G kinematic end state (high termination criteria) vs. a target executing a range turn and run maneuver (x G max-performance slide turn with x feet of altitude lost, acceleration to x M post maneuver); YELLOW: x M or x G kinematic end state vs. a target executing a x G x degree turn at Blue platform trigger squeeze; RED: unable to meet GREEN or YELLOW criteria)
  WPNKIN-2. (Within Visual Range: < x miles):
  (scoring criteria—GREEN: obtain first launch opportunity; YELLOW: obtain weapon engagement zone employment opportunity; RED: unable to meet GREEN or YELLOW criteria)

- Weapon lethality includes the ability to ignite, fly, guide, engage, and achieve a mission kill. The scoring criteria for the weapon will determine if the performance is sufficient against the threat. Lethality is the weapon's ability to guide to the intended target and to achieve the desired damage criterion. The lethality criteria are weapon dependent and are different for each. Critical measures:

Weapon Lethality-3. Does the missile damage the assigned target?
(scoring criteria—GREEN: probability of damage > x; YELLOW: y < probability of damage < x; RED: probability of damage < y)

Assessment Scoring (ASMT). ASMT represents the determination of "Effect/Kill" or "No Effect/Kill" against a target and the need/ability to re-engage. The critical measures are:

- ASMT-1. Is the outcome of each engagement correctly assessed?
(scoring criteria—GREEN: yes; RED: no)
- ASMT-2. Can the platform provide attributes that contribute to a fused assessment at a range that supports the TTP or concept of operation?
(scoring criteria—GREEN: yes, re-engagement possible by TTP desired engagement range; YELLOW: yes, re-engagement possible by a TTP minimum engagement range; RED: no)

Reconstitute Scoring (RCNST). RCNST represents the ability of the platform to recover.

- RCNST −1. Ability to reconstitute the platform capability within required timeframes for re-deployment/re-engagement.
(scoring criteria—GREEN: yes; RED: no)

9. **Apply the Scoring Metrics for Success Criteria to Mission Engineering Thread/Web for Assessment.** The capability of a platform or system to support a given mission thread task will be assessed against the task's critical measures using the scoring criteria presented in earlier sections. Test data is preferred as the basis of the assessment.
   (a) Score each test event GREEN, YELLOW, or RED based on the appropriate scoring criteria. It is expected that the test data will not all fall into one of the three categories, but will be spread across the three.
   (b) Unless the scoring criteria explicitly call out the proportion of tests that are required to satisfy GREEN or YELLOW criteria, the principal investigators and other subject-matter experts will analyze the test results in greater detail to determine the overall score.
   (c) The final output will present the findings and document the rationale behind each. Figure 8.23 is a sample of scoring applied to the system mapping.

10. **Manage the Mission Engineering Thread/Web for Assigned Mission Areas as Part of Capability Portfolio Management.** Manage the end-to-end mission engineering thread/web, for specific mission areas, through the implementation of the effects web framework. The goal is to emphasize the effectiveness of the mission engineering thread/web by managing the seams of the programs form-

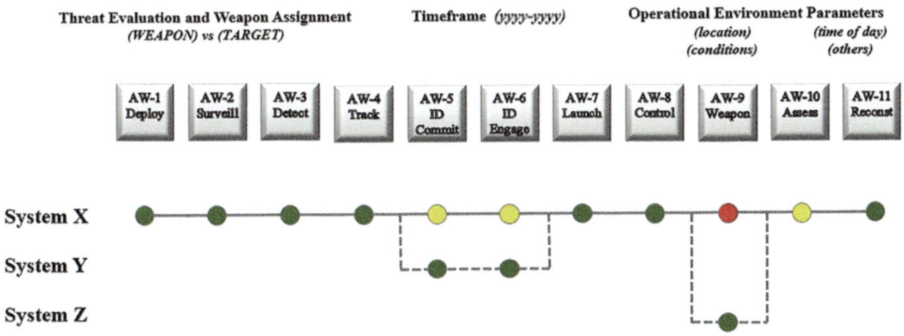

Fig. 8.23 Scoring criteria applied to mission engineering threads/webs

ing the capability (a SoS context) by driving mission criteria/requirements. Mission integration management review boards should evaluate the performance of the mission areas as part of performing Capability Portfolio Management.

Table 8.6 offers general scoring criteria for mission area assessments using critical measures to determine a platform's effectiveness score for particular effects chain/web. This critical measures table can be developed for the mission engineering thread to identify the key operational and performance criteria associated with each function and establish the basis for determining whether the required proficiency levels are achieved. The criteria address critical aspects of end-to-end mission execution, including search and detection, tracking and identification, engagement and weapon control, assessment, and post-mission performance, thereby helping identify functional deficiencies that could affect mission success.

8.11 Concept of Operations (CONOPS) Development

To execute MEI effectively within a complex organizational environment, the mission engineer must develop a Concept of Operations (CONOPS). A CONOPS serves as a foundational document that outlines the overarching strategy, objectives, and operational framework for a mission or system. In the context of MEI, a well-defined CONOPS provides a clear understanding of the mission's scope, requirements, and dependencies. It serves as a blueprint for integrating diverse capabilities and resources to achieve mission success. By delineating roles, responsibilities, and operational procedures, a CONOPS facilitates alignment and coordination among stakeholders, ensuring unity of effort and minimizing misunderstandings or conflicts during mission execution. Furthermore, a CONOPS enables iterative planning and adaptation, allowing for flexibility in response to evolving mission requirements or changing operational environments. Development of a robust CONOPS is

Table 8.6 Effects chain scoring methodology

Effects-Chain Functions		Critical Measures
Prepare/Configure		Preparations and selected configurations allow successful completion of the effects chain?
Search		Can provide full search coverage of tactical situation (TACSIT) threat sector?
		Can provide Desired Persistence Level?
		Collects, integrates, correlates, and disseminates surveillance information with acceptable operator workload?
Detect		Detects potential target at TACSIT-required ranges?
		Sufficient cue accuracy for tracking platform?
Track		Tracks multiple targets with sufficient accuracy to enable the weapon to meet weapon lethality requirements?
		Each target is maintained in one and only one track?
		Sufficient tracking support pre- and post- weapon launch for weapon lethality?
		Provides tracks at TACSIT-required ranges?
		Simultaneously tracks the TACSIT-required number targets?
Identify	**Commit**	Classifies targets to TACSIT-required level for decision making?
		Sufficient accuracy to assign Commit-Identify (ID) attribute to tracked target?
		Classification occurs at or beyond the TACSIT-required range?
	Engage	Identifies targets to the level required by the TACSIT?
		Sufficient accuracy to assign Engage-ID attribute to tracked target?
		ID occurs at or beyond the TACSIT-required range?
Defend		Can kill chain function(s) be performed without additional risk?
	Launch	Successful scheduling and employment of the weapon?
		Weapon integrated so all TACSIT-required capabilities are available?
	Control	Sufficient messages between controller and all supported weapons?
		Provides post-launch data with sufficient timelines and accuracy to support the weapon in a successful engagement?

Table 8.6 (continued)

Engage	**Weapon**	**Range Acquisition**	Sufficient kinematics or range to arrive at target with enough energy to produce TACSIT-required lethality?
			Employment range meets TACSIT-required re-engagement criteria?
			Employable at sufficient range to ensure debris does not endanger defended point?
			All rounds guide to the intended target?
		Lethality	Weapon is effective to TACSIT-required levels against the target?
Assess			The outcome of each engagement is correctly assessed?
			Assessment is timely enough to support re-engagement?
Post-Mission Scoring			Post-mission functions allow for successful kill chain completion when it is executed again?

therefore essential for enhancing mission effectiveness, promoting interoperability, and fostering collaboration across organizational boundaries.

Developing a CONOPS for specific missions involves a comprehensive approach and integrates various elements such as technology, resources, and operational procedures required to achieve mission success. MEI begins with a thorough analysis of mission objectives, constraints, and requirements to include the identification of key stakeholders, understanding the mission environment, and assessing potential risks. Once the foundational analysis is complete, the mission engineer uses this MEI approach to conceptualize the operations by developing potential mission engineering threads. This involves defining the overall mission architecture, outlining the roles and responsibilities of each component, and specifying the interactions between different elements. Factors considered include communication protocols, data exchange, and interoperability, all necessary to ensure seamless coordination during the mission. The example shown in Fig. 8.24 illustrates the constituent elements and communication events linked to the Joint Close Air Support CONOPS. The solid-line arrows, identified by communication event numbers, serve as instances of communication need lines. Subsequently, the information exchange requirements for each communication event are outlined in a standardized pattern.

Incorporating SoS engineering principles, the mission CONOPS addresses the integration of technology into the mission framework. This includes selecting appropriate technologies, designing interfaces, and ensuring the compatibility of systems. Other factors considered are reliability, maintainability, and scalability. Continuous refinement and validation of the CONOPS are essential steps in the MEI approach. This may involve simulations, tabletop exercises, and iterative reviews to identify potential issues and optimize the mission design. Additionally,

Joint Close Air Support

Mission Event No.	Description
1	Unit detects target
2	Commander decides to request CAS
3	Unit notified TACP
4	TACP passes request to ASOC < 5 min
5	ASOC coordinates with senior ground HQs which approve request
6	ASOC assigns on-call aircraft
7	CRC send aircraft to contact point (CP)
8	AWACS passes critical updates to aircraft > 95% Acrcy
9	JTAC briefs aircraft < 3 min
10	Aircraft depart initial point (IP)
11	JTAC controls CAS aircraft
12	Bombs on target > 98.9 % PK
13	Assessment

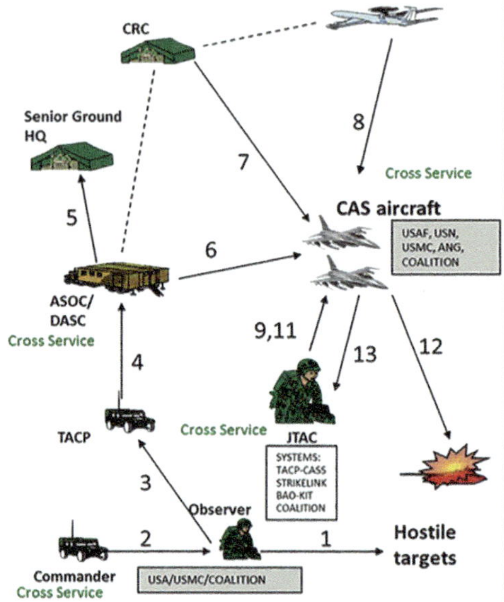

Fig. 8.24 Concept of operations for joint close air support example

feedback from stakeholders is crucial to ensure the concept aligns with expectations and requirements. Ultimately, MEI aims to create a robust and flexible CONOPS that maximizes mission success while minimizing risks. The MEI approach is holistic, considering not only the technical aspects but also the human and organizational factors that contribute to mission effectiveness. Through thoughtful analysis and collaboration, MEI sets the foundation for the successful execution of specific missions.

As we refine the **Mission Engineering and Integration (MEI) approach**, it becomes clear that every decision carries inherent risks that can impact mission success. Effectively assessing and managing these risks is essential to ensuring that proposed "to-be" SoS configurations are both feasible and resilient. The next chapter explores **risk assessment and management** and focuses on how to identify, analyze, and mitigate risks associated with critical decision-making within the MEI approach, ultimately strengthening the reliability and effectiveness of operational solutions.

References

1. Heisenberg, W. (1927). On the illustrative content of quantum theoretical kinematics and mechanics. *Journal of Physics, 43*(3–4), 172–198.
2. Moreland, J., Jr., et al. (2014). *Mission engineering integration and interoperability* (pp. 4–15). NSWCDD Leading Edge Technical Digest, NSWCDD-MP-15-00096 01/15. Retrieved from https://www.navsea.navy.mil/Home/Warfare-Centers/NSWC-Dahlgren/Resources/Leading-Edge/I-I-Leading-Edge/
3. Department of Defense. (2020). *Mission engineering guide v1.0*. Office of the Under Secretary of Defense for Research and Engineering. Retrieved from https://ac.cto.mil/wp-content/uploads/2020/12/MEG-v40_20201130_shm.pdf
4. Department of Defense. (2023). *Mission engineering guide v2.0*. Office of the Under Secretary of Defense for Research and Engineering. Retrieved from https://ac.cto.mil/wp-content/uploads/2023/11/MEG_2_Oct2023.pdf
5. Moreland, J., Jr. (2019). *Mission engineering and integration guide*. Retrieved from https://www.ise.vt.edu/content/dam/ise_vt_edu/adjunct/MEI%20Guidebook_Nov%202019.pdf
6. Moreland, J. D., Jr. (2009). Structuring a flexible, affordable naval force to meet strategic demand into the 21st century. *ASNE Journal, 121*(1), 35–51.
7. Peters, B. G. (2015). *Pursuing horizontal management: The politics of public sector coordination*. University Press of Kansas.

Chapter 9
Risk Assessment and Management

The level of critical thinking and judgment applied to risk assessment and management is pivotal in determining whether a program or mission can consistently meet its objectives. Managing these aspects requires strategic foresight, beginning with foundational decisions on program and operational means that consider the specific uncertainties and risks unique to each situation. The mission engineer must exercise rigorous analysis and well-informed judgment to successfully identify and control risks as the core to sound planning and management.

Risk is an inevitable part of life, woven into nearly every aspect of human experience and integral to both public and private sector activities. While definitions of risk vary across industries and organizations, they all share a fundamental concept: the uncertainty of outcomes. This uncertainty highlights that despite our best efforts to predict and prepare, we must navigate the unknowns that accompany any endeavor or decision.

In this chapter, the terms *risk*, *issue*, and *opportunity* are defined as follows [RISK223]:

- ***Risks*** are potential future events or conditions that could negatively impact the achievement of program or mission objectives related to success measures with uncertainty in the outcomes. Risks are characterized by (1) the probability (between 0 and 1) or likelihood of an undesired event or condition occurring and (2) the consequences, impact, or severity of the event if it were to take place, as depicted in Fig. 9.1.
- ***Issues*** are events or conditions that created a negative impact, such as risks that have materialized.
- ***Opportunities*** are potential future events that could positively impact the program's and/or mission's cost, schedule, and/or performance baseline.

The risk reporting matrix is shown in Fig. 9.1. While commonly used in risk assessment, it presents notable limitations. First, poor resolution affects risk matrices, as they often only compare a small subset of hazard pairs accurately, leading to "range

J. D. Moreland, Jr., *Mission Engineering and Integration*,
https://doi.org/10.1007/978-3-032-32132-9_9

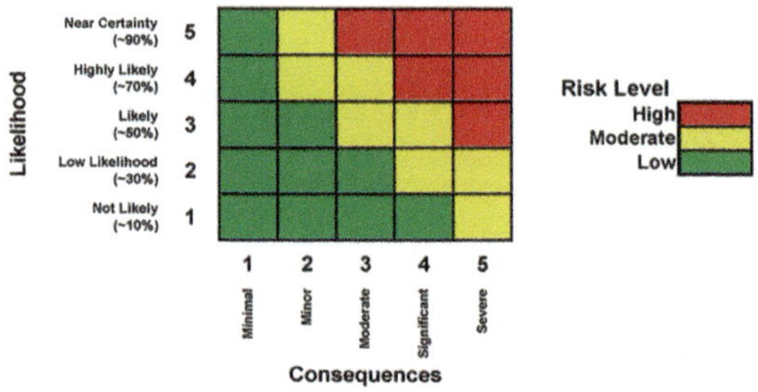

Consequence Level	Technical Performance/Life Cycle Sustainment	Schedule	Cost
1	Minimal or no consequence to technical performance.	Minimal or no impact	Minimal or no impact
2	Minor reduction in technical performance or supportability; can be tolerated with little or no impact on program.	Able to meet key dates	Budget increase or unit production cost increase (< 1% if Budget)
3	Moderate reduction in technical performance or supportability with limited impact on program objectives.	Minor schedule slip; able to meet key milestones with no schedule float	Budget increase or unit production cost increase (< 5% if Budget)
4	Significant degradation in technical performance or major shortfall in supportability; may jeopardize program success.	Program critical path affected	Budget increase or unit production cost increase (< 10% if Budget)
5	Severe degradation in technical performance; cannot meet Key Performance Parameters or key technical/supportability thresholds; will jeopardize program success.	Cannot meet key program milestones	Exceeds Acquisition Program Baseline threshold (> 10% if Budget)

Fig. 9.1 Risk reporting matrix

compression." Range compression results when risks of different magnitudes receive identical ratings. This typically happens when the number of categories available in the matrix is limited, which may reduce complex, continuous risk data into just a few broad classifications (e.g., "low," "moderate," "high"). Two risks that are quantitatively quite different may be assigned the same qualitative rating, reducing the matrix's effectiveness in clearly distinguishing higher-priority risks from lower ones. Distinguishing between varying levels of risk then becomes challenging. Second, the risk matrices can lead to errors in prioritization; matrices may mistakenly assign higher ratings to quantitatively smaller risks. In cases where risk frequency and severity are inversely related, matrices can even lead to decisions worse than those made randomly, undermining reliable risk assessment. Third, risk matrices are problematic for resource allocation. The categorical structure of a risk matrix does not support optimized allocation to mitigate specific risks because it does not account for nuances required for strategic resource deployment. Finally, ambiguity in inputs and outputs hinders objective use of risk matrices; consequence and likelihood often require subjective judgment. This subjectivity can lead to varying interpretations and conflicting ratings for the same risk, depending on the user. Due to these issues, users should approach risk matrices cautiously with a clear understanding of their limitations and explanations of the inherent assumptions.

In the context of MEI, risk refers to the potential for adverse outcomes that could impact the achievement of mission objectives due to uncertainties or challenges encountered throughout the lifecycle of the SoS capability. This includes risks associated with designing, acquiring, integrating, and sustaining the individual components and the collective SoS needed to support mission execution. Risk in this context is shaped by multiple factors, such as the technical performance of systems, reliability of integrations across systems, scheduling uncertainties, cost limitations, and operational readiness of both standalone systems and interconnected SoS architectures. In acquisition, risk also encompasses the challenges of aligning system capabilities with evolving mission requirements, managing interoperability across legacy and new systems, and ensuring that systems meet the operational availability targets to minimize downtime and enhance mission success. Effectively managing risk in MEI is essential to delivering systems and SoS that meet mission demands, are sustainable over time, and provide the necessary agility to adapt to dynamic mission scenarios and emerging threats. To meet the demands of a competitive environment and maintain an edge over near-peer adversaries, mission engineers need to shift beyond existing approaches to acquisition and risk management. Historically, acquisition processes have been encumbered by excessive procedures aimed at mitigating risk, even though achieving a completely risk-free system is impractical. Recognizing this challenge, risk must be treated as a critical factor in design, development, and testing. This amounts to a fundamental shift, focusing on streamlining high-risk acceptance processes. Eliminating redundant tasks and adopting a tailored, risk-based framework require new constructs—adaptive acquisition framework, contractor owned, and experimentation—to align with program-specific needs:

- **Adaptive Acquisition Framework (AAF):**
  - The AAF construct consolidates risk acceptance into a single milestone decision authority event before the initial system test. This approach minimizes paperwork and timeline churn while maintaining oversight through ongoing system performance assessments.
  - Example: The T-7A Red Hawk trainer aircraft leverages the AAF construct to expedite its development and testing phases.
- **Contractor Owned (CO):**
  - Tailored for programs involving contractor-operated systems, the CO construct distinguishes between civil operations and public operations. Risk acceptance responsibilities are aligned with operational authorities and tailored based on the regulatory framework of the operational environment.
- **Experimentation:**
  - Focused on experimental systems, this construct adapts risk assessment rigor based on available data and program objectives. It aligns decision-making authority with the operational stakeholder decision hierarchy rather than acquisition pathways.

The goal of the risk assessment framework and process is to ensure ongoing development and testing activities, enabling programs to more effectively identify and understand the risks being accepted while simultaneously working to minimize those risks for the operational communities. Before a risk can be assessed or managed, the risk manager needs to determine the risk attitude of the key decision-makers. Understanding their risk attitudes is essential for making informed decisions that align with mission goals as well as tolerance for uncertainty.

9.1 Risk Attitude

Risk attitude refers to a stakeholder's disposition or approach toward risk. It encompasses how one perceives, evaluates, and reacts to uncertain situations that involve potential losses or gains. Risk attitude is an important element in decision-making processes and can significantly influence choices in personal finance, business strategies, investments, operational decisions, and other areas where uncertainty plays a role. There are three main categories of risk attitudes:

- ***Risk-averse:*** Denotes individuals or entities that prefer to avoid risk and are willing to accept lower returns or benefits to minimize the possibility of loss. They prioritize stability and security over potential higher gains that come with higher risks.
- ***Risk-seeking:*** Includes those who are willing to take on higher levels of risk in pursuit of higher rewards. Risk-seekers are more comfortable with uncertainty and potential losses and may actively seek out high-risk, high-reward opportunities.
- ***Risk-neutral:*** Describes individuals or organizations that make decisions based on expected outcomes without a strong preference for or against risk. Those who are risk-neutral evaluate potential gains and losses objectively and choose options that maximize expected value, regardless of the risk involved.

In each category, risk attitude is influenced by various factors, including personal experiences, financial situation, cultural background, psychological traits, operational environment and conditions, and the specific context of the decision at hand. Managing risk effectively requires alignment with stakeholders' risk attitude, which can vary from risk-averse to risk-seeking, with a neutral stance falling between. A risk-averse attitude typically leads to a cautious risk management strategy, one based on prioritizing thorough identification and assessment of even low-probability risks, extensive planning for contingencies, and closely monitoring to avoid negative outcomes. While this approach can safeguard against unexpected setbacks, it may also slow progress and limit innovation. Conversely, a risk-seeking attitude favors pursuing higher rewards despite greater uncertainty, leading to faster decision-making, often with a focus on addressing only critical risks and fewer preventive measures. Risk-seekers may place more emphasis on adaptive responses to unforeseen challenges than on detailed planning. Meanwhile, those with a risk-neutral attitude aim for balance, making decisions based on expected outcomes without a strong preference for minimizing or embracing risk. This balanced approach typically results in objective risk prioritization and flexible strategies that adjust based on emerging data and changes in the environment. Tailoring risk management to align with risk attitude ensures that an organization's strategy reflects its overall mission and values.

9.2 Risk Assessment

Risk assessment in the context of program development, execution, and MEI is a systematic process that identifies, evaluates, and prioritizes potential risks that may impact achieving objectives throughout the lifecycle of a project or mission. In program development and execution, risk assessment involves examining technical, schedule, cost, and operational uncertainties that could hinder progress or result in failure to meet performance goals. By analyzing these factors, stakeholders prioritize which risks need immediate attention and develop mitigation strategies to address them proactively. In MEI, risk assessment extends to evaluating how complex systems and SoS operate under real-world conditions, considering both current and future operational environments. This includes assessing the interaction between systems, integration challenges, and evolving competition that could impact mission success. A comprehensive risk assessment enables decision-makers to balance potential gains with anticipated challenges, optimize resources, and ensure readiness, adaptability, and resilience in complex operational scenarios. Ultimately, the effectiveness of risk assessment in both program development and MEI hinges on its ability to provide insights that inform stakeholder decisions, guiding both preventative measures and contingency planning to safeguard mission outcomes.

Essentially, risk assessment aims to provide a comprehensive understanding of risks to inform key decision-making. Risk management encompasses the broader process of addressing those identified risks within risk assessment. While risk assessment lays the groundwork for understanding and quantifying risks, risk management actively engages in protecting from potential adverse outcomes through actionable measures.

9.3 Risk Management

Risk management constitutes a significant aspect of program management and mission execution. It draws from all disciplines, including SE, use of models and simulation, requirements definition, developmental and operational test, production planning, quality assurance, and logistics. Risk management needs to be both top-down (leadership) and bottom-up (working-level) to be successful. Effective managers encourage all stakeholders to take ownership of risk management. This responsibility includes the identification of risks, issues, and opportunities and, as appropriate, support of analysis, mitigation, and monitoring activities. Making risk management successful depends on process, but more importantly on people with knowledge and experience in the disciplines relevant to the product or mission activity. It is also critical to have the resolve to identify and address the risks that could influence objectives. An organizational climate open to external perspectives for independent reviews strengthens risk management effectiveness. Well-understood requirements, an integrated schedule coupled with success metrics, an independent cost estimate, and the tenacity to pull on the activities that reveal problems all contribute to prospects for success. While the process described in this chapter enables risk management, the risk mitigation plans for individual risks (output of the processes) are significantly more important. In the end, what matters most is not the process itself, but the quality and effectiveness of the risk mitigation plans and their implementation in reducing the risks to realizing objectives.

The steps of the risk management process are generally applicable to the management of risks and issues in multiple acquisition lifecycle or mission execution phases. At the same time, the nature of the specific actions taken for each step typically will differ depending on the phase of the implementation of the risk mitigation activity. The differences in the specific actions are driven by the changing types of individual risk, the information and tools available, the outcomes required, the degree of maturity and stability that must be demonstrated, and the residual risks that are tolerable following mitigation efforts. The risk management process must address planning, identification, analysis, mitigation, and monitoring of risks and issues.

9.4 Issue Management

Issue management is a crucial aspect of program and mission oversight, focusing on identifying and addressing events or conditions that have already occurred or are certain to occur in the future. Issues may arise from previously identified risks that have been realized or emerge independently without prior recognition of an antecedent risk. Regardless of origin, consequences of issues must be addressed promptly to avoid hindering operational progress. As the probability of identified risks increases, stakeholders should proactively anticipate their realization as issues and develop contingency plans to mitigate their impact. In cases where an issue arises from an unrecognized risk, stakeholders must quickly assess the consequences and determine the appropriate timing for resolution to formulate effective plans.

Moreover, stakeholders should evaluate whether these emerging issues could create additional risks, so that potential threats are appropriately assessed and managed. Issue management works in tandem with the risk management process, and programs benefit from leveraging common practices between the two while acknowledging their distinctive characteristics. Depending on the complexity and scale of activities, creating a separate issue-specific board can be beneficial. However, many entities can efficiently integrate issue management within existing structures. The primary goal is to maintain a balanced focus on both issues and risks, ensuring that immediate challenges do not overshadow ongoing risk management efforts. To facilitate effective issue management, stakeholders should establish a structured process that ensures issues are systematically identified, analyzed, addressed, and tracked to resolution. This process should empower the program to develop effective strategies for resolving critical or high-priority issues, validate these strategies at the appropriate decision authority levels, and allocate necessary resources for execution, thereby ensuring sustained progress toward stakeholder objectives.

9.5 Opportunity Management

Opportunity management is a strategic process that seeks to capture potential future benefits for a program or mission. Opportunities, unlike risks, focus on achieving positive outcomes, typically through proactive steps and resource allocation to attain objectives. Opportunity management is complementary to both risk and issue management. It aims to improve operational or programmatic outcomes by planning, identifying, analyzing, managing, and monitoring initiatives that can yield cost reductions, schedule improvements, and performance gains. In some cases, opportunity management also helps offset the impacts of identified risks.

To begin managing opportunities, stakeholders should implement a systematic identification and evaluation process. This starts with actively searching for potential enhancements that align with the operational objectives and stakeholder goals. As opportunities are identified, each is evaluated based on its potential benefits, the likelihood of achieving those benefits, and any associated risks. Thorough analysis enables stakeholders to weigh the costs, benefits, and potential risks before moving forward with an opportunity. Once an opportunity is approved, an opportunity management plan is developed to guide its realization while ensuring ongoing attention to risks and issues. Documenting the opportunity management process is crucial for consistency and transparency. With a structured approach to opportunity management, programs improve resource efficiency, which leads to meeting or exceeding cost-saving targets and enhances performance outcomes—factors driving overall mission success.

9.6 Risk and Issue Management Framework

Risk and issue management are interconnected processes that play a vital role in the success of program and mission execution. All initiatives inevitably encounter both risks and issues, necessitating continuous anticipation and proactive addressing. As defined earlier in this chapter, risks are identified by their probability of occurrence and the potential consequences should they materialize. Through effective risk

Fig. 9.2 Risk and issue management framework

management, programs allocate resources strategically to reduce the likelihood of a risk event or the impact if such an event were to occur. As the probability of specific risks increases, activities must anticipate the realization of these identified risks transforming into issues and implement mitigation plans early to manage the associated consequences. In contrast, issues differ fundamentally from risks in that their occurrence is certain rather than probable. Issues are characterized solely by their consequences, and issue management focuses on deploying resources to address and alleviate the negative outcomes linked to a past, present, or imminent event. Issues can arise from previously identified risks that have been realized, but they can also emerge without any prior risk recognition. Moreover, the occurrence of issues may generate new risks that need to be identified and managed.

To systematically manage both risks and issues, an adaptable framework can be applied across various phases of a program or mission lifecycle as shown in Fig. 9.2. This framework consists of five overarching steps for consistency; however, context can change depending on when these steps are applied to program or mission activities resulting in different specific actions. The next sections elaborate on the steps of this framework (Sects. 9.6.1, 9.6.2, 9.6.3, 9.6.4, and 9.6.5). By integrating risk and issue management, programs and missions enhance their resilience and responsiveness, ensuring effective navigation through uncertainties and challenges.

9.6.1 Step 1: Risk Management Process Planning

Effective process planning is essential for establishing a robust framework for proactive management of risks throughout the lifecycle of a program or mission. This process begins by assigning clear roles, responsibilities, and authorities to ensure that all stakeholders understand their specific contributions to risk management. Establishing a well-defined structure fosters accountability and enhances communication among team members. The concept of strategic risk management is not new. Businesses are constantly assessing the risks they face and taking steps to adjust to changing circumstances, whether by selling or purchasing new assets, taking on or reducing debt, or increasing or reducing their workforce. On a micro level, families are risk managers as well. They are constantly assessing risks faced in daily operations and respond by taking steps like fixing an old roof or getting more exercise to mitigate risks to property or personal health. Certain risks are accepted, like the risk of driving to work or allowing an old tall tree to remain right next to our home. The range of choices made in our lives is, in a sense, a form of strategic risk management.

Next, it is crucial to select and document an overall approach that encompasses the processes and procedures to be followed. This includes defining risk analysis criteria for assessing both the likelihood of potential risks and their possible consequences. Additionally, stakeholders should outline risk mitigation procedures that specify how identified risks will be addressed, thereby minimizing their impact on program objectives. Establishing traceability of identified risks to technical requirements and overall success goals is another key component of risk process planning. This alignment ensures that risk management efforts are directly connected to the success criteria and that strategies for risk management achieve stakeholder objectives. Furthermore, aligning the roles, responsibilities, tools, and information exchange across all stakeholders fosters collaboration and ensures a unified approach to risk management.

Determining the necessary resources for risk management is essential for effective execution. This includes establishing a budget for risk-related activities, allocating appropriate facilities, assigning qualified personnel, and creating a schedule that integrates risk management activities into the overall execution timeline. Finally, defining a risk management "battle rhythm" helps to establish regular intervals for risk reviews and updates. In this way, risk management remains an ongoing priority and adapts to changing circumstances.

9.6.2 Step 2: Identification

Identification is a critical step in the risk management process, providing the foundation for addressing potential challenges throughout the lifecycle of a program or mission. It begins with a comprehensive understanding of the operational context and specific requirements that shape it. By grasping the context in which the operation will occur, teams can more accurately identify risks that may affect performance, cost, or schedule. Various methodologies are available to facilitate robust risk identification. For example, independent assessments offer unbiased perspectives on potential risks, while careful analysis of the requirements helps to pinpoint specific obligations and constraints that may introduce risks. Engaging in brainstorming sessions with subject-matter experts generates diverse insights and encourages risk identification from different angles. Additionally, conducting interviews with stakeholders and operators across competencies often yields valuable information regarding potential issues based on operational experience.

A review of similar or historical programs is useful for providing lessons learned that highlight risks that previously impacted activities with similar characteristics. Trade studies or case studies also help in identifying risks associated with various design alternatives, ensuring that decision-makers are aware of the implications of their choices. Moreover, analyzing performance measures, resource data, lifecycle cost information, and trends in data is essential. These analyses provide insights into the health and performance of execution activities, helping to identify potential risks that may arise as the mission progresses. Finally, assessing technical performance at all levels, ranging from individual systems to SoS, integrated products, mission execution success criteria, and external interfaces, is crucial for uncovering risks that may affect interoperability or overall capability performance. By employing a comprehensive approach to risk identification that leverages these methodologies and analyses, stakeholders enhance their ability to proactively address risks, ultimately contributing to successful outcomes.

A well-constructed risk statement is essential for effective risk management and contains two key elements: (1) the potential event that may occur and (2) the associated consequences of that event. When applicable, the statement should include a third element: the existing contributing circumstance or cause of the risk. These statements should clearly define events that could negatively impact the ability to meet mission success criteria, e.g., cost, schedule, and performance objectives. This structured approach to specifying and communicating risks is critical to avoid ambiguity and inconsistency. One effective approach is the "if–then" format, which articulates the risk event or condition in the "if" clause and outlines the potential consequences in the "then" clause.

Risk Statement Format

"If <event X> happens then there is a risk <consequence> that the project could be impacted in <Y way>"

Example: If the new HR software is not delivered by October 1, then we may have to extend the contracts for the HR software testing team, meaning an increase in budget would be required.

Broken down: If [EVENT] the new HR software is not delivered by October 1, then we may have to [CONSEQUENCE] extend the contracts for the HR software testing team meaning an [IMPACT] increase in budget would be required.

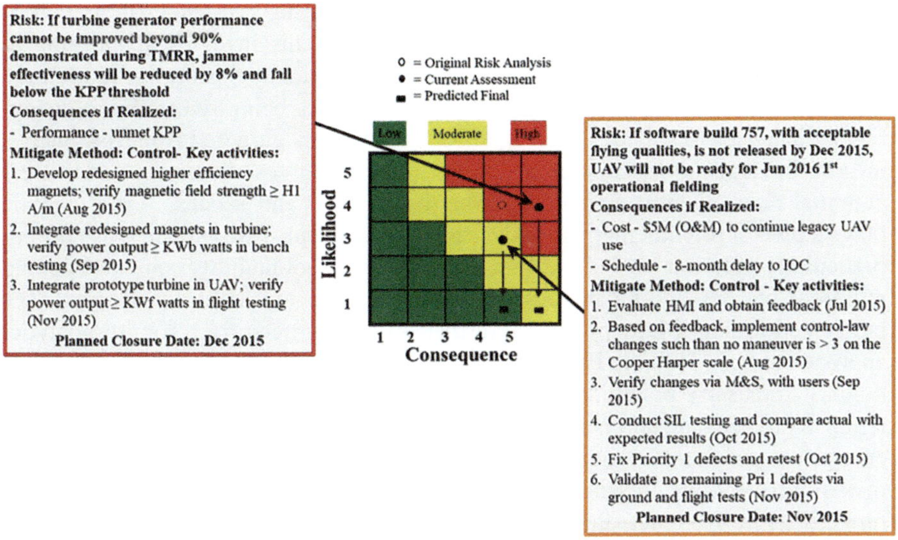

Fig. 9.3 Correct format for risk identification

For consistency, stakeholders are encouraged to adopt a single format for writing risk statements and to ensure that each statement is clear and concise, without including potential risk mitigation strategies or extraneous information. Figure 9.3 provides an example of strong risk statements and shows the linkage to the risk matrix for scoring.

9.6.3 Step 3: Analysis

Analysis is a critical component of the risk management process and focuses on answering two fundamental questions: (1) What is the likelihood of a risk event occurring and (2) What are the potential consequences? During this step of the process, estimates of the probability that a risk event will happen and assessment of the possible outcomes in terms of the success criteria are determined. This analysis culminates in determining the overall risk level, which is essential for prioritizing risks and guiding mitigation efforts. By estimating each risk's likelihood and consequence, decision-makers use risk analysis to provide valuable insights that enable stakeholders to effectively manage and prioritize their response strategies. Utilizing consistent, predefined criteria for likelihood and consequence creates a structured approach for evaluating risks and allows decision-makers to make objective comparisons. It is crucial for stakeholders to adopt a common framework for risk analysis to ensure alignment and facilitate communication.

While risk analysis often includes a degree of subjectivity and qualitative assessment, stakeholders should aim to bolster their analysis with quantitative data whenever feasible. A practical approach for conducting risk analysis is to focus initially on the potential impact or consequences. This involves quantifying the impacts on various factors across all phases of operational execution. Additionally, stakeholders must assess the likelihood of these risks being realized and conduct regular analyses to support ongoing tasks. The insights garnered from risk analysis inform the decision-maker's ability to prioritize resources effectively and mitigate potential risks, thereby enhancing overall resilience and success. Data analytics plays a pivotal role in enterprise-wide risk management (ERM), which has become particularly relevant and significant for modern accountability and transparency requirements demanding increased attention to risk and controls. Recent, well-publicized financial collapses have further driven the need for a robust, analytical approach to risk. Holistic ERM benefits from data analytics by enabling the stakeholders to classify potential events as either risks or opportunities, a process that requires precise data and comprehensive governance. By integrating data analytics, enterprises gain tools to quantify the impact of potential risks, allowing decision-makers to see both present risks and forecasted trends. This quantifiable insight turns ERM from a reactive process into a proactive framework. Accurate data, coupled with modern analytical techniques, provides organizations with the reliable measurements needed to make informed, strategic decisions. Without such metrics, risk management remains uncertain and reactive. Thus, a data-driven ERM approach not only bolsters risk control but also empowers informed decision-making that strengthens organizational resilience.

Once the analysis of likelihood and consequence is completed, stakeholders should utilize a risk matrix as shown in Fig. 9.1 to effectively communicate the results of their assessment. This matrix combines the likelihood of a risk occurring with the maximum consequence scores related to established success criteria, ultimately categorizing each risk into one of three levels: low (GREEN), moderate (YELLOW), or high (RED). This color-coded risk-level scoring provides a clear and concise method for presenting top-level risk analysis to stakeholders, making it easier to identify and prioritize areas requiring attention. While these risk-level ratings are essential for initial assessment, several additional factors should be considered when prioritizing risks. A primary consideration is the cost-effectiveness of potential risk mitigation options, as this plays a crucial role in determining how stakeholders allocate their limited resources among competing risks. Next is the frequency of occurrence of the risks, which should also be evaluated on a regular basis. Finally, stakeholders should appraise the timeframe in which they may materialize and any interrelationships between different risks that could compound their impact. Incorporating these factors into the risk reporting process establishes a better understanding of the risk landscape and ensures the most significant threats are addressed efficiently and effectively. This comprehensive approach to risk reporting enhances decision-making and supports the overall success of operations.

9.6.4 Step 4: Mitigation/Correction

Mitigation leading to correction is a critical component of risk management, outlining the options or combination of options and the specific implementation approach designed to address identified risks. It answers pivotal questions such as, "What is the plan to address the risk?" and "Should the risk be accepted, avoided, transferred, or controlled?" Following a thorough risk analysis, stakeholders should develop a strategy that evaluates the four mitigation options discussed in the strategies below to determine the best course of action or a hybrid of options based on the analysis, prioritization, and potential for risk reduction. These strategies are essential for managing identified risks effectively, allowing stakeholders to minimize adverse impacts or capitalize on beneficial opportunities.

Risk Response Strategies

Avoidance involves altering plans or requirements to eliminate the risk entirely, often by removing risky elements or changing the program approach. For example, a program might choose a proven technology over an untested one to avoid associated technical risks.

Transfer shifts the risk to another party, such as through insurance, outsourcing, or contractual agreements, making the third party responsible for managing the risk. This approach is often used when a program lacks the capacity or expertise to control the risk internally.

Mitigation reduces either the likelihood of a risk occurring or its potential impact, typically through proactive measures such as additional testing, design redundancies, or incremental development stages. Mitigation strategies can be complex, as they require detailed plans to control the risk within acceptable levels and continuous monitoring to assess effectiveness.

Acceptance, meanwhile, applies when a risk falls within an acceptable range or when it is impractical to eliminate or reduce further. Here, the program acknowledges the risk and monitors it without active intervention, sometimes establishing contingency plans if the risk materializes.

Each response strategy should be carefully evaluated based on cost, schedule, impact, and the program or mission objectives. The goal is to select a strategy or combination of strategies that align with the stakeholder's risk tolerance and resources while addressing potential adverse effects as effectively as possible. A well-documented treatment plan, clear accountability, and periodic reassessment ensure the chosen strategies remain relevant and effective throughout the program's lifecycle.

The chosen strategy for managing risks must be documented, ensuring it is presented at all significant decision points and milestones. This strategy should detail the specific actions to be taken, the timeline for implementation, the individuals responsible, and the anticipated impacts on the established success criteria, as well as the resources required to execute the individual risk mitigation plan, see Fig. 9.4 for an example.

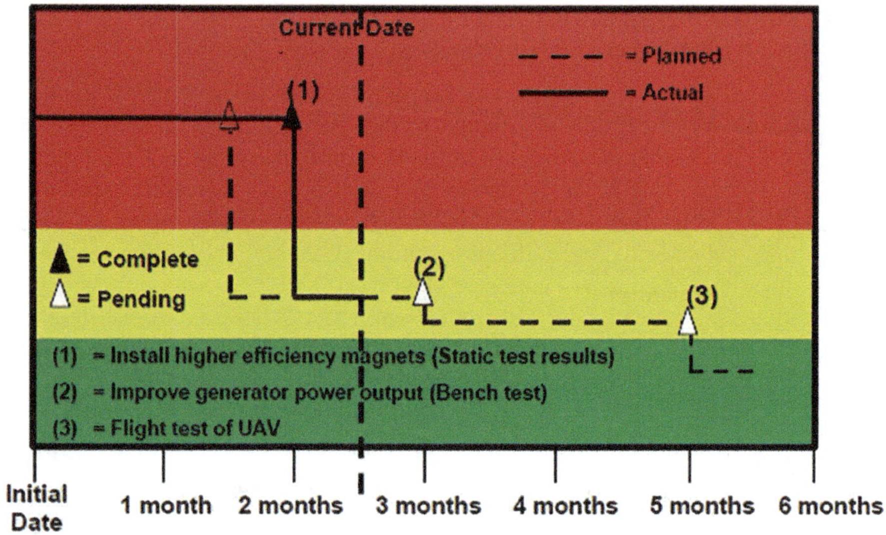

Fig. 9.4 Mitigation or "burn-down" plan

Key aspects of risk mitigation should include designing systems or SoS with features that enhance resiliency, such as safety margins, growth provisions, modularity, cybersecurity measures, graceful degradation, and avoidance of single points of failure. These provisions are established through careful analysis to maintain reasonable levels of risk avoidance without incurring excessive costs or compromising capabilities.

Some risk mitigation activities are executed as contingency plans activated by specific triggering events, emphasizing the importance of readiness. The level of detail in risk mitigation planning varies according to the operational execution phase and the nature of the risks involved, yet it must provide enough granularity to estimate the effort and technical scope needed, particularly in complex systems. In addition to considering the accept, avoid, and transfer options, stakeholders should prioritize the most effective mitigation strategies and select the optimal implementation approach for those strategies. It is crucial to keep appropriate peers and stakeholders informed about high-risk items, escalating issues as necessary, and including interdependency risks to assess the broader impact of risk mitigation actions across interrelated activities. This comprehensive approach ensures that risk mitigation is effective and aligned with overall program objectives.

9.6.5 Step 5: Monitoring

Monitoring is an important aspect of the risk management process that answers the essential questions: "How has the risk changed?" and "How effective are the risk mitigation plans?" This continuous process involves systematically tracking and evaluating the performance of risk mitigation strategies against established metrics throughout the project or mission lifecycle. It is imperative to recognize that not all risk mitigation efforts will yield successful results. Therefore, stakeholders must regularly reevaluate the effectiveness of the risk mitigation approach and associated activities to determine if additional actions are necessary. Potential decision points and actions must be identified to facilitate proactive responses to changing risk conditions. This monitoring step entails thorough recording, maintenance, and reporting of risks, risk analyses, risk mitigation efforts, and tracking results.

When a risk changes significantly, the stakeholders must adjust the risk mitigation strategy accordingly. If a risk is assessed to be lower than previously analyzed, the team may consider reducing or canceling risk mitigation activities, thereby reallocating resources to other areas. Conversely, if the severity of a risk increases, appropriate and timely risk mitigation efforts must be developed and implemented to address the heightened concern. All changes to the risk mitigation strategy should be documented and archived for future reference and historical analysis. Successful risk monitoring includes establishing timely and specific reporting procedures to ensure effective communication among the stakeholders.

Effective risk monitoring involves not only tracking the implementation and progress of risk mitigation activities but also ensuring that success measures are integral to the monitoring process after selecting the appropriate risk mitigation strategy. Regular status updates are critical to assessing changes in the likelihood and consequences of risks. Documenting risks that can be retired, as well as those that remain under mitigation, is also essential to prevent an unnoticed relapse of previously retired risks. Moreover, keeping lines of communication open is vital, enabling timely notifications to stakeholders regarding any ineffectiveness in mitigating risks.

9.7 Risk Categories

Risk categories and types help structure risk management by grouping risks into commonly encountered areas, facilitating more focused assessment, mitigation, and monitoring. Risks often fall into categories such as technical, programmatic, operational, and business (external).

Technical risks include challenges related to design, technology readiness, or integration, where innovations or dependencies on unproven technologies might fail to meet performance standards.

Programmatic risks stem from resource limitations, contractual obligations, or schedule dependencies, where unexpected shifts in funding, delays, or contract-related uncertainties could impact program execution.

Operational risks involve aspects related to real-world deployment or use, addressing factors like usability, maintainability, and alignment with mission requirements. These can become critical when considering the eventual user environment and potential changes during operational deployment.

Business (external) risks, not considered core with the proceeding categories, arise from broader factors outside the immediate control of the program team. These include economic shifts, regulatory changes, or supplier issues that can disrupt timelines and costs.

Within each category, risks can vary by type, such as strategic, financial, regulatory, or environmental risks. Distinguishing between risk categories and types allows stakeholders to better prioritize and apply resources across the range of potential risk sources and ensure high-impact areas receive focused attention. This organized approach aids decision-makers by clearly identifying the nature and origin of risks which in the end foster more effective and comprehensive risk management across the program's lifecycle.

Risk categories are essential in risk management, especially within the context of MEI, as they help organize and prioritize diverse risks across SoS and multi-mission environments. This operational landscape requires an in-depth understanding of risks that can affect mission success, which often means addressing interdependencies, technical challenges, and cross-functional factors that vary greatly in nature and impact. Organizing these into categories offers several key benefits as discussed below.

- **Improved risk visibility and focus**: By categorizing risks (such as technical, operational, cybersecurity, logistical, and safety risks), stakeholders gain a clearer picture of potential issues, making it easier to identify which types are most likely to affect mission success. For MEI, where risks span multiple systems and operational domains, categories enable risk teams to focus on the most critical risk types specific to mission success objectives.
- **Enhanced resource allocation**: Risk categories help prioritize resources and mitigation efforts. For example, risks related to integration challenges across different systems may require unique approaches and resources other than cybersecurity risks. Having defined categories allows risk teams to allocate resources efficiently, addressing the most impactful risks with appropriate and tailored mitigation strategies.
- **Streamlined communication and reporting**: Using categories to classify risks creates a common framework and language that can improve communication among stakeholders, including engineers, operators, and decision-makers. Multiple teams and disciplines are involved in MEI, so clear communication about risks is crucial for timely and coordinated responses. Categorized risks also simplify reporting by providing a structured view of risk across the operational mission landscape.
- **Systematic identification and analysis**: Categorizing risks supports a structured identification process, making it more likely that all relevant risks are identified early on. Risks are complex and may evolve as the SoS comprising the mission engineering threads interact or when operational conditions change.

Risk categories guide systematic reviews, helping to ensure comprehensive risk assessment and ongoing monitoring for each type of risk.

- **Effective integration planning and interdependency management**: MEI typically involves multiple interdependent systems forming mission engineering threads, where risks in one area can cascade into other areas. Risk categories help teams anticipate and manage these interdependencies by grouping related risks, resulting in more strategic integration planning and better handling of cascading risks that could jeopardize mission success objectives.
- **Informed decision-making and risk mitigation strategies**: Risk category structures facilitate the creation of targeted mitigation plans specific to each type of risk. This allows for more informed decision-making as stakeholders better understand the potential impacts of each risk category and prioritize efforts in alignment with mission goals and constraints.

9.8 Mission Execution Risks Considerations

Risk planners assess and articulate risks and identify potential mitigation. Identification, communication, and management of risk are fundamental in the MEI approach to understand the consequences of executing mission webs. In general, risk entails the probability and severity of loss linked to hazards. The severity of risk is determined by both likelihood of occurrence and magnitude of damage if the risk manifests. In most cases, risk and resources have an inverse relationship. As the level of resource commitment to a plan increases, the level of risk to that plan decreases. The identification and assessment of the mission web identify alternative mission engineering threads and resources that would reduce risk associated with the plan. However, resources are often inherently finite, and their use, or even planned use, may weaken overall security by creating opportunity costs and elevated risk elsewhere.

Assumptions are suppositions taken as true in the absence of proof. They are unavoidable in planning but incur risk. Planners must identify the role of assumptions in their plans and the impact if key assumptions prove to be invalid. Similarly, planners must also identify the impact of constraints and restraints on the operation. Any assumption that is not validated, to include assumptions from strategic guidance or objectives, becomes a risk to either the mission, force, or both. There is no magic formula for quantification of risk. Planners must provide decision-makers with an assessment of the expected risks, costs, and benefits, as well as the potential consequences of proposed actions. This assessment facilitates decisions to reduce, control, or accept risk through a shared knowledge of potential consequences.

From a military perspective, the decision to adopt a distributed force strategy to complicate an adversary's targeting introduces significant operational risks by reducing efficiency and increasing logistical complexity. High-tech weapon systems, such as modern fighter aircraft, rely on specialized maintenance equipment, which is traditionally centralized to maximize efficiency. Dispersed operations necessitate either the risky acceptance of limited maintenance capability or the costly acquisition, transport, and support of sufficient equipment at each operational site. In some regions with difficult physical landscapes, the logistical challenge is compounded by difficult terrain, poor infrastructure, and the need for multiple transportation modes, each vulnerable to adversary interdiction. Additionally, dispersed operations require minimizing centralized stockpiles, which further strains logistics over vast distances. The absence of intermediate hubs, critical for equipment reorganization and responsiveness, can undermine operational effectiveness. For example, the British reliance on Ascension Island during the Falklands War highlights the strategic importance of such nodes, without which large-scale dispersed operations may falter under logistical strain.

Risk to Mission

Risk to mission encompasses the ability of a force to effectively carry out current operations while managing acceptable levels of human, material, and financial costs. This risk assessment considers not only present capabilities but also the force's readiness to perform against evolving or anticipated challenges posed by competitors or threats. Evaluating risk to mission involves gauging whether the current capabilities and resources can meet operational demands without overextending personnel, depleting essential materials, or exceeding financial limits. Furthermore, this assessment incorporates expectations from critical stakeholders, ensuring that the force's performance aligns with strategic objectives and prepares for potential shifts in the threat landscape.

Mitigating risks to mission regarding logistics in contested environments requires a balance between efficiency and resilience. One approach is shifting from a "pull" logistics model, which relies on just-in-time delivery using predictive analytics, to a "push" logistics model, which operates on scheduled logistics packets based on anticipated consumption rates. While the push model is less efficient, as it may deliver unneeded supplies and strain transportation capacity, it is inherently more resilient to disruptions than the lean pull system. However, the reliance on intermediate storage nodes in push logistics introduces vulnerabilities, particularly in contested environments where these nodes are likely targets. Another risk mitigation strategy involves prepositioning stockpiles of essential materiel at likely dispersal sites. This reduces the strain on long-range supply chains during active operations but introduces its own set of challenges. Historically, cached materiel has often been abandoned due to hasty relocations during conflicts, and maintaining these caches incurs significant costs in terms of upkeep, security, and the opportunity cost of tying up resources. Additionally, cached sites may create detectable signatures, revealing operational plans and potentially inviting enemy targeting. Despite these drawbacks, prepositioning stockpiles can serve as a deterrent or provide critical support in certain scenarios.

A comprehensive approach to addressing risk to mission challenges at all levels of conflict is critical for ensuring operational effectiveness and resilience. In the case of the DoD, leadership has increasingly focused on contested logistics, examining vulnerabilities from the strategic to the tactical level. Strategically, it assesses risks to the industrial base and inter-theater transportation capabilities, such as sealift and airlift. At the theater level, the focus shifts to aerial refueling, intra-theater mobility, and emerging innovations like space-based supply chains. However, the most pressing challenges arise closer to the front lines, where new service concepts for dispersed operations face severe logistical hurdles as discussed in the previous paragraph. These land-based, dispersed-site strategies must contend with operating in heavily contested operational environments, presenting risks that could undermine their feasibility and effectiveness. Without a holistic view of risks spanning all conflict levels, these concepts can impose self-defeat or disproportionate costs on joint operations, undermining broader strategic objectives. Comprehensive risk assessment and mitigation planning as captured in the MEI approach are therefore essential to ensure sustainability and success across the spectrum of conflict.

Risk to Force

Risk to force refers to the potential dangers and vulnerabilities faced by personnel and assets in the execution of a mission. This risk encompasses the likelihood and impact of harm to individuals, damage to equipment, and other losses incurred during operations. Managing risk to force involves assessing factors such as the physical and environmental hazards of the mission environment, the operational readiness and protection of personnel, and the adequacy of resources like protective equipment, medical support, and extraction plans. Additionally, risk to force includes the psychological toll on personnel due to exposure to high-stress situations, as well as the strategic implications of potential casualties or loss of critical assets on morale and mission sustainability. Ensuring that risk to force is kept at a manageable level is essential for mission success and for maintaining the long-term health, safety, and effectiveness of the force.

In some cases of operational planning, risk to force is transferred to risk to mission introducing significant challenges in the context of contested logistics and operational effectiveness. Agile employment of forces, while designed to enhance resilience through dispersed and austere operations, exemplifies this trade-off. Historically in defense scenarios, achieving air superiority depends on high sustained sortie generation rates. However, the inefficiencies inherent in dispersed operations can severely constrain sortie rates, diminishing the force's ability to achieve operational objectives. These dynamics shift the burden of risk from protecting the force to compromising the mission. This underscores the importance of comprehensive risk assessments that account for the trade-offs between protecting the force and achieving the mission, ensuring that operational concepts like agile force employment align with strategic and logistical realities.

Risk to National Security

More applicable to the military sector, risk to national interests involves threat or competition to high-level priorities, including security, economic stability, alliances, and global influence. This type of risk reflects the potential for adversarial actions, regional conflicts, or international instability to disrupt key national goals or to compromise access to critical resources, trade routes, and strategic locations. It also considers how military and diplomatic decisions affect relationships with allies and how these actions are perceived on the global stage. Managing risk to national interests requires balancing immediate operational needs with long-term objectives, such as preserving national sovereignty, promoting economic security, and upholding commitments to allies and international norms. Failure to address this risk can lead to diminished global standing, weakened alliances, and an increased vulnerability to external influence or economic coercion, all of which can negatively impact a nation's overall strength and resilience.

Temporal Nature of Risk
Risk assessment in defense planning is inherently influenced by temporal considerations, requiring trade-offs across short-term, mid-term, and long-term horizons. For any given decision, defense planners must navigate the balance between immediate operational readiness, ongoing development of capabilities, and the overall capacity to sustain future missions. Investing heavily in current readiness ensures that the force is prepared to address immediate threats, yet it may limit resources available for developing new technologies or expanding force capacity, which are crucial for addressing emerging threats. Conversely, prioritizing long-term capability and capacity investments can leave immediate readiness gaps, potentially compromising response effectiveness in the short term. This temporal trade-off demands a strategic approach that aligns current needs with future objectives, balancing near-term operational demands with the resilience and adaptability required for enduring national security.

Operational Availability
Risk and operational availability (A_O) are closely linked when executing missions, as they both directly impact the ability of a force to perform effectively and respond to threats. A_O refers to the readiness of systems, equipment, and personnel to carry out missions as planned, while risk considers the probability and severity of potential negative events that could disrupt mission success. High risk often correlates with reduced A_O. For example, if equipment fails or personnel are injured, the mission may be delayed or require additional resources, lowering overall availability. Conversely, ensuring high A_O can reduce certain risks by maintaining adequate readiness levels. This involves prioritizing maintenance, resource allocation, and personnel preparedness to minimize vulnerabilities that could otherwise increase the likelihood of mission failure or compromise. Therefore, a balance must be achieved: minimizing risk enhances A_O by preparing for unforeseen challenges, while optimizing availability ensures that the force can sustain high performance, even under adverse conditions.

A_O indicates the percentage of time that a system or SoS within a unit are operationally capable of performing an assigned mission and can be expressed as (uptime/(uptime + downtime)). Determining the optimum value for A_O requires a comprehensive analysis of the system and its planned use as identified in the concept of operations, including the planned operating environment, operating tempo, reliability alternatives, maintenance approaches, and supply-chain solutions. A_O is fundamentally a probability function that incorporates the interdependent factors of reliability, maintainability, and supportability.

- **Reliability** is measured by **Mean Time Between Failure (MTBF)**, which assesses the average time a system operates before experiencing a failure. Higher MTBF values indicate better reliability.
- **Maintainability** is measured by **Mean Time To Repair (MTTR)**, which represents the average time required to repair a system after a failure. Lower MTTR values suggest that repairs are faster, enhancing availability.
- **Supportability** is measured by **Mean Logistics Delay Time (MLDT)**, which captures the time taken for logistical elements, such as spare parts and personnel, to become available for repairs or maintenance tasks. Lower MLDT values indicate an effective logistical support system, which minimizes delays.

By calculating and integrating these three components (MTBF, MTTR, MLDT), A_O predicts the likelihood that a system will be operational and ready for use at any given time, balancing readiness needs against resource constraints.

$$A_O = \mathrm{MTBF} / (\mathrm{MTBF} + \mathrm{MTTR} + \mathrm{MLDT})$$

High A_O values indicate a system that is both reliable and easy to maintain and well-supported logistically. Reliability and ease of maintenance and logistical support are essential for minimizing downtime and maximizing mission availability. This probabilistic approach allows defense planners to assess and enhance system readiness at each stage, from acquisition through operational deployment, aligning system design and supportability with the mission's operational demands. Additionally, for systems using Commercial Off-The-Shelf (COTS) hardware and software, it is crucial that logistics support is designed or selected concurrently to ensure that the program's supportability requirements are met.

Operational availability is a critical component when considering suitability, as it reflects a system's readiness to perform its intended functions under real-world conditions. Suitability evaluates whether a system can be deployed, operated, and sustained effectively, and operational availability captures the practical aspect of this by quantifying the percentage of time the system is available for use. This relationship highlights the dependence of operational availability on factors such as reliability (how often the system fails), maintainability (how quickly it can be repaired), and supportability (the adequacy of logistics and resources). These factors, in turn, are central to suitability, as they determine whether the system can reliably meet its Measures of Performance (MoPs) parameters while being accessible and functional for mission-critical operations. By integrating operational availability into suitability, decision-makers gain a comprehensive view of a system's capacity to consistently achieve its Measures of Effectiveness (MoEs) in operational settings. In this way, operational availability serves as a tangible expression of how well a system's suitability enables it to align its technical performance with mission success. Although MoPs quantify a system's technical capabilities, such as speed, accuracy, or capacity, these metrics alone are insufficient to guarantee mission success if the system is not suitable for sustained operation in real-world scenarios. This is where suitability becomes essential, as it ensures that the system can maintain its MoPs in the context of operational challenges, thereby enabling it to achieve mission objectives defined as MoEs. This availability is critical for meeting MoEs, as a system that is frequently out of service will fail to contribute to mission success, regardless of its inherent performance capabilities.

9.9 Development of Risk Measures

Developing effective risk measures involves a structured process to quantify the likelihood and consequence of potential risks on stakeholders' objectives. First, clearly defining the scope and goals of risk assessment is essential, as these will guide which metrics are relevant. Risk measures are typically built on two core components: likelihood (or probability) and consequence (or impact), which were defined and discussed in the beginning of this chapter. Likelihood assesses the chance of a risk event occurring, while consequence evaluates the potential effects on cost, schedule, performance, or reputation. To develop these measures, data collection is key because it provides the objective foundation needed for analyzing potential risks. Without robust data, assessments would be speculative and could lead to underestimating or overestimating risks, either of which can have significant consequences for decision-making and resource allocation. Sources of this data are drawn from historical data, expert judgment, statistical models, and scenario analysis to accurately estimate likelihood and impact. Quantitative risk measures can include expected value (probability multiplied by consequence), Value at Risk (VaR), which estimates potential losses within a given confidence interval, and conditional value at Risk (cVaR), for scenarios with extreme losses. Qualitative measures, such as high, medium, or low rankings, may be used when precise data is unavailable. Standardized scoring systems and matrices can facilitate comparison and prioritization of risks. Additionally, defining threshold levels for acceptable risk helps determine when mitigation actions are necessary. Regular reviews and updates to risk measures are crucial to ensure these measures reflect the current risk environment, as both internal and external factors affecting risk are subject to change over time.

9.10 Uncertainty in Risk Management

Incorporating uncertainty into risk management is essential for building resilience in decision-making and preparing for a range of possible outcomes. Since risk management often deals with incomplete information, especially in complex or dynamic environments, it is important to acknowledge and manage uncertainties to minimize potential negative impacts effectively. Here are some key approaches for integrating uncertainty into risk management:

- **Probabilistic Risk Assessment (PRA)**: PRA uses probability distributions to represent the likelihood and impact of different risks, allowing organizations to model the variability and uncertainty inherent in risk scenarios. Techniques such as Monte Carlo simulations provide a range of possible outcomes based on random variations in key parameters, enabling risk managers to visualize a spectrum of risks instead of just single-point estimates.
- **Sensitivity and scenario analysis**: Sensitivity analysis assesses how sensitive an outcome is to change in specific risk factors, helping identify which uncertainties have the greatest influence on risk exposure. Scenario analysis, on the other hand, evaluates risks under various hypothetical situations (e.g., best-case, worst-case, and baseline scenarios). Both methods help organizations understand how different factors and assumptions might affect outcomes, enabling more robust risk planning and contingency preparation.
- **Bayesian updating**: Bayesian analysis is a statistical approach that allows for updating risk estimates as new information becomes available. This approach is particularly useful in dynamic environments where risks evolve over time. Bayesian updating provides a structured way to refine risk estimates based on incoming data, allowing for continuous improvement in managing uncertainties.
- **Use of confidence intervals and ranges**: Rather than focusing on single-value estimates, risk assessments can express uncertainty by using confidence intervals. For example, a project's cost or timeline might be expressed as a range (e.g., "Cost expected to fall between $10 million and $15 million with 90% confidence"). This approach provides a more nuanced view of risk and helps stakeholders understand the level of certainty associated with different estimates.
- **Expert judgment and Delphi technique**: When quantitative data is scarce, expert judgment can be invaluable in assessing uncertain risks. Techniques like the Delphi method, a process of gathering and refining expert opinions, allow organizations to incorporate qualitative insights and consensus from specialists, providing a balanced perspective on risks that are hard to quantify.
- **Fuzzy logic and qualitative approaches**: Fuzzy logic enables organizations to manage ambiguities by categorizing risks with descriptive terms rather than precise numbers. This is useful when risks are difficult to quantify or involve complex, subjective factors. For instance, instead of assigning a precise likelihood to a risk, fuzzy logic might describe it as "moderately likely," which accommodates ambiguity and offers a flexible approach to uncertain risks.
- **Continuous monitoring and feedback loops**: Establishing a continuous risk monitoring process is essential for dealing with uncertainties that may shift over time. By keeping track of real-time data and continuously evaluating the effectiveness of risk mitigation measures, organizations can adjust their strategies promptly as uncertainties evolve. Feedback loops allow for the dynamic adaptation of risk management plans based on changing conditions.
- **Decision trees and expected value analysis**: Decision trees help organizations visualize different decision paths and their associated probabilities and impacts. By analyzing the expected value of each path, organizations can weigh the costs

and benefits of various options under uncertain conditions, helping to make informed decisions even when outcomes are uncertain.

Incorporating uncertainty into risk management is about building flexibility and adaptability into the risk management framework. By acknowledging uncertainty, organizations can move beyond binary risk assessments and embrace a spectrum of possibilities, allowing for more realistic planning and better alignment of resources. This helps organizations react to risks as they materialize and proactively prepare for a wider range of potential scenarios, increasing resilience in the face of the unknown.

9.11 Entering the Future of Digital Risk Management

The future of digital risk management is set against a backdrop of unprecedented technological advancement and an ever-expanding digital landscape. The rapid increase in computing power and the plummeting costs of technology since the 1970s have paved the way for an era where data generation is at an all-time high. As nearly every human activity is now digitally recorded, from health metrics to consumer behavior, it has become a daunting challenge to manage the vast amounts of data produced while ensuring robust risk management practices. For example, look at the banking sector, where customer expectations are rising sharply to enhance operational efficiency. Today's consumers demand instant service and personalized offerings, pushing banks to innovate rapidly. The implications for risk management are significant; financial institutions must adapt quickly to maintain competitiveness and meet new demands. A critical trend emerging is the stringency of the regulatory environments. In the case of banks, this will lead them to allocate substantial resources toward compliance and will serve as a catalyst for banks to digitize their risk functions to improve transparency and efficiency. Thus, digitization emerges as a strategic lever to manage regulatory burdens effectively, suggesting a synergistic relationship between compliance and technological advancement.

Several challenges impede the swift digitization of risk management. Legacy information technology systems and data quality issues rank as significant barriers, with many risk managers reporting that outdated technologies hinder their ability to access and utilize high-quality data effectively. Furthermore, the conservative nature of risk management, rooted in the necessity for cautious decision-making, may stifle the adoption of innovative digital practices. To counteract this inertia, risk leaders must cultivate a culture that embraces agility and iterative design, which appear to be at odds with traditional risk management practices. Additionally, talent shortages in the areas of digital technologies and analytics further complicate efforts to modernize risk functions, necessitating a concerted focus on training and recruitment. Looking ahead, the drive to digitize risk management will require a comprehensive transformation across organizational structures and processes. Risk management functions will need to broaden their data sources, leveraging nontraditional information to enhance their risk assessments. Automating routine processes and employing advanced analytics will be essential for improving accuracy and consistency. This holistic approach will enable organizations to remain agile in an ever-changing digital environment, ensuring that they can meet both regulatory requirements and customer expectations effectively. The future of digital risk management is promising, but significant challenges must be overcome, embracing a culture of innovation and collaboration across stakeholders.

Some general steps to follow in building a digital risk management ecosystem:

- **Data management** is crucial for a successful digital risk management program. Improved data governance and operational models will enhance data quality, leading to more consistent risk and business decisions and ensuring the stakeholders respond effectively to data needs. A key enhancement involves recognizing data risk as a fundamental component of the risk taxonomy, linking it to a defined "risk appetite" statement and a comprehensive data-control framework.
- **Process and workflow automation** is vital in enhancing a digital risk program. By automating tasks like collateral data entry, often utilizing robotic process automation (RPA), stakeholders can streamline various activities into smart workflows. These workflows consist of integrated sequences that involve both humans and machines throughout the entire process. This automation not only boosts efficiency but also creates a more seamless and timely experience.
- **Advanced analytics and decision automation** are transformative elements in modern risk management. Utilizing sophisticated risk models, such as those based on machine-learning algorithms, enables organizations to identify complex patterns. These advanced analytics enhance the accuracy of predictions related to defaults and other risk events. This potential for rapid decision-making streamlines processes and enhances the overall effectiveness of risk management strategies in a fast-paced digital environment.
- **Real-time and flexible infrastructure** is essential for supporting the various components of a digital risk program. This evolving risk infrastructure will facilitate the integration of innovative data-storage solutions, new user interfaces, and streamlined access to the vendor ecosystem. Employing techniques such as application as a service, which can be sourced from application service providers, enhances the adaptability of the infrastructure.
- **Smart visualization and interfaces** are poised to revolutionize how risk insights are communicated and utilized. By leveraging advanced technologies, risk management presents information in more intuitive, interactive, and personalized formats. This includes the development of risk dashboards or control towers, augmented-reality platforms for customer engagement, and other innovative interfaces designed to enhance user experience. These advancements will facilitate better understanding of risk data, thereby empowering stakeholders to make informed decisions swiftly, ultimately improving risk management outcomes.
- **External partnerships** are becoming increasingly vital for enhancing risk management capabilities. By collaborating with external providers, stakeholders significantly improve processes. This partnership approach allows critical team

members to leverage specialized expertise and resources across organizational boundaries that streamline operations and enhance efficiency. By integrating external resources, organizations can enhance their risk management frameworks and adapt more effectively to the evolving challenges of the digital landscape.
- **Culture and talent** are crucial components for evolving risk management in the digital age. Organizations will increasingly need personnel who are both digital-savvy and fluent in risk management and mission operations. This dual proficiency is essential for fostering an agile culture that embraces innovation and experimentation. In this context, new roles such as data scientists and modeling experts will be pivotal in a digitized risk function. This approach encourages a deeper understanding of risk while ensuring that team members can effectively leverage advanced analytics and technology to enhance decision-making. By prioritizing talent development in alignment with a forward-thinking culture, organizations can better navigate the complexities of modern risk management and capitalize on new opportunities.

Transforming risk management into a digital ecosystem is a complex and multifaceted endeavor that requires focus on some fundamental themes:

- ***Establishing a vision for digital risk:*** Define a clear vision to understand the future landscape of risk management for stakeholders. Articulate the key activities that the risk function will undertake the evolving mandate and role of risk, and the metrics for measuring success.
- ***Identifying digitization opportunities:*** Prioritize the application of digital tools to the most promising activities, supported by a robust business case that estimates the total potential impact. Take decisive steps to implement digital initiatives, even while simultaneously developing the necessary assets and skills to support these transformations in the future.
- ***Implementing a cohesive initiative framework:*** Run a series of coordinated efforts or pilots to achieve strategic goals and seize identified opportunities in small defined phases. This requires a well-thought-out governance structure and operating model, as well as the adoption of innovative techniques such as agile sprints and digital factories. Business rules or operating procedures need to be updated immediately as the new digital process is implemented. If not, stakeholders can easily revert to the old way of doing business and culture to accommodate their comfort level. By taking a balanced approach, organizations can successfully navigate the complexities of digital risk transformation while minimizing potential pitfalls.

Having explored **risk assessment and management**, we now turn our attention to the financial implications of the decisions made within the MEI approach. In Chap. 10, we will focus on **cost estimation of potential solutions**, examining the expenses involved in building, integrating, and fielding solutions to address operational gaps. This process is crucial for ensuring that the proposed solutions remain within affordability caps and economic constraints. It also helps balance effectiveness with fiscal responsibility throughout the lifecycle of a system. Considering both risks and costs promotes informed decisions to ensure the success of the mission while maintaining economic feasibility.

Reference

1. Department of Defense. (2023). *Risk, issue, and opportunity (RIO) management guide for defense acquisition programs*. Defense Acquisition University. Retrieved from https://www.dau.edu/sites/default/files/2023-10/DoD%20RIO%20Management%20Guide%2009-2023.pdf

Chapter 10
Cost Estimating Potential Solutions

10.1 Introduction

Synthesis and analysis have opposite meanings, but both describe a cost estimating process, from rough estimates to detailed approaches. An in-depth review of a cost estimating process reveals that cost estimating is indeed an assemblage of separate, distinct, and subordinate parts. This inward look would also reveal that each subordinate part is treated as a whole to be separated into its elements for examination, modification, and/or updating before becoming part of a total estimate. The challenge with force-level cost estimating is how to work across many systems at different maturity levels with a good understanding of the interfaces for integration and interoperability. Most cost estimating processes are both analytical and synthesizing in nature and are an effort directed toward accurately calculating the future costs of systems to SoS with varying levels of complexity. Many organizations are consumers of cost estimates and analyses. Cost estimates formulate and implement phases, non-advocate cost estimates, source selections, what-if exercises, affordability studies, economic analyses, and Analysis of Alternatives (AoA), as well as support numerous types of decisions related to projects. System cost must be a design variable to assure focus on major cost drivers during design and to challenge estimates that deviate strongly from history. The cost estimating process provides the decision-maker with a clear understanding of the cost risk inherent in the project, cost of alternatives within the project, and information to make resource allocation decisions. [DAU20] Once the decision is made to proceed with the project, cost estimates provide management with critical cost-risk information to improve the control of resources in the present and the future as well as provide insight into the impact of project changes on the program budget. The cost estimating process must therefore be adaptable and flexible while holding firm to the principles, objectives, and practices of cost estimating.

J. D. Moreland, Jr., *Mission Engineering and Integration*,
https://doi.org/10.1007/978-3-032-32132-9_10

This chapter provides descriptions of the cost estimating environment, details on the cost estimating process, and information on techniques that support the cost estimating process such as economic analysis and other cost estimating support applications for the cost estimating community. Where appropriate, past experiences are presented to build knowledge into the cost estimating process and to pass on corporate cost history capturing the cost estimating process's evolution to its current state.

Cost estimating and analysis demand a forward-looking approach that relies in part on past acquisition and ownership cost history but with adjustments to best predict the mission under consideration for the future program. Therefore, this chapter will discuss proactive and effective methods of keeping lines of communications open for estimating and assessing cost, schedule, and associated risks to address cost issues effectively while leveraging the experience and knowledge of all stakeholders both present and past. Many management reform initiatives are in place across the industrial sectors to control costs, improve performance, and achieve mission goals and objectives. With each reform, the goal has been ensuring affordability and getting the best value for the dollar. Program managers must put their programs through performance reviews (i.e., objective measures to determine if the program is performing as anticipated based upon baseline costs, schedule, and risks). Budget submissions must be integrated with performance-based budgeting. The cost estimator must be aware of and respond appropriately to these and the other environmental drivers as they support the programmatic health. This process provides the basis for making trade-off decisions considering affordability versus needs and translates those needs into concepts and eventually into system designs.

A formal budget process drives the daily operations of programs and is dependent on collaboration among the various stakeholders, each working toward the fulfillment of the vision set forth by their organization's strategy. The four interrelated and overlapping phases of this budgeting process are **planning**, **programming**, **budgeting**, and **execution** (PPBE). The planning and analysis that constitutes the first phase of the PPBE process is the foundation for the formulation of the organization's yearly budget. To support the budgeting process effectively and comply with all regulatory and statutory requirements associated with receiving program funds, continuous, accurate, and forward-focused investment planning and analysis are required. As a current year budget is executed, the next year's budget must be formulated and planned. PPBE is a resource allocation process for making informed, affordability assessments, and resource allocation decisions on acquisition programs. The PPBE system integrates end-user requirements with resource requirements: the natural outflow of this consolidation is the budget, and ultimately, the programs that produce systems for the commercial and government markets. The PPBE process examines capabilities horizontally and relies on the future year planning strategy to plan funding in short term based upon long-term consequences.

10.2 Cost Estimating Enablers

Cost estimating is a highly structured process that has evolved over many years into the formal framework. It is an essential element to effective program management and is required for realistic program planning and decision-making. Following the cost estimating process and framework presented in this chapter permits developers and program managers to optimize the use of limited financial resources. Cost estimators must be proficient and aware of the financial management, performance measurement, schedule analysis, acquisition management, as well as the technical aspects of a program to support the cost estimating process effectively. Improving performance is one of the key tenets in any organization; therefore, program proponents need to demonstrate with adequate technical rigor that programs they advocate actually accomplish their goals.

Accurate cost estimating that considers risks and benefits as a quantifiable systematic process puts a program or project on a solid foundation, as does the consistent and continuous application of system engineering and program management. [GAO20] Cost estimating is extremely important to all organizations and associated program managers as it is a key function in determining costs at the onset of a program, assuming proper budgets are in place, and in managing and controlling cost throughout the program's lifecycle. Disciplined and independent cost engineering is an essential aspect of the systems engineering (SE) process and a required capability contributing to improving the efficiency and effectiveness of organizations' programs and operations. Technical decisions and subsequent program direction depend on a clear understanding of the cost and cost implications of various technical options. Credible cost estimates based on rigorous cost estimating standards and processes must be in place for organizations to meet assigned responsibilities. The goal is to provide a solid cost estimate for the programs and organizational budgets to provide a tracking metric for programmatic health following a formal and rigorous, well-documented process. This increases the chances of program success and builds credibility in affordability outlooks.

Governance and authorities with respect to authoritative data and analysis are very important with respect to holding individuals accountable for execution of programs. An authoritative decision-maker needs to be selected to be responsible for providing and making cost engineering decisions. This individual and their respective pyramid structures of subject-matter experts in their field need to provide technical authority with bedrock credibility and allow for essential independent functioning. Under each of these pyramid structures of engineering managers and lead engineers, experts who in their field provide technical authority with bedrock credibility and allow for essential independent functioning. These individuals need to have authority, responsibility, and accountability to establish, monitor, and approve technical products and policy in conformance to higher tier policy. In addition, these experts need to be entrusted and empowered to oversee cost engineering processes and make technically sound engineering decisions and must do so with integrity and discipline. This way decisions are made in a timely and responsive manner, not requiring excessive review and oversight.

10.3 Cost Estimating for Program Lifecycle

This section provides an overview of cost estimating activities in a representative project lifecycle to include the milestone events, the types of data needed/available, and the estimating methodologies to support these estimating activities during the program lifecycle. Depending on the complexity of a program and where the program is in the program lifecycle, data availability and management, milestone events, and cost estimating methods may vary in difficulty. For example, early in a program lifecycle, the estimator may have very little available. Late in the program lifecycle, large amounts of data may be available but need to be verified for consistency in program content or normalized before use in an estimate. Figure 10.1 provides a representative overview of a general program lifecycle.

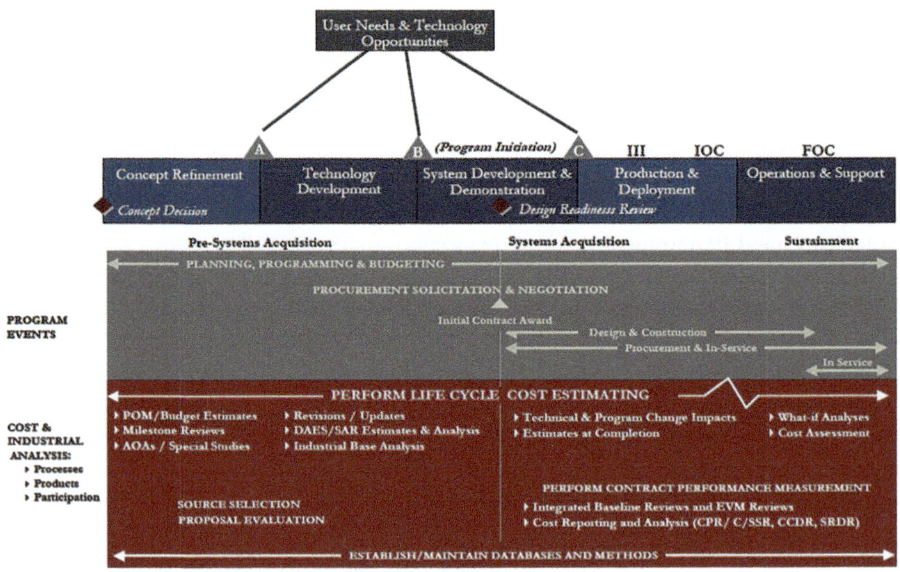

Fig. 10.1 Cost estimating activities

10.3.1 Lifecycle Cost Estimate

Lifecycle Cost Estimate (LCCE) is the total cost to the customer of a program over its full life. A cost estimate provides a projection for each lifecycle cost element for every aspect of the program required to respond to the needs identified in the requirements. These costs can also be grouped by the following lifecycle cost categories:

- **Research and development**: Includes development and design costs for the lead system development, engineering and design, test and evaluation, and other one-time costs for the system design features. It includes the nonrecurring costs for development, design, startup, prototypes, software, initial spares, live fire test and evaluation, special tooling and test equipment, facility changes, etc.
- **Procurement investment**: Includes the total production (or low-rate production) and deployment costs of the prime system and its related support equipment and facilities. It includes any related initial spares and repair parts.
- **Operation and support:** Includes all direct and indirect costs incurred in using the prime system (i.e., manpower, fuel, maintenance, and support) through its entire lifecycle. Manpower costs should be based on estimates and expressed in terms of the Manpower Estimate Report (MER).
- **Disposal (inactivation)**: Includes the costs of disposing (inactivating) the prime equipment after its useful life.

LCCEs provide an exhaustive and structured accounting of all resources necessary to identify all cost elements including project feasibility, project definition, system definition, preliminary and final design, fabrication and integration, deployment, operations and sustainment (O&S), and disposal effort. The LCCE encompasses all past ("sunk"), present, and future costs. Lifecycle costing enhances the decision-making process, especially during the early planning/concept formulation phase of the acquisition cycle. Design trade-off studies conducted during this period can be evaluated on a total cost basis as well as on a performance/technical basis. These estimates support budgetary decisions, milestone reviews, and investment decisions. An LCCE can also be referred to as a Program LCCE (PLCCE) if the estimate is created in support of a program. A PLCCE is developed to ensure that costs are fully accounted for across a program. The lifecycle of a program equals its total life, beginning with mission feasibility and extending through O&S and disposal. The PLCCE should be comprehensive and structured to identify all cost sustainment elements and should capture all costs necessary to design, develop, deploy, field, operate, maintain, and dispose of a system over its lifetime. As members of the product or program design team, cost estimators prepare a PLCCE by translating the technical and design parameter characteristics and schedules into cost estimates using established cost estimating methodologies. Iterative and ongoing reviews are conducted with the technical team during the design process until the cost estimator and the project management team are confident that the cost estimate credibly reflects the baseline program's design requirements, technical capabilities, management structure, and operational scenarios. Then, the PLCCE becomes the basis for the program's budget baseline.

10.3.2 Cost Estimating Classifications

Every estimate described in this section follows a specific process, yet every estimate is different. Many factors go into preparing an estimate. Time and data are the most critical resources when preparing an estimate. If either of these resources is missing when preparing an estimate, the estimate could be affected, prompting the estimator to make assumptions when developing the estimate. Using cost classifications is useful in standardizing the issues that affect an estimate and communicating this information to the users of the estimate.

Over the years, cost estimating has been a challenge largely because cost estimates tend to be presented on an equal basis but vary in quality. That is, some budget estimates are sometimes submitted prematurely and lack acquisition strategy plans, so a classification is necessary to characterize the quality of the cost estimate. The cost estimator also must factor the cost classification of the cost estimate into the risk assessment for the estimate. The cost estimate classification system uses letters of the alphabet to designate estimate quality (a description follows in the next subsections). It is important to note that the cost estimate classification can be no higher than the level of fidelity of the technical inputs that are provided.

10.4 Class A: Budget Quality (New Construction)

The budget quality cost estimate is the ultimate goal of the cost estimating process, since this classification indicates that the estimate is of budget quality. A budget quality rating is a commitment to stakeholders that additional funds will not be required in the future to complete the scope of work for the program. The only exception allowed on this commitment involves out-year inflation rates assumed in the estimate. The costs driven by these rates can be adjusted plus or minus in future budgets as based on forward pricing rate agreements and approved material indices. If items of uncertainty do exist, appropriate growth factors must be included, and the cost estimate and the documentation fully noted. The Cost Estimating Relationships (CERs) used to calculate the cost estimate should be based on: (1) an accepted weight estimate using bid information, (2) current weight estimate when using cost data from the contractor's latest Cost Performance Report (CPR), or (3) similar construction data of the prospective manufacturing plant(s) where new designs are costed. Risk considerations have significant influence in determining budget quality cost estimates. If major equipment has not met the requirements of "approved for full production," an additional cost allowance for an alternative fallback position may be justified for a budget quality cost estimate. If the contract is planned to include a standard compensation escalation clause, projected manufacturing plant escalation cost calculations should be based on: (1) a program manager's developed contract award, start of construction and delivery schedules, and (2) global insight projections of the appropriate labor and material indices. Additionally, if the estimate is base dated, escalation calculations should be made using the approved escalation model. If technical design, program planning, or economic (cost) information lacks credibility or, in the opinion of the cost estimator, significant information is questionable or not up-to-date, the cost estimate should be classified as either a feasibility study quality or rough order of magnitude quality, as appropriately defined below.

10.5 Class B: Budget Quality (Conversion)

Uncertainties related to system conversions, modernizations, and system life extensions at times cannot be resolved until after contract award. Therefore, for these types of estimates, a Class A budget quality classification is not appropriate. For a conversion, modernization, or system life extension cost estimate to conform to this classification, the detailed scope of work requirements shall include: (1) the description and weights of equipment or systems to be removed, relocated, or added and (2) a list of proposed alterations and an adequately defined repair package.

Additionally, the scope of work must include costs for similar repairs and modernization items. Comparable conversion or system life extensions should be available, or an initial manufacturing plant cost estimate of projected repairs, and modernizations improvements should be made by the prospective manufacturing plant. The potential interface and interdependency considerations of these items should be determined as related to the existing design. An allowance should be made to recognize that the system condition may differ considerably 2 years after the budget is submitted, i.e., when the system is actually worked on.

Other items that should be included are: (1) the status of the current manufacturing plant workload and additional workload projections for the prospective conversion, modernization, or system life extension manufacturing plant; (2) assessment by the system cost estimator for productivity considerations; (3) realistic projection of labor rates at the manufacturing plant; and (4) the expected use of premium pay for overtime if the schedule so requires.

10.6 Feasibility Quality Cost Estimates

Feasibility quality cost estimates are those costs prepared by using design information resulting from system feasibility studies. The feasibility study usually produces at least one-digit Work Breakdown Structure (WBS) system weights and only general guidance with respect to major electronics and equipment. Cost estimates that fit this classification also involve those derived by inflating to current dollars from a previous cost for a similar system and then making rough or gross adjustments for expected changes in design, program requirements, or program cost factors. Any cost estimate derived from a current budget year estimate by deflating or inflating to some other year by the application of a labor and material manufacturing index will be designated as a feasibility quality cost estimate. The manufacturing plant type (private or government) and number of systems to be built in a single plant are often not known when deriving feasibility quality cost estimates. Escalation calculations are based either on inflating the escalation cost contained in the total cost estimate used as the base (reference) estimate or on using a flat percentage of the manufacturing plant portion of the cost estimate based on an approximation calculation.

10.7 Rough Order of Magnitude (ROM)

The rough order of magnitude quality cost estimate results when design information does not meet the standards equivalent to a system, other platform, or systems feasibility study. The design study may produce rough calculations of system weights and the basis for the weights; other system design parameters may not be founded on sufficient technical information and analysis to support high reliability in the design. Some examples include (1) a new design of an unprecedented system, (2) a system that is initially designed to carry many developmental equipment, or (3) a system designed beyond the current state of the art.

10.8 Directed or Modified Cost Estimate

Directed or modified quality cost estimates apply to estimates that are (1) not developed through the normal estimating process, (2) provided by other outside groups, or (3) directed by higher authority. Directed cost estimates are generally a total cost limitation established without the benefit of a fully developed design concept and related cost estimate such as a cost goal or cost target. Moreover, a directed estimate also applies to any previous cost estimate performed at a higher classification level that was changed to conform to budget cuts or restrictions on a total cost that is not based on scope decisions.

10.9 Cost Estimating Process

Phase 1 of the cost estimating process shown in Fig. 10.2 is called the Project Definition. [NAVSEA20] [NASA15] During this phase, the estimator clarifies the reason for the estimate, defines expectations, and begins to understand the project

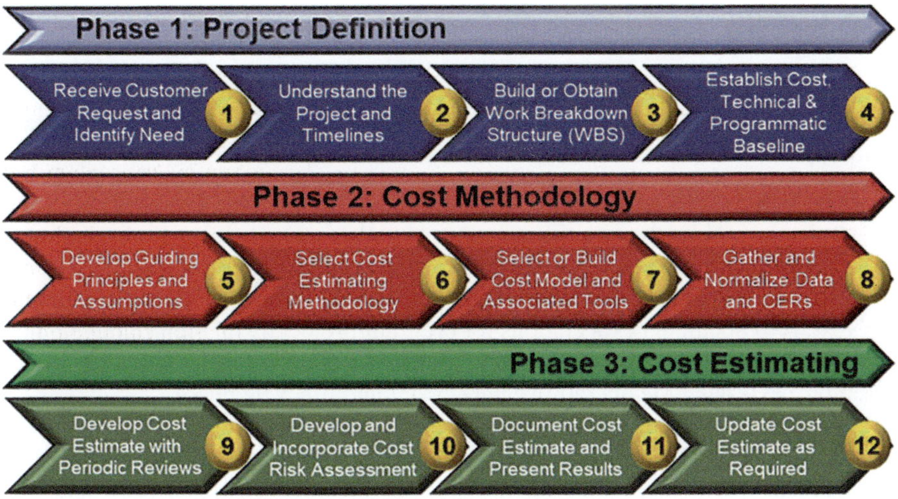

Fig. 10.2 General cost estimating process

and timelines for the cost estimate tasking. As the cost estimate is defined and project information gathered, a WBS and cost, technical, and programmatic baselines are obtained. These items help define the project and form the foundation for the cost estimate. As the estimator continues through the estimating process, these steps may be revisited as new information is obtained.

Phase 2 of the cost estimating process, Cost Methodology, includes four tasks (detailed in Sect. 10.9.1) that create the approach and framework for the estimate. Developing the guiding principles and assumptions will be the most revisited task in this part of the cost estimate process. As methodologies are selected and data are gathered, the guiding principles and assumptions, methodologies, and even the cost model may be refined with different cost tools considered, as appropriate. The final challenging task in this phase involves the collection of critical data and CERs. The estimator needs to recognize that data, once collected, may need to be normalized to support a particular cost model or estimating method so adjustments can be made for inflation. This makes the raw data set consistent and fit for use in CERs, models, or estimates. Phase 3 of the cost estimating process, Cost Estimating, has four tasks identified that include the actual conduct with periodic reviews, incorporation of risk, presentation, and maintenance of the cost estimate (See Sect. 10.9.2). All these tasks are important in their own right, and together they become critical for a defensible and complete estimate. As noted above, the cost estimating process is shown as a series of successive steps. It is in practice often an iterative process that is rarely linear. For example, the initial estimating results may be revised one or more times because of changes to the technical and programmatic baselines, questions about the sensitivity of results, changes in the initial guiding principles and assumptions, or indications that the estimate is incomplete. The revised results and findings are then documented and, along with the estimating models and data, made available for subsequent estimating and analyses.

10.9.1 Project Definition Tasks (Phase 1)

To properly estimate the cost of a project, it is vital that the project be thoroughly defined and understood by the estimators. Therefore, the initial part of the cost estimating process is dedicated to the following tasks.

- **Task 1: Receive Customer's Request and Identify Need**: The goal is to communicate sufficiently with the customer to gather enough project information to generate a quality cost estimate. The first step involves the identification of the customer(s) and stakeholder(s) who will use the results of the cost estimate. Next, is to document expectations for program/project by (a) identifying the purpose of the cost estimate; (b) specifying mission needs, objectives, and goals; and (c) assessing the operating environment and lifecycle phase.
- **Task 2: Understand the Project and Timelines**: The goal focuses on gathering and reviewing all relevant project information for evaluation (e.g., an existing technical baseline, previous estimates, lessons learned, customer feedback, budget data, and programmatic data such as schedules). In addition, this is the time to discuss schedule, data, expectations, and resource requirements with the requesting customer. In these early lifecycle phases of a project, there will be many unknowns. It is the role of the cost estimator to ask insightful questions that help the project management staff make decisions regarding key aspects not normally considered in an early stage (e.g., maintenance concept and testing strategy) and to address issues such as personnel, schedule, technologies, and cost drivers that can have a major impact on risk. Information gathering is no less important in later phases of a project when more is known and overlooking any element could affect the estimate's outcome.
- **Task 3: Build or Obtain Work Breakdown Structure**: The goal of this task is to provide a consistent structure that includes all elements of the project that the cost estimate will cover. It is critical to develop or receive the WBS as this element is key to performing good project management. The purpose of a WBS is to divide the project into manageable pieces of work to facilitate planning and control of cost, schedule, and technical content. A WBS ensures that all work to be performed on the project is organized and aligned in accordance with the total scope of a program, using a hierarchical structure. Each element in the cost estimate represents the cost to do that work. This structure becomes the cost estimator's framework for ensuring full coverage (without double counting) of the project's objectives. The WBS is also used as a communication tool to present the project's scope in an understandable form that can be easily communicated to the project team and other stakeholders. The type of estimating product varies by program lifecycle phase. In early phases, the cost estimator will either obtain a high-level project WBS from the project staff or work with them to develop one. A project WBS is the comprehensive WBS comprising all lifecycle phases, recurring and nonrecurring costs, and components including the hardware for the product. Other items such as mission assurance, SE and Integration (SE&I), Integration and Test (I&T), mission operations, data analysis, outreach, science

team, and project management comprise the WBS. At a higher level WBS (e.g., WBS Levels 1 and 2), it is important to standardize the WBS not only to facilitate data (and data analysis), but also to enable apples-to-apples comparisons to heritage programs, see Fig. 10.3 showing a Level 4 WBS example.

- **Task 4: Establish Cost, Technical, and Programmatic Baseline**: This task focuses on establishing a common baseline document that thoroughly describes the project to be used by the project team and project estimators in developing their estimate(s). The project cost, technical, and programmatic baseline defines and provides quantitative and qualitative descriptions of the project characteristics from which cost estimates will be derived. As such, this baseline ensures that cost projections individually developed by the project manager and the independent review organizations are based on a common definition of the system and project. The goal of this step is to ensure that the estimating team understands and can obtain the documents that thoroughly describe the project to be estimated. This project baseline should identify any area or issue that could have a major cost impact (e.g., risks) and, therefore, must be addressed by the cost estimator. Several activities associated with understanding the complete program characteristics, including:
  - Gather and review all relevant project data (e.g., existing technical baseline, previous estimates, lessons learned and customer feedback, and budget data as well as other programmatic data such as schedules)
  - Collect system characteristics, configuration, quality factors, security, operational concept, and the risks associated with the system
  - Obtain the system's (or the project's) milestones, schedule, management strategy, implementation/deployment plan, including launch, test strategy, security considerations, and acquisition strategy. Work closely with the organization sponsoring the cost estimate for access to pertinent documentation describing the system being estimated. Every estimate, regardless of size,

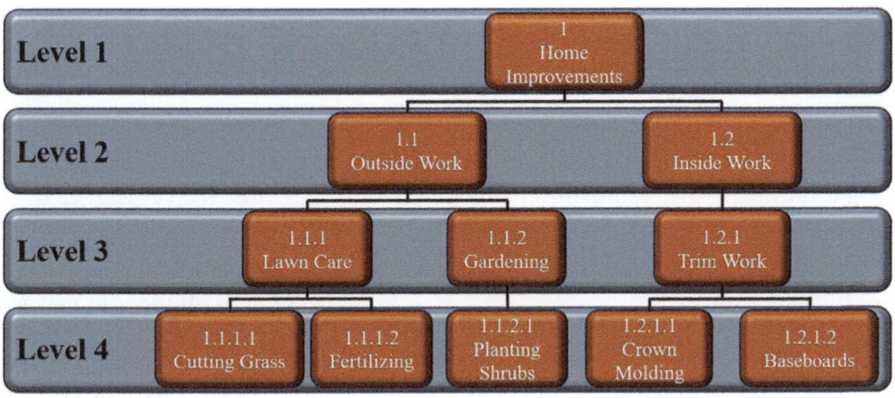

Fig. 10.3 Work breakdown structure example

needs to define what is being estimated. Projects that are smaller in size or earlier in their project's lifecycle may only require a simple data sheet with technical requirements provided by the project to support developing a Rough Order of Magnitude (ROM) cost estimate.

10.9.2 Cost Methodology Tasks (Phase 2)

The next four tasks of the cost estimating process relate to selecting and administering the cost methodology, which will guide the development of the cost estimate. These four tasks are listed and detailed below.

- **Task 5: Develop Guiding Principles and Assumptions**: The development of guiding principles and assumptions is intended to communicate the scope, context, and environment within which the cost estimate is developed. The cost estimator works with the program manager and members of the technical team to establish and document a complete set of guiding principles and assumptions that are necessary to provide definition to the project and the estimate and to bound its scope. These guiding principles and assumptions show all involved what costs are included and what costs are excluded in the current estimate. This promotes easy comparisons to future estimates and those conducted by independent agencies. Guiding principles and assumptions provide cost estimators with the means to bound the estimate, focus attention on the most important elements, and provide temporary resolution of undefined technical and programmatic questions. Since it is impossible to know every technical or programmatic parameter with certainty before and into the design phase of a project, a complete set of realistic and well-documented guiding principles and assumptions adds to the soundness of a cost estimate. Descriptions of relevant missions and system characteristics, personnel, maintenance, support, and logistics policies are generally included in the guiding principles and assumptions.
- **Task 6: Select Cost Estimating Methodology**: This task focuses on selecting the most appropriate cost estimating methodology (or combination of methodologies) for the data available to develop a high-quality cost estimate. The three basic cost estimating methods currently used across a system's lifecycle include analogy, parametric, and engineering build-up, as well as extrapolation from actuals using Earned Value Management (EVM). There is a high probability that the prevalent cost estimating methodology changes over the project lifecycle because the amount of data available to conduct the project estimate typically changes over time.
  Analogy Cost Estimating Methodology—The analogy method uses the cost of a similar system, adjusts for differences, and estimates the cost of the new system. This technique identifies a currently fielded system (comparable system) similar in design and/or operation to the proposed system. Analogy cost estimates are based on a comparison and extrapolation to similar items or efforts. Cost data

from a past program that is technically representative of the program to be estimated serves as the basis of estimate. Cost data is then subjectively adjusted upward or downward, depending upon whether the subject system is more or less complex than the analogous program. Clearly, subjective adjustments compromise the validity and defensibility of the estimate and should be avoided, while historical data and analysis are considered more credible and defensible. This estimating approach is typically used when an adequate amount of program and technical definition is available to allow for the proper selection, and adjustment, of comparable program costs. An analogous approach is also used when attempting to estimate a generic system with very little definition. The analogy system approach places heavy emphasis on the opinions of "experts" to modify the comparable system data to approximate the new system and is therefore increasingly untenable as greater adjustments are made.

- ***Parametric Cost Estimating***—Parametric cost estimates are a result of a cost estimating methodology using statistical relationships between historical costs and other program variables such as a system's physical or performance characteristics, contractor output measures, or personnel loading. Generally, an estimator selects parametric cost estimating when only a few key pieces of data are known, such as weight and volume. The implicit assumption in parametric cost estimating is that the same forces that affected cost in the past will affect cost in the future. The data that relate to these estimates are weight characteristics and design complexity, respectively. The major advantage of using a parametric methodology is that the estimate can usually be conducted quickly and be easily replicated. As part of this approach, an estimator performs a regression analysis to develop one or more CERs. The main objective of regression analysis is to generate a CER in the form of an equation that represents a relationship between a "dependent" variable to one or more "independent" variables. The dependent variable is called that because it responds to changes in the independent variable where:

  - The value of the dependent variable, designated by the symbol Y, is calculated.
  - The value(s) of the independent variable(s), designated by the symbol X, are known.
  - The resulting relationship between Y and X can be described mathematically.

  When working with CERs, the dependent Y variable typically represents cost, while the independent X variable(s) represents the parameter(s) of the system or effort being estimated. Independent variables or "cost drivers" should be chosen because there is a strong correlation between these variables and cost (Y) and because there are sound principles for the relationship being investigated. For example, the assumption may be made that the weight of a component drives its total cost. Figure 10.4 provides an example of how this may be portrayed as a standard Cartesian coordinate graph. The dependent variable is the Y-axis (Cost), and the independent variable is the X-axis (Weight). By using historical data that compare cost to an independent variable and plotting, we can establish whether there is a relationship between the variables. From these data points, a "line of best fit" can also be plotted (depicted as the dashed blue line in Fig. 10.4). The line of best fit to the data can be tested and used for a CER.

CERs established early must be periodically examined to ensure that they are current throughout the life of a cost estimate and that the input range of data being estimated is applicable to the system. All CERs should be detailed and documented. If a CER is improperly applied, a serious estimating error could result. Microsoft Excel™ or other commercially available modeling tools are most often used for these calculations.

Engineering Build-Up Methodology—Sometimes referred to as "bottom-up" estimating, the engineering build-up methodology produces a detailed project cost estimate computed by estimating the cost of every activity in a Work Breakdown Structure, summing these estimates, and adding appropriate overheads. This costing methodology involves the computation of the cost of a WBS element by estimating at the lowest level of detail (often referred to as the "work package" level) wherein the resources to accomplish the work effort are readily distinguishable and discernable. This is often referred to as the Cost Breakdown Structure (CBS) or the Cost Estimating Structure (CES). In most cases, the labor requirements are estimated separately from material requirements. Overhead factors for cost elements such as other direct costs, general and administrative expenses, materials burden, and fee are applied to the labor and materials costs to complete the estimate. An experienced technical team member typically works

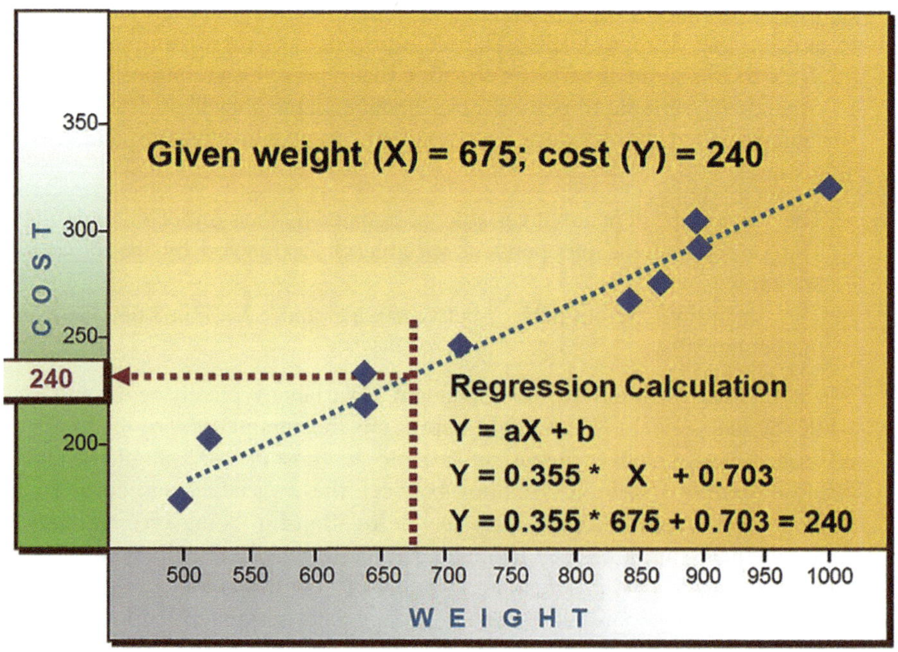

Fig. 10.4 Scatter plot and regression line of cost (Y) versus weight (X)

with the cost analyst, who prepares the engineering build-up estimates. The cost estimator's role is to review the bottom-up estimate for reasonableness, completeness, and consistency with the project guiding principles and assumptions. It is also the cost estimator's responsibility to test, understand, and validate the knowledge base and data used to derive estimates. In some situations, the engineering community provides "professional judgment," but only in the absence of empirical data. Experience and analysis of the environment and available data provide latitude in predicting costs for the estimator. This method of engineering judgment and expert opinion is known as the Delphi method. When relying on the Delphi method, the cost estimator's interview skills are important to capture and properly document the knowledge being shared from an engineer's expert opinion. The Delphi method usually involves assembling a group of experts to converge on a value by iterating estimates using varying amounts of feedback. During this process, individuals are generally not identified to the outside and, in some experiments, not identified to each other.

Figure 10.5 illustrates the relationship between cost estimating methods (on the y-axis) and acquisition lifecycle phases (on the x-axis). This heatmap visually represents which methods are used during each phase, with darker shades indicating active use.

- **Task 7: Select or Build the Cost Model and Associated Tools:** The objective of this task is to select the most appropriate model/tool or to create a model to estimate the cost. Factors that influence the selection process include data and resource availability, schedule, and cost. A variety of estimating tools and models are available for cost and schedule analysts to use in developing an estimate,

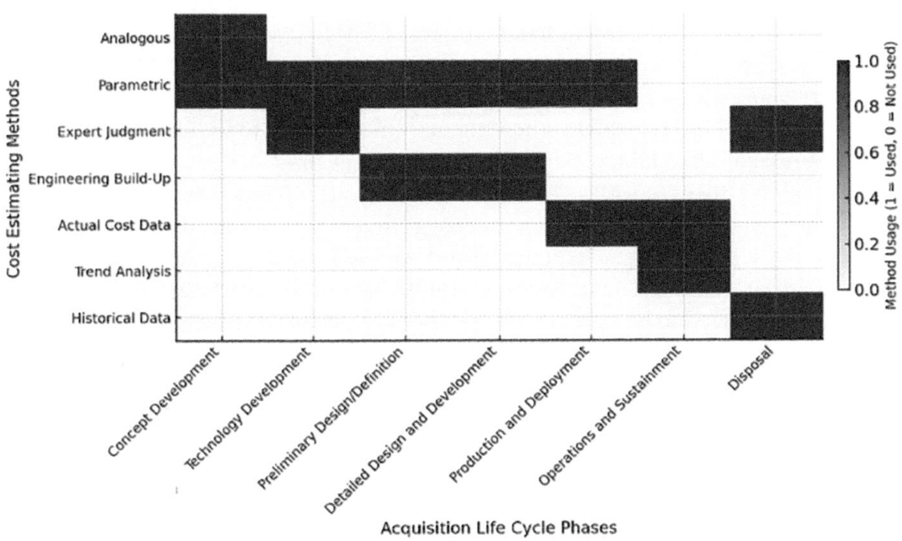

Fig. 10.5 Cost estimating methods across acquisition lifecycle phases

including tools that are newly developed and other commercially available tools for which a license can be obtained. Cost estimating models can vary significantly in terms of data availability and automation. Some models require the user to create datasets and develop CERs, while other models already include datasets and automate many functions for the cost estimator. A model can also use a variety of estimating methodologies and direct inputs to complete a full estimate. Many commercially available models are parametric models that generate estimates based on specific parameters that drive an estimate's cost. The cost drivers include items such as weight, volume, quantity, and schedule. These models can be used when only a few input parameters are known to generate a high-level estimate. If many cost drivers have been identified and many known technical input parameters exist, these models can also be used to generate highly detailed and complex cost estimates. Commercially available parametric models use normalized industry data sets in generic and sometimes proprietary algorithms. For best results, these models should be calibrated to historical data that is comparable to the system being estimated to ensure that the estimate considers factors such as the project environment. If an estimator chooses to build his or her own model, following a disciplined process will ensure a credible product. Once the estimator has identified the need for a model and determined the model type, the model design can begin. The importance of spending time up front designing and understanding the model cannot be underestimated. The model developer needs to define the scope of the model, its ultimate use, and the approach for integrating the data collected and new CERs developed. While planning the development, it is important to document the model guiding principles and assumptions. After the model is developed and populated with at least preliminary cost data, it must be validated before the estimator uses it. Once validated and any corrections or updates have been incorporated, the model is ready to generate estimates.

- **Task 8: Gather and Normalize Data and CERs:** This task is focused on arming the cost estimator with as much information as possible so that the most accurate and defendable cost estimate can be developed. Data collection and data normalization are two cost estimating activities critical to the development of a credible and defendable cost estimate. Sufficient time and resources should be planned for these efforts. Data collection is one of the most difficult, time-consuming, and costly activities in cost estimating. Data needs are not always clear at the assignment's beginning, and data requirements often evolve during an estimate's development. An estimator needs to recognize that data, once collected, may need to be normalized to support a particular cost model or estimating method [OMB92]. Because uncertainty is the underlying driver in a cost-risk analysis, it is critical to concurrently collect risk data to support the cost-risk assessment. Many experts that will be interviewed and data that will be reviewed in this effort will not only support the cost estimate, but may assist in identifying risks early and potentially save time by reducing data collection during the cost-risk assessment later in the process. Therefore, it is imperative that the cost estimator meet with experts early and often to discuss uncertainty in technology, design, requirements, etc. As previously noted, data collection can occur in earlier steps, such as collecting data for regression analysis to support a methodology or even earlier in the process when the estimator is understanding the project. A well-documented set of project requirements ensures that the cost estimators are estimating the same product being designed by the technical team. If some of the cost model inputs are not explicitly contained in the requirements document, the cost estimator will have to coordinate with the cognizant technical points of contact to obtain the needed data through interviews and/or by survey mechanisms. Often, the cost analyst is estimating conceptual designs, so a requirements document does not exist. However, often conceptual design data such as master equipment lists and functional block diagrams are available. Collecting programmatic data, such as schedules, is also key to data collection efforts.

- Normalizing data for cost includes adjusting for inflation, which makes the raw data set consistent and fit for use in CERs, models, or estimates. Data may be adjusted for inflation again later when it has been incorporated into the cost estimate and the estimate as a whole is adjusted for inflation. The full estimate may be adjusted for inflation to show the results in Base Year (BY), Constant Year (CY), or Real Year (RY) dollars. Through escalation, inflation adjusts costs to reflect the decrease in the purchasing power of money over time. The inflation factor is the "multiplier" used to account for the change in the price of a product or service over time. Escalation factor (or weighted inflation) is the multiplier that accounts for inflation plus the normal occurrence of allocating money in 1 year and incorporates outlay rates over a number of years. While inflation is the most common data normalization technique to improve consistency in a data set, other normalization techniques can be just as important. Adjustments for learning or cost improvement curves may apply. Production rate (units produced over a time period) may also affect the raw data set, thereby calling for adjustment. In the case of production rates, patterns or influences in the production of the item, such as facilities or personnel may affect the data. Possible adjustments needed to normalize data include the following:

  - Scope consistency between the historical data of a product and the product being estimated
  - Unusual events or anomalies in a project's life, such as extra testing, failures, or labor anomalies
  - Technology improvements and advancements, where the data may need to be adjusted by engineering judgment
  - Raw data from reporting system anomalies or changes, such as a change in rates, factors, or hours for standard reporting
  - Reporting system differences that may require mapping accounting classifications or categories of data to WBS elements

The estimator should review and validate the normalized data to ensure that a consistent data collection methodology, data collection formats, and procedures to identify data anomalies are in place. Considerations such as data sufficiency to support the estimating methodology selected and documentation to ensure the traceability of adjustments made to the data are also critical. These documented factors assist the estimator with the validation of the data, lead to data reliability, and ultimately contribute to estimate credibility. If an estimator considers each of these elements when identifying and collecting data, analyzing schedules, and normalizing data, the repeatability and credibility of the data supporting the estimate will be improved.

10.9.3 Cost Estimating Tasks (Phase 3)

- **Task 9: Develop Cost Estimate with Periodic Reviews**: The primary objective of this task is to create an initial LCC point estimate by following a structured process. Eight activities are key in developing the estimate:
  - Verifying the Ground Rules and Assumptions (GR&A)
  - Populating the cost model with normalized data
  - Ensuring the estimate complies with full-cost accounting principles
  - Running the model to calculate costs
  - Time phasing the estimate
  - Adjusting for inflation
  - Performing cross-checks or reconciling estimates
  - Developing or updating the cost track to compare with previous or independent estimates.
- Once the appropriate cost model has been selected or built and relevant data gathered, the model is populated based on the GR&A and used to generate a point estimate. While producing a point estimate is a critical milestone, it is only the starting point in the broader cost-risk analysis process.
  - **Time Phasing Estimate**: The cost estimator must time-phase the data to align with the planned development schedule, utilizing techniques such as Beta Curves, historical spreads, engineering judgment, or budget constraints. Similar to normalizing data for inflation, the estimate must also be adjusted for inflation throughout its lifecycle. Baseline cost estimates, typically generated in CY dollars, are first phased by fiscal year and then converted into annual RY or then-year dollars. These inflation-adjusted RY dollars serve as inputs for the annual project budget. When using the estimate for budget development, adjusting the cost phasing to reflect budget authority is essential because it ensures alignment between the timing of budget authority and the actual expenditure of funds.
  - **Develop Probabilistic Cost Estimate**. This method involves quantifying risks and uncertainties that could lead to delays or increased costs, which enables the creation of more credible cost and schedule baselines. Probabilistic techniques allow for the effective communication of the impact of resource changes, providing quantified probabilities of meeting planned targets. Over the past decade, the cost estimation community has shifted from deterministic

estimates, which lack uncertainty quantification, to probabilistic estimates, which provide deeper insights into key cost and schedule drivers. These estimates empower management to actively address risks and develop strategies to mitigate costs while offering valuable information for reporting and providing analysis to stakeholders, leadership, and partners. A probabilistic cost estimate often begins with a deterministic "point estimate" but enhances it by incorporating risk and uncertainty using methods such as Monte Carlo simulations or analytical techniques. These methods are classified as input-based, output-based, or scenario-based and provide a more accurate range of probable outcomes for cost, schedule, or both. Unlike summing individual "most likely" estimates, which tends to understate total costs, probabilistic analysis accounts for aggregated uncertainties. The process involves identifying and quantifying risks, with iterations often necessary as data, models, and methodologies evolve. Cost estimators may choose to develop a technical baseline estimate before conducting probabilistic analysis, but this step can also be performed in parallel. Ultimately, probabilistic analysis ensures decision-makers have a clear view of project risks and provides a comprehensive range of probable outcomes for improved decision-making and planning.

- **Task 10: Develop and Incorporate Cost-Risk Assessment**: Developing a cost-risk assessment involves six key activities designed to evaluate a project's confidence level and estimate the unallocated future expenses (UFE) required to achieve a desired confidence level. These steps include:

  - Identifying the project's cost drivers and risks with input from project/program managers and staff
  - Developing probability distributions for technical and schedule cost drivers
  - Addressing cost model uncertainty
  - Running the risk model
  - Determining the probability that the actual cost is at or below the point estimate
  - Recommending sufficient UFE to meet the desired confidence level

  This process identifies and analyzes critical project risks within the context of cost, schedule, and technical objectives and constraints. Cost-risk analysis captures estimating errors, uncertainties in technical design and requirements, and project risks to produce a probabilistic range of potential outcomes. This approach not only quantifies the UFE needed to reach a specified confidence level but also allows cost estimators to document risks, facilitating proactive cost management. Risk dollars are typically allocated to phases of the project where issues are most likely to arise, such as between Preliminary Design Review (PDR) and Critical Design Review (CDR), and during Integration and Test. At this point, conducting a sensitivity analysis is strongly recommended to identify major cost drivers that are design parameters having the greatest impact on cost when they change. This analysis helps determine the sensitivity of the cost estimate to variations in assumptions, technologies, or system designs. This offers decision-makers valuable insight into how program choices influence cost model outputs. By providing a range of estimates alongside an understanding of the likelihood of outcomes within that range, sensitivity analysis proves more practical than relying solely on a single-point estimate. Sensitivity analysis can occur at any stage of cost estimation, but it typically starts with an unconstrained solution that fulfills all mission objectives. Adjustments are then made to reduce costs, ensuring that the changes do not compromise the core benefits of the solution. This iterative process highlights the interplay between design decisions and costs, which then enables informed trade-offs that balance budgetary constraints with mission objectives.

- **Task 11: Document Cost Estimate and Present Results**: The purpose of cost documentation is to provide a detailed written justification for a program's cost estimate, often referred to as the Basis of Estimate (BoE). This document outlines the ground rules, assumptions, and key drivers used in developing the cost and schedule estimates, including model inputs, analogies, and supporting details. Given the significance and complexity of programs, preparing thorough and transparent documentation is a substantial effort. The documentation should be detailed enough to allow independent cost analysts or review teams to reproduce the estimate accurately. While it is unrealistic to standardize the format across all parts of an SoS enterprise, maintaining internal consistency within each stakeholders' organization promotes completeness, quality, and transparency. Cost estimators document LCC results throughout the entire estimating process, not just at its conclusion. In addition, the final documentation must include detailed estimates for each cost element, inputs for cost-risk estimates, and analyses of cost-risk results. It should also fully explain how each part of the estimate was derived and identify or provide the databases used.
- **Task 12: Update Cost Estimate as Required**: Updating a cost estimate ensures its accuracy and relevance as better information becomes available, enhancing its credibility over time. Regular updates enable cost estimators to defend their estimates, shorten turnaround times for revisions, and provide decision-makers with clearer insights for major decisions or "what-if" scenarios. This process involves two key activities:

  - Incorporating customer feedback and lessons learned into the next estimate version.
  - Updating the estimate as the project evolves through its lifecycle phases and milestone reviews.

  Maintaining an accurate project cost baseline is crucial for identifying potential cost overruns early. In addition, cost estimates should be revised whenever project content changes and reconciled with the baseline to assess the cost impact of proposed program alternatives. One effective approach for updating estimates is to adjust them based on the project's actual performance. Earned value management is widely recognized as a best practice for making such adjustments on a regular basis, such as monthly or quarterly, to ensure estimates remain aligned with project progress.

10.10 Cost Estimating Deliverables

In this section, an overview of the cost estimating representative deliverables that can be requested and developed across the lifecycle of programs and organizations. Many of these deliverables occur on a specific timeline, and others occur at the request of program managers creating three major categories of deliverables: Cost Estimating and Analysis, Economic Analysis, and Industrial Base Analysis. The first category of deliverables is Cost Estimating and Analysis, which includes deliverables such as the LCCE or PLCCE. The LCCE or PLCCE is the most common deliverable created in the programmatic domain. The second category of deliverables is Economic Analysis, which includes deliverables that support special studies as well as ongoing program activities. Some economic analysis deliverables include labor/overhead rate sheets, economic analyses, material inflation rate projections by WBS, and Business Case Analyses. The third category of deliverables is Industrial Base Analysis, which addresses activities that provide cost analysts and programs with continuously updated information on the industrial base. Examples of these deliverables include material indices, inflation data factors, and industrial base sector studies. A detailed listing of these representative deliverables grouped by these three categories can be found in Table 10.1, with information provided on each deliverable including the formal title, description of deliverable, and deliverable purpose.

10.11 Cost Risk and Uncertainty Methodologies

Cost risk and uncertainty exist through all phases of a project's lifecycle. It is important for a cost estimator to identify and distinguish between risk and uncertainty, as they are distinct and consequential inputs to the analysis. A cost analyst must be able to defend the uncertainty and identify the risk assessments built into the cost estimate. Cost-risk assessments are employed on missions to understand risks and help ensure resources and plans are adequate to deliver projects on time and within budget. Cost risk must be carefully and quantitatively assessed in developing and presenting any cost estimate for several reasons. First, when trade studies are conducted, a single cost estimate, such as an expected cost, may mislead the trade team by not revealing the potential for overruns. Second, at technical reviews and before authorities proceed with decision recommendations, the cost estimate must include an appropriately chosen level of unallocated future expense to achieve a desired confidence level. The objective of a cost-risk analysis is to produce a credible project cost cumulative distribution function (CDF, or "S-curve") for the range of costs of the project.

There are six activities associated with developing a cost-risk assessment that are needed to understand the current confidence level of the project and estimate the amount of unallocated future expense necessary to achieve a desired confidence level:

Table 10.1 Cost estimating deliverables

Cost Estimating and Analysis		
Deliverable	**Description**	**Purpose**
Life Cycle Cost Estimate (LCCE) or Program Life Cycle Cost Estimate (PLCCE))	An estimate accounting for the total cost to a customer of acquisition and ownership of a system over its full life. Identification of all cost elements pertaining to the total life of a project, beginning with mission feasibility and extending through operations and support and disposal or conclusion. A PLCCE is specifically conducted for a program.	Facilitates supporting an acquisition process, the investment decision-making process and supports budgetary decisions by providing an exhaustive and structured accounting of all costs related to a program. Becomes the project budget baseline, ensuring that costs are fully accounted for, and that each project's life cycle cost is minimized.
Independent Cost Estimate (ICE)	An independent estimate conducted by an organization outside the acquisition chain, using the same detailed technical and procurement information. Other estimates to support modifications may include estimates for conversion, activation, Modernization, and service life extension.	Serves as a comparison to the program manager's estimate to assist in determining the fairness and reasonableness of an estimate. Results are reconciled with the PLCCE.
Independent Cost Assessment (ICA)	An outside evaluation of a cost estimate, taking into account both the quality and accuracy of the estimate in question, looking for specific cost and technical issues. Also, a process used to determine whether the current program estimate reflects the program of record.	An ICA can be requested by a program manager or an outside source to gain perspective on the quality and accuracy of an estimate.
Estimate at Completion (EAC)	An EAC is conducted using Earned Value data. The EAC is the actual cost of work completed to date plus the predicted costs and schedule for finishing the remaining work. It can also be the expected total cost of an activity, a group of activities, or of the project when the defined scope of work is completed.	An EAC is conducted to provide an estimated actual cost of remaining work to assist in determining the overall cost of a program.
Participation in Source Selection	Provide industrial base, production, producibility, cost estimating and analysis and schedule expertise to source selection team.	Provide industrial base technical support to source selection team.
Basis of Estimate (BOE) Reviews	Provide independent technical review and assessment of selected design and production related cost and manhour estimates.	Provide the program manager with in-depth government and industrial assessments of selected cost estimates
Economic Analysis		
Deliverables	**Description**	**Purpose**
Business Case Analysis (BCA)	A BCA provides the justification for a proposed investment. It combines strategic analysis with quantitative cost/benefit/risk analysis.	A business case has three primary functions: to clarify/structure the planning and analysis required for effective decision-making, to determine the value of an investment or business initiative, and to guide on-going investment management and evaluation. A thorough business case details acquisition, implementation, and performance measurement strategies to create a foundation for detailed program/asset management plans.
Material Inflation Rate Projections by Work Breakdown Structure (WBS)	Annual survey of manufacturers of systems and equipment supporting programs.	Used to support program cost models in aggregate and for specific systems.
Material – Bureau of Labor Statistics/Global Insight by WBS Group	Material inflation indices by WBS groupings using representative Bureau of Labor Statistics indices for historical values, and Global Insight projections.	Used to inflate material costs or Cost Estimating Relationships from one date to another.
Labor – Bureau of Labor Statistics Projection Tables	Monthly actual/projected Bureau of Labor Statistics Labor and Material inflation indices.	Used to calculate contract escalation for contracts with compensation adjustment clauses.
Inflation Data Sheet/Factor Program	Cost conversion factors for major appropriations across industrial sectors.	Used to convert program costs in terms of one fiscal year to another fiscal year (move budget dollars between years). Convert "then-year" dollars to "constant-year" dollars (or the reverse).

(continued)

Table 10.1 (continued)

Unique Inflation Indices	Actual/projected inflation indices for specific products, services, or labor categories.	Used to evaluate reasonableness of labor/material inflation in contractor proposals. Used by cost community to inflate unique labor or material items (mission systems, technical support, IT support, etc.).
Industrial Base Analysis		
Deliverables	**Description**	**Purpose**
Workload Chart	Graphical representation of the resources necessary to design, construct, maintain, and repair systems under contract and future potential system contracts at a single manufacturing facility. Charts can be aggregated to regional and total level.	Used to analyze potential resource issues at a specific manufacturing location and to evaluate the resource impact caused by changes to baseline system assumptions at a single manufacturer.
Manufacturer Labor/Overhead Rate Sheets	Projected direct labor and overhead rates for the major manufacturing plants per sector, based on Forward Pricing Rate Agreements data adjusted for workload assumptions.	Used in cost estimates - applied to man-hour estimates to estimate manufacturing labor costs.
Private Sector Government Manday Rate for Repair and Modernization	Private sector government labor-day rate for repair facilities conducting repairs or modernization work on systems. Includes rate for the previous year as well as projections through multiple years using either forward priced rates or escalated rates using appropriate indices. Provides rates for individual facilities.	Used in development of system maintenance and modernization budgets, development of estimates for installation costs associated with proposed system change documents.
Industrial Base Sector Studies	Analytical reports assessing specific manufacturers or industry sectors such as propulsion shafting, batteries, etc.	Used by various program managers to evaluate diminishing sources, foreign restrictions, sole source, corporate mergers, corporate viability and other issues that may represent risk to manufacturing programs.
Special Feasibility Studies	Provide feasibility review of candidate manufacturing plants for alternatives proposed by program managers.	Used by program managers to support Analysis of Alternatives and market studies.

- Determine the project's cost drivers and risks with input from program managers
- Develop probability distributions for the technical and schedule cost drivers
- Develop probability distributions for the cost model uncertainty
- Run the risk model
- Identify the probability that the actual cost is less than or equal to the point estimate
- Recommend sufficient unallocated future expenses to achieve the desired percent confidence level

All projects should develop plans and budgets based upon a quantification of risk and uncertainty that could cause the project to take longer or cost more than initially anticipated. Program managers should request budget allocations that reflect that information with a certain probability that the project will be completed at or below this amount (i.e., probability is usually 70%). Project management's cost-risk roles and responsibilities, as well as program and project cost-risk requirements by life-cycle phase, include the following activities:

- Risk assessments
- Risk evaluations
- Risk mitigation

- Identification of margin and UFE (formerly known as reserves)
- Associated oversight and approval processes

By adhering to these guidelines, cost estimators and analysts will improve the quality and accuracy of systems cost estimates, help to generate realistic budget plans, and provide decision-makers with accurate and realistic cost data to better inform the decision-making process.

10.11.1 Difference Between Risk and Uncertainty

There is an important distinction between the terms "risk" and "uncertainty." The taxonomy and definitions of risk and uncertainty have been defined by multiple sources in different ways. For this discussion, risk and uncertainty are differentiated using the following definitions:

- **Risk** is the chance of loss or injury. In a situation that includes favorable and unfavorable events, risk is the probability that an unfavorable event will occur.
- **Uncertainty** is the indefiniteness of the outcome of a situation. It is assessed in cost estimate models to estimate the risk (or probability) that a specific funding level will be exceeded.

Cost-risk management integrates a risk management process which includes cost estimating, cost-risk assessment/analysis, and EVM. Supporting disciplines include procurement, source selection, cost data collection, and cost data analysis. Integrating cost-risk assessment throughout the project lifecycle enables decision-makers to manage the cost risks and analysts to continually identify and quantify cost risks throughout the program lifecycle. A summary of the three-step cost-risk management process follows.

- **Identify and Quantify Cost Risk**—First, identify and assess risk. Examples of risk can include items that have low Technology Readiness Levels (TRLs), components used for the first time or for new operating environments, schedules that are too short, requirements that change, etc. Next, translate risk assessment into cost impact, and perform risk-based cost analysis (i.e., S-curve). The term S-curve can also be used to indicate an S-shaped chart resulting from a cumulative likelihood distribution. In this function, an S-curve is a tool of quantitative risk analysis that project management would use to determine the possible dangers of any given course of action; see Fig. 10.6.
- **Establish Cost-Risk Reporting**—Develop the Request for Proposal, EVM, Data requirements description, and equivalent project plan requirements. Evaluate EVM and LCC based on the defined requirements in proposals/project plans.
- **Manage Cost Risk Using Reported Data and Analysis**—Perform EVM performance measurement and S-curve analysis. Compile end-of-contract cost-risk data for database updates, data evaluation and analysis, and cost-risk algorithm updates.

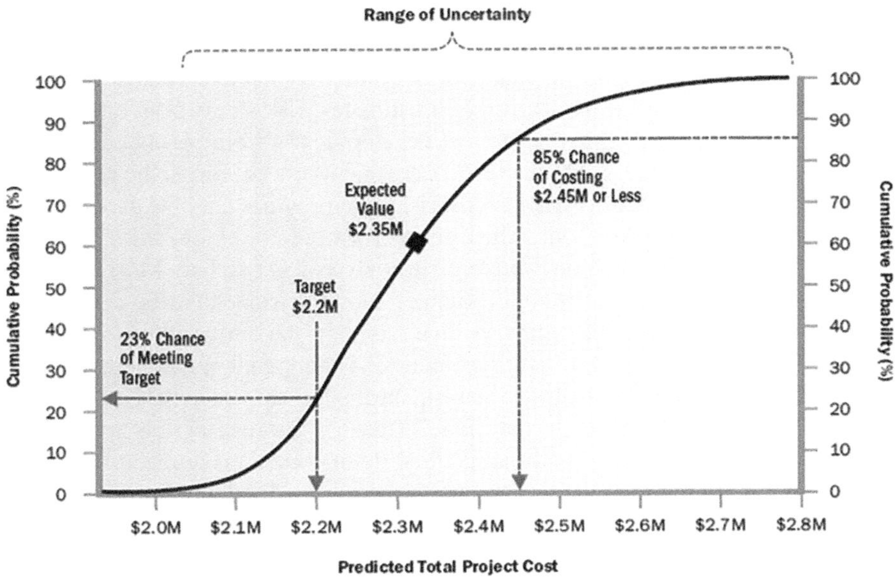

Fig. 10.6 S-curve example

10.11.2 Cost-Risk Assessment

Cost-risk assessment is the process of identifying and analyzing critical project risks within a defined set of cost, schedule, and technical objectives and constraints. Cost-risk assessment balances the probability of failing to achieve a particular outcome against the consequences of failing to achieve that outcome. Assessing cost risk also allows the cost estimator to document risks that accommodate proactive management of project costs. The purpose of cost-risk assessment is to capture uncertainty in areas such as cost estimating methodology, technical risk, schedule risk, and programmatic factors to go from a deterministic point estimate to a probabilistic estimate. A credible baseline estimate is the key starting point in generating a cost-risk-adjusted estimate and developing confidence intervals. Note that a point estimate is usually based on most likely/current-best-estimate inputs. Historically, on large-scale projects, possible impacts of risks have been addressed by establishing contingencies added to a base cost estimate. Contingencies were typically estimated budget allowances set using simple rules of thumb, such as 10% of the base cost.

Risk analysis provides an analytical basis for establishing defensible cost estimates that quantitatively account for likely project risks. It is important to keep in mind that this analysis should be continually reviewed and updated as more data become available. Projecting future outcomes by undertaking a certain course of action (or inaction) allows credible risk analysis. This analysis utilizes various methods of modeling, analysis, and evaluation and thus contains various types of uncertainty. In general, uncertainties may be attributable to a number of factors including (1) the statistical nature of data, (2) the insufficient understanding of physical and biological phenomena, and/or (3) unpredictable events (e.g., natural, biological, and human behavior). For cost estimates, risk stems from uncertainties encountered during the course of project development—from concept exploration to operational sustainment. The cost-risk assessment process forces the cost estimator and program manager to consider cost risks and provides tangible data for use as the basis for informed decisions. The estimator should not be limited in analyzing just the data that have been reviewed (e.g., just the project risk list). Most important, when estimating project costs, the estimator must consider that the costs are an uncertain quantity and the point estimate is not the only possible estimate. Figure 10.7 demonstrates that point estimates of individual WBS elements using triangular and normal distributions can be quantified as "most likely" (mode), "50th percentile" (median), or "expected value" (mean). The use of this terminology implies that costs are statistical in nature and defined by their probability distributions. When the number of WBS elements increases, the distribution of the total cost of the WBS elements approximates the normal distribution (Fig. 10.8). This is known mathematically as the Central Limit Theorem. This theorem states that the average of the sum of a large number of independent, random variables with finite means and variances converges to a normal random variable. Three of the more commonly used probability distributions for assessing risk are triangular, normal, and lognormal distributions.

As shown in Fig. 10.9, triangular distributions require defining three inputs (most likely, lowest, and highest). They can be determined quantitatively, for example, by using three cost estimates for a single project element to define the distribution of the range of costs for that element. They can also be determined subjectively, as when experts determine the range of possible values for a model input or risk magnitude. Note that analysts and engineers tend to overestimate best-case outcomes and underestimate worst-case outcomes. Therefore, cost analysts frequently do not treat their minimum and maximum values as the end of points of the triangular distribution. Instead, they adjust the extremes of the triangular distribution to allow for a wider range of potential outcomes.

The normal and lognormal can be used when the mean and standard deviation are known (see Fig. 10.10). This approach eliminates the need to specify minimum and maximum values. Sometimes the normal distribution can produce negative values that, obviously, do not make logical sense for describing cost. Furthermore, the normal distribution's inherent symmetric characteristic cannot characterize the common tendency for costs to go up rather than down over time. For these two reasons, many cost analysts prefer the lognormal distribution, which cannot go below zero and is nonsymmetrical (e.g., can be right-skewed), is typically preferred over the normal distribution.

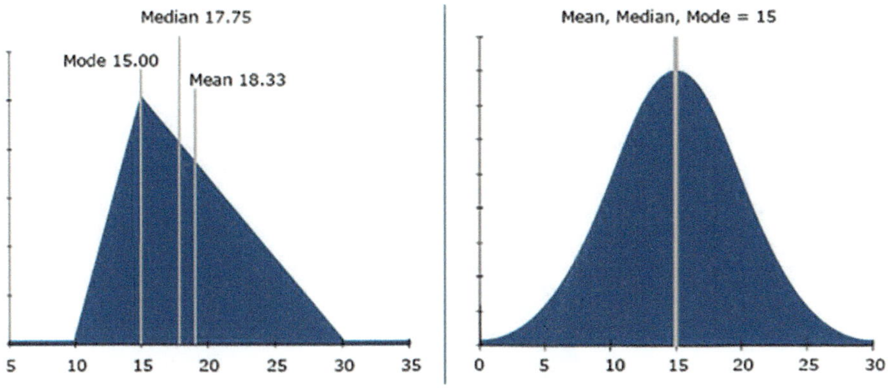

Fig. 10.7 Statistics of triangular and normal distributions

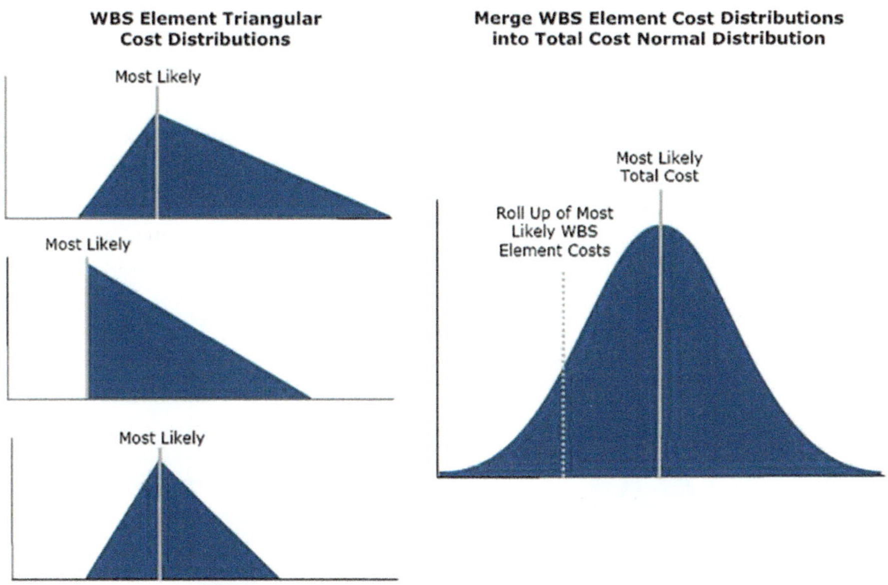

Fig. 10.8 Central limit theorem

10.11.3 Cost Estimating Risk as Part of the Cost Estimating Process

Cost estimating risks include economic factors such as rate uncertainties, cost estimating errors, and statistical uncertainty inherent in the estimate. Cost estimating risk is dependent upon other fundamental risk dimensions (technical, schedule, and programmatic), all of which must be assessed to arrive at an accurate picture of project risk. Cost estimating risk assessment considers the cost, schedule, and

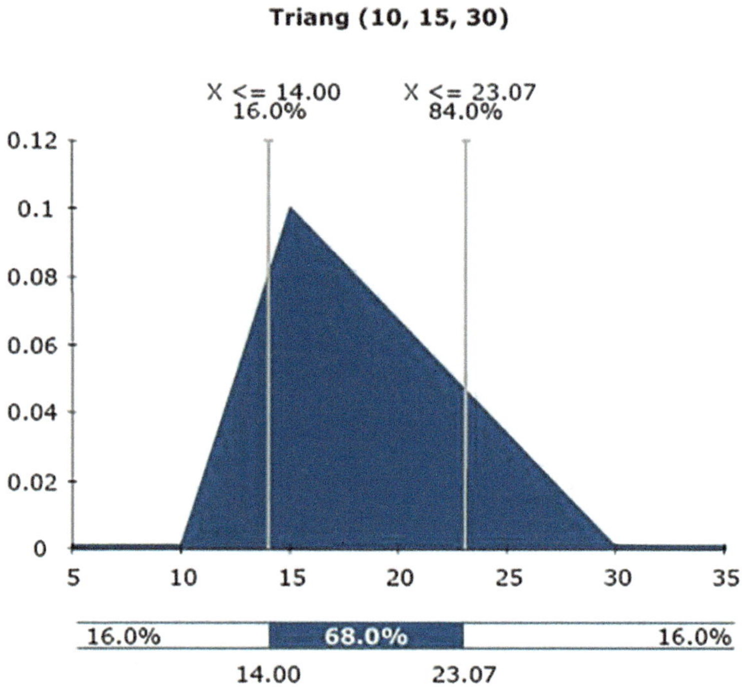

Fig. 10.9 Triangular distribution example

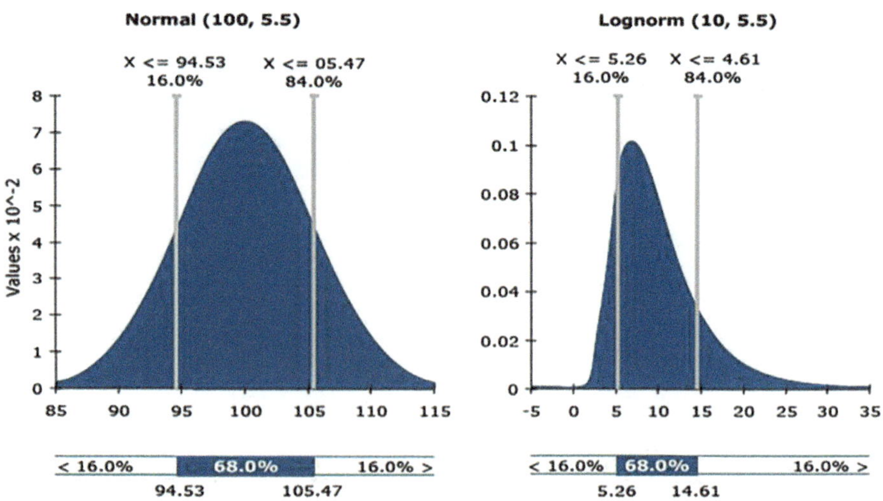

Fig. 10.10 Normal and lognormal distributions

technical risks that are then factored back into the cost estimate. To quantify the cost impacts due to risk, sources of risk need to be identified. Cost analysts should be

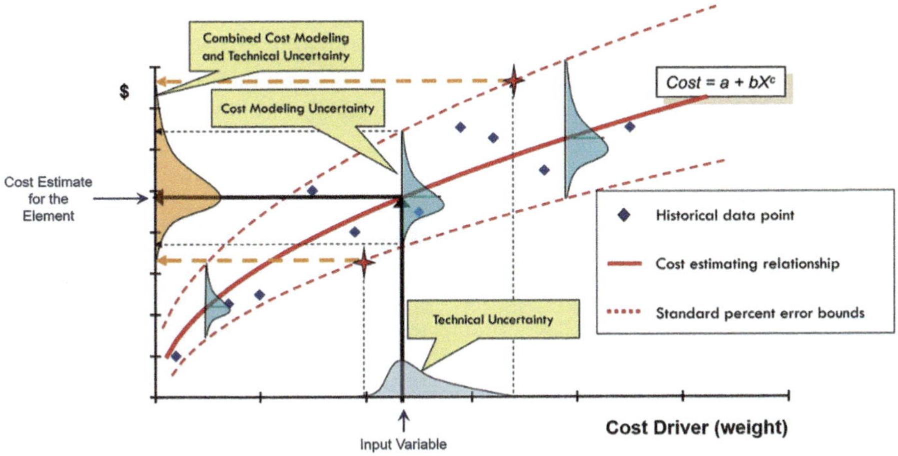

Fig. 10.11 Cost model and technical input risks

concerned with three sources of risk and ensure that the model calculating the cost also accounts for these risks:

- **Risk inherent in the cost estimating methodology**. For example, using a regression-based CER includes an associated Standard Error of the Estimate, confidence intervals, and prediction intervals. The cost estimator can use any of these to include cost estimating methodology risk in the estimate.
- **Risk inherent in the technical and programmatic aspects of the systems being developed**. The technology's maturity (TRLs are good indicators of this risk source), design/engineering, integration, manufacturing, schedule, and complexity, etc., fall into this risk category. Quantifying the cost impacts due to these kinds of risks is not as statistically based as CER risk (Fig. 10.11), which graphically displays the effects of cost estimating methodology risk and technical input risk.
- **Risk inherent in the correlation between WBS elements**. Correlation assessment determines to what degree one WBS element's change in cost is related to another and in which direction. For example, if the cost of System A goes up and the cost of System B goes up, then a positive correlation results between both systems' costs. Many WBS elements within an SoS have positive correlations with each other. The cumulative effect of this positive correlation tends to increase the range of the possible costs.

Types of risk may be classified as known-knowns, known-unknowns, and unknown-unknowns. The known-knowns are the discrete risks for which likelihood and consequence are understood. The known-unknowns are discrete risks for which likelihood and consequence are not understood. Unknown-unknowns are, by definition, unknown–but they are not unknowable. The unknown-unknown risks are generally quantifiable only through subjective assessment. The potential exists for reducing the effect of this class of risk by interviewing more subject-matter experts or by waiting for additional information to become available. Lastly, for risks that are unknown and for which prior assessment is not possible, allowances can still be made for contingency. The challenge for cost estimators and schedule analysts is determining how best to accommodate these risks, uncertainties, and unexpected surprises without over- or understating their effects and without negatively impacting the project budget.

Having examined the **cost estimation of potential solutions**, we now shift our focus to the ongoing management of mission capabilities. In the next chapter, we explore **mission capability portfolio management**, which is crucial for maintaining awareness of the health of mission capabilities and ensuring their effective execution. This process involves monitoring established mission engineering threads and webs to assess their effectiveness, thereby allowing for real-time detection of any capability failures. Proactively managing these capabilities ensures that mission objectives are continuously met and, when necessary, adjusted to keep pace with or outpace adversaries and competitors.

References

1. Defense Acquisition University (DAU). (2020). *Cost estimating and analysis guide*. Defense Acquisition University Press.
2. U.S. Government Accountability Office (GAO). (2020, March). *GAO cost estimating and assessment guide: Best practices for developing and managing capital program costs*. GAO-20-195G.
3. Naval Sea Systems Command (NAVSEA). *NAVSEA cost estimating handbook*. NAVSEA, U.S. Department of the Navy.
4. National Aeronautics and Space Administration (NASA). (2015, February). *NASA cost estimating handbook*, Version 4.0. NASA.
5. Office of Management and Budget (OMB). (1992). *Circular A-94: Guidelines and discount rates for benefit-cost analysis of federal programs*. Executive Office of the President.

Chapter 11
Mission Capability Portfolio Management

Mission Engineering and Integration (MEI) supports the function of capability portfolio management as outlined in DoD Directive 7045.20 [CPM08]. Mission engineering threads/webs identify operational needs based on the way the forces plan to fight with the "as-is" force architecture following Combatant Commanders' operational plans and contingency plans. These mission engineering threads/webs also inform the deployment of the systems needed to accomplish a mission within a system-of-systems (SoS) context.

The relationship between portfolio management and MEI is crucial for organizations aiming to align their strategic objectives with operational capabilities effectively. Portfolio management focuses on the centralized management of various projects, programs, and operations to achieve overarching strategic goals. It ensures that resources are allocated efficiently, risks are managed effectively, and initiatives are executed in alignment with stakeholder strategies. On the other hand, MEI involves the systematic approach to designing and integrating complex systems to accomplish specific mission objectives. By integrating portfolio management practices with MEI principles, organizations better ensure that their portfolio of projects and programs is tailored to support mission-critical objectives by producing the right systems and platforms to form SoS capabilities for mission execution. The coalescence of these holistic approaches provides the necessary mechanisms for managing resources, addressing interdependencies, and optimizing the delivery of capabilities needed to fulfill mission requirements.

Through the review of strategic, tactical, and operational capabilities and gaps, portfolio management offers valuable feedback essential for resource planning and management. By assessing these capabilities and identifying gaps, portfolio managers gain insights into where resources should be allocated to address deficiencies and enhance strengths. Moreover, this provides feedback for monitoring the overall health of the portfolio and allowing proactive adjustments for better alignment with strategic objectives. Additionally, when portfolio managers collaborate with program and project management teams, their guidance and support help ensure that

J. D. Moreland, Jr., *Mission Engineering and Integration*,
https://doi.org/10.1007/978-3-032-32132-9_11

projects and programs are executed effectively, contributing to the achievement of desired outcomes and the overall success of the portfolio. This integrated approach fosters synergy across the organization, enabling resources to be utilized efficiently and decision-making to drive strategic initiatives forward effectively.

Portfolio management practitioners operate within a complex environment characterized by overlapping and sometimes conflicting interests as many stakeholders are involved in the development and acquisition of the required SoS for missions. In such conditions, they often face challenges associated with information overload and a lack of actionable intelligence. To effectively navigate this environment and anticipate behavior, portfolio manager needs to study the portfolio as a complex SoS. Programs and projects serve as temporary mechanisms for delivering change within this environment. Therefore, adopting an SoS perspective enables a deeper understanding of the objectives that drive change initiatives and the means through which these objectives are achieved. By viewing the portfolio as a complex SoS, practitioners can better analyze the interconnectedness of various elements, anticipate potential challenges, and make informed decisions to optimize portfolio performance and achieve strategic goals. This holistic approach fosters a more comprehensive understanding of the dynamic interactions within the portfolio ecosystem, ultimately enhancing the practitioner's ability to drive successful outcomes.

Adopting a holistic approach to portfolio management involves integrating systems thinking alongside prevailing linear thinking practices. This approach goes beyond merely examining and understanding portfolio elements individually; it also entails analyzing the portfolio as a whole and focusing on the effects of interactions between these elements. While linear thinking traditionally emphasizes sequential cause-and-effect relationships, systems thinking considers the interconnectedness and interdependencies within the portfolio ecosystem. By applying systems thinking in parallel, practitioners gain insights into how changes in one element can ripple throughout the entire portfolio, affecting outcomes and performance. This broader perspective allows for a more nuanced understanding of complex dynamics, facilitating better decision-making and more effective management of portfolios.

11.1 Defining Portfolio Management

11.1.1 Portfolio Defined

A portfolio is a collection of projects, programs, subsidiary portfolios, and operations managed collectively to achieve strategic and mission objectives. The elements within a portfolio, such as programs and projects, which ultimately produce systems and platforms, are quantifiable through processes like identification, categorization, evaluation, prioritization, and authorization. These elements are a collection of SoS forming a mission capability and have diverse objectives. Portfolio elements compete for a share of limited resources, with the allocation driven by operational mission strategies and capabilities. Therefore, decision-makers must examine their unique circumstances to determine the best way to optimize and balance the portfolio elements, so that the allocation of resources maximizes the strategic impact and aligns with organizational goals.

A portfolio exists to achieve mission capability strategies and goals, comprising a set of current and future portfolio elements. Like programs and projects, portfolios have a lifecycle. However, unlike programs and projects, which have a more limited duration, portfolios often have greater longevity and require sustained management attention. At any given time, a portfolio represents a collection of its selected elements and reflects one or more stakeholder strategies and objectives for that period. Therefore, a functioning portfolio should represent an organization's intent, direction, and progress at any given moment, aligning with its strategic aims and evolving needs.

11.1.2 Portfolio Lifecycle

The portfolio lifecycle encompasses the ongoing processes and functions that occur within a set of portfolios, programs, projects, and operations over a continuous timeframe. In a dynamic world, this lifecycle needs to be adaptive and flexible to accommodate constantly changing internal and external influences, ensuring stakeholders remain competitive and relevant in their capabilities. The controls required to manage a portfolio must be flexible to adapt to the complexities of various portfolio scenarios. Adjustments to portfolio elements are likely needed when changes in the operational context and conditions arise. Strategic objectives, performance metrics, and product and service design are critical inputs into the portfolio lifecycle. Understanding these inputs thoroughly is essential to grasp the portfolio's value to stakeholders. Defining the value derived from the portfolio's functions, products, and services from all stakeholder perspectives helps eliminate non-value-added programs, projects, or activities. Mapping the value stream is key to identifying which value-added programs, projects, and activities are essential to strategic objectives, and conversely, which ones offer little or no benefit.

A portfolio's continuous lifecycle consists of several stages: initiation, planning, execution, and optimization. As a portfolio progresses through its lifecycle, information and decisions flow within and between each of these stages in an iterative, rather than strictly sequential, manner. All stages, including initiation, are adaptable, flexible, and fluid. The stages are defined as follows:

- **Initiation**: The critical first stage, triggered by the formal release of stakeholder objectives and goals. Initiation sets the approach and principles for the major processes that will define how the portfolio and its elements are managed throughout the lifecycle. The primary goals of this stage include 1) establishing an operational strategy, 2) identifying the portfolio elements within the scope, and 3) defining a long-term roadmap with financial goals, performance metrics, and stakeholder governance. This foundational phase ensures that the portfolio aligns with strategic objectives and provides a clear framework for subsequent planning, execution, and optimization stages.

- **Planning**: The primary goals of the planning stage are to develop the portfolio management plan and reach agreement on several key aspects:
  - Manage portfolio elements within scope
  - Identify interdependencies between portfolios and their elements
  - Identify risks and issues along with the development of response plans
  - Determine resourcing requirements
  - Prioritize portfolio elements based on mission essential tasks mapping
  - Confirm the governance framework and stakeholder accountability
  - Establish portfolio metrics to measure success
  - Define the scope of elements within the portfolio
  - Detail product and/or service requirements and specifications

This comprehensive planning ensures that all components are aligned and prepared for effective execution and optimization. During periodic portfolio strategic planning, the portfolio manager revisits the model to check that it aligns with stakeholder values, benefits, and operational strategic goals. This phase initiates the initial optimization of the portfolio, as it is formally aligned with the strategy during this review. Portfolio managers scrutinize metrics to ensure they correspond to realistic goals and objectives across all stakeholders. Existing portfolio elements, initially agreed upon during the initiation stage, are examined in greater detail, including their scope, timing, budget, associated risks, resourcing requirements, and interdependencies. Additionally, the portfolio management team reassess the prioritization criteria based on the performance of new elements introduced since the last strategic review. Changes are then proposed and evaluated based on these updated criteria to maintain alignment and optimize the portfolio's effectiveness.

- **Execution**: The primary goals of the execution stage are to:
  - Lead the delivery of all elements within each portfolio
  - Actively manage and resolve risks and issues across interdependencies and within the portfolio elements
  - Reprioritize and modify subsidiary portfolios as needed
  - Monitor benefits-realization potential based on element delivery
  - Managing portfolio assets and resources specific to the portfolio

Portfolio execution is carried out through its various elements and operations. The health of the portfolio is reported through regular status updates and by reviewing the performance metrics established during the planning phase. The portfolio team evaluates proposed changes based on ongoing stakeholder needs, which may necessitate the introduction of new elements into the portfolio. These new elements are then assessed as required, based on unplanned critical needs or the positive outcomes of proofs of concept, pilots, or feasibility studies.

- **Optimization:** Is the process of making a portfolio as effective as possible by maximizing available conditions, constraints, and resources. During this stage, the portfolio manager facilitates discussions with stakeholders to ensure that all intended benefits for the remaining elements within the portfolio are realized.

Critical supportive activities for maintaining optimization include monitoring portfolio performance and recommending changes to the portfolio element mix, as well as ensuring portfolio element performance and compliance to control optimization. The purpose of monitoring and controlling is to understand when changes need to be made to the portfolio to maintain alignment with strategic objectives and maximize overall effectiveness.

11.1.3 Portfolio Management

Portfolio management is the centralized management of one or more portfolios aimed at achieving strategic objectives. It involves the application of activities to align the portfolio and its elements with stakeholder strategies. Additionally, portfolio management can be perceived as a dynamic activity through which stakeholders invest their resources to attain strategic objectives. This process encompasses a range of activities including identifying, categorizing, monitoring, evaluating, integrating, selecting, prioritizing, optimizing, balancing, authorizing, transitioning, controlling, and terminating portfolio elements. Through these actions, portfolio management ensures that resources are allocated effectively, risks are managed appropriately, and the portfolio remains aligned with the organization's strategic vision.

Operational and stakeholder strategies, along with their respective objectives, serve as guiding principles for portfolio management, shaping decisions, direction, purpose, and resource allocations aimed at achieving targeted values. These values may vary significantly, ranging from overarching organizational objectives to more specific goals aligned with individual stakeholder missions and visions. However, numerous challenges complicate the process of realizing these objectives, with strategic alignment and execution emerging as the most critical. Obtaining and maintaining support from decision-makers, balancing feasibility with necessity, and determining both short- and long-term benefits and goals also present significant hurdles. Moreover, effective portfolio management necessitates adeptly managing resources, including capacities and capabilities, while simultaneously ensuring the ability to execute and sustain initiatives over time. Addressing these challenges requires a comprehensive and strategic approach that prioritizes alignment, stakeholder engagement, resource optimization, and the ability to adapt to evolving circumstances.

Portfolio management embodies an organized approach aimed at achieving strategic results, anchored by fundamental principles that guide its operations:

- Pursue excellence in executing strategic plans
- Ensure that initiatives are carried out efficiently and effectively to attain desired outcomes
- Emphasize the importance of enhancing transparency, responsibility, accountability, sustainability, and fairness throughout the portfolio management process
- Foster ethical practices and stakeholder trust and confidence
- Seek to balance portfolio value against overall risks, meticulously assessing and managing risks to optimize the balance between potential rewards and consequences
- Ensure investments in portfolio elements are aligned with stakeholder strategies and operational objectives
- Exercise active and decisive leadership to optimize resource utilization
- Foster a culture that embraces change and risk, and adeptly navigates complexity to enable successful outcomes

11.2 Evaluation of Mission Capability Portfolios

The evaluation process in portfolio management is not a one-time event but a continuous endeavor, made necessary due to the dynamic nature of environments where operational conditions or programmatic factors can change at any moment. This ongoing evaluation involves reviewing the current set of elements which comprise systems and platforms included in the portfolio, and selecting and assessing those that significantly impact the portfolio's overall results. Defining these key portfolio elements demands a deep understanding of strategic objectives and the portfolio's role in achieving them. Both internal and external environmental factors must be considered. Therefore, identifying these key elements relies on regularly reviewing and updating the main strategic objectives for all stakeholders and the portfolio itself. This iterative process ensures that the portfolio remains aligned with evolving strategic priorities and is optimally positioned to deliver value to stakeholders amid changing circumstances. The evaluation process within portfolio management can be prompted by various activities, with several common triggers:

- Change becomes necessary in stakeholder strategies and objectives because of internal or external factors.
- An operational gap forms between the overall expected portfolio results and the actual results.
- An ongoing key element is delayed, closed, or has a major change on its validated plans.
- A new portfolio element is validated.
- A new opportunity (e.g., new technology) has emerged.

Portfolio optimization is the ongoing practice by which benefits, risks, and resources are balanced and enhanced. It should be an integral part of any effective planning process to ensure alignment between supply (e.g., assets and intellectual resources) and demand (e.g., threats and regulatory changes). With multiple projects underway or in the pipeline at any given time, managing the balance of priorities, resource alignment, allocation, and overall bandwidth is crucial to avoid overruns of time and funding and prevent overutilizing human and technological resources. This practice involves examining all components of the portfolio, highlighting dependencies and constraints, and identifying the most effective bundling of projects to define synergies and leverage points. By doing so, portfolio optimization ensures that resources are used efficiently, and strategic objectives are met effectively. Some potential challenges encountered as part of executing portfolio optimization include:

- Misalignment of portfolios and resources with objectives and strategies
- Lack of transparency or access to reliable, accurate authoritative source of truth data
- Resources being exhausted due to time spent on numerous low-value projects
- Difficulty in rapidly adjusting portfolios in response to operational factors and adversarial threats
- Evaluating the impact of changes in one individual element (e.g., system and platform) on the other elements
- Evaluating the cyclical effects of reprioritization or optimization

Managing strategic alignment allows the portfolio manager to respond to changes in organizational strategy and improve the capacity to accept and act on significant strategic changes that impact portfolio performance. As strategy evolves, it is crucial to compare the current state with the future state, which may involve evolutionary or incremental changes. This comparison may reveal a gap that necessitates realignment of resources or adjustments in the portfolio element mix to support the strategic shift. Changes in portfolios are common, and depending on their significance, some reworking may be necessary to ensure continued strategic alignment. This process involves identifying the gap between the current and future states and analyzing the impact of and response to strategic changes, including changes in resources such as people, processes, and technology. Portfolio managers examine strategic objectives and goals to determine which parts of the portfolio structure will address specific strategic goals. They then review the existing portfolio or inventory of work to decide which work or elements should continue, be closed, or be newly addressed. Techniques used include analyzing the drivers of strategic change to determine the desired contribution of initiatives to strategic objectives and assessing benefits and measurable contributions while considering all potential risk constraints.

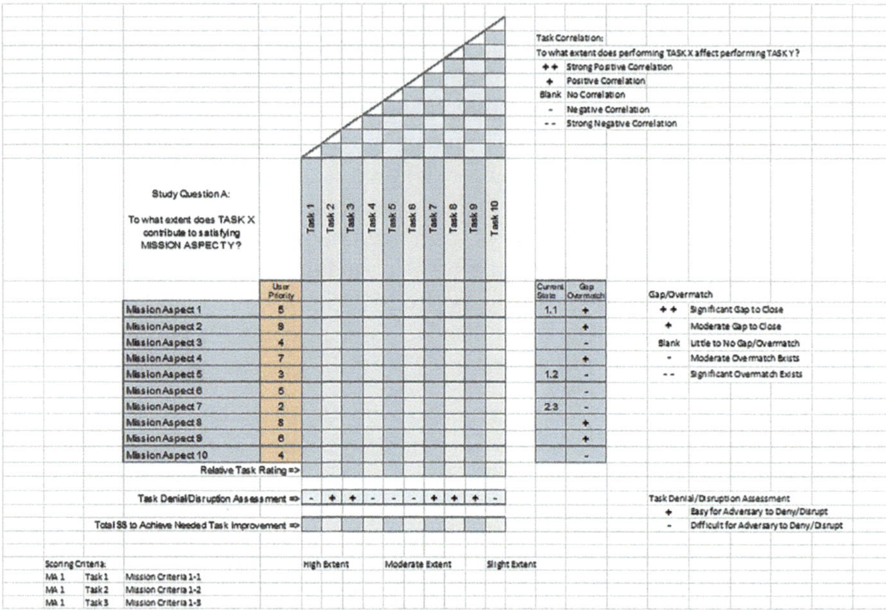

Fig. 11.1 QFD general framework for capability portfolio management

11.2.1 Using Quality Function Deployment (QFD) Matrix

The Quality Function Deployment (QFD) matrix is the outcome of a structured methodology used to translate stakeholder requirements into specific technical or operational characteristics of a system, product, or process. Originally developed in the manufacturing and product development industries, QFD systematically ensures that the final deliverable aligns with the customer's needs and expectations. It achieves this through a matrix-based approach, often referred to as the House of Quality (HoQ). The HoQ maps out relationships between high-level requirements and specific design features. QFD involves multiple phases that progressively break down broad mission or operational objectives into technical specifications, implementation details, and process controls. The methodology is highly effective for complex, multi-layered decision-making environments, making it suitable for capability portfolio management in MEI. This general framework is illustrated in Fig. 11.1.

In MEI, capability portfolio management involves selection, development, integration, and optimization of systems and technologies to meet strategic defense or security objectives. Given the complex, interdependent nature of mission capabilities, QFD provides a structured way to align portfolio decisions with mission needs by systematically evaluating requirements and technical trade-offs. Key steps in applying QFD to capability portfolio management are:

- **Identify Mission Needs and Stakeholder Requirements**
  This step establishes how specific tasks contribute to broader mission aspects by defining mission-level objectives and requirements. It identifies stakeholder needs, strategic goals, and key operational tasks required to achieve mission success. This step serves as the foundation for linking high-level mission objectives to system functions, operational tasks, and technical capabilities. Mission-level analysis typically occurs in this first step of the QFD process, often referred to as the Mission Needs Analysis (WHATs Identification) phase, which includes the following actions.
  - Capture and categorize stakeholder inputs, including user requirements, operational needs, and national security objectives.
  - Define mission-specific capability goals, such as intelligence gathering, surveillance, targeting, logistics, or cyber warfare resilience.
  - Break down operational tasks: Identifies specific tasks (e.g., surveillance, target tracking, data fusion) that directly support the mission.
  - Map task contribution to mission success: Uses a structured framework (e.g., QFD matrix) to show how different tasks align with mission-critical aspects.
  - Prioritize mission-critical tasks: Helps decision-makers focus on tasks that provide the most operational value.
- **Develop the House of Quality (HoQ) Matrix**
  - Create a QFD matrix where mission requirements (WHATs) are mapped against technical or system attributes (HOWs) following the general framework in Fig. 11.1. In the QFD matrix, this step is represented on the left side ("WHATs"), listing mission needs that will be mapped to technical solutions and capabilities in subsequent steps.
  - Evaluate the strength of relationships between requirements and system capabilities using weighting criteria.
- **Analyze System and Capability Interdependencies**
  - Determine how different systems contribute to the overarching mission capabilities, see Fig. 11.2.
  - Identify conflicting or competing requirements that need prioritization.
- **Prioritize Capability Investments**
  - Use QFD results to rank and prioritize portfolio elements based on their alignment with mission-critical objectives.
  - Identify gaps where new system development, technology insertion, or system upgrades are needed. Figure 11.3 depicts the analytical flow.

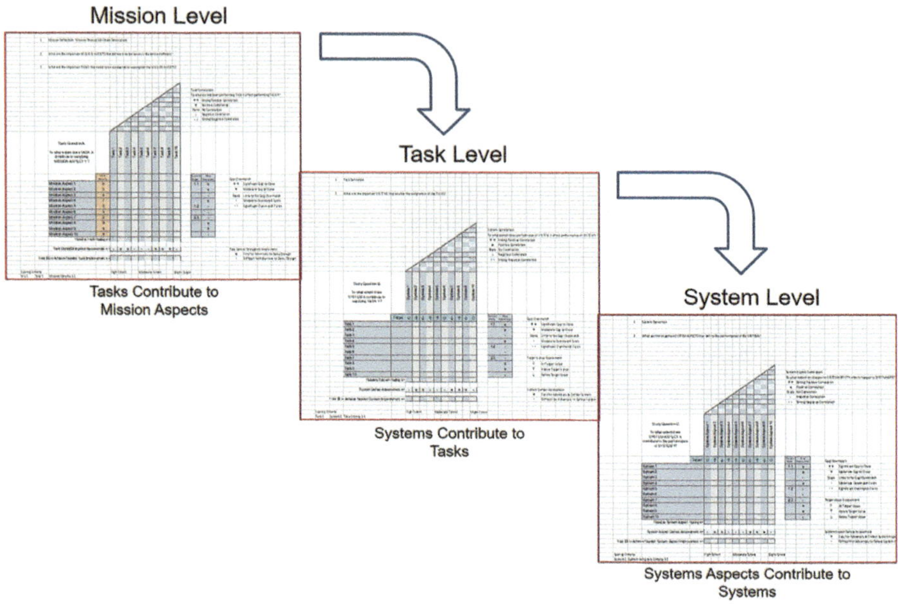

Fig. 11.2 Mission to tasks to systems mapping

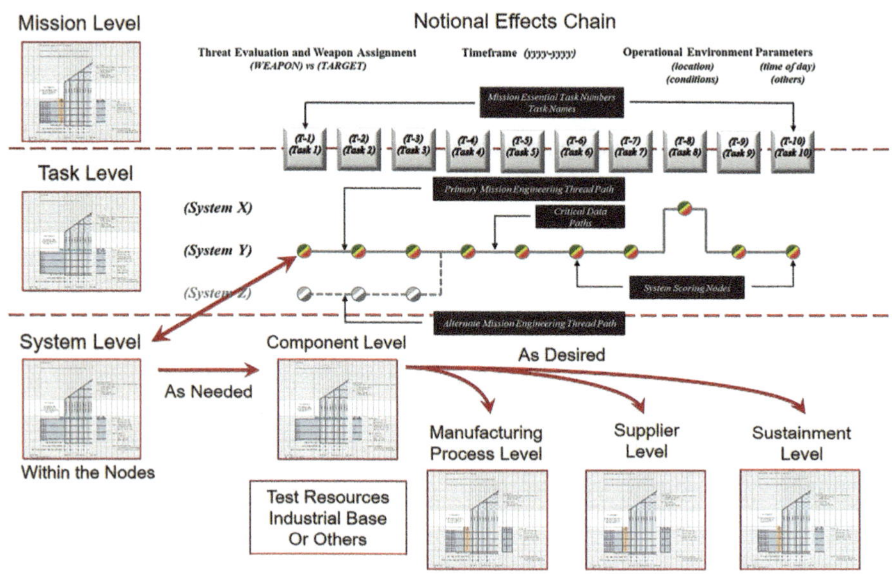

Fig. 11.3 Analytical flow

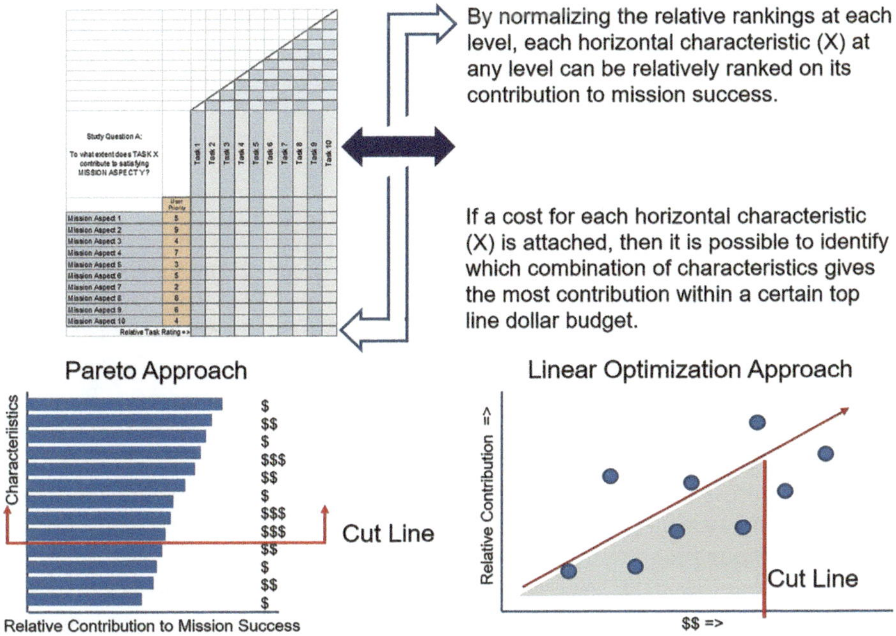

Fig. 11.4 Optimize capability portfolio for budget decisions

- **Optimize Portfolio Decisions and Resource Allocation**
  - Conduct trade-off analysis to balance mission effectiveness, risks, costs, and resource availability as illustrated in Fig. 11.4.
  - Adjust capability roadmaps based on emerging threats, technological advances, or shifting mission priorities.
- **Monitor and Update Portfolio Performance.**
  - Implement continuous evaluation loops to assess how well-deployed systems meet evolving mission demands.
  - Use QFD matrices dynamically to adapt to changes in operational conditions or stakeholder expectations.

QFD is critical in performing capability portfolio management because it provides a structured and systematic approach to aligning mission needs with technical capabilities, resource allocation, and strategic priorities. By translating high-level mission objectives into actionable system requirements, QFD ensures that every capability investment directly contributes to mission success. It facilitates decision-making by identifying key operational tasks, assessing interdependencies, and prioritizing initiatives based on mission impact. Additionally, QFD enhances transparency, stakeholder alignment, and risk management by systematically evaluating trade-offs and optimizing resource utilization. Ultimately, it enables a mission-driven approach to capability development, ensuring that investments are both effective and adaptable to dynamic operational conditions.

11.2.2 Balanced Scorecard for MEI

In the complex and dynamic landscape of MEI, effective capability portfolio management requires a structured approach to align resources, technologies, and operational strategies with overarching mission objectives. The Balanced Scorecard (BSC) provides a comprehensive framework to evaluate and optimize portfolio elements by integrating multiple performance perspectives such as mission impact, operational effectiveness, resource utilization, and innovation readiness. By applying BSC to MEI, decision-makers can ensure that capabilities are strategically aligned, performance is continuously assessed, and resources are allocated efficiently to address emerging threats and operational demands. This approach enables a data-driven, mission-focused methodology for enhancing readiness, interoperability, and promoting long-term mission success.

Implementing the BSC in capability portfolio management for MEI offers a strategic approach to aligning organizational objectives with mission-driven outcomes. The BSC framework, traditionally used in performance management, provides a balanced view across four key perspectives including financial, customer, internal processes, and learning and growth. By applying this framework to capability portfolio management, organizations ensure that their investments in MEI capabilities are not only efficient but also aligned with broader strategic goals. This integration enables organizations to evaluate performance across multiple dimensions, optimize resource allocation, and ensure that their MEI initiatives deliver maximum value. Some general steps in this implementation are:

- **Define Strategic Objectives and Mission Alignment**:
  - Identify key mission success criteria and align them with stakeholder priorities.
  - Define capability goals, such as system interoperability, force readiness, or technological superiority.
  - Develop a mission-centric strategy map that illustrates how different portfolio elements contribute to mission objectives.

  Example: If the mission objective for defense is rapid deployment capability, the portfolio should prioritize systems that enhance mobility, logistics efficiency, and force projection.
- **Establish Key Performance Indicators (KPIs) for Each Perspective**: Each BSC perspective requires specific, quantifiable KPIs to measure success.
  - **Mission and Stakeholder Perspective (Strategic Impact)**

  Mission Success Rate—Percent of completed missions meeting all objectives.

Stakeholder Satisfaction Score—Rating from military leaders, policymakers, and operational commanders.
Capability Effectiveness Index—Measurement of how well portfolio systems fulfill mission roles.

- **Operational Effectiveness Perspective**

System Availability & Reliability—Mean Time Between Failures (MTBF) and readiness rates.
Operational Efficiency—Percent of reduction in operational bottlenecks.
Interoperability Score—Degree of seamless integration between systems.

- **Resource and Financial Perspective**

Cost-Benefit Ratio of Portfolio Elements—Investment efficiency for each capability.
Budget Utilization Rate—Percent of allocated funds effectively spent.
Workforce & Skill Readiness—Percent of personnel trained for new system operations.

- **Innovation & Readiness Perspective**

Technology Readiness Level (TRL) and Capability Readiness Level (CRL)—Maturity of emerging technologies.
Time-to-Deployment for New Capabilities—Speed of fielding innovative solutions.
Adaptability Index—Organization's ability to pivot in response to threats.

- **Align Capability Portfolio Decisions with BSC Insights:**
  - Use the BSC results to prioritize and optimize portfolio elements.
  - Allocate resources dynamically based on mission-critical capabilities and operational gaps.
  - Identify underperforming elements and restructure, reallocate, or phase them out.

  Example: If a major logistics system is found to have low operational efficiency and high costs, BSC analysis can trigger an evaluation to replace, upgrade, or retire the system.
- **Monitor and Adjust Portfolio Strategies Based on BSC Reports:**
  - Continuously track performance using BSC dashboards and scorecards.
  - Adjust portfolio elements as mission priorities, threats, and technologies evolve.
  - Conduct regular reviews to refine capability investments and mitigate emerging risks.

  Example: If intelligence reports indicate a new cyber warfare threat, the portfolio may reprioritize investments in cyber defense systems based on BSC insights.

By leveraging the BSC, mission engineers and capability portfolio managers can systematically assess, prioritize, and optimize portfolio elements to achieve strategic mission success with maximum efficiency and effectiveness.

11.3 Portfolio Risk Attitude

A stakeholder's strategic risk "attitude" or "appetite" refers to the level and type of risk they are willing to accept to achieve their strategic goals. As discussed in Chap. 9, this appetite varies depending on factors such as the operational environment, organizational culture, strategic drivers, and the personal preferences of senior decision-makers. Different risk categories may have varying levels of acceptable risk, and these appetites can shift over time. Portfolio managers must navigate these complexities by taking calculated risks, avoiding certain threats, and seizing select opportunities while disregarding others. Establishing performance standards is key to articulating these risk appetites in a meaningful way. While defining risk thresholds is challenging, it is essential, as stakeholders have different tolerances for risk, often referred to as their risk attitudes. A common understanding of risk and preparation for the likelihood and impact of known threats is vital. Stakeholders must agree on the maximum level of risk tolerance for each risk category before proceeding with any action.

To manage risk effectively, stakeholders must determine an acceptable level of exposure in pursuit of their strategic objectives. This is guided by measurable risk thresholds that reflect the risk appetite of both the lead organization and other stakeholders. These thresholds define the acceptable variation around a strategic objective. The risk appetite statement should be informed by the perspectives and concerns of all stakeholders and take into account the implications of current organizational strategies and practices.

Having explored **mission capability portfolio management**, we now turn our attention to the critical aspect of **mission resilience and cybersecurity**. The next chapter examines how resilience plays a pivotal role in maintaining operational effectiveness, particularly when facing cyberattacks and their associated attack vectors. As we execute the MEI approach, understanding the different pathways through which adversaries may target systems is essential for safeguarding capabilities. Focusing on resilience and addressing potential attack vectors better prepares the mission engineer to detect and respond to cyber threats, ensuring that mission objectives remain intact even in the face of disruptive challenges.

Reference

1. U.S. Department of Defense. (2008, September 25). *Capability portfolio management*. DoD Directive 7045.20. Available through the U.S. Department of Defense website.

Chapter 12
Mission Resilience and Cybersecurity

Mission resilience refers to the ability of an organization or system to maintain essential functions, services, and capabilities in the face of disruptions, challenges, or adversities. To plan for mission resiliency and ensure mission success, the mission engineering stakeholders must perform risk management to determine potential alternative mission engineering threads. Using a Mission Engineering and Integration (MEI) approach, organizations can dissect the mission into fundamental components, or mission threads, and identify associated risks. By evaluating these risks in terms of their consequence impact and probability of occurrence, organizations can develop alternative threads and mitigate them. Each alternative is thoroughly assessed to determine its potential impact on mission success and its likelihood of effectively managing identified risks. This comparative analysis allows for the selection of the most resilient mission thread, striking a balance between risk reduction and the likelihood of mission success. Implementing the chosen thread with continuous monitoring and adaptive strategies further strengthens mission resiliency, ensuring that the mission can withstand and recover from potential disruptions.

In the context of cybersecurity, mission resilience encompasses the capacity to withstand and recover from cyber attacks, breaches, or disruptions that may threaten the integrity, availability, or confidentiality of critical mission functions and data. Achieving mission resilience involves implementing a combination of proactive measures, risk management strategies, and response capabilities to mitigate the impact of cyber incidents and ensure the continuity of essential operations. This may include deploying robust cybersecurity controls, establishing backup and recovery mechanisms, conducting regular risk assessments, and developing incident response plans.

A trustworthy computing systems perspective becomes the foundation for developing, constructing, and managing a cyber infrastructure that is more resistant to attacks and possesses greater resilience. The approach can be customized to accommodate diverse technical and policy needs, encompassing considerations such as

J. D. Moreland, Jr., *Mission Engineering and Integration*,
https://doi.org/10.1007/978-3-032-32132-9_12

privacy and accountability. The extensive scope of this endeavor encompasses various research methodologies, spanning theoretical exploration to practical experimentation, with potential involvement of human subjects. Theoretical frameworks, models, cryptographic techniques, algorithms, methodologies, system architectures, programming languages, software solutions, tools, and evaluation frameworks are all areas of interest for potential research efforts.

Mission resilience also fosters a culture of preparedness, adaptability, and collaboration within an organization to effectively respond to unexpected events or emergencies. This may involve training personnel, conducting exercises and drills, and establishing clear communication channels to coordinate response efforts and minimize downtime. Furthermore, mission resilience extends beyond technical measures to encompass broader organizational resilience principles, such as leadership commitment, stakeholder engagement, and organizational agility. By integrating cybersecurity into broader resilience frameworks, organizations enhance their ability to anticipate, withstand, and recover from a wide range of disruptions, including cyber threats.

12.1 Science of Cybersecurity

The term *science* has undergone a transformation in meaning since Aristotle first used it to describe a body of knowledge. Initially, it simply referred to accumulated knowledge. However, in contemporary understanding, many associate the term science with knowledge derived from systematic experimentation, making experimentation the defining characteristic of a science. This definition finds its satisfaction in the natural sciences, where experimentation is integral to forming and validating theories or laws that offer verifiable predictions about both natural and man-made phenomena. Consequently, the natural sciences remain unaffected by this evolving definition of science.

With the advancement of disciplines like computer science, the notion of science expands beyond mere experimentation. Computer science, for example, delves into the study of processes that can be efficiently automated. In this context, laws governing specifications (problems) and implementations (algorithms) serve as a means to encapsulate knowledge about what processes can be automated effectively. Thus, computer science transcends the traditional experimental model and adopts a framework where laws and principles guide the understanding and implementation of automated processes. This evolution demonstrates the adaptability and broadening scope of the scientific endeavor to encompass emerging fields and methodologies beyond traditional experimentation. The construction of secure systems is essential for protecting digital assets and ensuring the integrity of information in an interconnected world. To establish a firm foundation for secure system development, developers need a body of laws that enables them to predict the consequences of design and implementation choices accurately. These laws should possess several key attributes:

- ***Transcend Specific Technologies and Attacks***: The laws should not be tied to specific technologies or attack vectors but should instead offer general principles and guidelines that are applicable across diverse technological environments and threat landscapes. This ensures the relevance and longevity of the laws in real-world settings where technologies and threats evolve rapidly.
- ***Introduce New Models and Abstractions***: By introducing new models and abstractions, the laws not only offer predictive power but also contribute to the pedagogical value of secure system development. These models and abstractions provide developers with conceptual frameworks for understanding security principles and practices, thereby enhancing overall knowledge and expertise in the field.
- ***Facilitate Discovery of New Defenses***: The laws should facilitate the discovery of new defense mechanisms and strategies by identifying common patterns, vulnerabilities, and attack vectors. By elucidating the underlying principles of security, the laws empower developers to innovate and devise novel approaches to mitigate emerging threats.
- ***Describe Non-obvious Connections Between Attacks, Defenses, and Policies***: The laws should describe non-obvious connections between attacks, defenses, and policies, offering a deeper understanding of the security landscape. By elucidating the relationships between different elements of security, the laws enable developers to make informed decisions and design robust defense mechanisms that address complex and multifaceted threats.

In the realm of natural sciences, laws govern the relationships between quantities found in nature, such as $E = mc^2$ and $PV = nRT$, with continuous mathematics often used to specify these laws. However, continuous mathematics is not inherently tied to the concept of scientific laws; rather, predictive power is paramount. Similarly, laws governing digital computations, pertinent to cybersecurity, are typically expressed using discrete mathematics and logical formulas. The subject matter of these laws in cybersecurity revolves around ensuring the security of systems despite potential attacks, wherein systems must adhere to prescribed policies dictating permissible actions and defenses are employed to prevent compromises. Therefore, the development of laws in cybersecurity should aim to relate attacks, defenses, and policies. For maximum applicability, laws should ideally relate classes of attacks, defenses, and policies, categorizing them based on essential characteristics such as the nature of the attack (e.g., cyber, physical, or hybrid), defense strategies (e.g., preventive, detective, or corrective), and policy types (e.g., regulatory, strategic, or operational). This approach allows for the formulation of generalized laws such as "Defenses in class X enforce policies in class Y despite attacks from class Z."

12.2 Ensuring Resilience Against Cyber Threats

Ensuring resilience against cyber threats is a multi-faceted challenge that requires a comprehensive approach to safeguarding digital assets and critical operations. Resilience in the face of cyber threats entails the ability to detect, respond to, and recover from cyber incidents effectively and efficiently, minimizing their impact on organizational functions and objectives [NIST21]. To achieve resilience, organizations must adopt proactive measures, develop robust response capabilities, and cultivate a culture of cybersecurity awareness and readiness. One key aspect of ensuring resilience against cyber threats is by implementing robust cybersecurity measures that mitigate the risk of successful attacks. This is achieved by deploying advanced security technologies, such as firewalls, intrusion detection systems, and encryption protocols, to protect networks, systems, interfaces, and data from unauthorized access and malicious activities.

A system's interfaces serve as the primary conduit through which its environment can interact with and perceive the effects of system execution. While some interfaces are directly tied to hardware components, such as keyboards and a mouse for input and graphic displays, or printers for output, others may be less conspicuous or even obscure. Software interfaces are also prevalent within the operating system. Offering various ways for programs to communicate, either overtly through system calls and shared memory or covertly through side channels like battery level or execution timings. Since interfaces are the exclusive means for influencing and observing system execution, they inherently constitute the sole avenues for conducting attacks against a system. Consequently, the set of interfaces and their specific operations form a natural basis for defining classes of attacks.

Furthermore, resilience against cyber threats requires organizations to establish effective incident response plans and procedures to detect and respond to cyber incidents promptly. This involves developing clear escalation protocols, assigning roles and responsibilities to incident response teams, and conducting regular training and exercises to test the effectiveness of response plans. By having well-defined incident response processes in place, organizations can minimize the impact of cyber incidents and restore normal operations swiftly. In addition to these proactive measures and incident response capabilities, a culture of cybersecurity awareness among employees is essential for ensuring resilience against cyber threats. This involves providing cybersecurity training and education to employees at all levels of the organization, promoting best practices for data protection and safe computing habits, and encouraging employees to report any suspicious activities or security incidents promptly.

Mission resilience is the cornerstone of organizational readiness, particularly in the execution of mission essential tasks. It embodies an organization's ability to adapt, withstand, and recover from adversities or disruptions while maintaining operational continuity and effectiveness. Central to mission resilience is identifying critical tasks necessary for mission accomplishment through comprehensive risk assessments and clearly identified potential threats and vulnerabilities that may impede task execution. By implementing strategies for risk mitigation, such as redundancy and contingency planning, organizations bolster their capacity to persevere in the face of challenges. Continuous monitoring and evaluation ensure that organizations remain vigilant and identify areas for improvement and reinforcement of resilience capabilities.

12.3 Policies Driving Cybersecurity

Traditionally, the cybersecurity community has approached policy formulation through the lens of three fundamental requirements: confidentiality, integrity, and availability. Confidentiality pertains to determining which principals or entities are permitted to access specific information. Integrity concerns the permissible changes to the system, encompassing alterations to stored information, resource usage, and outputs, ensuring data remains accurate and unaltered. Availability focuses on the timing of inputs and outputs, delineating when data must be accessible, or services rendered. However, this classification may present challenges when establishing the foundational laws that underpin the scientific direction of cybersecurity into the future. While these requirements have served as fundamental principles in cybersecurity policy development, they may not fully capture the complexities and nuances of modern cyber threats and defense strategies. As cybersecurity landscapes evolve and threats become increasingly sophisticated, a more comprehensive and adaptable framework is needed to address emerging challenges and reflect the multifaceted nature of cybersecurity. Consequently, the traditional classification of confidentiality, integrity, and availability may need to be refined or expanded to accommodate the evolving cybersecurity landscape and support the development of robust and effective cybersecurity laws and practices.

Contrasting the conventional approach with trace properties reveals a distinct framework where safety and liveness (the property that a system continues to make progress toward an expected outcome) are treated as separate and independent classes. Safety is paramount so that no detrimental events occur during system execution. Liveness guarantees the eventual occurrence of specific desired events [ALPERN85]. This delineation offers added value as safety and liveness are each associated with distinct proof methods, enhancing the rigor of requirement formulation. However, trace properties, while offering this unique classification, fall short in adequately specifying all confidentiality and integrity policies. They are effective for sequential programs; however, they prove insufficient for concurrent programs due to their characterization of execution as a relation. Safety ensures the absence of adverse events throughout execution, while liveness asserts the occurrence of positive events. Traces represent sequences of states and form the basis of trace properties, whereby each trace satisfies a characteristic predicate associated with the property. Yet not all sets of traces constitute trace properties. For instance, information flow, which requires a correlation between variable values across all traces, lacks a characteristic predicate dependent solely on individual traces, thus not qualifying as a trace property. Notably, every trace property can be expressed as the conjunction of is either a safety property and a liveness property. Furthermore, the safety-liveness classification provides a foundation for reasoning about and verifying system behavior, extending the distinction beyond conceptual definitions to the analysis and proof of specified properties [ALPERN87]. Thus, the safety-liveness classification for trace properties encompasses proof methodologies that go beyond mere conceptual distinctions.

Any classification of policies within the realm of cybersecurity is inherently tied to the system model being employed, particularly with regard to the interfaces defined by the model. These interfaces dictate the operations available to potential adversaries seeking to exploit vulnerabilities within the system. For instance, one approach to modeling a system involves representing it in terms of the set of possible indivisible state transitions it undergoes during operation. This model is useful for expressing laws pertaining to classes of policies enforced by various operating system mechanisms, such as reference monitors or code rewriting, which regulate permissible changes to the system state. Alternatively, a system may be modeled as a black box that processes information streams from certain channels and produces outputs on others. In this case, stream models are often utilized to quantify information leakage or corruption in output streams. It is important to acknowledge that the science of cybersecurity is unlikely to revolve around a single model or a solitary classification of policies. Rather, it will encompass a diverse range of models and classifications tailored to address the multifaceted nature of cybersecurity challenges. These models and classifications will evolve alongside advancements in technology and threat landscapes, ensuring that cybersecurity practices remain robust and adaptable in the face of emerging risks and vulnerabilities.

12.4 Defense Techniques for Cybersecurity

A broad and diverse array of defenses populates the cybersecurity literature, each offering unique approaches to improving system security. Among these, program analysis and rewriting emerge as a distinct class characterized by the allocation of effort toward deploying the defense primarily before execution. This class of defenses, known as language-based security, encompasses strategies that involve analyzing and potentially modifying program code to enhance security measures. While program analysis and rewriting defenses aim to proactively identify and address security vulnerabilities before program execution, run-time defenses focus on detecting and mitigating security threats during program execution in real time. Both approaches play crucial roles in enhancing cybersecurity posture and protecting systems from various threats and attacks.

Run-time defenses are built upon a foundational framework comprised of a few fundamental mechanisms that include

- ***Isolation***: This involves preventing the execution of one program from accessing interfaces associated with the execution of other programs. Examples of isolation mechanisms include physically separated hardware components, virtual machines that create isolated environments, and processes with segmented memory.
- ***Monitoring***: A reference monitor intervenes whenever an operation in a specified set is initiated. It has the capability to block subsequent execution to ensure compliance with enforced policies. Monitoring mechanisms encompass hardware features like memory mapping, processor modes that restrict certain instructions, operating system kernels, and firewall systems.
- ***Obfuscation***: This technique involves transmitting or storing code or data in a form that can only be deciphered with knowledge of a secret. By keeping this secret inaccessible to attackers, the protected content becomes unintelligible and resistant to manipulation. Examples include data encryption, digital signatures, and program transformations that increase the complexity of crafting attacks.

This taxonomy of mechanisms provides a basis for classifying run-time defenses. Additionally, another perspective on defenses involves the concept of trust relocation, where security measures aim to shift the burden of trust away from potentially vulnerable components or interfaces to more secure environments or entities. For instance, when an application is executed under the supervision of a reference monitor, trust in the application is effectively shifted to trust in the reference monitor. This perspective encourages exploration of overarching principles that govern the substitution of trust from one component to another; therefore, designing security mechanisms that allow for the transfer of trust between components, thereby strengthening the overall security posture of systems and applications.

While the ideal scenario for establishing cybersecurity laws would involve logical deduction from axioms, reality often requires working with axioms that embody approximations or models, akin to practices in the natural sciences. For instance, Newton's laws of motion simplify reality by disregarding factors like friction and relativistic effects. To validate the accuracy of these approximations and models, experimentation becomes essential. Just as laboratories support experimentation in the natural sciences, test beds are indispensable for conducting controlled experiments in cybersecurity. Experimentation in computer science diverges from traditional experimental methods. Instead of conducting experiments in controlled environments, computer scientists often validate their ideas by building prototypes. These prototypes serve to unearth hidden assumptions about reality and demonstrate the feasibility and scalability of new system designs. Additionally, prototypes help identify missing functionality and assess the utility of novel features in artifacts like programming languages and systems. In the context of cybersecurity where novel approaches to building systems and defenses are continually sought, the validation of these proposals may necessitate building prototypes. Note, however, that prototyping in cybersecurity is not tantamount to engineering a secure system. Prototypes are crafted explicitly to validate assumptions and observe emergent behaviors, thereby supporting the development of a science of cybersecurity. As a result, this field encompasses both investigative computer science and experimentation, with prototypes serving as critical tools for validating theoretical concepts and exploring new ideas.

12.5 Zero Trust Architecture

Zero Trust Architecture (ZTA) is a crucial approach in modern cybersecurity, especially given the increasingly sophisticated nature of cyber threats. By not relying on traditional perimeter-based security models, a ZTA emphasizes continuous authentication and authorization, regardless of the user's location or the network accessed. This results in a more robust defense against potential breaches and unauthorized access. The core principle of ZTA, "never trust, always verify," underscores the importance of validating every access attempt, even if it originates from within the network perimeter. This shift from a perimeter-based approach to a more granular, identity-centric security model helps organizations better protect their sensitive data and resources, reducing the risk of insider threats and lateral movement by malicious actors. Implementing a ZTA involves careful planning and integration across various components of the information technology infrastructure, including:

- ***Identity, Credential, and Access Management (ICAM) System***: Manages user identities, credentials, and access rights, allowing access to network resources based on user information such as roles, permissions, and authentication factors. This system may employ advanced authentication methods like multi-factor authentication or biometric verification to ensure that only authorized users can access specific resources at designated times.
- ***Security Analytics***: Plays a vital role by analyzing vast amounts of data, including threat intelligence feeds, activity logs, and network traffic patterns. By leveraging data analytics and artificial intelligence techniques, security analytics can detect anomalous behavior and potential security threats in real time. This proactive approach enables organizations to identify and respond to security incidents promptly, minimizing the risk of breaches and data loss.
- ***Endpoint Protection***: Another essential component focused on securing individual devices (endpoints) and the data they store or access. Endpoint protection solutions monitor devices for signs of intrusion, known vulnerabilities, and malware infections, helping to prevent unauthorized access and data breaches. By ensuring the security of endpoints, ZTA can mitigate risks associated with compromised devices and unauthorized access attempts.
- ***Encryption***: Plays a critical role in ZTA by safeguarding sensitive data from unauthorized disclosure, tampering, or access. By encrypting data both in transit and at rest, organizations maintain the confidentiality and integrity of their information assets, even in the event of a security breach or unauthorized access attempt. Encryption technologies help organizations comply with data protection

regulations and industry standards, enhancing overall security posture within a ZTA framework.

A ZTA is paramount to MEI and mission execution, serving as a foundational element for ensuring mission success in dynamic and high-stakes environments to foster a security posture built on continuous authentication and stringent access controls. Critical systems and sensitive information are central to mission accomplishment, involving intricate networks, diverse stakeholders, and rapidly evolving threats. A key benefit of a ZTA lies in its ability to mitigate insider threats and unauthorized access attempts, which can pose significant risks to mission integrity. By treating every access request as potentially malicious and requiring continuous verification of user identities and device trustworthiness, mission organizations can effectively thwart both internal and external threats before they compromise critical systems or data. Moreover, a ZTA enhances the resilience and adaptability of MEI operations by enabling real-time threat detection, rapid response capabilities, and seamless integration with emerging technologies. In an environment where mission requirements evolve rapidly, having a security framework that dynamically adjusts to changing conditions is essential for maintaining operational effectiveness and mission readiness.

12.6 Cyber Kill Chain

The Cyber Kill Chain® (CKC) framework, developed by Lockheed Martin in 2022, elucidates how attackers navigate networks to identify and exploit vulnerabilities from the perspective of intrusions. The core of an intrusion involves the aggressor creating a payload to penetrate a trusted boundary, establishing a foothold within a secure environment, and then executing actions to achieve their goals. These goals may include lateral movement within the environment or compromising the confidentiality, integrity, or availability of a system.

The CKC framework outlines the sequential steps attackers follow during offensive cyber operations against their targets. By understanding these steps, organizations can better anticipate, detect, and mitigate cyber threats. As shown in Fig. 12.1 [LM24], the CKC is divided into seven stages: Reconnaissance, Weaponization, Delivery, Exploitation, Installation, Command and Control (C2), and Actions on Objectives. These steps are defined with an outline of preventive measures that network defenders can take at each stage:

1. **Reconnaissance**:
   - **Description**: Attackers gather information about the target to identify potential vulnerabilities.
   - **Preventive Measures**: Network defenders use intrusion detection systems (IDS) and monitor network traffic for unusual patterns, employ honeypots

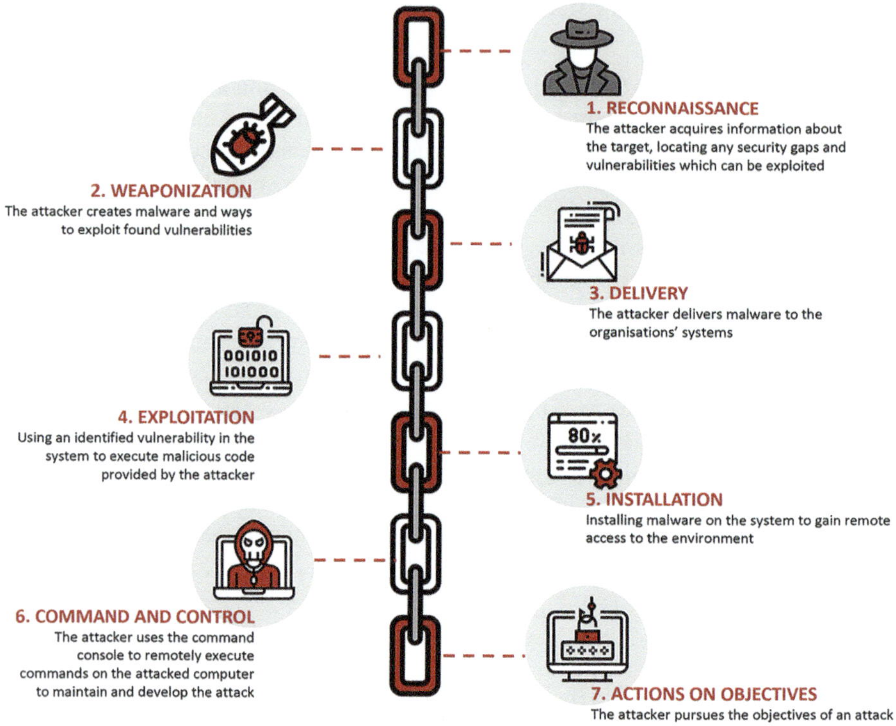

Fig. 12.1 Lockheed Martin Cyber Kill Chain. (Adapted from [LM24])

(decoys) to detect reconnaissance activities, and limit publicly available information about their network architecture.

2. **Weaponization**:
   - **Description**: Attackers create a malicious payload, such as malware, tailored to exploit identified vulnerabilities.
   - **Preventive Measures**: Defenders implement robust email and web filtering, ensuring software and systems are regularly updated and patched, and use threat intelligence to stay informed about emerging threats to prevent weaponization.
3. **Delivery**:
   - **Description**: The payload is delivered to the target through methods like phishing emails, malicious websites, or compromised devices.
   - **Preventive Measures**: Training system users on recognizing phishing attempts, using email and web gateways with advanced filtering capabilities, and deploying endpoint protection solutions can reduce the risk of successful delivery.

4. **Exploitation**:
   - **Description**: The payload is executed, exploiting the vulnerability to gain initial access to the target system.
   - **Preventive Measures**: Cyber security teams conduct regular vulnerability assessments and penetration testing, implement application whitelisting, and employ behavioral analysis tools to help identify and block exploitation attempts.
5. **Installation**:
   - **Description**: Malicious software or tools are installed to establish persistent access within the compromised system.
   - **Preventive Measures**: Using endpoint detection and response (EDR) solutions, implementing least privilege access controls, and maintaining a secure configuration of all systems can prevent installation.
6. **Command and Control (C2)**:
   - **Description**: The attacker sets up communication channels to remotely control the compromised system.
   - **Preventive Measures**: Network defenders monitor for anomalous outbound traffic, use Domain Name System filtering to block known malicious domains, and deploy network segmentation to limit the spread of C2 channels.
7. **Actions on Objectives**:
   - **Description**: The attacker carries out their goals, which may include data theft, system disruption, or lateral movement within the network.
   - **Preventive Measures**: Network defenders regularly back up critical data, use data loss prevention (DLP) tools, implement network segmentation, and continuously monitor for suspicious activity to quickly detect and respond to malicious actions.

By implementing these preventive measures at each stage of the CKC, organizations can enhance their defensive posture and mitigate the risk of cyber intrusions.

At a strategic level when the campaign analysis is executed, analyzing multiple intrusion kill chains over time reveals commonalities and overlapping indicators. For instance, the correlation of two intrusions across multiple kill chain phases can identify highly dimensional correlations. Through this process, defenders can recognize and define intrusion campaigns, linking together years of activity from a persistent threat. Key indicators emerging from the most consistent correlations provide focal points for defenders to prioritize the development and implementation of effective courses of action. These stable indicators are less prone to volatility and offer reliable predictions about the characteristics of future intrusions when observed frequently. Leveraging an adversary's persistence becomes an asset for defenders to fortify their posture.

The primary goal of this campaign analysis is to discern patterns and behaviors of intruders, including their tactics, techniques, and procedures (TTP). The focus is on understanding "how" they operate rather than specifically "what" they do. While positively attributing the identity of intruders is less crucial, evaluating their capabilities, doctrine, objectives, and limitations is paramount; intruder attribution may naturally arise as a by-product of this comprehensive analysis. As defenders examine new intrusion activities, they can either link them to existing campaigns or identify entirely new sets of behaviors indicative of previously unknown threats, tracking them as new campaigns. Evaluating defensive postures on a campaign-by-campaign basis enables defenders to develop strategic courses of action to address any identified gaps in security. Another key objective of campaign analysis is to comprehend the intruders' intent. By identifying technologies or individuals of interest, defenders can better understand the adversary's mission objectives. This involves analyzing trends in intrusions over time to assess targeting patterns and scrutinizing any exfiltrated data. This analysis helps prioritize targeted security measures to safeguard individuals, networks, or technologies effectively.

12.7 Risk Management and Cyber Resiliency

Organizations are tasked with managing various types of risks associated with their systems, which encompass cyber resources within a broader risk portfolio. These risks include:

- Financial and reputational risks
- Programmatic or project-related risks (such as cost, schedule, and performance)
- Security risks linked to the organization's mission or business activities
- Cybersecurity risks arising from legislative, regulatory, policy, or standard requirements

A proposed cyber resiliency solution aims primarily to mitigate mission, business, or operational risks but can also have ancillary benefits such as reducing security, reputational, supply chain, and performance risks. However, like any risk management solution, it may inadvertently elevate other risks such as financial, cost, or schedule risks.

In a collaborative effort, systems engineering (SE) and risk management professionals articulate the potential adverse impacts of alternative solutions. These professionals evaluate whether the impacts align with the organization's risk tolerance and assess whether adopting a proposed solution aligns with the organization's risk management strategy. Ultimately, these professionals communicate risk trade-offs to the organization's risk executive for informed decision-making. At the organizational level, adopting a cyber resiliency perspective in risk management involves analyzing and managing risks associated with programs and initiatives across various levels. These initiatives may entail investments in, transitioning to, using, or transitioning away from different cyber technologies. The environment in which a system is engineered is seldom static. Organization-wide programs and initiatives, often driven by directives, may include transitioning to a zero trust architecture, mitigating software supply-chain risks, or migrating from network protocols.

These overarching programs and initiatives can impact efforts at lower levels, such as specific acquisition programs for systems or services, or initiatives aimed at redefining missions or business processes. By taking a holistic approach to risk management from a cyber resiliency standpoint, organizations better navigate the dynamic landscape of cyber technologies and initiatives while ensuring the effectiveness and security of their systems and operations. In the realm of cyber resiliency, various techniques are applied to enhance the resilience of an SoS. While many techniques are generally applicable, some offer greater benefits in certain contexts such as the following examples:

1. **Contextual Awareness via Mission Dependency and Status Visualization**:
   - This technique involves understanding the interdependencies between systems within an SoS and visualizing their status. By employing this technique, organizations can predict potential mission impacts resulting from cyber effects of adversary activities on individual systems or system elements.
2. **Calibrated Defense-in-Depth and Consistency Analysis for Coordinated Protection**:
   - Calibrated Defense-in-Depth and Consistency Analysis approaches ensure that the various protections across constituent systems operate with consistency and in coordination. This helps prevent or delay the advancement of adversaries across systems within the SoS.
3. **Realignment**:
   - In situations where constituent systems within an SoS were not originally designed to work together and have different missions, functions, and risk frames, realignment techniques are beneficial. Specifically, approaches such as Offloading and restriction ensure that core system elements are appropriately aligned with the overall mission of the SoS.

By strategically applying these techniques and approaches, organizations can increase the resilience of their SoS, mitigating the impact of cyber threats and ensuring continued mission success. In the realm of systems security engineering, cyber resiliency analysis serves to evaluate whether the cyber resiliency properties and behaviors of a system, regardless of its lifecycle stage, are adequate for the organization's needs. This assessment considers the organization's mission assurance, business continuity, and other security requirements in the face of threats, including Advanced Persistent Threats (APTs).

APTs use sophisticated methods like spear-phishing, custom malware, and network infiltration to establish and maintain a long-term presence within target systems. One well-known operation involved the group's breach of the United States Democratic National Committee in 2016, where they reportedly exfiltrated sensitive data. More recently, in 2020, they targeted organizations working on COVID-19 vaccine research, aiming to gather intelligence on vaccine development.

The goal of cyber resiliency analysis is to provide insights that support SE and risk management decisions regarding the system under consideration. Depending on factors such as the system's lifecycle stage, programmatic considerations, the analysis may recommend various actions. These actions may include architectural changes, integration of new products or technologies into the system, adjustments in how existing products or technologies are utilized, or modifications in operating procedures or environmental protections. These recommendations align with and are aimed at implementing the organization's risk management strategy effectively.

Having explored **mission resilience and cybersecurity**, we now shift our focus to how to assess and validate mission effectiveness through comprehensive **test and evaluation in mission context**. Chapter 13 discusses testing from a mission-based perspective, emphasizing the importance of leveraging various testing approaches throughout the system lifecycle. By employing "Test and Evaluation as a Continuum" and incorporating design of experiments, the mission engineer can ensure that every stage of the MEI approach is thoroughly evaluated, allowing for continuous refinement and optimization of systems. This methodology helps validate resilience, performance, and effectiveness, ensuring that systems are mission-ready and able to adapt to evolving challenges.

References

1. Alpern, B., & Schneider, F. B. (1985). Defining liveness. *Information Processing Letters, 21*(4), 181–185. https://doi.org/10.1016/0020-0190(85)90056-0
2. Alpern, B., & Schneider, F. B. (1987). Recognizing safety and liveness. *Distributed Computing, 2*(3), 117–126. https://doi.org/10.1007/BF01782772
3. Lockheed Martin Cyber Kill Chain. https://www.lockheedmartin.com/en-us/capabilities/cyber/cyber-kill-chain.html
4. NIST. (2021, December). National Institute of Standards and Technology Special Publication 800-160, Vol. 2, Rev. 1, 310 p.

Chapter 13
Test and Evaluation for Mission Effectiveness

13.1 Traditional Test and Evaluation Process

Test and Evaluation (T&E) constitutes a systematic process through which systems or system-of-systems (SoS) undergo testing and subsequent evaluation to various aspects such as design progression, performance levels, and supportability. The T&E effort provides knowledge to manage risks in developing, producing, operating, sustaining, and modernizing systems and capabilities to both the acquisition and operational communities. The T&E effort is conducted across and throughout the lifecycle of systems. Traditionally, the T&E process has followed a linear trajectory and is characterized by discrete test events that progress through several key phases, including Contractor Testing (CT), Developmental Testing (DT), Operational Testing (OT), and Live Fire Testing (LFT). This sequential approach aims to actualize the capabilities of the delivered system effectively. During CT, the system is tested by the contractor responsible for its development to ensure that it meets specified requirements and performs as intended. This phase often involves rigorous testing of individual components to verify their functionality and interoperability. DT serves multiple purposes within the T&E process. It aids in refining the engineering design and development process, validates manufacturing processes, ensures quality control, and verifies the achievement of technical performance specifications and objectives. Subsequently, OT focuses on assessing the operational effectiveness and suitability capabilities of the system. It aims to estimate the system's performance in real-world scenarios, identify necessary modifications or improvements, and provide valuable insights into tactics, doctrine, organization, standard operating procedures, and personnel requirements. One of the most important goals of OT is to characterize a system's or SoS' end-to-end mission effectiveness over the operational envelope. Such characterization of performance informs the stakeholders and the system operators of capabilities and limitations in the various conditions that will be encountered during operations. The LFT plays a vital

J. D. Moreland, Jr., *Mission Engineering and Integration*,
https://doi.org/10.1007/978-3-032-32132-9_13

role in assessing the survivability, lethality, and operational effectiveness of systems to ensure they meet the rigorous standards and requirements of critical operations and missions. This phase typically occurs later in the developmental testing phase, following CT, DT, and OT.

The overall T&E process provides confidence in the capabilities and readiness of systems, supporting informed decision-making and successful acquisition, deployment, and sustainment of complex systems, such as military and aerospace. T&E's significance is maintaining appropriate levels of performance and safety assurance for the specific application. In addition to this assurance of system readiness, the goal is to help the program manager reduce costs and risks.

Integrating across systems involves assessing individual systems within SoS and their contributions to delivering integrated capabilities across various industrial sectors and end-user communities. Requirements should be developed not just for individual systems but also for the larger SoS, characterizing the relationships between different mission elements using integrated architectures and model-based systems engineering (MBSE) tools. Additionally, policy and guidance on tools and T&E are necessary to verify systems, stakeholders, and end users' ability to send, receive, and accept data for executing designated capabilities and missions. A challenge lies in addressing internal and external interfaces for the SoS, including those provided by programs affiliated across organizations and industrial sectors. Integration and interoperability testing are crucial for verifying satisfactory data transmission across these interfaces to ultimately assure that these systems deliver an integrated mission capability to tackle industrial challenges and execute designated mission effects.

13.1.1 Importance of Rigorous Testing and Evaluation

Rigorous testing is paramount in Mission Engineering and Integration (MEI) for several reasons. The complexities and high stakes involved in missions, especially in fields such as space exploration, defense, and critical infrastructure, demand a robust testing process. These key reasons highlight the importance of rigorous testing in MEI:

Risk Mitigation

- **Explanation**: Rigorous testing helps identify and mitigate potential risks and uncertainties before and during the mission.
- **Importance**: Reducing risks is critical to ensure the safety of personnel, protecting valuable assets, and increasing the likelihood of mission success.

Validation of Systems and Components

- **Explanation**: Testing validates the functionality and performance of individual systems and components, ensuring they meet the specified requirements.
- **Importance**: Validation is essential to confirm that the mission-critical components operate as intended, reducing the likelihood of failures during the mission.

Integration Testing

- **Explanation**: Integration testing assesses the interoperability of different systems to ensure they work cohesively as an integrated SoS.
- **Importance**: Coordinated functioning of systems is crucial for mission success; integration testing helps identify and resolve compatibility issues.

Environmental Testing

- **Explanation**: Testing under environmental conditions simulates the challenges that the mission may encounter, i.e., extreme temperatures, radiation, or atmospheric conditions.
- **Importance**: This testing ensures that systems can withstand and perform under various environmental conditions encountered during the mission execution.

Operational Testing

- **Explanation**: Operational testing assesses the performance of systems in realistic scenarios, simulating the actual conditions of a representative mission environment.
- **Importance**: Identifying operational challenges and refining operating procedures optimizes performance during the mission.

Cybersecurity Testing

- **Explanation**: As cyber threats are a significant concern, testing the robustness of cybersecurity measures ensures the mission's resilience against potential cyber-attacks.
- **Importance**: Protecting sensitive data, communication systems, and mission-critical infrastructure from cyber threats is paramount.

Human Factors Testing

- **Explanation**: Assessing the impact of human factors, including operator interfaces, workload, and decision-making, helps optimize human systems interactions.
- **Importance**: Human elements of the mission engineering thread for mission execution are better integrated so that operators can effectively manage and respond to mission essential tasks.

Compliance with Standards and Specifications

- **Explanation**: Testing ensures compliance with industry or government standards, specifications, regulations, and safety protocols.
- **Importance**: Adherence to standards and specifications is crucial for regulatory approval, certification, and maintaining the integrity of the mission.

Training

- **Explanation**: Testing allows operators and personnel to familiarize and train themselves with mission systems, procedures (i.e., Tactics, Techniques, and Procedures, Rules of Engagement), and potential challenges.
- **Importance**: Fully prepared and trained personnel contribute to efficient and effective mission execution and rapid response to unexpected events.

The requirement for rigorous testing as part of MEI is not only a quality assurance measure but also a fundamental component of risk management and operational readiness. A thorough T&E Plan provides an investment in the reliability, safety, and effectiveness of mission engineering threads and mission execution.

13.1.2 Validation of Critical Systems in Mission Context

Traditional T&E and mission-based testing represent distinct approaches to assessing the capabilities of systems or SoS. Traditional T&E follows a structured process of evaluating individual system components through predetermined test events, such as CT, DT, OT, and LFT. These tests focus on verifying technical specifications and performance metrics, aiming to ensure systems meet predefined standards and requirements. In contrast, mission-based testing emphasizes evaluating systems within the context of their intended operational missions [MORELAND14]. Rather than assessing components in isolation, mission-based testing and experimentation examines how systems contribute to achieving mission objectives and operational outcomes within specific operational scenarios. Test scenarios and conditions are tailored to simulate real-world missions, allowing for a dynamic assessment of system performance in mission execution. In this approach, test and experimentation scenarios are carefully crafted to simulate actual mission conditions, including the environment, threats, and operational objectives. These scenarios are designed to "stress-test" the critical systems under realistic conditions, leading to a thorough evaluation of their capabilities and performance. Evaluation criteria are aligned with mission objectives and operational requirements, focusing on overall effectiveness, suitability, and performance in fulfilling mission tasks. While traditional T&E provides insights into system capabilities and limitations against technical requirements, mission-based testing and experimentation offers a broader perspective, assessing system readiness for deployment in real-world scenarios and their ability to support mission success, illustrated in Fig. 13.1.

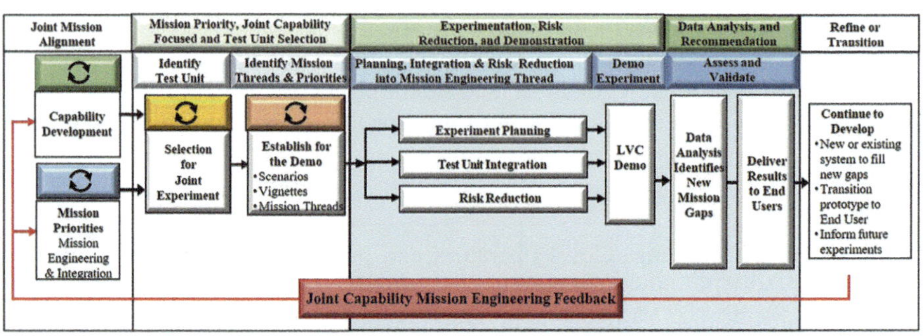

Fig. 13.1 Generic mission-based experimentation approach

The *Mission Priority, Joint Capability Focused, and Test Unit Selection Phase* monitors new technologies through a joint selection board to identify technologies for demonstrations. The next phase, *Experimentation, Risk Reduction, and Demonstration*, introduces prototypes, systems, SoS, and modernization priorities to establish an operationally relevant vignette for assessment of the capability within a representative operational environment during an exercise event. The goal of this phase is to monitor and experiment, determine risk reduction, and provide insights on integration prior to a demonstration. Finally, the *Data Analysis, and Recommendation Phase* performs the capability assessment to determine if a prototype or system is ready for transition at the correct maturity level with an accompanying report on newly defined gaps.

Three key aspects of mission-based test design are (1) the identification of mission-critical functions and objectives, (2) the creation of realistic test scenarios that replicate operational conditions, and (3) the execution of tests to validate the systems' ability to meet mission requirements. This approach ensures systems are tested in context, providing valuable insights into their effectiveness and suitability for deployment in actual missions. By validating critical systems within mission context, organizations can gain confidence in the reliability and performance of these systems under real-world conditions. This approach helps to mitigate risks and uncertainties associated with system deployment, ultimately enhancing mission success and operational effectiveness.

13.1.3 Design of Experiment

Design of Experiment (DoE) centers on developing rigorous, adequate, and defensible test programs and evaluating their results. In recent years, most test designs have focused exclusively on verifying threshold requirements, relying too heavily on hypothesis tests for test sizing, and often fail to embrace the statistical tenets of DoE. Furthermore, some organizations have not updated their data analysis practices to capitalize on the benefits of using DoE. DoE is a strategic approach where deliberate alterations are made to input variables of a system or process, while the resulting impacts on response variables are carefully observed and measured. This methodology finds utility across both physical processes and computer simulation models. Notably, experimental design serves as a potent tool for optimizing the quantity of information garnered from a study, while minimizing the volume of data necessitated. In particular, factorial experimental designs stand out by scrutinizing the effects of multiple factors concurrently, rather than modifying one factor in isolation. By adopting factorial designs, analysts can ascertain the sensitivity of individual factors as well as their combined effects, offering a comprehensive understanding of system dynamics. In the fiercely competitive global landscape of T&E, these efficient methods for examining multiple factors concurrently are indispensable.

The foundation of experimental design lies in addressing the challenges posed by nuisance factors, with the aim of enhancing experimental efficiency. These fundamental principles include:

- **Randomization**: This technique safeguards against unknown biases that could skew experimental results. For instance, in an experiment comparing a baseline procedure to a new one, randomizing the testing sequence mitigates the impact of factors like instrument drift. By conducting tests in a random order, such as baseline-new-new-baseline-new-baseline, any biases or drifts are averaged out, ensuring more accurate conclusions.
- **Replication**: Replicating experimental runs helps validate findings by assessing consistency and variability in results. By repeating experiments under identical conditions, researchers can gauge the reliability of observed effects and reduce the influence of random fluctuations or errors.
- **Blocking**: Blocking involves grouping experimental units into homogeneous subsets, or blocks, based on certain characteristics. Researchers can then account for systematic variations and enhance the precision of estimates by reducing variability within each block. For instance, in agricultural experiments, blocking involves grouping test plots based on soil type or environmental conditions.
- **Orthogonality**: Orthogonal designs ensure that experimental factors are independent of each other, facilitating clear interpretation of main effects and interactions. By selecting orthogonal factors, researchers can efficiently assess the impact of individual variables without confounding effects from other factors.
- **Factorial Experimentation**: Factorial designs enable the simultaneous investigation of multiple factors and their interactions. By varying factors at different levels, factorial experiments provide comprehensive insights into the effects of each factor and their combined influence on the response variable.

These fundamental principles in experimental design collectively contribute to the robustness and reliability of scientific investigations, enhancing researchers' ability to draw accurate conclusions and make informed decisions based on experimental data.

SoS interactions between performance variables are critical considerations as these variables can have profound effects on system effectiveness, efficiency, and resilience. An SoS comprises multiple independent systems that collaborate to achieve a common goal or functionality. Screening experiments facilitate understanding the interdependencies and synergies among system components. These systems interact in complex ways, influencing each other's performance and overall system behavior. Their interaction occurs in an experiment when the effect on the response variable is contingent upon the level of another factor. Simply put, the combined impact of two factors on the response variable cannot be accurately predicted based solely on individual effects. When an interaction exists between two factors, their joint effect on the response variable can either be greater (synergy) or less (interference) than what is anticipated when each factor acts independently.

In scenarios requiring a process assessment involving numerous input variables and measured output variables, screening experiments are valuable. This would be true of complex simulation models or a manufacturing process with inputs such as raw materials, temperature, and pressure. Screening experiments help identify which input variables (factors) predominantly contribute to the variability observed in the output (responses), thereby identifying the key factors, often referred to as the "drivers." Screening experiments typically involve only two levels of each factor and may also be termed as "characterization testing" or sensitivity analysis.

Test designs and integrated evaluation frameworks need to be developed by stressing the critical elements below:

- ***Create a clear test goal for each phase of test.*** As stated earlier, a successful test plan must identify the goal of the test. Goals need to be clearly identified in the test plan and should be specific. Future test plans must state clearly that data are being collected to measure a particular response variable (possibly more than one), to characterize the system or SoS' performance by examining the effects of multiple factors. Test plans must also clearly delineate what statistical model (e.g., main effects and interactions) is motivating the strategic factor variation of the test. An example of a clearly stated test goal is:

The purpose of the test is to characterize miss distance under a range of operationally realistic conditions. The characterization of mission performance through the primary response variable miss distance will be determined using a main-effects and second-order interaction model.

Ensure metrics are mission-oriented, relevant, informative, and do not rigidly adhere to the narrowest possible interpretation of definitions in requirements documents. Too often the goals of T&E are not captured by technical performance requirements. This is especially true when response variables are limited to the technical requirements of the system under test when, in fact, the mission capability requires an SoS to succeed. Ideal test metrics provide a measure of mission accomplishment (not technical performance for a single system), lend themselves to good test design (i.e., be continuous in nature), and in general cover the reasons for procuring the systems. Another pitfall is relying on binary metrics as the primary response variable on which to base the test design. Even if the requirements document defines a probability-based metric, great effort is expended to find a related continuous measurement for the test design. It is not frugal to test based solely on the evaluation of probability-based metrics, rather than on continuous metrics.

- ***Do not limit factors varied in the test to those defined in the requirements.*** End users will employ the system or SoS in conditions that are different from those identified for system development and specification compliance. Testers are responsible for evaluating systems across the conditions under which systems will actually be employed. For example, consider a threshold that is based only on one threat type; it is not acceptable to base the OT on that one use case, especially if performance is expected to vary against different use cases. The operational conditions and use cases must be considered as a factor in the design, and all operationally relevant use cases should be accounted for in the levels of that factor. Using DoE techniques allows for a fiscally sound, robust evaluation of the requirement *alongside* a relevant and feasible characterization across the operational envelope. Limiting the effort to replicating runs/events under one condition (such as that identified in the requirements document) is much less informative than a strategic spread of runs/events across all relevant operational conditions.

- ***Ensure important factors are strategically controlled in test design and execution.*** Although a mission-based test design process identifies factors, the proposed run plan often relegates them to recordable conditions. This is a risky test strategy because all levels of the factor of interest might not be observed in the OT. Significant factors must be controlled whenever possible; this will ensure coverage of the full operational envelope. As an additional benefit, controlling significant factors helps to better explain the variance in the data, leading to more meaningful test results. For factors that truly cannot be controlled, testers need to evaluate the effect on confidence and power for other factors (since choosing not to control a factor often decreases the power to observe other factor effects of the test). Moreover, testers should ensure the recordable conditions are included in the analysis.
- ***Avoid confounding factors.*** Some test plans vary factors in such a way as to confound the effects, even when it was eminently feasible to avoid doing so. For example, testing all low-altitude runs for a specific system using short-slant ranges and high-altitude runs at long-slant ranges. It becomes impossible in the analysis to determine whether altitude or slant range was the primary cause of the observed difference in performance. Test strategies that confound factors are uninformative, inefficient, eliminate the ability to predict performance in other conditions, and preclude the ability to determine which factor affects performance the most. Physical limitations, non-repeatable conditions, and test infrastructure might force a confounded test design in some cases; however, these cases should be strenuously avoided.
- ***Do not use single hypothesis tests to size the test program.*** Even when the right focus is made on the use of power and confidence to scale a test, sometimes the intent of the test is still not achieved. If the goal of the test is to characterize performance across the primary factors, then the number of runs/events necessary to accomplish this goal cannot be determined from a hypothesis test which compares a single average across all conditions to a single threshold requirement (or average historical performance). The hypothesis test is lacking in several respects:
  (a) Most important, a single hypothesis test fails to identify differences in performance across the operational envelope. This is essential for meeting one of the primary objectives of OT to characterize a system's end-to-end mission effectiveness across the operational envelope.
  (b) The average performance across all conditions is typically a meaningless analytical result; not only should this be avoided in the analysis of the data, but it should also be avoided in test planning. The goal of the test is not to "roll up" all the results across all conditions and compare to a single number derived from a requirements document (which was likely defined under a different set of conditions). An extreme example is the case when the perfor-

mance in one set of conditions is zero and in another is perfect; the average of 50% across the environments is meaningless.

(c) Similarly, the effect size selected for a single hypothesis test is not typically meaningful. This results in misleading characterizations of power and produces a false sense of adequacy for the test, especially when several factors are expected to result in performance differences.

(d) It is rare that the requirements document threshold, which is used as the null hypothesis, is a true threshold, in the sense that we must ensure our tests reject systems that cannot be proven to be above threshold.

In some cases, a single hypothesis test might be warranted: tests where it is impossible to vary factors, or in cases where the threshold requirement must be met, and confidence in correctly rejecting the null hypothesis is assured. In those cases, a series of operating characteristic curves for various test sizes is warranted to determine test risk, as opposed to the use of a single poorly defined effect size, fixed confidence level, and single power calculation. In all other cases, an estimate of the precision of the test must be provided for assessing the adequacy of characterizing the system's performance across conditions. This will typically be in the form of power estimates to detect factor effects; however, other methods may be acceptable provided they capture the accuracy of the test's ability to characterize performance.

- ***Understand that adding a condition does not necessarily increase the size of the test.*** Although this effect is true for factorial designs, many other design types exist to overcome these difficulties and ensure testing adequately covers the operational envelope without significantly increasing the necessary test resources. A strategic spread of resources across all conditions, particularly if factors are continuous in nature, ensures that the precision of the test remains acceptable without undue replication in every test condition. Therefore, important factors should not be eliminated from the test design to avoid increasing the test size. Instead, the test design should include all operationally relevant factors and capitalize on the statistical tenets of DoE which enable the test size to remain manageable while ensuring the operational envelope is covered adequately.
- ***Understanding removal of condition may not decrease size of the test.*** A test program based on DoE requires accomplishing all the test events across the various conditions to achieve the precision to determine operational performance. In many test designs (but not all), the factors are continuous in nature; testing in one condition (level) helps to determine the common variance of the response variable across all conditions. Eliminating runs in one set of the conditions, therefore, will reduce the precision of the measurement in other conditions when employing a statistical model (a key tenet of DoE analysis). In cases where it becomes apparent that a factor or level should be eliminated, the test design (including the originally determined sample size) must remain adequate to characterize performance across the remaining factors.

13.2 Advancing Traditional T&E to 'T&E as a Continuum'

T&E must be reimagined, placing increased emphasis on the operational and mission context in which the system under test is expected to perform. Therefore, the way in which we think about validation and how we test in support of validation must result in high levels of confidence in executing complex integrated solutions to form effects webs consisting of the primary and alternative paths required for resiliency and redundancy, featuring multiple (joint) SoS. As such, the evaluation of a system under test must go beyond discrete T&E blocks within the program lifecycle. Instead, it must be carried out from conception to adoption (and fielding), within the context of a capability-expected contribution to overall joint capabilities and mission execution.

Major emphasis continues to be placed on the development and restructuring of policies, forums, and processes to drive integration and interoperability across SoS' capabilities in accordance with the five pillars that support the desired end state for T&E are shown in Table 13.1.

The goal of this transformation is to address the challenges posed by adversaries' rapid capability development and the accelerating pace of innovation in emerging technologies such as artificial intelligence (AI), autonomy, quantum computing, and 6G. To effectively navigate this landscape, a new approach is available, i.e., shifting from traditional T&E practices to a more integrated framework termed *T&E as a Continuum*. This new framework embodies three key attributes:

- **Capability and Outcome Focused Testing**: Instead of conducting testing activities independently from systems engineering (SE) and MEI efforts, *T&E as a Continuum* emphasizes testing that is closely aligned with desired capabilities and outcomes. This ensures that testing activities are purpose-driven and contribute directly to the achievement of operational objectives.
- **Agile, Scalable Evaluation Framework**: This new framework prioritizes agility and scalability in evaluation processes. This means that testing methodologies and evaluation criteria are designed to be flexible and adaptable to changing requirements and evolving technological landscapes. By adopting agile principles, evaluation frameworks better accommodate the rapid pace of innovation and respond quickly to emerging threats and opportunities.
- **Enhanced Test Design**: A key aspect of the transformation involves improving the DoE or Design of Test scenarios and methodologies. This includes leveraging advanced modeling and simulation techniques, as well as incorporating real-world data and feedback loops from operational environments. By enhancing test design, T&E activities can generate more accurate and actionable insights, leading to better informed decision-making throughout the capability development lifecycle.

Overall, these attributes are essential for achieving capability delivery at the "speed of relevance," enabling the rapid fielding and evolving capabilities that effectively address emerging threats and challenges in today's dynamic operational

Table 13.1 Advancements in T&E capabilities

Pillar for desired end state	Driving Elements
Test the way we fight	• Ensure accurate representation of the Joint, multi-domain force and operating environment during test events • Establish processes, resources, and capabilities to evaluate joint capabilities and mission engineering threads (SoS' performance)
Accelerate delivery of operable weapons	• Maintain discoverable, accessible, and secure authoritative data repositories • Deliver near real-time test data analysis and assessment results • Establish tools and processes to optimize the execution of integrated T&E to include Design of Experiments (DoE) • Develop digital documentation and tracking of T&E strategies and plans for transparency
Improve survivability and effectiveness of capabilities in a contested environment	• Minimize mission-critical vulnerabilities and maximize SoS' capabilities in contested environment • Employ real-time tracking and mitigation of mission-critical vulnerabilities as systems and adversarial threats evolve
Continuous advancement of T&E capabilities to handle complex system-of-systems	• Use credible digital tools in T&E to include digital twins • Ensure adequate and operational assessments for operational performance for AI-enabled systems • Use predictive health analytics for continuous tracking and mitigation of system performance degradation in operational environments
Foster an agile and enduring T&E workforce across the enterprise	• Hire a highly skilled T&E workforce to handle complexity and challenges • Reinforce a robust recruitment/retention program by offering an effective continuous learning program • Perpetuate a flexible and innovative workforce operating model with cross-domain pollination

environment. *T&E as a Continuum* is built upon the current digital transformation foundations aimed at advancing engineering practices and decision-making processes. The goals driving this digital transformation include (1) formalizing the use of models to inform decision-making, (2) establishing authoritative sources of truth, (3) integrating technological innovation, (4) creating supporting infrastructure and environments, and (5) fostering a culture and workforce that embrace digital engineering principles [DESTRAT18]. Derived from these goals are three key enablers for implementing *T&E as a Continuum*:

- **Robust Live, Virtual, and Constructive (LVC) Testing Environments**: These environments provide the necessary infrastructure for conducting testing activities across a spectrum of scenarios, ranging from physical testing in live environments to virtual simulations and constructive modeling. By leveraging LVC testing environments, testing engineers can conduct T&E activities more efficiently and comprehensively, resulting in timely assessment of system performance and risk characterization.
- **Model-Based Environment**: A model-based approach to testing allows for the integration of models and simulations throughout the engineering lifecycle. By using digital models to represent system behavior and performance, T&E activities can be conducted in a more iterative and collaborative manner, enabling early identification of issues and opportunities for improvement.
- **"Digital" Workforce**: A workforce that is equipped with the necessary skills and tools to leverage DE practices is essential for the successful implementation of *T&E as a Continuum*. This includes training personnel to use digital tools and methodologies, while fostering a culture of innovation and collaboration within the workforce.

By combining these key enablers, *T&E as a Continuum* aims to provide timely and comprehensive information on system performance and risk characterization throughout the engineering lifecycle. This integrated approach not only supports decision-making on individual systems but also enables informed management of capability development portfolios, ultimately facilitating the rapid development and fielding of increasingly complex systems and SoS.

13.3 Joint Test Concept

To continue the advancement in the T&E process necessitates building a Joint Test Concept (JTC) with agility and flexibility to move away from the current isolated and sequential block-testing approach from CT to DT, and then OT. The JTC includes a new layering schema that moves away from vernacular and sequential behavior of testing levels. The JTC should include a focus on overlapping layers that ensure the system meets performance requirements in isolation, within the predefined SoS. By utilizing a mission engineering thread construct within a joint, multi-domain environment, the JTC will produce realistic adversarial threat representations and accurate depictions of the SoS capabilities. A potential example of this new layered test environment is depicted in Fig. 13.2.

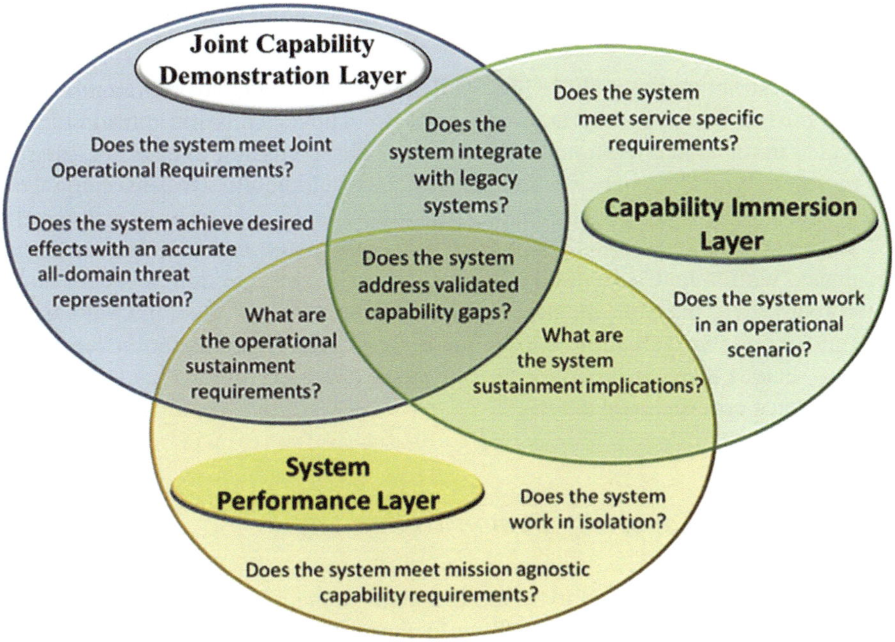

Fig. 13.2 Joint test concept foundation layers

13.3.1 System Performance Layer

The System Performance Layer (SPL) represents the most adaptable and dynamic component within the overall JTC evaluation framework. Its primary function is to evaluate how the system under test addresses identified capability gaps within a mission engineering thread, regardless of specific operational contexts. This layer is responsible for ensuring that the system meets predefined mission success criteria for performance both in isolation and within the larger SoS. The complexity of the mission thread assessed by the SPL is contingent upon the predetermined requirements for system performance. In the existing acquisition framework, SPL assessments are likely to be tailored to specific services or domains. While the SPL shares similarities with traditional T&E processes, it does not serve as the entry point to the evaluation process. Depending on factors such as system complexity, data availability, and performance requirements, the assessments conducted within the SPL layer could be fully addressed in either the Joint Capability Demonstration Layer (JCDL) or the Capability Immersion Layer (CIL). This underscores the non-hierarchical nature of the layers, despite their representation as a Venn diagram.

13.3.2 Capability Immersion Layer

The CIL is structured to evaluate how well the system tested performs against established mission success criteria within a set of validated mission threads that are executed in realistic operational environments. These mission threads are designed to represent typical scenarios where the system would be utilized, ensuring that its performance aligns with operational expectations.

In the current acquisition framework, it is anticipated that the mission threads evaluated within the CIL will be tailored to specific service requirements or have a joint focus. Like the SPL, the assessments conducted within the CIL could also be integrated into the evaluations carried out in the JCDL. This integration depends on factors such as the complexity of the SoS, the availability of data, and the specific performance criteria being assessed.

13.3.3 Joint Capability Demonstration Layer

The JCDL evaluates a portfolio of SoS in a multi-domain, joint operational environment that is representative of real-world theaters of operation. This evaluation includes incorporating elements such as a realistic adversary or threat and validated effects webs, which are networks of interconnected effects resulting from the interaction of various systems. The primary objective of the JCDL is to assess the performance parameters required for fielding, including Key Performance Parameters (KPPs) and Key System Attributes (KSAs). Because assessments are conducted in operationally representative scenarios, the JCDL can identify any operational performance gaps that need to be addressed prior to or early in the sustainment phase of fielded systems.

For example, a system may be deemed ready for fielding despite having certain limitations, such as occasional failures in extreme operating conditions or the need to quickly adapt to interoperability requirements with newly identified programs of record or legacy systems. A comprehensive joint capability assessment allows operators to develop and evaluate operational scenarios using more complete authoritative performance data across multiple complex SoS and effects webs, thereby ensuring readiness for real-world deployment.

The existing T&E process typically follows a linear and unidirectional path within an assessment. This, however, can pose a risk of catastrophic failure to the program if issues are not identified and addressed early. Recognizing this potential negative outcome, the JTC layering and overlap structure acknowledges the need for flexibility in the assessment process. The developer can interject systems into the formal JTC T&E assessment process at any phase of development, especially if the developer possesses historical assessment data that could be integrated into the JTC T&E formal assessment. This approach yields an iterative workflow that is neither linear nor unidirectional. For instance, a system may initially be deemed ready for fielding based on assessments conducted in the SPL or CIL. However, during sustainment, it may become apparent that further assessment is necessary to ensure effective integration into operations. In such cases, the system may need to return to SPL or CIL assessments to address any identified issues and optimize performance within the operational environment. This iterative approach enables continuous improvement and refinement throughout the lifecycle of the system with the ability to reframe the problem statement if necessary.

The JTC foundation is designed to complement the DoD's existing emphasis on SE, seamlessly integrating within its hierarchical structure. By aligning with SE principles, the JTC framework ensures compatibility with MEI, digital engineering, and other related disciplines within the SE domain. This alignment facilitates the smooth application of the JTC framework within the existing acquisition framework, particularly in the context of T&E. By embedding the JTC layers within the SE hierarchy, organizations effectively leverage the synergies between different disciplines and streamline the implementation of innovative approaches to testing and evaluation processes as illustrated in Fig. 13.3.

By leveraging the three-layer JTC foundation and adopting flexible workflows, the JTC structure moves away from the traditional practice of sub-optimizing systems. Instead, it embraces an end-to-end lifecycle (E2EL) T&E framework. This shift represents a departure from siloed approaches to T&E processes, allowing for a more holistic and integrated assessment of systems throughout the lifecycle. With the JTC structure in place, organizations can streamline T&E activities, enhance collaboration across different phases of development, and ultimately improve the effectiveness and efficiency of testing and evaluation efforts.

Derived from the principles of capability portfolio management, the concept of E2EL T&E shifts the focus from individual systems to overarching capabilities. By

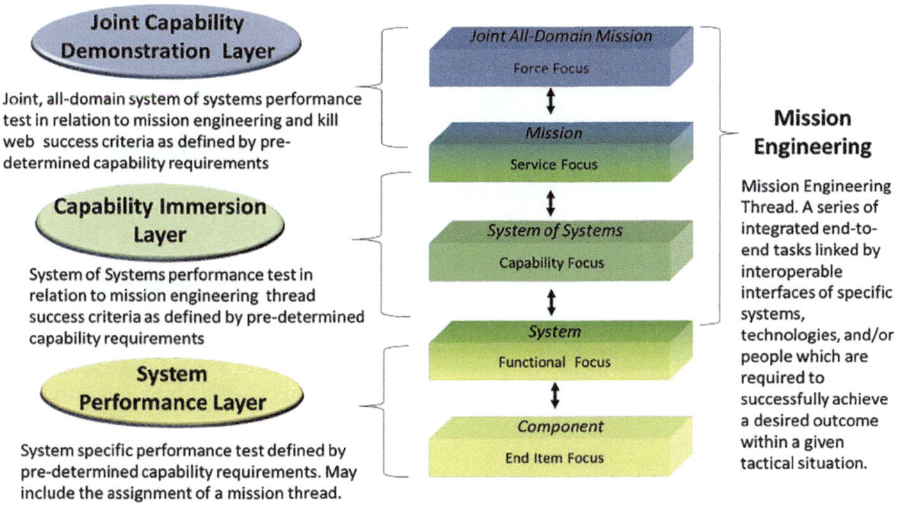

Fig. 13.3 Fitting the JTC to systems engineering hierarchy

prioritizing capabilities over specific systems, E2EL T&E enables a closer alignment between acquisition strategies and operational planning requirements. This approach ensures that T&E efforts are directed toward assessing the effectiveness and suitability of capabilities in meeting operational needs, rather than just verifying the performance of isolated systems. As a result, E2EL T&E enhances the integration of acquisition and operational considerations, leading to more informed decision-making throughout the lifecycle of developmental programs as depicted in Fig. 13.4.

The capability lifecycle begins by identifying capability gaps and conducting MEI to map systems to mission essential tasks within validated mission threads to identify the mission engineering thread. This initial phase is followed by the exploration and maturation of various solution approaches, which may involve research and development efforts as well as prototyping and experimentation. Once a solution approach is selected, the system lifecycle commences with the down-selection of the chosen solution and the commitment to design, develop, and deliver the system according to a defined acquisition pathway. However, the capability lifecycle extends beyond system deployment, necessitating ongoing T&E activities and performance monitoring in the field. This continuous assessment supports long-term sustainment and ensures operational resilience throughout the system's lifecycle.

In a complex modern operating environment, where SoS are often necessary to support a single capability, it is essential to move beyond a single-system capability concept to embrace a multi-system framework. Multi-system E2EL T&E processes ensure the presence of appropriate operational redundancies to support complex effects webs. Moreover, they facilitate the continuous integration of emergent requirements both post-fielding and throughout the sustainment phase across the entire capability portfolio. This approach acknowledges the interconnected nature of operational missions and the need for flexible and adaptable systems to effectively address evolving threats and operational challenges. The multi-system E2EL

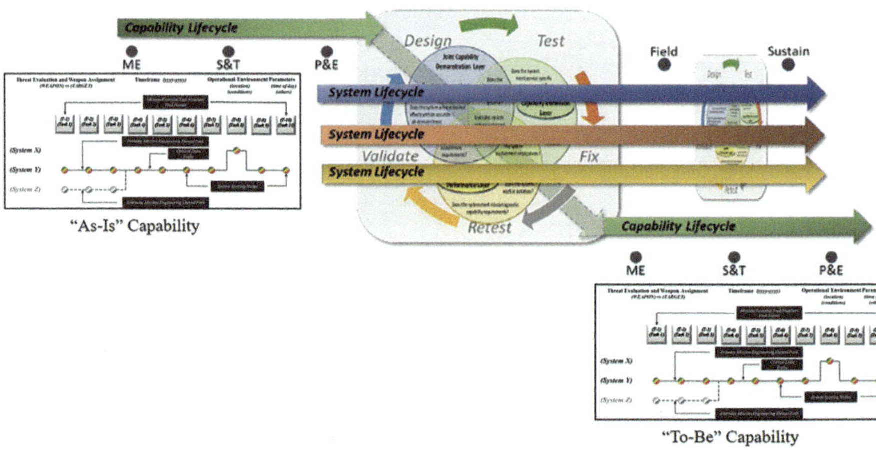

Fig. 13.4 E2EL T&E framework

approach is flexible and scalable, enabling a more focused assessment of multiple system acquisitions and a deeper comprehension of their individual and collective impacts on capabilities when considering capability portfolio management.

13.4 Integrated Test and Evaluation Techniques

In the preceding sections discussing the evolution of T&E (particularly with the concepts of *T&E as a Continuum* and the Joint Test Concept), it becomes evident that integrated testing is the fundamental cornerstone to realize the potential of these advanced techniques. Integrated testing, which combines DT and OT, represents a critical shift toward a more holistic and comprehensive approach to evaluating system performance across its entire lifecycle. The concept of *T&E as a Continuum* emphasizes the seamless integration of testing activities from the earliest stages of system development through operational deployment and sustainment. By conducting integrated testing, developers and operators more readily identify and address performance issues early in the development process, leading to more robust and reliable systems in the operational phase.

Similarly, the Joint Test Concept emphasizes the collaboration and coordination of testing efforts among multiple stakeholders, including government agencies, industry partners, and operational users. Integrated testing thus plays a pivotal role in this collaborative framework by enabling the evaluation of system interoperability, integration, and performance across diverse operational environments and mission scenarios. The future direction on this T&E vector allows stakeholders to assess system performance in realistic operational conditions, identify interoperability challenges, and validate system capabilities against operational requirements for dynamic operational environments.

Integrating DT and OT to evaluate SoS capabilities through the utilization of DoE methodologies presents a comprehensive approach to assess SoS performance across the entire lifecycle. Employing DoE principles offers the test developer a systematic variation of input variables and the measurement of their impacts on the systems' performance. By employing factorial designs, the developer can evaluate the main effects of individual factors and their interactions simultaneously. This provides a holistic understanding of SoS behavior under various operational scenarios.

Screening experiments identify the most influential factors affecting SoS performance efficiently, guiding subsequent testing efforts. Response surface methodology aids in modeling the relationship between input variables and system performance metrics, facilitating optimization of SoS performance. Statistical analysis of experimental data enables stakeholders to identify critical performance drivers, quantify their effects, and optimize SoS design and operation to meet operational requirements effectively. Applying this integrated approach that leverages DoE methodologies enhances the reliability, efficiency, and effectiveness of SoS testing, ultimately contributing to the successful deployment and operation of complex SoS.

13.4.1 Distributed Test and Evaluation

Executing distributed testing of complex SoS using network infrastructure capabilities, such as the Secret Defense Research and Engineering Network (SDREN), with test nodes across the United States, offers several opportunities for enhancing SoS evaluation and readiness. Leveraging the SDREN promotes secure and reliable communication and collaboration among distributed test nodes, ensuring the confidentiality and integrity of sensitive defense-related information. This network infrastructure provides a trusted platform for exchanging data, sharing test results, and coordinating testing activities across geographically dispersed locations. By distributing testing activities across multiple nodes located throughout the United States, organizations have the ability to replicate diverse operational environments and scenarios, providing a more comprehensive assessment of SoS performance under varying conditions. This distributed approach supports the evaluation of SoS interoperability, resilience, and scalability across different geographic regions and mission domains. Figure 13.5 illustrates the architecture and connectivity of the SDREN, highlighting its role as a vital platform for secure and collaborative testing and evaluation activities across multiple nodes within the United States.

Additionally, distributed testing using SDREN facilitates collaboration among various stakeholders, including government agencies, defense contractors, research institutions, and operational users. By bringing together diverse expertise and resources, organizations improve leverage of collective insights and capabilities to address complex SoS challenges. The Tactical Data Link Integration Exerciser (TENA) is a component of the SDREN, contributing to its capabilities for advanced T&E and allowing for collaboration across organizational boundaries and

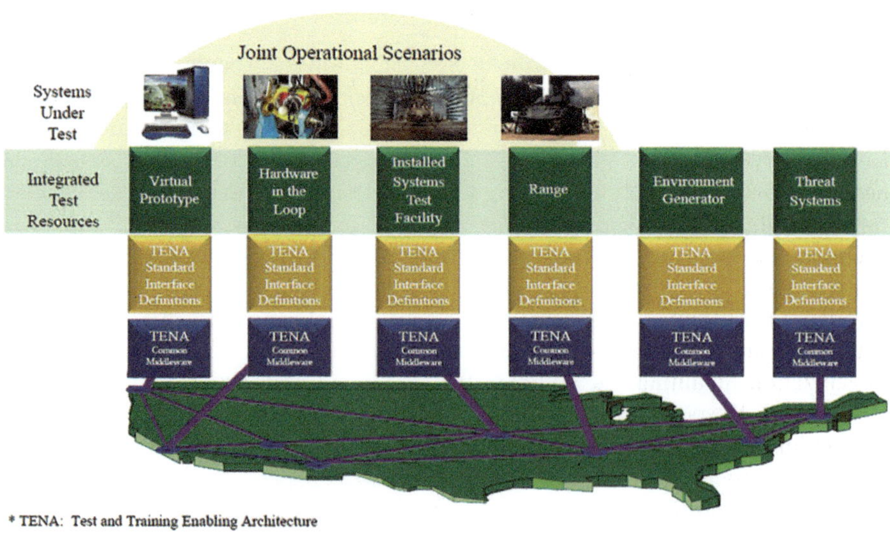

Fig. 13.5 SDREN infrastructure architecture

operational domains. TENA serves as a framework for integrating and exercising tactical data links (i.e., Link 16, Link 11, Tactical Radios, etc.) in complex operational environments. Within SDREN, TENA facilitates the simulation and emulation of tactical data links across distributed test nodes, enabling realistic and interoperable testing scenarios for SoS. By incorporating TENA into SDREN, users can assess the performance, interoperability, and resilience of tactical data links in a secure and collaborative environment. Furthermore, distributed testing enables the simultaneous conduct of parallel and concurrent testing activities across multiple nodes, reducing testing timelines and accelerating the feedback loop for identifying and addressing performance issues. This agile testing approach promotes rapid iteration and refinement of SoS capabilities, leading to faster development cycles and more timely deployment of operational systems.

13.4.2 Live, Virtual, and Constructive Environment

Establishing a Live, Virtual, and Constructive (LVC) T&E environment allows for a comprehensive test approach for SoS at different maturity levels to assess capabilities and readiness across the development lifecycle. In the early stages of SoS development, a virtual T&E environment is utilized to simulate system components and interactions in a controlled and repeatable manner. Virtual simulations enable rapid prototyping, experimentation, and concept validation, allowing developers to assess system architecture, interoperability, and performance in a cost-effective and risk-free environment. By leveraging virtual simulations, stakeholders readily identify design flaws, refine system requirements, and optimize SoS architectures before proceeding to physical testing.

As the SoS matures, live T&E environments become increasingly important for evaluating system performance in real-world operational conditions. Live testing involves conducting experiments and exercises using physical hardware and operational assets to assess SoS functionality, interoperability, and mission effectiveness. Live T&E events provide valuable insights into system behavior under realistic scenarios, allowing stakeholders to validate system capabilities, verify operational requirements, and assess readiness for deployment. Experimentation, evaluation, and testing are integral components of assessing and improving systems, products, or processes, each playing a distinct yet interconnected role. Experimentation involves the systematic manipulation of variables in controlled environments to gather data and understand underlying principles or hypotheses across the Doctrine, Organization, Training, Materiel, Leadership, Personnel, Facilities, and Policy (DOTMLPF-P) spectrum and Technology Readiness Levels (TRL). It serves as the foundation for generating new knowledge, exploring potential solutions, and informing the development of innovative technologies or methodologies. Evaluation, on the other hand, focuses on the systematic assessment of the effectiveness, efficiency, or quality of a system, program, or intervention. It occurs in real-world settings to measure outcomes, impacts, or performance against predefined criteria or objectives. Evaluation encompasses a range of methodologies, from qualitative assessments to quantitative analyses, to provide a comprehensive understanding of the subject under evaluation. Testing, meanwhile, involves the systematic examination or assessment of a system, product, or process to verify its functionality, performance, or compliance with specified requirements. Testing activities span from unit testing of individual systems to SoS-level integration testing and validation, aiming to identify defects, validate requirements, and ensure that the product or systems meet user needs. While experimentation, evaluation, and testing serve distinct purposes, they are interconnected processes that often inform and complement one another as characterized in Table 13.2. Findings from experimentation may guide the development of evaluation frameworks, while evaluation outcomes may inform testing strategies. Moreover, insights gained from testing may prompt revisions to experimentation hypotheses or inform the refinement of evaluation criteria. Together, these processes contribute to evidence-based decision-making, continuous improvement, and the advancement of knowledge and practice across various domains.

In parallel with live testing, constructive T&E environments are employed to supplement testing activities and explore a broader range of operational scenarios. Constructive simulations involve modeling and simulating SoS behavior using computational models, algorithms, and data-driven simulations. Constructive T&E

Table 13.2 Experimentation, test, and evaluation focuses

	Experimentation	Evaluation	Testing
Definition	An analytical activity founded upon observation or experience derived from unbiased trials conducted under controlled conditions within a representative environment to **discover something unknown, to test a hypothesis, or establish or demonstrate some knowledge within a specific context.**	Independent (of material developer) process to **determine if a system satisfies the approved requirements.**	Developmental: **Verify design risks are minimized; safety of the system is certified; achievement of system technical performance is substantiated; readiness for OT is certified.** Operational: **Field test under realistic operational conditions** with users who represent those expected to operate and maintain the fielded system.
Focus	Concept, Required Capability Across DOTMLPF-P Generally, TRL 7 or lower	System (Materiel) TRL 6 or higher	System (Materiel) TRL 7 or higher
Purpose	Should we pursue this idea/capability?	Should we continue to develop this system?	Should we buy this system?

enables the assessment of SoS performance across a diverse set of operational environments, mission scenarios, and threat conditions, providing stakeholders with a comprehensive understanding of system capabilities and vulnerabilities. By integrating LVC T&E environments, stakeholders conduct a range of testing activities at different maturation levels to assess SoS capabilities comprehensively. This integrated approach allows technical risks, operational limitations, and performance gaps to be identified early in the development process. This timely identification facilitates informed decision-making and risk mitigation strategies. Ultimately, leveraging LVC T&E environments throughout the SoS lifecycle enhances system resilience, interoperability, and readiness for operational deployment in dynamic and challenging environments.

13.5 Using Modeling and Simulation (M&S)

Modeling and Simulation (M&S) have emerged as indispensable tools in the arsenal of modern industries. Their importance is underscored by the growing reliance on complex SoS architectures across various sectors. From aerospace to healthcare, industries are grappling with intricate networks of interconnected systems that demand a thorough understanding of their behaviors and interactions. M&S provides a virtual sandbox for engineers and designers to explore these complexities, offering insights into emergent behaviors, dependencies, and performance characteristics. As traditional methods of development struggle to keep pace with the intricacies of modern systems, M&S offers a cost-effective and efficient alternative. By enabling virtual testing, risk mitigation, and performance optimization, M&S accelerates the development cycle while ensuring the robustness and resilience of new capabilities. Its cross-industry applicability further cements its role as a catalyst for innovation, driving progress across a wide spectrum of industrial domains. As industries continue to evolve and embrace ever more sophisticated technologies, the role of M&S in shaping the future of development and innovation is set to become even more pronounced.

M&S for T&E forms a fundamental component of Integrated Lifecycle Management. Its strategic application throughout a system's lifecycle yields significant advantages, notably in mitigating program risks. Such utilization offers manifold benefits for program managers, systems engineers, mission engineers, decision-makers, and system users alike. The essence of T&E M&S lies in:

- Enhancing confidence levels
- Reducing field test durations and associated costs
- Augmenting data availability for pretest predictions and posttest validation
- Facilitating test control to ensure safety
- Simulating scenarios that are otherwise challenging to replicate

The foundational underpinnings for M&S approaches require a sturdy intellectual and conceptual base firmly rooted in the laws of science. These foundational principles, encompassing disciplines such as physics, mathematics, and engineering, provide the fundamental framework for understanding and modeling complex systems and phenomena. By adhering to the laws of science, M&S practitioners ensure the accuracy, validity, and reliability of their simulations, enabling them to capture the underlying mechanisms and behaviors of the systems under study. This adherence also allows M&S engineers to develop the predictive models that can anticipate system responses under various conditions, thus supporting decision-making and problem-solving efforts. This, with their strong intellectual foundation in the laws of science which fosters innovation and creativity, allows these engineers to drive advancements in simulation methodologies and technologies for building robust, credible, and impactful simulations that can address a wide range of challenges across different domains and applications.

The criticality of M&S for MEI cannot be overstated when depending on quantitative data to assess the performance of systems and SoS from a verified and validated M&S tool. In the realm of mission planning and execution, where precision, efficiency, and adaptability are paramount, M&S serves as an indispensable toolset. It provides a virtual environment where the complexities of mission scenarios can be thoroughly analyzed, tested, and optimized before real-world implementation. Through M&S, mission planners simulate various operational scenarios, assess the impact of different factors, and refine strategies to achieve mission objectives. This capability is particularly crucial in scenarios involving SoS integration across operational domains and military services, where multiple interconnected components must seamlessly collaborate to accomplish overarching mission goals. M&S helps identify potential bottlenecks, vulnerabilities, and dependencies within the system, and this enables proactive mitigation strategies and ensures mission success. Furthermore, M&S supports interdisciplinary collaboration by providing a common platform where stakeholders from diverse domains can contribute their expertise and perspectives to the mission planning process. As missions become increasingly complex and dynamic, the role of M&S in MEI will continue to play a critical role.

13.5.1 Dimensions of Fidelity

Fidelity, or the level of realism and accuracy in a simulation, directly influences effectiveness and applicability across various domains [ALLAIRE14]. Fidelity encompasses several dimensions, including physical, functional, and behavioral fidelity, each contributing to the overall realism of the M&S tool. Physical fidelity refers to how closely the M&S replicates the physical attributes and characteristics of the real-world system. This includes factors such as geometry, materials, and environmental conditions. High-physical fidelity simulations provide detailed representations of the system's appearance and dynamics, enabling realistic visualizations and assessments of its performance. Functional fidelity pertains to the accuracy with which the M&S models the functionalities and operations of the real-world system. It involves replicating the behaviors, interactions, and responses of system components in accordance with their intended functions. M&S with high-functional fidelity represents the functions and behaviors of the system at a level of detail appropriate for assessing its performance under different conditions, allowing users to assess its behavior and performance under different conditions [PACE15]. Behavioral fidelity refers to the realism of the behaviors and decision-making processes exhibited by entities within the M&S. This aspect is particularly relevant in M&S involving human operators, autonomous agents, or complex systems with emergent behaviors. High-behavioral fidelity simulations accurately represent the actions, interactions, and decision-making processes of these entities, providing insights into their behaviors and performance in realistic scenarios.

Comparing fidelity between models representing the same target system component involves evaluating their outputs for identical inputs. In this context, a high-fidelity model demonstrates lower output error compared to a low-fidelity model when assessed against a reference output. This comparison typically involves measuring the discrepancy between the model outputs and the reference output. A high-fidelity model, characterized by its detailed and accurate representation of the system, is expected to yield outputs that closely align with the reference output, resulting in minimal error. Conversely, a low-fidelity model, which may sacrifice some level of detail or accuracy for efficiency or simplicity, is more likely to exhibit higher output error due to its limited representation of the system. As illustrated in Fig. 13.6, by quantifying output error and comparing it against a reference, analysts discern the relative fidelity levels of different models and ascertain their suitability for specific applications or analyses.

Choosing a fidelity level in M&S is a critical decision influenced by specific objectives and resource constraints. High-fidelity M&S provides detailed and accurate representations of systems but demands significant computational resources and development effort. Conversely, low-fidelity M&S offers scalability and efficiency at the cost of realism. Balancing fidelity levels involves trade-offs between realism, computational complexity, and resource requirements. Understanding the domain's requirements and constraints is vital for determining the appropriate fidelity level to achieve desired outcomes effectively. To address this challenge, mapping high-fidelity to low-fidelity counterparts allows applying physical constraints within the low-fidelity space and reducing computational burdens. Real-world scenarios

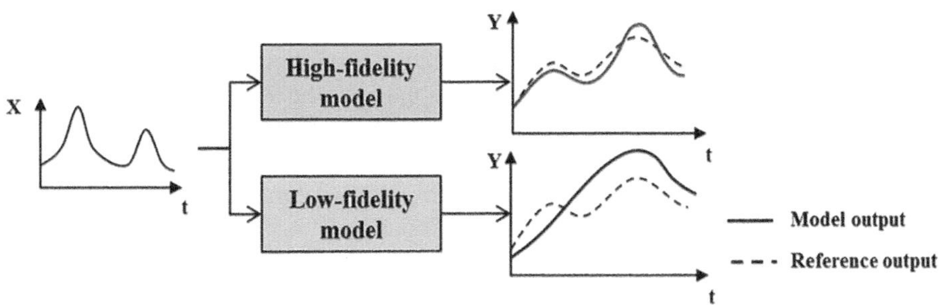

Fig. 13.6 Fidelity of model

involve data of varying fidelities from diverse sources. Efficiently integrating multi-fidelity data becomes crucial due to resource constraints associated with high-fidelity data acquisition. If high- and low-fidelity data can achieve the same representation, a method that utilizes all fidelity levels for performance assessment while guiding high-fidelity M&S with low-fidelity physical constraints ensures a balance between efficiency and accuracy. Using this approach maximizes data utilization while maintaining computational tractability, essential for advancing M&S capabilities in complex domains.

Fidelity in M&S is delineated across three fundamental dimensions: accuracy, repeatability, and resolution. Accuracy refers to the extent to which a parameter or variable, whether singular or a set, within an M&S faithfully aligns with reality or a designated standard. It signifies the precision of the representation, gauging how closely the model's outputs mirror real-world phenomena. Repeatability, on the other hand, assesses the consistency of outcomes generated by the same model (or referent) when subjected to identical input conditions. It measures the reliability of the model's predictions or outputs, emphasizing the degree of consistency in its performance. Lastly, resolution encapsulates the level of detail or granularity with which a parameter or variable can be discerned within the model. It characterizes the extent to which the model can capture fine-scale features or nuances within the system under study, highlighting its capacity for detailed analysis and representation. These dimensions collectively contribute to the overall fidelity of an M&S, reflecting its faithfulness, reliability, and granularity in depicting the real-world phenomena or systems it seeks to emulate.

Different methods exist for combining fidelities in M&S, with the goal of leveraging the strengths of each fidelity level to achieve both accuracy and computational efficiency. Two primary approaches are commonly considered: multi-fidelity surrogate models (MFSMs) and multi-fidelity hierarchical models (MFHMs). In MFSMs, algebraic surrogates enhance the performance of low-fidelity models (LFMs) by incorporating information from high-fidelity model (HFM) predictions. Conversely, MFHMs select fidelity levels based on specific criteria, which may vary depending on the context of the modeling task. Within MFSMs, these four correction paradigms are often employed:

1. **Multiplicative correction**: This approach involves adjusting the output of the low-fidelity model by scaling it with a correction factor derived from the high-fidelity model predictions.
2. **Additive correction**: Similar to multiplicative correction, but instead of scaling the output, an additive correction term is applied to the low-fidelity model's predictions based on the discrepancy with the high-fidelity model.
3. **Comprehensive correction**: This method integrates information from both high and low-fidelity models through a comprehensive correction function, which may involve more complex mathematical relationships.
4. **Space mapping**: Space mapping techniques involve mapping the low-fidelity model's input space to the high-fidelity model's input space, so that high-fidelity information can be incorporated into the low-fidelity predictions.

Historically, MFSMs have been predominant in the research landscape, constituting most research articles, while MFHMs have represented a significant minority. MFHMs distinguish themselves by selectively employing fidelity levels based on specific optimization criteria. This stands in contrast to the more inclusive architecture of MFSMs, which aim for a comprehensive integration of low- and high-fidelity data. However, recent observations suggest a growing preference for MFHMs with the shift driven and facilitated by advancements in computational capabilities and optimization algorithms. This enables the adoption of more sophisticated approaches that were previously computationally prohibitive. As such, the evolution of M&S techniques continues to reflect the dynamic interplay between computational resources, methodological advancements, and the quest for more efficient and accurate modeling paradigms.

13.5.2 M&S Data

The verification and validation of data and data sources is a critical process that scrutinizes the data utilized in the development and execution of M&S to make sure authoritative data from an authoritative source of truth is used. The credibility of data hinges not only on their production and maintenance processes but also on how they are transformed and integrated within the M&S framework. Data verification entails confirming that the selected data are the most suitable for the intended application and that they undergo appropriate preparation or transformation for integration into the M&S. This step ensures that the data are well-suited for their intended purpose and align with the requirements of the simulation. Data validation, on the other hand, involves verifying that the data accurately represent the aspects of the real world intended to be simulated within the M&S. This process ensures that the data faithfully capture the relevant characteristics and behaviors of the real-world phenomena being modeled, enhancing the accuracy and reliability of the simulation results. Together, data verification and validation play pivotal roles in ensuring the integrity and fidelity of M&S by confirming the appropriateness and accuracy of the data used throughout the simulation process.

Data requiring verification and validation are categorized into the following five main categories within the context of M&S:

- ***Data needed to verify M&S requirements***: These are the data used to ensure that the M&S system meets its specified requirements and objectives.
- ***Data needed to build the conceptual model***: These data are utilized in constructing the conceptual framework of the M&S tool, forming the basis for subsequent development and validation activities.
- ***Data needed to validate the M&S***: This category includes data used to assess the accuracy and reliability of the M&S outputs by comparing them to real-world observations or empirical data.

- ***Data needed to perform experiments***: These data are essential for conducting experiments or simulations within the M&S environment, providing inputs and parameters necessary for the simulation runs.
- ***Data needed to run M&S decision aids***: This category encompasses data used in decision-making processes facilitated by M&S tools, providing the information required to generate insights and recommendations.

It is imperative that the data used for developing and validating M&S are appropriate for their purpose. Integrating data verification and validation activities alongside other model development tasks ensures that the data utilized in the M&S system are relevant, accurate, and reliable. Incorporating these activities that are focused on the quality of data from an authoritative source throughout the M&S lifecycle increases the confidence level for the practitioners in the integrity and fidelity of the M&S results. The effectiveness and utility of the M&S system are ultimately improved for decision-making and analysis purposes.

13.5.3 M&S Verification, Validation, and Accreditation (VV&A)

M&S fidelity plays a crucial role in M&S assessments, including verification, validation, and accreditation (VV&A). With the growing reliance on M&S across various domains of science, engineering, and everyday life, establishing confidence in the correctness and applicability of M&S results is increasingly vital. This confidence hinges on the proper determination of M&S fidelity. Verification involves the identification and elimination of errors in logic, mathematics, or programming within M&S tools. This critical process aims to ensure that the M&S code and logic accurately perform their intended functions. Additionally, verification evaluates the extent to which M&S development activities adhere to state-of-the-practice software engineering techniques. Through scrupulous verification, practitioners then validate the correctness of M&S implementations and enhance confidence in their reliability and effectiveness.

Validity pertains to the relative appropriateness of M&S results for a specific purpose or application. While fidelity focuses on the accuracy of the simulation output, validity assesses whether those results are suitable for the intended use. Validity often incorporates information about fidelity as part of its assessment process. This means that fidelity serves as one of the factors considered when determining the suitability of M&S results for a particular purpose. By ensuring high fidelity, M&S practitioners can enhance the validity of their simulations, thereby increasing confidence in the reliability and relevance of the results for the intended applications. This validation process is integral to assessing the effectiveness and appropriateness of the conceptual model within an M&S application. It comprises two primary components: structural validation and output validation. Structural validation involves an internal scrutiny of M&S assumptions, architecture, and algorithms in relation to the intended use. This component delves into the coherence and suitability of the M&S framework and its components, ensuring they align with the objectives and requirements of the application. On the other hand, output validation focuses on evaluating the fidelity of M&S results compared to the perceived "real world." This component involves comparing the output generated by the M&S tool with empirical data or observations to gauge the accuracy and reliability of the model's predictions. Together, these two facets of validation play a crucial role in establishing confidence in the M&S application, confirming both its internal structure and its ability to accurately represent real-world phenomena.

The accreditation process evaluates several key factors, including the V&V status of a specific model version, the quality and source of its data support, and the qualifications of the analysts or users responsible for operating the model and interpreting its results. At the heart of the accreditation process is the accreditation authority, an individual who holds responsibility and accountability for decisions or actions based on the specific usage of M&S. The accreditation authority possesses the sole discretion to determine whether a model or simulation receives accreditation. This authority also dictates the level of effort required to support the accreditation decision, which may entail conducting additional V&V activities or reviewing existing M&S documentation and past VV&A history. Accreditation ultimately falls within the management responsibility of the requiring agency, with support from the designated V&V agent, who assists in ensuring that the accreditation process adheres to established standards and guidelines.

MEI with digital engineering as an enabler relies heavily on utilizing M&S, because its use verifies the trustworthiness of the models employed which is crucial for effective design of complex systems [AHNER21]. Validation, a pivotal aspect of ensuring model reliability, involves assessing the extent to which an M&S and its associated data accurately represent the real world concerning their intended uses. This process establishes whether a model can reliably depict the associated physical system. However, validation outcomes often oversimplify the assessment, categorizing models as either validated or not, which fails to capture the nuanced degree of trustworthiness. Moreover, this classification typically results from subjectively chosen validation methods combined to determine a model's validity. Current practices also lack continuous reassessment of model validity in response to evolving data availability or changes in the model's use case. A universal, objective validation method that can be readily and repeatedly applied across the diverse spectrum of M&S applications is notably absent. Addressing these challenges is essential to ensure the reliability and effectiveness of M&S in supporting MEI endeavors relying on digital engineering techniques.

13.6 Cybersecurity Test and Evaluation

Cybersecurity Test and Evaluation (T&E) refers to a comprehensive suite of activities integral to assessing the security posture and resilience of systems against cyber threats. It encompasses a range of methodologies and techniques to evaluate various aspects of cybersecurity to ensure the robustness and effectiveness of security measures. These activities include vulnerability assessments, which identify weaknesses and susceptibilities within systems or networks, and security controls testing, which evaluates the functionality and effectiveness of implemented security measures. Penetration testing is another critical component of cybersecurity T&E, involving simulated cyber-attacks to uncover vulnerabilities and weaknesses that could be exploited by malicious actors. Adversarial testing goes further by simulating real-world attack scenarios to assess the system's ability to withstand sophisticated cyber threats and attacks. Additionally, cybersecurity T&E encompasses testing related to a system's operational resilience, evaluating its ability to maintain essential functions and services in the face of cyber incidents or disruptions.

Moreover, cybersecurity T&E evaluates a system's cyber survivability capabilities within the context of its operational mission. This involves assessing the system's ability to continue operating effectively despite cyber-attacks or compromises. Adopting the phrase "Prevent, Mitigate, and Recover" (PMR) facilitates consistency in describing the assessments of system cyber survivability and operational resilience, as opposed to the varied terminology used by different organizations in this domain. This standardizing of terminology provides clarity and coherence in discussing these critical concepts among practitioners, policymakers, and stakeholders. The use of consistent terminology also helps align assessment frameworks, methodologies, and practices across organizations, enabling more effective collaboration and coordination in addressing cyber threats and ensuring the robustness and resilience of systems against cyber-attacks and disruptions. These phases of cybersecurity captured in the PMR phrase are defined as:

- ***Prevent***: The capability of safeguarding critical mission functions from cyber-threats. This aspect of cybersecurity focuses on implementing proactive measures and controls to mitigate the risk of cyber-attacks and unauthorized access to systems or data. By employing preventive measures such as firewalls, intrusion detection systems, access controls, and encryption, organizations aim to thwart potential cyber-threats and prevent unauthorized activities that could compromise the integrity, confidentiality, or availability of critical mission functions. The goal of prevention is to create a resilient security posture that minimizes the likelihood of successful cyber-attacks and ensures the continuous operation and integrity of essential systems and services.
- ***Mitigate***: Encompasses the capacity to detect and respond effectively to cyber-attacks, while also assessing resilience to withstand such attacks and ensure mission completion. This aspect of cybersecurity emphasizes the importance of proactive monitoring, incident detection, and rapid response capabilities to mitigate the impact of cyber threats on organizational operations. By promptly

identifying and containing security incidents, organizations can limit the damage caused by cyber-attacks and minimize disruptions to critical functions. In addition, the concept of mitigation involves assessing the resilience of systems and networks to withstand cyber-attacks and maintain operational continuity. This includes evaluating the effectiveness of security controls, redundancy measures, and disaster recovery plans in mitigating the impact of cyber incidents.
- ***Recover***: Denotes the resilience of systems to recover from cyber-attacks and prepare mission systems for subsequent conflicts or operational challenges. This facet of cybersecurity focuses on restoring operations, data, and functionality following a cyber incident, and ensuring the rapid recovery and full restoration of critical mission systems. Effective recovery involves implementing robust incident response plans, backup and restoration procedures, and continuity of operations strategies to minimize downtime and restore affected systems to their pre-incident state. By promptly recovering from cyber-attacks, organizations can mitigate the impact of disruptions, minimize losses, and maintain operational readiness for future conflicts or challenges.

Given the practical challenges of addressing all vulnerabilities, susceptibilities, and exploitable attack paths before a system is deployed, organizations often rely on a mission-based approach to prioritize cybersecurity T&E efforts. This mission-focused process involves identifying, estimating, assessing, and prioritizing risks based on their potential impacts on operational missions resulting from cyber effects on the employed system(s). Aligning T&E activities with mission objectives aids stakeholders in effectively prioritizing resources and in focusing testing efforts on the most critical areas of concern. This approach acknowledges that not all vulnerabilities can be eliminated but instead seeks to manage risks and maximize the protection of mission-critical functions. Mission-based planning process helps organizations to systematically evaluate the potential consequences of cyber threats on operational missions and allocate resources accordingly. By prioritizing risks based on their impact on mission success, stakeholders make better informed decisions about risk acceptance, mitigation strategies, and resource allocation, ultimately enhancing the overall cybersecurity posture of the system(s) in question.

13.6.1 Dedicated Cyber Range Infrastructure for Cybersecurity Testing

A cyber range for executing cybersecurity T&E is paramount due to several critical factors. First, cyber ranges offer a controlled yet realistic environment conducive to simulating a wide array of cyber threats, attacks, and scenarios. By replicating authentic network configurations and attack vectors, the cyber ranges enable T&E teams to thoroughly assess the efficacy of cybersecurity measures in defending against diverse cyber threats. Moreover, cyber ranges serve as invaluable platforms for training and skill development, providing cybersecurity professionals with hands-on experience in detecting, responding to, and mitigating cyber-attacks. This hands-on training is instrumental in improving incident response capabilities and enhancing the overall cybersecurity posture of organizations. Additionally, cyber ranges foster experimentation and innovation by offering a sandbox environment for testing new cybersecurity technologies, tools, and techniques. They facilitate collaboration and information sharing among T&E stakeholders, enabling collective efforts in addressing common cybersecurity challenges and improving cybersecurity capabilities. Furthermore, cyber ranges aid in risk reduction by allowing organizations to identify and mitigate cyber risks proactively, before they impact operational systems and missions. In essence, cyber ranges play a pivotal role in enabling realistic, collaborative, and proactive cybersecurity T&E activities, ultimately strengthening the resilience and readiness of organizations against evolving cyber threats.

Positioning a cyber range as a common infrastructure resource offers a transformative approach to delivering integrated cybersecurity solutions across diverse domains and organizations. By centralizing the cyber range as a shared resource, various stakeholders can access a collaborative platform for testing, training, and validating comprehensive security measures. This shared infrastructure promotes cross-domain collaboration and information sharing. Organizations are then able to collectively address cybersecurity challenges and enhance their overall cyber resilience. The cyber range serves as a neutral ground where organizations from different sectors and industries can come together to conduct joint exercises, share best practices, and collaborate on cybersecurity initiatives. Through this collaborative approach, organizations can leverage collective expertise and resources to address common cybersecurity threats and vulnerabilities effectively. Moreover, a centralized cyber range enables organizations to maximize the value of their cybersecurity investments by avoiding duplicative efforts and leveraging shared resources and expertise. By pooling resources and capabilities, organizations can access advanced technologies, tools, and expertise that may be cost-prohibitive to acquire individually. Additionally, a centralized cyber range provides a standardized testing environment that promotes consistency and interoperability across different cybersecurity solutions and technologies.

For flexibility purposes, cyber ranges need to have scalability as well as elasticity. Scalability in a cyber range refers to its ability to accommodate the intended user population effectively, while elasticity pertains to the time required to expand the range's capacity to support additional users. Ideally, a cyber range should support its full potential user population immediately and quickly increase capacity upon demand. Cyber ranges based on local hardware infrastructure face limitations due to finite random-access memory and hard drive space, allowing them to scale only until local resources are exhausted, thus lacking elasticity. Expanding capacity beyond these limits involves purchasing and configuring additional hardware and software, which can take weeks or even months to accomplish. Conversely, public cloud-based ranges generally excel in scalability, tapping into additional cloud provider resources as needed. They also offer high elasticity when automated, drawing on public cloud infrastructure to provision system resources for extra users. In addition to computer and storage infrastructure, adequate server-side bandwidth is crucial for scaling to handle high user access during peak times. Limited scalability or elasticity can lead some commercial ranges to restrict simultaneous access, requiring stakeholders to reserve time slots or wait for resource availability.

13.6.2 Integrated Cybersecurity Testing

Integrated cybersecurity T&E facilitates sharing a single test point or mission across multiple test events and enables data to satisfy multiple objectives concurrently. This approach enhances operational realism by replicating real-world scenarios and providing opportunities for early identification of system design improvements. By aligning test events with a common test point or mission, organizations optimize resources, streamline testing activities, and maximize the value of collected data. Additionally, integrated T&E enables stakeholders to assess cybersecurity capabilities in a holistic manner, considering various aspects such as vulnerability assessments, security controls testing, and operational resilience within a unified framework. This comprehensive approach furthers a more thorough evaluation of cybersecurity posture and readiness, which in turn leads to the early detection of weaknesses or gaps in system design. Furthermore, integrated T&E fosters collaboration and information sharing among stakeholders, facilitating cross-functional analysis and decision-making to address identified issues and drive continuous improvement.

This type of testing encompasses not only technical vulnerabilities but also examines the effectiveness of response protocols, defense layers, and team readiness. To execute integrated cyber testing effectively, organizations should start with clear objectives that outline the scope of testing, such as identifying specific vulnerabilities or assessing team responses under simulated attacks. Proper execution involves creating a realistic network environment within a cyber range, mirroring the organization's actual infrastructure, including network configurations, data assets, and operational systems. Once the environment is set, security teams can introduce different threat scenarios, varying attack types, and incident simulations to gauge how systems and personnel respond under pressure. Regularly reviewing and adjusting the testing scenarios to reflect evolving threats ensures that the insights remain relevant. Finally, analyzing results and implementing improvements based on the findings completes the cycle, creating a continuous feedback loop for refining cybersecurity practices.

Having discussed **test and evaluation in a mission context**, we now look ahead to the tools and technologies that can drive future advancements in MEI. Chapter 14 explores **technological and transformational enablers**, focusing on the cutting-edge tools and capabilities that can be applied within the MEI framework. By integrating transformative technologies into the SoS, the mission engineer can enhance both the execution of the MEI approach and the overall effectiveness of SoS capabilities. These innovations will not only improve operational performance but also help shape the future landscape of mission readiness and adaptability.

References

1. Ahner, D. K., Jones, N., Key, M., Adams, W., Burke, S., & Weeks, C. (2021). *A conceptual framework for the establishment of model readiness levels*. White Paper, Scientific Test & Analysis Techniques Center of Excellence (STAT COE).
2. Allaire, D. L., & Willcox, K. E. (2014). A mathematical and computational framework for multi-fidelity design and analysis with computer models. *International Journal for Uncertainty Quantification, 4*, 1–20.
3. U.S. Department of Defense. (2018). *Digital engineering strategy*. U.S. Department of Defense. Retrieved from https://ac.cto.mil/wp-content/uploads/2019/06/2018-Digital-Engineering-Strategy_Approved_PrintVersion.pdf
4. Moreland, J. D., Jr. (2014). *Mission level test and experimentation: Drive mission level performance during early development*. NSWCDD Leading Edge Technical Digest, NSWCDD-MP-15-00096 01/15, pp. 126–133.
5. Pace, D. K. (2015). Fidelity, resolution, accuracy, and uncertainty. modeling and simulation in the systems engineering life cycle, simulation foundations. In *Methods and applications* (p. 369). https://doi.org/10.1007/978-1-4471-5634-5_10

Chapter 14
Technological and Transformational Enablers

The advancement of new technologies presents an opportunity to revolutionize mission engagement strategies by focusing on gaining a decisive advantage in decision-making over adversaries. These technologies include AI-enabled decision aids, autonomous unmanned systems in military defense, enhanced passive sensors, smaller weapons, and capabilities in electronic and cyber warfare. By leveraging these tools, military forces introduce complexity and uncertainty, execute targeted attacks on critical targets, and create confusion among opponents. However, merely integrating new technologies into existing operational frameworks limits the potential to conventional thinking of the past and risks rapid proliferation among adversaries. Instead, maximizing the benefits of new technologies requires innovative operational concepts that harness the full potential of the new tools. Commercial innovation studies have shown that the effectiveness of a new technology often depends on its specific application or use case, which can determine whether it leads to incremental improvements or revolutionary advancements that redefine the entire market. Therefore, a strategic approach to combining emerging technologies with novel operational concepts is essential for achieving transformative effects in executing operational missions. Figure 14.1 illustrates the dynamic relationship between technologies and operational concepts in shaping future strategies.

In this model, sustaining technologies represent advancements that offer incremental improvements to existing systems and employ similar mechanisms to achieve operational objectives. Conversely, disruptive technologies utilize novel mechanisms to deliver value, often measured using different metrics compared to previous technologies. These disruptive technologies have the potential to significantly alter the mission landscape by introducing new capabilities and methods of achieving strategic goals. Furthermore, these new technologies can be integrated into either evolutionary or revolutionary operational concepts. Evolutionary concepts involve incremental changes to current tactics and strategies, leveraging new technologies to enhance existing approaches. In contrast, revolutionary concepts seek to fundamentally redefine operations by adopting entirely novel approaches to

J. D. Moreland, Jr., *Mission Engineering and Integration*,
https://doi.org/10.1007/978-3-032-32132-9_14

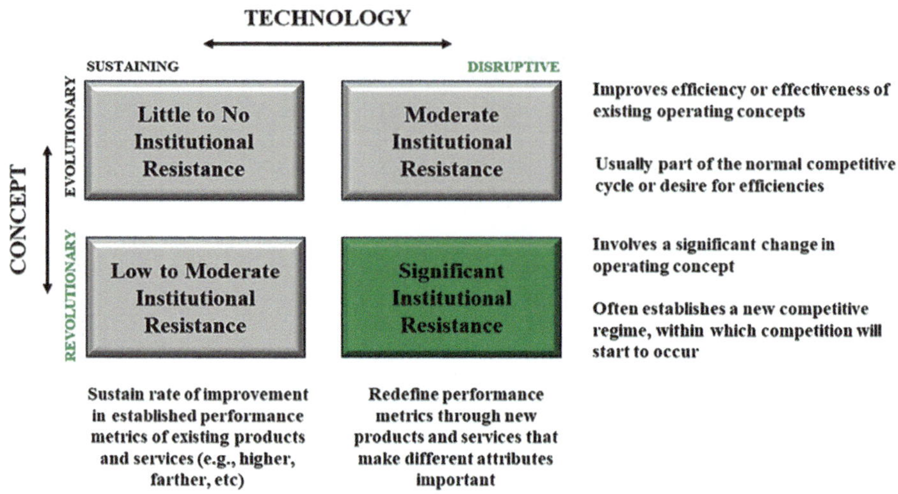

Fig. 14.1 Combining technologies with new operational concepts

achieving objectives. By combining emerging technologies with innovative operational concepts, entities can adapt to evolving threats and gain a competitive edge. Some combinations of disruptive technology and revolutionary operational concepts have led to significant innovations that established new metrics for success and ushered in new competitive regimes. It's important to note that the emergence of these new metrics is often only recognized in hindsight. Consequently, some of these innovations have been retrospectively characterized as revolutions, especially in military affairs. These revolutions signify fundamental shifts in the way warfare is conducted, often resulting from the successful integration of disruptive technologies with visionary operational concepts.

14.1 Artificial Intelligence and Machine Learning

Artificial intelligence (AI) and machine learning (ML) have been discussed in previous chapters where relevant; however, this section focuses on the fundamental concepts and key areas of advancement. The discipline of AI involves the simulation of human intelligence by machines, enabling reasoning, learning, and decision-making. Machine learning, a subset of artificial intelligence (AI), uses algorithms to learn patterns from data, enabling predictions, decisions, and automation without explicit programming for each specific task. Among ML techniques, deep learning (DL), which utilizes neural networks, has driven significant breakthroughs in fields like image recognition, speech processing, and autonomous systems. The convergence of high-performance computing, vast datasets, and ML/DL frameworks has accelerated AI advancements, solving complex problems without the need for traditional coding. Learning paradigms in AI/ML are:

- **Supervised Learning**: Supervised ML relies on labeled datasets, where inputs are paired with corresponding outputs (e.g., spam vs. non-spam emails). Algorithms learn from this labeled data to predict outcomes for unseen examples. Applications include fraud detection, medical diagnosis, and recommendation systems.
- **Unsupervised Learning**: This paradigm uncovers patterns and relationships within unlabeled data. It is frequently used for clustering, dimensionality reduction, and anomaly detection, helping to reveal hidden structures in large datasets.
- **Self-Supervised Learning**: A more recent approach, self-supervised learning generates labels from unlabeled data by predicting missing or hidden information, such as completing sentences or reconstructing images. It powers advancements in natural language processing and generative AI models.
- **Reinforcement Learning (RL)**: RL algorithms interact with environments to achieve specific goals, optimizing actions through iterative feedback. This paradigm is commonly applied in robotics, game development, and autonomous navigation systems, where adaptive decision-making is essential.

14.1.1 AI/ML Applications

AI/ML has transformed industries worldwide by automating processes, optimizing decision-making, and uncovering new insights from data. By leveraging advanced algorithms, these technologies have enabled breakthroughs across diverse fields, driving innovation, efficiency, and precision. From healthcare and finance to manufacturing and scientific research, AI/ML applications are reshaping the way organizations operate and deliver value.

Healthcare: The introduction of AI has revolutionized healthcare by enabling more precise disease diagnosis, accelerating drug discovery, and personalizing treatment plans. DL algorithms interpret medical imaging, such as X-rays and magnetic resonance imaging, with remarkable accuracy, assisting in early detection of diseases like cancer. Additionally, AI helps identify genetic markers for rare conditions and predicts patient outcomes based on complex datasets, empowering clinicians with data-driven decision-making tools.

Finance: In the financial sector, AI optimizes investment strategies, enhances fraud detection, and powers algorithmic trading. Machine learning models process vast datasets to forecast market trends, identify anomalies, and assess credit risks, providing financial institutions with predictive insights and stronger safeguards against cybercrime.

Manufacturing: In modern manufacturing, AI enables predictive maintenance, which minimizes equipment downtime and ensures operational continuity. Computer vision systems driven by ML enhance quality control by identifying defects on production lines with precision, reducing waste, and improving product reliability.

Customer Experience: Customer experiences are transformed using AI/ML which provides personalized recommendations and real-time support. Recommendation systems suggest products or services tailored to individual preferences, while AI-powered chatbots provide human-like interactions, resolving customer inquiries and enhancing satisfaction.

Scientific Research: In scientific research, AI accelerates innovation by processing and analyzing massive datasets. It facilitates advancements in material science, climate modeling, and protein folding, unlocking insights previously unattainable through traditional methods. These capabilities are driving progress in areas such as renewable energy, environmental sustainability, and medical advancements.

Defense: In the defense sector, AI/ML enhances operational capabilities by enabling intelligent surveillance, autonomous systems, and advanced decision support. AI-driven tools analyze vast amounts of data in real time for threat detection, mission planning, and resource optimization. ML models power predictive maintenance for military equipment, while autonomous systems, such as drones and robotic vehicles, improve situational awareness and reduce risks to personnel. Additionally, AI strengthens cybersecurity measures and supports training simulations, ensuring readiness in complex and dynamic environments.

AI/ML applications continue to expand, offering transformative solutions across industries and redefining how organizations approach complex challenges. By leveraging these technologies, businesses and researchers can unlock new opportunities and achieve unprecedented levels of efficiency and precision.

14.1.2 Challenges in Artificial Intelligence/Machine Learning

While AI/ML has the potential to revolutionize industries, they also present significant challenges that must be addressed to ensure their safe, fair, and effective use. From reliance on high-quality data to the need for transparency and ethical considerations, these challenges highlight the complexities of implementing AI/ML at scale. Addressing these obstacles is critical to unlocking the full potential of AI while mitigating risks and unintended consequences.

Data Dependency: The effectiveness of ML models is heavily dependent on the availability of high-quality, labeled datasets. Insufficient, biased, or unrepresentative data can lead to inaccurate predictions, perpetuating biases or resulting in unethical outcomes. Ensuring data diversity and fairness is crucial for creating models that generalize well across various applications and populations.

Computational Requirements: Training large-scale AI models requires vast computational resources, including high-performance graphics processing units and large-scale cloud infrastructures. These resource demands often create barriers for smaller organizations, limiting their ability to access and deploy advanced AI/ML capabilities. Developing more efficient algorithms and accessible computing solutions is essential to democratizing AI.

Interpretability: DL models, such as neural networks, often function as "black boxes," making it difficult to understand how decisions are made. Difficulty in interpretation is especially problematic in sensitive domains like healthcare, law, and finance, where transparency and accountability are critical. Enhancing explanation capacity in AI systems is necessary to build trust and ensure ethical use.

Ethical and Social Implications: AI's rapid adoption has raised concerns about data privacy, job displacement, and algorithmic bias. Misuse or improper implementation of AI systems leads to unfair outcomes, erosion of trust, and significant societal impacts. Addressing these issues requires a focus on fairness, accountability, and robust governance frameworks to guide the ethical development and deployment of AI technologies.

Overcoming these challenges requires collaboration across disciplines, innovative solutions, and strong regulatory frameworks. By addressing these issues, organizations can harness the transformative power of AI/ML while ensuring its responsible and equitable use.

14.1.3 Future Direction and Opportunities

The field of AI/ML is rapidly evolving, with transformative advancements poised to reshape numerous industries and society at large. As we look ahead, several emerging trends are particularly promising in terms of their potential to revolutionize AI systems' capabilities and the ways they interact with the world. These innovations offer new opportunities to solve complex problems, enhance human-AI collaboration, and create more resilient, adaptable systems.

Multimodal Learning: AI systems that can process and integrate multiple types of data, such as text, images, and audio, are set to unlock exciting new possibilities. This multimodal approach will enable the development of more context-aware applications that can understand and respond to human input in more natural and intuitive ways. For example, digital assistants powered by multimodal learning better comprehend complex queries, providing more accurate and personalized responses. Similarly, autonomous vehicles can rely on multimodal data to improve navigation and decision-making in dynamic environments, further enhancing safety and efficiency.

Real-Time Adaptation: One key challenge for AI systems is their ability to adapt to real-time changes in their environment. Advancing real-time learning will allow AI models to continuously evolve and adjust to new information, enhancing their resilience in critical applications. This capability could be crucial in fields like disaster response, where AI systems need to quickly adjust to shifting conditions and provide decision support in high-pressure situations. Real-time adaptation will also empower industries like healthcare, where AI could help monitor patients and respond to sudden changes in their condition, improving outcomes and patient care.

Quantum Artificial Intelligence: The integration of quantum computing with AI is another groundbreaking opportunity that could exponentially accelerate problem-solving in several domains. Quantum AI holds the potential to revolutionize fields such as cryptography, enabling secure communications systems that are far more robust than those traditionally available. Additionally, quantum-powered AI models could accelerate molecular modeling, drastically improving research in drug discovery and materials science. The combination of quantum computing and AI could bring about breakthroughs that are currently unimaginable with classical computing alone.

AI-Enabled Automation: AI-driven automation is already transforming industries by taking over repetitive, time-consuming tasks, allowing human workers to focus on more strategic and creative endeavors. This shift is unlocking new efficiencies and creating opportunities for humans to engage in higher level roles while it drives innovation and economic growth. In manufacturing, AI can streamline supply chains and optimize production processes, while in customer service, AI can manage routine inquiries, leaving humans to handle more complex issues. The widespread adoption of AI in automation is empowering businesses to innovate faster, while also fostering new industries and job opportunities that leverage AI to solve complex challenges.

14.1.4 Artificial Intelligence/Machine Learning Role in Mission Engineering and Integration

AI/ML is reshaping MEI by enabling the analysis, prediction, and optimization of complex systems and system-of-systems (SoS) in dynamic operational environments. At the core of AI-enabled applications are learning algorithms, which allow systems to model real-world phenomena and improve over time without explicit programming. These technologies offer unprecedented capabilities for understanding, simulating, and optimizing mission-critical systems. Machine learning has become a cornerstone for enabling autonomous and predictive decision-making in MEI. Likewise, DL has seen significant commercial success and plays a vital role in addressing complex mission challenges. Despite their origins in early computational research, ML and DL have only recently achieved widespread application due to advances in computational power and data availability. These tools are now integral in scenarios requiring real-time feedback, rapid prototyping, and iterative development and are enabling systems to adapt to evolving mission requirements. By incorporating ML and DL into MEI, organizations can move beyond traditional methods and embrace data-driven strategies for improved situational awareness, resource optimization, and mission planning.

First, let us digress to explain what is needed to develop an effective, well-integrated, interoperable system. Effective system integration and interoperability hinge on well-designed interfaces at system boundaries. Interfaces facilitate seamless communication and data exchange between components, subsystems, and systems within a broader SoS framework. To support integration and interoperability in MEI, interfaces must be designed with the following principles. Note the cases in which generative AI can assist.

Standardization: Utilize widely accepted standards and protocols, such as the Open Systems Architecture (OSA) or Application Programming Interfaces (APIs), to ensure compatibility across diverse systems. Standardized data formats and communication protocols help systems interoperate efficiently without requiring extensive customization.

Modularity: Design interfaces to support modularity, enabling individual components to be upgraded, replaced, or modified without disrupting the entire system. This approach enhances adaptability and scalability in response to evolving mission requirements.

Data Interoperability: Implement data transformation layers that translate information between systems with differing data structures or formats. Generative AI can assist in creating middleware solutions that facilitate real-time data translation and synchronization.

Semantic Alignment: Ensure that systems share a common understanding of data semantics, such as meaning, relationships, and context. Generative AI can help develop shared ontologies and simulate interactions to verify semantic alignment across systems.

Security and Resilience: Incorporate secure communication protocols, encryption, and authentication mechanisms to safeguard interfaces against cyber threats. Generative AI can simulate potential attack vectors to test and improve the resilience of system interfaces.

Generative AI plays a critical role in enhancing system interfaces for integration and interoperability. By analyzing existing system interactions, generative AI identifies inefficiencies and recommends optimized interface designs. Additionally, it simulates the behavior of integrated systems under various conditions, ensuring robust performance in dynamic mission environments. Generative AI also enables the creation of synthetic data to test interfaces at system boundaries, allowing engineers to validate functionality, identify bottlenecks, and refine communication protocols without relying on costly real-world testing. By supporting iterative development and validation, generative AI reduces the risk of integration failures and accelerates the deployment of interoperable systems.

Integrating AI Microservices into Hard-Real-Time SoS to Ensure Trustworthiness of Digital Enterprise Using Mission Engineering by A. C. Murphy and J. D. Moreland, Jr. provides research results on embedding AI algorithms directly into microservice applications to leverage machine learning, reasoning, and decision-making capabilities within each functional microservice [MURPHY21]. This research and new capability allowed for the processing of complex data, management of interacting constraints, and integration of heterogeneous knowledge sources. In these systems, AI microservices serve as key enablers for creating an integrated development environment that is designed around ontologies, rules, and reasoning, using semantic technology to structure and represent knowledge. These ontologies and rules form a formalized model of the system's domain, which can be expanded or adjusted incrementally as new data or integrations are introduced. The flexibility of this approach ensures that new knowledge is incorporated without disrupting the existing system and provides significant economic and operational advantages by reducing the risk of system overhaul and making adaptation more efficient as needs change. The use of functional programming principles in the AI microservices ensures that business logic is captured within the system's type structure, where inputs and outputs are rigorously defined. AI microservices can be deployed independently or integrated into existing enterprise-level systems via Web APIs. This facilitates seamless integration with other application services or client-server architectures. Containers and Kubernetes are instrumental in enabling the deployment and orchestration of these microservices, ensuring they can be scaled, monitored, and maintained with minimal overhead. Containers provide lightweight, isolated environments that improve the portability and efficiency of microservices. On the other hand, Kubernetes automate the deployment, scaling, and management of these containerized services, ensuring high availability and reliability across distributed systems. Together, these technologies provide a powerful framework for developing, deploying, and managing AI-driven applications that can reason about and solve complex problems in real time. The result is a SoS that can operate with greater confidence, adaptability, and reliability. Therefore, organizations can better meet mission-critical requirements while continually evolving their systems with minimal risk and maximum scalability.

14.2 Application of Category Theory

Category Theory (CT) is a branch of mathematics that focuses on relationships and structures rather than the elements themselves. It provides a high-level, abstract framework for studying how objects (systems, data, or concepts) and their relationships (processes, transformations, or mappings) interact. At its core, CT helps engineers understand not just the objects in isolation but the "edges" or connections between them. In a category, objects represent entities (e.g., systems, datasets, or functions), and morphisms (arrows) represent the relationships or interactions between those objects. The strength of CT lies in its ability to formalize these connections, making it easier to model and reason about complex interactions in a structured, systematic way. This abstraction allows engineers to focus on the relationships and dependencies between systems, ensuring that they function cohesively as part of the SoS. Some of the key concepts and terms in CT include:

Objects: Abstract representations of entities (e.g., systems or subsystems).

Morphisms (Arrows): Connections or relationships between objects that describe how one object transforms into another.

Composition: The ability to combine morphisms, creating a chain of transformations that still respects the underlying structure.

Identity: A morphism that maps an object to itself, preserving its structure.

Composition: Process that ensures that when multiple interfaces or transformations occur sequentially (e.g., Sensor → Processor → Decision System), the overall interaction remains consistent and preserves the desired behavior.

Functional Mapping: A functor maps one category (e.g., the data structure of System A) to another category (e.g., the data structure of System B) while preserving the relationships between objects and morphisms. This ensures transformations between systems remain consistent and well-defined.

Natural Transformations: These describe how one mapping (e.g., data conversion) can transition to another in a structured way, providing a blueprint for adapting interfaces as systems evolve.

Limits and Colimits: These concepts describe how systems can combine into larger SoS or how a system can be decomposed into simpler subsystems while maintaining a consistent structure.

Formalizing integration and interoperability in SoS requires a structured approach to aligning diverse systems that may have different data formats, interfaces, and operational constraints. CT provides a powerful mathematical framework for describing these relationships and ensuring compatibility across interconnected systems. Performing a functional mapping using CT guarantees data, and functional mappings between systems remain consistent and well-defined. In addition, incorporating natural transformations provides a formal blueprint for managing interoperability as systems change, ensuring that integrations remain robust and adaptable. By leveraging these principles, CT enhances the ability to integrate complex SoS, enabling seamless data exchange, modularity, and long-term system evolution without disrupting mission-critical operations.

In the context of MEI, CT provides a robust framework for modeling architectures, analyzing interfaces, and ensuring seamless interoperability. By focusing on the relationships between constituent systems rather than just individual components, CT allows engineers to design SoS architectures that support overall mission capability. One key application is formalizing system interfaces. Engineers are then better able to detect inconsistencies, redundancies, or potential points of failure that could disrupt operations. Additionally, functors and natural transformations provide a structured approach to mapping data formats and processes across different systems, ensuring smooth integration and communication. This mathematical foundation also enhances adaptability, as it allows systems to be added, upgraded, or replaced without compromising the integrity of the SoS. By preserving the structure of interactions, CT safeguards cohesive and effective mission-critical operations even as technologies and operational requirements evolve over time. The authors in "The Application of Applied Category Theory to Quantify Mission Success"

[GARRETT23] introduced the application of CT to MEI as a framework for modeling and integrating complex SoS. Their work highlights the unique challenges of "wicked problems," where traditional engineering approaches struggle due to unstructured, context-dependent, complex missions combined from a set without regard to order.

The current emphasis on CT stems from its ability to address the increasing complexity, interdependencies, and dynamic nature of modern SoS. Traditional systems engineering (SE) methods that focus on optimizing individual systems using linear models and decision trees are proving inadequate for contemporary SoS environments across industrial sectors. These contemporary environments require decentralized control, and their evolving interactions demand a more abstract and flexible mathematical framework. Recent advancements in computational tools, including graph databases and symbolic computation platforms, now enable the practical application of CT for modeling and simulating SoS at scale. Additionally, the shift from system-level optimization to edge-focused interoperability has driven interest in CT's ability to rigorously model interactions between constituent systems, ensuring seamless integration and adaptability. The emergence of wicked problems, such as climate modeling and global supply-chain disruptions, has further highlighted the limitations of traditional engineering approaches, making CT a valuable tool for addressing uncertainty and integrating diverse mathematical frameworks. Moreover, the need for modular, scalable, and interoperable solutions in SoS aligns with CT's use of functors, natural transformations, and composition to create flexible, reusable models. Finally, interdisciplinary collaboration across engineering, mathematics, computer science, and other soft sciences has accelerated CT's adoption, as it provides a universal language for bridging domain expertise. With these factors converging, CT is now recognized as an essential tool for managing the dynamic and interconnected nature of SoS, offering a formalized approach to integration, adaptation, and mission success.

14.2.1 Advancement of Category Theory

The Defense Advanced Research Projects Agency (DARPA) is exploring the application of CT to address the challenges associated with complex adaptive systems and SoS. One notable initiative in this area is the Complex Adaptive System Composition and Design Environment (CASCADE) program. This program was launched by DARPA to enhance the modeling and design of intricate interconnected systems. CASCADE aims to develop novel mathematical techniques that provide a deeper understanding of component interactions and a unified view of system behaviors while seeking to create a formal language for composing and designing complex adaptive systems with a focus on:

- Applied mathematics, particularly category theory, algebraic geometry, topology, and sheaf theory

- Operations research, control theory, and planning, especially in stochastic and non-linear control
- Modeling applications responsive to challenges in medicine logistics, adaptive logistics, reliability, and maintenance
- Search and rescue platforms and modeling
- Adaptive and resilient urban infrastructure

By leveraging CT, CASCADE aims to develop comprehensive response models that account for the many components and interactions inherent in complex missions, whether in urban or austere environments. This approach could inform the design of future forward-deployed surgical capabilities by making sure the functions, structures, behaviors, and constraints of the medical system, composed of surgeons, helicopters, communication networks, transportation, time, and blood supply are accurately modeled and understood. In addition, DARPA has funded research projects that apply CT to various domains. For instance, in 2004, mathematicians were awarded grants to relate the geometric Langlands program to quantum field theory, reflecting DARPA's interest in exploring the potential applications of advanced mathematical theories in complex systems. The Langlands program established a profoundly intricate network of relationships across different branches of mathematics, offering new perspectives and innovative solutions to long-standing problems. Through these initiatives, DARPA is at the forefront of integrating CT into the design and analysis of complex adaptive systems, aiming to enhance resilience and adaptability in dynamic and unpredictable environments.

14.2.2 Category Theory Serves as Glue Across Tools

Solving integration and interoperability problems within the MEI trade space requires the combination of multiple tools to decompose and structure complex SoS with an emphasis on the interfaces. Functional Flow Block Diagrams (FFBDs), N^2 matrices, and CT, as described herein, are complementary tools that help design, analyze, and manage complex SoS. Each provides a distinct perspective on system behavior and interactions, and their relationships can be understood as follows:

1. **Functional Flow Block Diagrams: Behavioral Sequencing**

   Functional Flow Block Diagrams (FFBDs) are used to represent the sequential functional flow of a system, illustrating the logical progression of tasks or actions required to achieve mission objectives. They are especially useful for visualizing workflows and dependencies between functions.

   - **Purpose**: In MEI, FFBDs focus on capturing *what functions need to occur and in what order*. They help engineers analyze operational sequences and dependencies within a mission thread.
   - **Limitations**: While FFBDs excel at representing behavior and sequence, they do not explicitly capture system interfaces or interactions between subsystems.

- **Functions**: Each function in an FFBD corresponds to an object in CT.
- **Arrows**: In the FFB, arrows between functions represent morphisms in CT, capturing the flow of data or control.
- **Composition**: Composition in CT mirrors the sequential flow of functions in FFBDs, ensuring that a sequence of actions forms a consistent whole.

2. **N^2 Matrix**: **Interface Mapping**

   The N^2 matrix complements FFBDs by focusing on interfaces and interdependencies between system components. It is a square grid where:

   - **Purpose**: In MEI, the N^2 matrix provides a clear picture of how systems interact at their boundaries, highlighting communication, dependencies, and integration points.
   - **Limitations**: A lack of a temporal or sequential aspect, which is critical for understanding how functions evolve during a mission.
   - Each system element in the N^2 matrix corresponds to an object in CT.
   - The interactions or interfaces between elements (off-diagonal cells) are morphisms in CT.
   - The CT framework ensures that these interfaces are mathematically consistent, enabling modularity and adaptability.

3. **Category Theory**: **Abstract Structural Representation**

   Category Theory (CT) provides a mathematical framework to unify and analyze the relationships captured by FFBDs and N^2 matrices. It abstracts the systems and interactions into objects and morphisms:
   - **Purpose**: In MEI, CT enables engineers to reason about both the functional flow (FFBD perspective) and interactions/interfaces (N^2 matrix perspective) within a single unified framework. By focusing on relationships and composition, CT helps identify patterns, ensure consistency, and formally validate system behaviors.
   - Use of FFBDs emphasizes sequential behavior, while N^2 matrices focus on interactions at boundaries. Application of CT bridges these perspectives by allowing engineers to formalize both the flow of actions and the relationships between system interfaces in a cohesive structure.
   - Using CT, engineers can represent a mission thread as a graph or string diagram, where nodes are functions (FFBDs), edges are interactions/interfaces (N^2 matrix), and the entire graph adheres to compositional rules.

Application in MEI

In MEI, the design of SoS often requires:

1. **Behavioral Modeling (FFBDs)**: Used to understand how functions flow across mission threads and mission engineering threads.
2. **Interface Analysis (N^2 Matrix)**: Ensures systems communicate effectively and interoperate seamlessly.
3. **Formal Validation (CT)**: Validates mathematically that the functions, flows, and interfaces align with mission objectives and constraints.

By integrating FFBDs, N^2 matrices, and CT:

- Engineers gain a holistic view of mission dynamics, including both sequential operations and system interactions.
- Mathematical rigor is provided by CT to ensure the system's design is modular, scalable, and adaptable, reducing risks during integration and deployment.

MEI Example

Imagine a disaster response SoS involving drones, ground sensors, and a command center:

- **FFBD**: Models the sequence of actions (e.g., drone reconnaissance → sensor data collection → command decision).
- **N^2 Matrix**: Captures interfaces (e.g., data transfer between drones and sensors, communication between sensors and command) in network graph.
- **CT**: Provides a unified framework to represent the flow of data (from FFBD) and interfaces (from N^2 matrix). This ensures compositional integrity across the SoS. For example, it can validate that sensor data matches the command center's required input format for a seamless interaction.

Aggregating multiple network graphs from monoplex to multiplex structures enables a more comprehensive representation of complex systems by integrating different layers of interactions. A monoplex network is a single-layered graph where nodes represent entities of interest, and edges capture a single type of relationship between them, often aggregating multiple interactions into a unified structure, which can lead to a loss of nuanced information about the system's complexity. In contrast, a multiplex network preserves multiple distinct relationship types by organizing them into separate layers, where each layer maintains its unique topology. This allows inter-layer edges to support integrated analyses without collapsing different interaction modalities into a single network. Networks are now fundamental across diverse fields such as biology, social sciences, and transportation because they provide an intuitive data structure for storing relationships between entities. For example, in genomics, a network might represent genes as nodes, with edges indicating functional or regulatory relationships based on gene expression patterns. However, biological systems often involve multiple types of interacting entities, such as genes, proteins, and metabolites, forming a heterogeneous network rather than a standard monoplex structure. Heterogeneous networks contain multiple node types and edge types, better capturing the complexity of real-world interactions. The power of networks lies in their ability to facilitate advanced analyses, such as network exploration, ML, and DL algorithms. Centrality measures can identify key nodes, and propagation algorithms can infer node similarities. Graph neural networks can leverage network topology for predictive tasks. However, many real-world systems involve multiple types of relationships between the same entities, necessitating more sophisticated models.

A simple approach to handling multiple relationship types is to merge separate networks into a single monoplex network by combining the edges across layers.

This can be represented as a weighted network where edges may be assigned weights based on their frequency or average strength across layers. However, this aggregation process loses the distinct topology of each individual layer, potentially obscuring meaningful patterns. A more effective approach is to maintain the integrity of each network layer within a multiplex network, where each layer retains its unique structure but is connected to other layers through inter-layer edges. These edges are typically identity connections between the same node across layers. They act as tunnels facilitating information flow while preserving layer-specific topology. This design enables integrated analyses using algorithms like Random Walk with Restart on supra-adjacency matrices. In contrast to multiplex networks, heterogeneous networks explicitly model different node types, making them more suitable in cases where different entity classes (e.g., genes, proteins, and diseases) interact in complex ways.

The relationship to N^2 matrices emerges as these network graphs structure the interactions between system components, aiding in visualizing dependencies and integration points. When initially treated as a monoplex, an N^2 matrix can be translated into adjacency matrices or directed acyclic graphs, where nodes represent components and edges define functional, informational, or dependency relationships. These graphs often resemble hierarchical networks, bipartite graphs (when distinguishing between different types of entities, such as components and interactions), or flow networks (when processes and dependencies are directional). However, treating an N^2 matrix as a monoplex may obscure the layered nature of interactions, making it difficult to distinguish between different types of dependencies such as structural, functional, and informational links.

CT provides a rigorous mathematical framework for overcoming these limitations by modeling multi-layered interactions explicitly. Through categorical constructs such as functors and natural transformations, the engineer can systematically map entities and relationships across network layers, ensuring structural consistency and preserving relational properties. Applying CT to multiplex and heterogeneous networks allows development of more expressive models for complex systems. The result is enhanced decision-making and interoperability in fields such as MEI, biological network analysis, and AI-driven knowledge discovery.

14.3 Digital Engineering

The history of digital transformation is deeply intertwined with the evolution of industrial revolutions, each marked by technological breakthroughs that reshaped industries and economies, as depicted in Fig. 14.2. The First Industrial Revolution was driven by the advent of mechanization, steam power, and the rise of factories, marking the transition from handcrafted production to machine-based manufacturing. The Second Industrial Revolution introduced mass production powered by electrification, significantly enhancing manufacturing efficiency and scalability. This period laid the groundwork for modern industrial processes. The Third

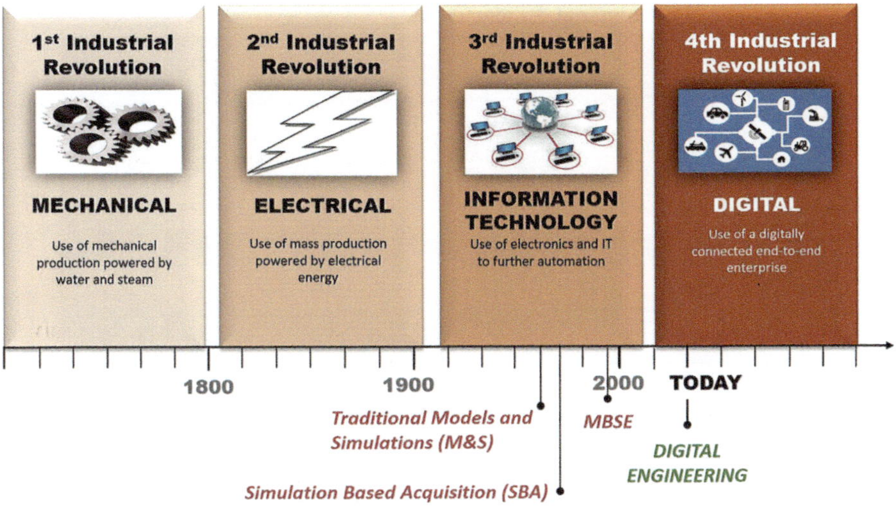

Fig. 14.2 Industrial revolution history

Industrial Revolution emerged with the advent of computers and information technology, enabling automation, digital data processing, and early forms of networked communication, which revolutionized businesses and scientific research. Now, we are entering the Fourth Industrial Revolution (Industry 4.0), where digitalization fundamentally transforms innovation and operations through interconnected smart systems, AI, and cyber-physical integration. The DoD has recognized the transformative power of digital technologies, with its M&S initiatives evolving into digital engineering (DE) which is an integrated, model-based approach that enhances engineering practice across the entire lifecycle of systems. However, challenges persist, particularly in Simulation-Based Acquisition (SBA), where an overemphasis on simulation alone has proven misleading and confusing, much like challenges in Model-Based Systems Engineering (MBSE). Despite these obstacles, DE represents a paradigm shift, leveraging technological innovation to optimize complex systems and enhance decision-making.

Digital engineering (DE) leverages the use of computer models and data to integrate engineering tasks across disciplines, virtually, before entering production. The aim is to modernize and achieve efficiency and consistency in engineering to reduce product development cycle time, improve product quality, increase customer satisfaction, and enable growth. DE is a major thrust within both the government and industry to transition away from the traditional document-centric process of maturing a system from initial conception through sustainment into a much more dynamic, efficient, and flexible DE process driven by the authoritative source of truth data as illustrated in Fig. 14.3.

This digital transformation moves away from using documents and spreadsheets as the technical baseline in favor of linked models and digital artifacts. The ability to assess practices, processes, and product concepts in a virtual environment has

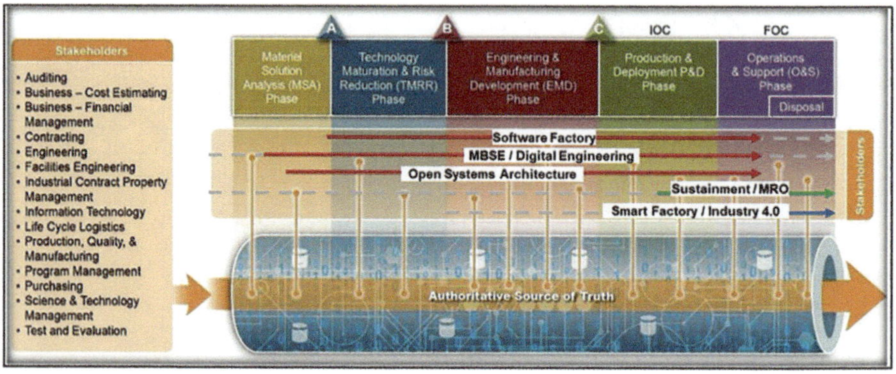

Fig. 14.3 Digital transformation across the acquisition lifecycle

generated the use of new terms, i.e., "digital twin" and "digital thread" (extensive discussion of these terms is in Sect. 14.3.3). Simply stated, the digital twin is the current representation of a product or system representing an organization's machines, controls, workflows, and systems. The digital thread in a relation is a record containing digital artifacts of a product or systems lifeline from its creation to its removal. Both potentially offer tremendous future benefits for operating models, revenue streams, and relationships. Implementing the digital twin within Model-Based Engineering (MBE) inclusive of MBSE, Model-Based Design (MBD), Model-Driven Software Development (MDSD), Model-Based Manufacturing (MBM), and Model-Based Sustainment (MBS) enables authoritative technical data, software, information, and knowledge bringing decision-makers the right information to make critical decisions. This positive shift in the development capability drives down the cost needed for "re-work," which then exponentially increases the ability to discover subsequent problems later in the lifecycle. The ultimate vision of a DE strategy should be to realize a single digital representation of a system under design, where each component is accurately represented using analytical and descriptive models that can be accurately and effectively traced to the initial set of mission and requirement definitions across a digital thread, as shown in Fig. 14.4.

Development of complex SoS requires a structured process using an MBD approach to capture the interdependencies as well as the behaviors of the constituent systems and the networks interconnecting these systems. Whether for the development of controls for a flight system, an industrial robot, a wind turbine, a production machine, an autonomous vehicle, an excavator, a sensor, or an effector, the manual effort of writing code and using document-based requirements capture is laborious and inadequate. This approach leads to answering critical questions for design through a trial-and-error approach or testing using a physical prototype. The downfall to this approach occurs when a single requirement changes causing the entire system to require recoding and rebuilding. This may delay the project by days or even weeks with rippling effects on the SoS. Rather than using handwritten code and documents, a model-based design model can incorporate the physical model, the control algorithms, and the operational environment. With this digital model, a

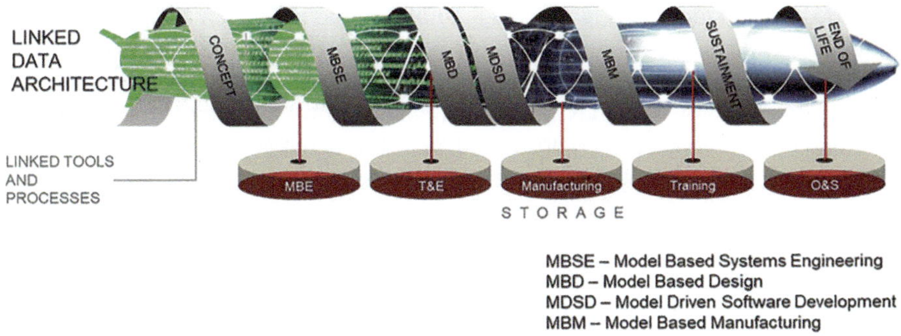

Fig. 14.4 Digital flow of authoritative data across tools and processes

simulation can be run at any point to acquire an instant view of system behavior. The engineer can then test out multiple "what-if" scenarios and trade-off analyses without risk, without delay, and without reliance on costly hardware.

In MBD, rich text requirements are captured, analyzed, and managed with custom attributes by linking them to designs, code, and tests. In addition, requirements can also be imported and synchronized from external sources such as requirements management tools. When design requirements change, the specific part of the design or the test directly affected by the change is highlighted with notification so appropriate action can be taken to address the change. To enhance this MBD approach, dynamic analysis is performed by using executable models composed from static architectural views on multi-domain operations across SoS for requirements validation and verification. MBD thus provides insights into system-level behavior and performance that cannot be obtained by static analysis alone. The number of ideas that can be explored is virtually limitless. Requirements, system components, intellectual property, and test scenarios are all captured in the model. Because the model can be simulated, the engineer can investigate design problems and questions long before building expensive hardware. Rapid evaluation of multiple design ideas, exploring trade-offs, and employing sensitivity analysis amplify insights on design change effects on the individual systems as well as the SoS. Likewise, MBD allows the engineer to generate code directly from the design model instead of writing thousands of lines of code by hand with the model serving as a bridge between the software engineers and the control systems engineers. The generation of real-time code is used for rapid prototyping or production. Rapid prototyping in turn provides a fast, inexpensive way to test algorithms on hardware in real time and perform design iterations in minutes rather than weeks. With the same rapid prototyping hardware and design models, this enables engineers to conduct hardware-in-the-loop testing and other test and verification activities to validate hardware and software designs before production. Production code generation converts the model into the actual code that will be implemented on the production's embedded system. The generated code is then optimized for specific processor architectures and integrated with handwritten legacy code.

MBM is also identified in Fig. 14.4 within digital ecosystems as a path toward the Fourth Industrial Revolution, Industry 4.0. Industry 4.0 is the ongoing automation of traditional manufacturing and industrial practices, using modern smart technology. It generally includes sourcing (extending to the supply base) through manufacturing and delivery. Industry 4.0 provides manufacturing and test-data feed to the product lifecycle digital thread and consumes models and the data from DE in addition to providing data for engineering feedback. Besides elements of full automation, Industry 4.0 includes substantial increases in human-assisted tasks, from data analysis, delivery and visualization to collaborative robotics, and human-assisted digital and physical automation to improve the efficiency and human experience of factory operators and leaders. Lastly, in the operations and support phase of the system lifecycle, digital sustainment is being implemented to modernize and develop common, scalable solutions for maintenance, prognostics, and training as it drives increased operational availability and lower lifecycle costs for fielded products.

The resulting model-based acquisition package, including digital artifacts that link mission operations and capabilities to enterprise-level architectures, will improve a user's ability to visualize, work, develop, share, and use additional model features through improved digital artifact portability. As identified in Fig. 14.4, other areas of digitization include Model-Driven Software Development (MDSD) and Model-Based Manufacturing (MBM). Software uses the same principles of digital sharing to program critical code as part of MDSD. This allows for the development of architectures with the ability to scale to large applications like SoS much like interconnecting Lego blocks. It is especially important to quickly compose a solution that is compliant with open standards while using commercial development practices (such as DevSecOps) in cloud-based environments. This allows developers to sit and develop anywhere. DevSecOps is an organizational software engineering culture and practice that integrates development (Dev), security (Sec), and operations (Ops) into a unified, continuous process, see Fig. 14.5 [DOD19]. It is a preferred approach for delivering secure and resilient software, emphasizing automation, integration, and continuous learning to improve software quality and velocity. By embedding security throughout the development pipeline and leveraging advanced technologies such as AI, machine learning, and digital twins, DevSecOps enhances cybersecurity, facilitates continuous integration and delivery, and supports rigorous SE practices.

These model-based artifacts enable increased traceability and transparency to authoritative sources of data and truth about systems and serve as a mechanism to allow for the detection of errors, inconsistencies, and broken linkages earlier in the development lifecycle. Digital transformation spans the entire product lifecycle, revolutionizing how capabilities are conceived, developed, tested, deployed, and sustained as illustrated in Fig. 14.6. Within the framework of a digital acquisition strategy, this transformation enables an integrated, data-driven approach that enhances decision-making, accelerates development timelines, and optimizes resource allocation. By leveraging digital engineering, MBSE, and digital twins, organizations can simulate, analyze, and refine designs before physical production.

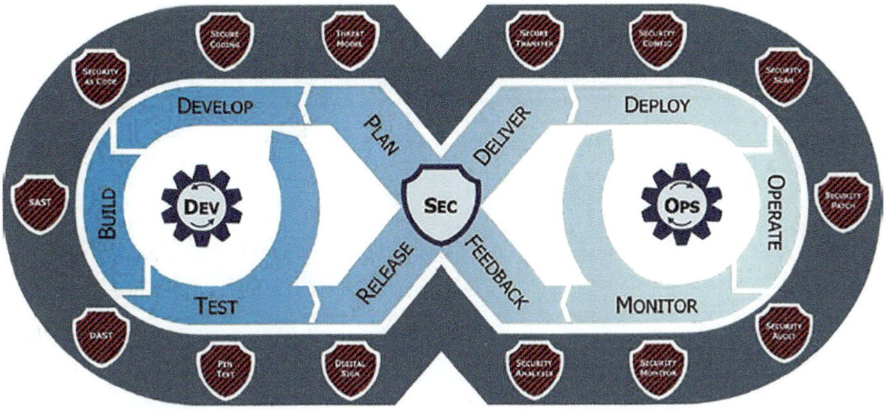

Fig. 14.5 DevSecOps process for continuous integration and deployment

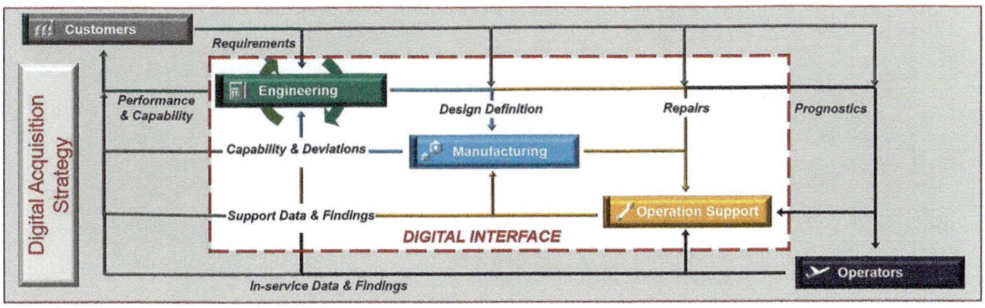

Fig. 14.6 Digital transformation spans the product lifecycle

Throughout production and sustainment, digital threads ensure continuity of data, enhancing interoperability and lifecycle management while supporting predictive maintenance and agile updates.

This end-to-end digitalization fosters greater flexibility, responsiveness, and mission effectiveness, ensuring that systems remain adaptable to evolving operational needs and technological advancements. Table 14.1 provides characteristics of today's state and the future desired end state of this digital ecosystem.

14.3.1 Digital Engineering Strategy

In general, a DE strategy redefines the approach to SE by integrating advanced digital technologies and practices. Digital engineering is described as an integrated digital approach using authoritative sources of system data and models as a continuum across disciplines to support lifecycle activities from concept through disposal. This strategy builds upon and modernizes traditional SE by leveraging computational technology, modeling, analytics, and data science to address the growing

Table 14.1 Characteristics of today and future state digital ecosystem

From		To
Disparate, disconnected tools and platforms	►	Integrated Digital Thread where data can easily flow from one tool to another
Manual data transfers and conversions throughout the lifecycle	►	Automated data transfer and conversions to maintain semantics
Data is hard to find, stale, and not under configuration management	►	Data easily accessible to all and users notified of all updates
Lack of consistent model of entire system that includes all disciplines	►	Federated models from complete system model which is rigorously managed

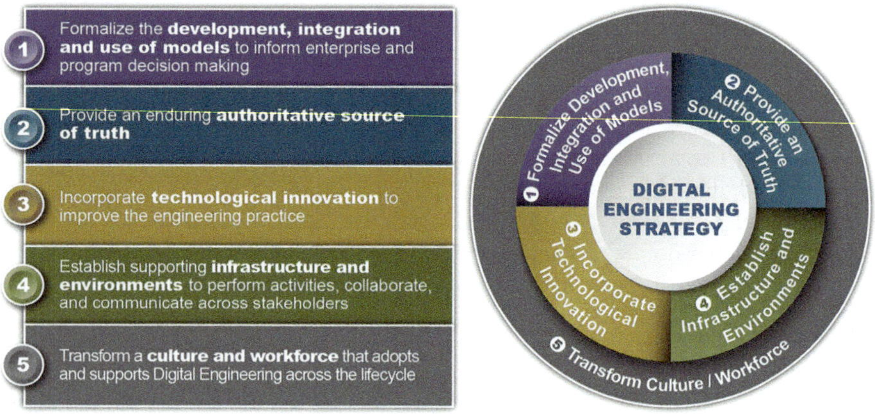

Fig. 14.7 Five general goals for digital engineering strategy

complexity in acquisition and operations. It provides a foundation for more agile, informed, and efficient decision-making by enabling the seamless integration and reuse of digital artifacts throughout the system lifecycle. Across the industrial sectors, the shift to DE reflects recognition of its necessity to meet the demands of modern acquisition and sustainment processes, ensuring that society remains prepared and adaptable in an increasingly complex world. Common goals for the implementation of DE are captured in Fig. 14.7 [DESTRAT18].

Goal #1: Formalize the development, integration, and use of models to inform enterprise and program decision-making

The first goal aligns closely with the core principles of model-based approaches where models (1) serve as an integral part of the technical baseline; (2) evolve continuously throughout the project lifecycle; (3) are integrated across various project disciplines; and (4) can be shared and reused across projects, including all stakeholders.

These elements form a foundational aspect of DE, with models serving as powerful tools for representing systems, phenomena, entities, or processes. In the early phases of the lifecycle, models allow for the exploration of a system within a virtual environment before physical construction begins. As the lifecycle progresses, high-fidelity physics and engineering models closely mirror their physical counterparts. These models enable the analysis, simulation, characterization, and visualization of systems, processes, and phenomena across disciplines. When kept up to date with the project's development, they serve as the most accurate and current record of the project design. Digital engineering extends beyond traditional model-based approaches that typically focus on a particular activity or aspect (e.g., MBD, MBM, and model-based X) of the lifecycle. From MBSE, the focus is on formal systems modeling across the lifecycle, regardless of the activity being performed. The vision for DE is to encompass the broad spectrum of models as a continuum across the lifecycle of activities. Some challenges to overcome for the implementation and accomplishment of this goal are:

- **Model Integration**: Models are not developed or used across domains, acquisition phases, and programs.
- **Model Curation**: Models are not curated such that information can be preserved, discovered and used across the lifecycle.
- **Model Credibility**: Traditional verification, validation, and accreditation approaches do not account for model credibility and trust in the digital age.

Goal #2: Provide an enduring authoritative source of truth

The second goal focuses on establishing an authoritative source of truth that enables stakeholders across organizations and locations to access, manage, analyze, utilize, and distribute models effectively. Building on Goal #1, this objective integrates digital models into a unified source of authoritative information. It transitions the primary mode of communication from static, disconnected documents to dynamic digital models and data, fostering greater consistency and collaboration. Some challenges to overcome for the implementation and accomplishment of this goal are:

- **Authoritative Data**: Vast amounts of data are scattered across multiple stovepiped systems and organizations in various forms.
- **Governance**: Managing and controlling data sources are fragmented or ad hoc.
- **Digital Artifacts**: Digital artifacts are exchanged in a document-based culture.

Goal #3: Incorporate technological innovation to improve the engineering practice

The third goal moves beyond traditional model-based approaches by embracing technological innovations that enable a seamlessly connected, end-to-end digital enterprise. Models play a pivotal role in driving advancements in digital technologies, unlocking transformative capabilities. As technology continues to evolve rapidly, this goal emphasizes leveraging and integrating new innovations as they emerge to enhance and modernize engineering practices. Some challenges to overcome for the implementation and accomplishment of this goal are:

- **End-to-End Solutions**: DE activities are disjointed across the lifecycle.
- **Engineering Practice Innovation**: The way engineers leverage technology is transformed to be responsive to change.

Goal #4: Establish a supporting infrastructure and environments to perform activities, collaborate, and communicate across stakeholders

The fourth goal aims to advance and integrate DE computing infrastructures and environments to support the preceding goals effectively. Currently, organizations and programs operate diverse, mission-specific infrastructures that are often complex, costly, and challenging to manage, secure, and support due to their activity-specific nature. To address these challenges, efforts should focus on transitioning toward a more unified, collaborative, and trusted environment. This evolution must deliver capabilities that keep pace with technological advancements, strengthen cybersecurity and intellectual property protections, and enhance information sharing among stakeholders. Some challenges to overcome for the implementation and accomplishment of this goal are:

- **IP and Critical Technology Protection**: Strategies for protecting and securing the integrity of classified and proprietary digital data are limited.
- **Information Technology (IT) Infrastructure**: These infrastructures are not designed for complex digital model-based engineering activities.
- **Methods, Tools, and Processes**: Current methods, tools, and processes do not holistically support the DE activities from design to disposal.

Goal #5: Transform a culture and workforce that adopt and support digital engineering across the lifecycle

To excel in DE, organizations must foster a culture and workforce that actively embrace digital approaches. Digital engineering transforms how solutions are prototyped, experimented with, and tested in virtual environments before being delivered to the customer. This shift involves adopting an authoritative source of truth, transitioning from static documents to dynamic models, and leveraging former digital methods, processes, and tools. Furthermore, DE impacts related business functions such as contracts, legal compliance, and financial management. By enabling greater sharing of digital models both within and across industrial sectors, DE enhances collaboration and drives innovation. Some challenges to overcome for the implementation and accomplishment of this goal are:

- **Workforce Skills Training**: There are limited incentives for workforce skills development, insufficient training capacity, and resources to meet the demand.
- **Policy, Guidance, and Standards**: Few policies, guidance, and standards exist to comprehensively address DE activities.
- **Metrics**: A common set of metrics is lacking to serve as leading indicators of adoption and effectiveness.

14.3.2 Digital Transformation

Digital transformation signifies a fundamental shift in how organizations leverage technology, people, and processes adapt to a rapidly changing digital environment. It is a comprehensive integration of digital technologies across all areas of an organization, resulting in profound changes in operations, culture, and value delivery to stakeholders. At its core, digital transformation extends beyond simply adopting new tools; it requires rethinking traditional business models to unlock the full potential of data, automation, and connectivity. This transformation is essential in an era of accelerating technological innovation and heightened global competition, where agility, efficiency, and resilience are critical for success.

Digital transformation also embraces strategically leveraging technology to achieve information superiority and enhance mission performance. This includes applying DE principles across systems and programs to holistically support mission execution. The transformation is guided by fundamental principles that emphasize the use of digital models as central tools for system development and management. Digital models enable various program uses, such as requirements analysis, architecture development, design and cost trades, system evaluations, and sustainment planning, among others. By employing approved models and simulations, organizations can address challenges such as interoperability, integration, and SoS complexity across the acquisition lifecycle.

To ensure successful digital transformation, organizations must establish a model-centric approach that integrates authoritative technical data and artifacts throughout a system's lifecycle. These digital system models serve as collaborative products of mission and design engineering efforts, ensuring that data generated and consumed during program activities is interconnected and readily accessible. Embracing digital transformation promotes the ability to adapt to present and future challenges, unlocking new opportunities for innovation and operational excellence.

14.3.3 Digital Engineering Capability

A digital engineering (DE) capability represents a transformative approach to designing, developing, and managing complex systems across their entire lifecycle. Leveraging cutting-edge technologies, data-driven methodologies, and secure ecosystems enables seamless integration of digital tools, processes, and collaboration among stakeholders. This capability not only bridges the phases of the acquisition lifecycle but also ensures the fidelity and relevance of digital models through continuous updates and validation. By fostering innovation and operational efficiency, a DE capability empowers organizations to optimize performance, enhance security, and drive informed decision-making, ensuring success in the dynamic and demanding environments of modern defense and industry.

A commitment to fostering the successful implementation of DE across programs by iteratively developing a robust digital engineering capability has become a necessity in today's complex environment. Objectives for the design of this digital engineering capability include:

1. **Enhance Accessibility and Usability**: Enable individuals and organizations throughout the ecosystem to access and utilize DE resources. This includes personnel responsible for requirements development, architecture, capability design, testing, evaluation, operation, training, production, and sustainment of systems across their lifecycles.
2. **Integrate Acquisition Phases**: Connect the various phases of the acquisition lifecycle, facilitating the seamless flow of information and feedback across acquisition activities and processes.
3. **Ensure Model Fidelity and Management**: Support the development, validation, and maintenance of technically accurate digital systems and models. These models must have appropriate fidelity levels to simulate operational environments effectively during testing activities.
4. **Incorporate Operational Security and Automation**: Include processes, infrastructure, and software for the integration of security into all phases of the software development lifecycle to include operations and automated testing to streamline the following functions:
    (a) Verification and validation of systems and operational requirements.
    (b) Automated security testing, such as vulnerability scanning and penetration testing, adhering to advanced data security practices like zero trust.
    (c) Timely, repeatable, and iterative software distribution to operational environments.
5. **Secure the Digital Ecosystem**: Operate within a secure ecosystem that leverages policies, standards, and best practices to maintain confidentiality, integrity, and availability of the DE capability and data. This includes:
    (a) Advanced security measures to mitigate risks associated with aggregating and sharing DE information.
    (b) Compliance with privacy regulations for environments handling personally identifiable information.
6. **Meet Operational Security Standards**: Adhere to operational security requirements.
7. **Align with the Data Strategy**: Support the data strategy by ensuring data is visible, accessible, understandable, linked, trustworthy, interoperable, and secure.

The DE capability is built upon a cohesive set of integrated elements that collectively support the development, deployment, and sustainment of complex systems. These elements are outlined below composed of infrastructure, tools, processes, and collaborative ecosystems necessary to enable a robust DE environment:

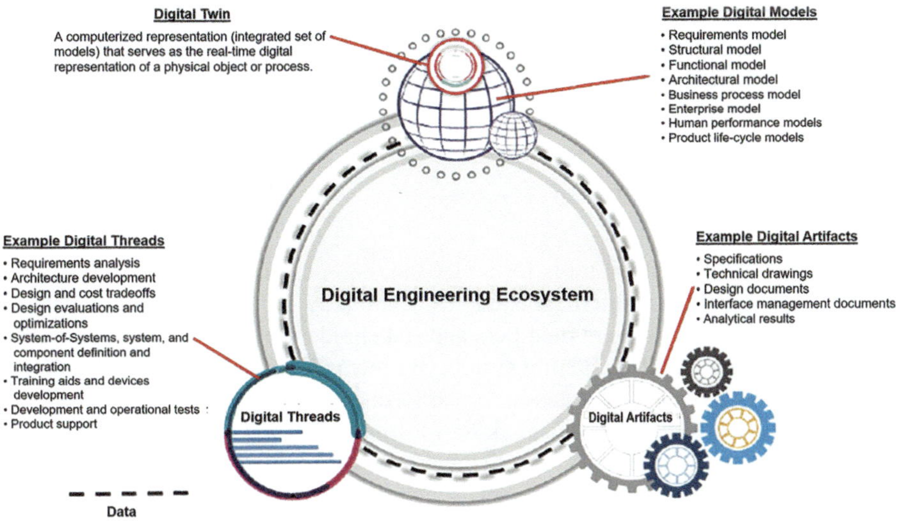

Fig. 14.8 Digital engineering ecosystem

1. **Infrastructure and Architecture**: The digital engineering ecosystem relies on a robust infrastructure that supports automated system development, design, testing, evaluation, production, operation, training, and sustainment throughout the defense acquisition process. These infrastructure elements reside in the central circle of Fig. 14.8 and are a necessity in making the ecosystem function. Key assets include:

   (a) Hardware
   (b) Software
   (c) Networks and Cloud Services
   (d) Tools
   (e) Skilled Workforce

2. **Processes and Methods**: The ecosystem incorporates stakeholder approaches to DE, including processes, methods, and practices that facilitate the extraction of knowledge through digital threads and artifacts derived from digital models. These processes and methods also reside in the central circle of Fig. 14.8 and are foundational to making the ecosystem function.
3. **Collaboration and Integration**: A digital engineering ecosystem fosters collaboration across stakeholders, including government entities, contractors, suppliers, and operators. These collaborative digital environments enable the development of models, execution of simulations, and optimization of digital models and twins, forming a comprehensive digital thread.
4. **Feedback Mechanisms**: The ecosystem integrates feedback loops to maintain an authoritative source of truth, ensuring the continuous improvement of digital models and systems.

As Fig. 14.8 depicts, with well-established and operational elements, the DE capability will empower the enterprise to innovate, optimize processes, and achieve superior outcomes across systems' lifecycles.

14.3.4 Digital Models

Digital models are critical tools for understanding, designing, and managing complex systems and their interdependencies. They play a central role in facilitating communication among team members and stakeholders, ensuring clarity and alignment throughout the lifecycle of a system. By integrating simulations and analyses, digital models provide a foundation for exploring concepts, evaluating system characteristics and alternatives, and making informed decisions. When enriched with real-world data, these models form the authoritative source of truth, enabling accurate assessments of overall system performance.

Effective configuration control is essential for maintaining the integrity and reliability of digital models and digital twins. These models, along with their associated information and data, must be traceable across all phases of the system lifecycle, including operational capabilities, requirements, design constructs, production, testing, training, and sustainment. Ensuring traceability not only supports long-term functionality and sustainability but also facilitates upgrades and cost-effective management. Programs should carefully consider data usage and rights during planning, acquisition, and contracting phases to secure the necessary access and ensure that the system will be affordable and upgradable. Verifying and validating digital model baselines prior to technical milestones is an additional critical step in maintaining accuracy and system alignment.

Digital models come in various forms, each tailored to specific aspects of system design and management. Key types of digital models and their functions include:

- Requirements Models: Capture and refine operational and technical requirements.
- Structural Models: Define the physical and organizational architecture of a system.
- Functional Models: Depict operational processes and behaviors.
- Business Process Models: Represent workflows and organizational processes.
- Architecture Models: Define the framework for system integration and interaction.
- Enterprise Models: Illustrate organizational structures and strategies.
- Physics-Based Models: Simulate physical phenomena for system validation.
- Human Performance Models: Evaluate human interactions with the system.
- Threat Models: Analyze potential risks and vulnerabilities.
- Product Lifecycle Models: Manage information across the lifecycle of a product.

A powerful evolution of digital models is digital twins. Initially developed during research and development phases, digital models can evolve into fully functional digital twins once physical counterparts are deployed, enabling ongoing refinement and performance improvements. They serve as dynamic, virtual counterparts to physical products, systems, or processes. These twins integrate advanced models, real-time sensor data, and inputs to replicate and predict system behavior throughout the lifecycle of their physical counterparts. They ensure continuous data synchronization, allowing them to remain up-to-date and reflective of changes in the physical system. Digital twins meticulously replicate all aspects of their real-world counterparts, including instruments, sensors, environments, and components, providing a controlled platform for testing concepts and conducting experiments before real-world implementation. Unlike traditional simulations, digital twins incorporate machine learning, enabling iterative self-improvement and enhanced accuracy over time. The concept dates back to the 1990s and was formally introduced to manufacturing by Dr. Michael Grieves in 2002. Digital twins then gained prominence through NASA's use in space exploration missions. Advancements in data management, high-performance computing, and sensor technology have expanded digital twins' scalability and applicability across industries. Notably, digital twins differ from simulations in their ability to perform concurrent analyses, offering a holistic view of interconnected processes with real-time updates and two-way feedback. These capabilities empower organizations to streamline decision-making, discover optimal solutions, adapt to changing conditions, and ensure continuous improvement throughout a product's lifecycle.

Applications of digital twins now span various disciplines, such as aerospace and mechanical engineering, where the high costs of building and testing physical systems make virtual experimentation highly advantageous. By enabling conceptual analysis and validation in a digital environment, digital twins save time, reduce production costs, mitigate risks, and maintain safety and performance. Furthermore, they are invaluable in testing complex scenarios and edge cases where extreme situations consisting of multiple parameters push operational limits and make it difficult or impossible to replicate in real-world conditions. This capability is particularly crucial for tasks in hostile environments, such as training autonomous vehicles or robotic systems for austere missions. To manage these complexities, digital twins often employ advanced control systems such as model predictive control algorithms, designed to handle unexpected challenges in extreme conditions. Their use accelerates design processes, reduces costs and risks, and ensures safe and efficient operations. Digital twins hold the potential to revolutionize operations across industries, paving the way for more sustainable and secure environments for future exploration and activities.

14.3.5 Digital Threads

The digital thread is a foundational, data-driven architecture that connects and integrates information throughout a product's lifecycle, ensuring seamless traceability, accessibility, and consistency. Acting as the authoritative communication and data platform for an organization's products, the digital thread enables engineers to access a system's comprehensive definition, including design details, performance data, supply-chain information, and software components, while preserving the mission context. It provides an extensible and configurable analytical framework that facilitates the controlled interplay of technical data, software, and knowledge within the digital engineering ecosystem. By linking authoritative data and orchestrating digital models across a system's lifecycle, the digital thread transforms raw data into actionable insights, supporting proactive maintenance, optimizing processes, and reducing unplanned downtime. This architecture allows diverse audiences to extract, integrate, and utilize data for a plethora of purposes. These include requirements analysis, architecture development, design evaluation, system integration, cost estimation, training, testing, sustainment, airworthiness, and nuclear certification. Moreover, the digital thread supports a continuous feedback loop, enabling iterative improvements and informed decision-making across all phases of a system's lifecycle. For example, in vehicle malfunctions, the digital thread's traceability pinpoints root causes by examining the original design and subsequent modifications, making way for repairs to be more efficient and vehicle performance better.

In manufacturing, the integration of digital twins with their corresponding digital threads has transformed traditional processes. For example, in smart factories, cyber-physical systems use digital twins to replicate physical systems, providing real-time status updates and enabling precise control of production processes. These systems leverage digital threads to ensure seamless data flow across the product lifecycle, which improves efficiency and enables innovation. By providing real-time measurements and enabling iterative simulations, this approach optimizes manufacturing systems and enhances business processes across the product-value chain. The combined use of digital twins and digital threads is transforming industries by enabling comprehensive system analysis, optimizing performance, and reducing risks. Their integration represents a significant leap forward in leveraging digital technologies to drive innovation, enhance efficiency, and unlock new possibilities in engineering, manufacturing, and beyond. In addition, performance feedback and health monitoring of the system can be performed real time, as the real system becomes operational in that real world.

14.3.6 Digital Reasoning Thread

The Digital Reasoning Thread (DRT) represents an emerging capability within digital transformation that extends the traditional digital thread by capturing and maintaining the logic, assumptions, intent, and rationale underlying decisions made across the system and mission lifecycle. While conventional digital threads excel at managing structured data flowing through engineering and operational activities, they do not record the reasoning behind decisions or the contextual factors that shape mission outcomes. DRT addresses this gap by creating a persistent reasoning layer that carries context across tools, agents, stakeholders, and mission phases, enabling both humans and AI systems to understand not only what decisions were made but why they were made. This introduces a new class of digital continuity centered on reasoning rather than data, transforming decision logic into a reusable enterprise asset.

In complex mission environments, reasoning continuity is essential because decisions rarely occur in isolation. MEI requires the coordination of diverse domains such as operations, intelligence, logistics, cyber, modeling and simulation, and SoS engineering. Each of these domains carries unique assumptions, constraints, and heuristics that shape how mission threads are constructed and evaluated. The DRT supports this by maintaining a coherent representation of decision-making logic across heterogeneous environments, ensuring that mission alternatives, capability allocations, and operational choices remain consistent as these contexts evolve. Through this continuity, DRT enhances the MEI emphasis on adaptability and mission resilience by allowing planners and systems to trace how mission logic changes over time and how those changes influence downstream engineering and operational activities. A core advantage of the DRT is its ability to unify reasoning across heterogeneous models, tools, and disciplines, which has long been a challenge in both systems engineering (SE) and mission planning. Industrial and mission environments rely on multiple specialized tools, domain-specific languages, inconsistent formats, and evolving semantics. Similar to the difficulties encountered when transitioning from SysML v1 to SysML v2, implementing operational modeling in the Unified Architecture Framework (UAF), and managing domain fragmentation across planning tools, reasoning is often trapped inside documents, email exchanges, or the knowledge of individual subject-matter experts. The DRT provides a structured mechanism to extract, synthesize, and preserve reasoning, allowing AI systems, decision-support tools, and human operators to work from a shared contextual foundation. This capability is essential for MEI, where mission engineering threads depend on integrating disparate domains into coherent operational constructs.

The DRT also provides foundational support for explainable and governed mission decision-making. Because mission decisions carry operational and safety implications, they must be auditable, traceable, and compliant with governance policies. DRT includes built-in mechanisms for oversight, risk-tiered decision controls, role-based constraints, and a governance layer that ensures autonomy remains bounded and accountable. This capability aligns directly with MEI's need for decision transparency and mission-level accountability, particularly when integrating human-machine teaming and autonomous agents into mission execution. By preserving assumptions, constraints, causal relationships, and decision paths, the DRT ensures that mission planners can replay, validate, and justify decisions across the mission lifecycle. One of the most powerful contributions of the DRT is its support for causal and *What-If* analysis. The ability to evaluate alternatives, understand impacts, and reason about mission outcomes is central to MEI, especially as missions increasingly rely on dynamic force composition and effects webs. DRT enables reasoning engines and operators to explore alternative mission paths and evaluate SoS interactions using consistent logic, thereby strengthening mission agility and resilience. Its integration with digital twins extends this capability into simulation-based forecasting and predictive mission analysis, providing a rigorous method to evaluate the implications of decisions before executing them in real operational contexts.

The alignment between the DRT and MBSE, particularly within the UAF, is another critical contribution. UAF's operational, capability, resource, and service views capture the structural and functional characteristics of missions and systems, yet they do not capture the reasoning that drives design or operational decisions. DRT complements these models by maintaining the rationale, goals, constraints, and assumptions that inform architecture choices, mission thread structure, and capability alignment. This enhances architectural coherence and strengthens the relationship between mission intent and system design, ensuring that systems and operations remain aligned throughout the mission lifecycle. In this way, the DRT enriches UAF/MBSE workflows by linking decision logic to authoritative models, improving cross-domain consistency, and enabling more robust mission analysis and evaluation. In addition, the DRT extends the MEI framework by filling a long-standing gap between the engineering perspective and the operational decision cycle. MEI emphasizes the creation of mission threads and webs that can adapt dynamically to changing threats, contexts, and operational constraints. DRT enables this agility by preserving the reasoning that underpins mission decisions, supporting rapid reconfiguration and adaptation of mission constructs while maintaining logical integrity. As missions become increasingly data-rich, multi-domain, and AI-enabled, the DRT provides the mechanism to integrate reasoning into the mission lifecycle, enabling faster, more coherent, and more trustworthy mission decisions.

14.3.7 Digital Artifacts

Digital artifacts are dynamic outputs generated directly from digital models, serving as key deliverables within the digital engineering ecosystem. These artifacts are created using the standards, rules, tools, and infrastructure embedded in the ecosystem to ensure consistency and alignment across all phases of a system's lifecycle. Digital artifacts enable seamless communication and collaboration among stakeholders by providing precise and up-to-date information. Some examples of digital artifacts include:

- **Design specifications** that define system parameters and requirements.
- **Technical drawings**, such as authorization boundaries and data flows, for visualizing system architectures.
- **Design documents** capturing detailed system designs.
- **Interface management documents** facilitating integration across subsystems.
- **Analytical results** providing insights into system performance and optimization.

- **Bills of material** enumerating components and materials for production.
- **Software source code** developed as part of system functionality.
- **Work breakdown structures** detailing project deliverables and tasks.
- **Production or machining instructions** for manufacturing processes.
- **Test planning and cases** guiding verification and validation activities.
- **Schedules** for tracking progress and timelines.
- **Product support strategies** ensuring long-term sustainment.
- **Data flow diagrams** illustrating information exchange within the system.

These artifacts reflect the authoritative source of truth and ensure traceability. This allows organizations to maintain alignment with requirements and objectives while adapting to evolving needs. Through their integration within the digital engineering ecosystem, digital artifacts play a critical role in fostering efficiency, innovation, and informed decision-making.

14.3.8 Digital Engineering Enables Mission Engineering and Integration

Digital engineering (DE) is a transformative enabler for MEI, leveraging MBE to virtualize real-world systems and enhance their development, deployment, and sustainment. At the heart of this linkage is the ability to use digital models and simulations to represent complex systems and processes. This fosters a seamless integration of engineering disciplines and diverse engineering efforts across the lifecycle of physical systems. This approach enables organizations to reap the benefits typically associated with digitally native domains, such as rapid experimentation, iterative development, and continuous feedback loops. Digital models serve as authoritative sources of truth, allowing stakeholders to collaboratively develop, analyze, and refine mission objectives and capabilities in a virtual environment. These models enable engineers to evaluate trade-offs, identify risks, and optimize solutions without the time and resource constraints of traditional physical testing.

While this approach prioritizes disciplines that focus on creating virtual real-world systems, software tools, processes, and integrations remain fundamental to achieving its objectives. Software serves as the backbone for creating, managing, and leveraging digital models, providing the computational power and analytics required for MBE to function effectively. For example, advanced simulation tools and integrated digital environments facilitate the creation of high-fidelity models that replicate physical systems'behaviors. These tools also support interoperability, which is essential for mission integration. Moreover, the implementation of DE within the context of MEI through MBE provides the agility required to address dynamic operational challenges and emerging threats. Virtualized environments enable rapid prototyping and iteration, allowing engineers to explore alternative solutions and refine mission capabilities more quickly. This accelerates the development timeline and reduces cycle times while improving the accuracy and

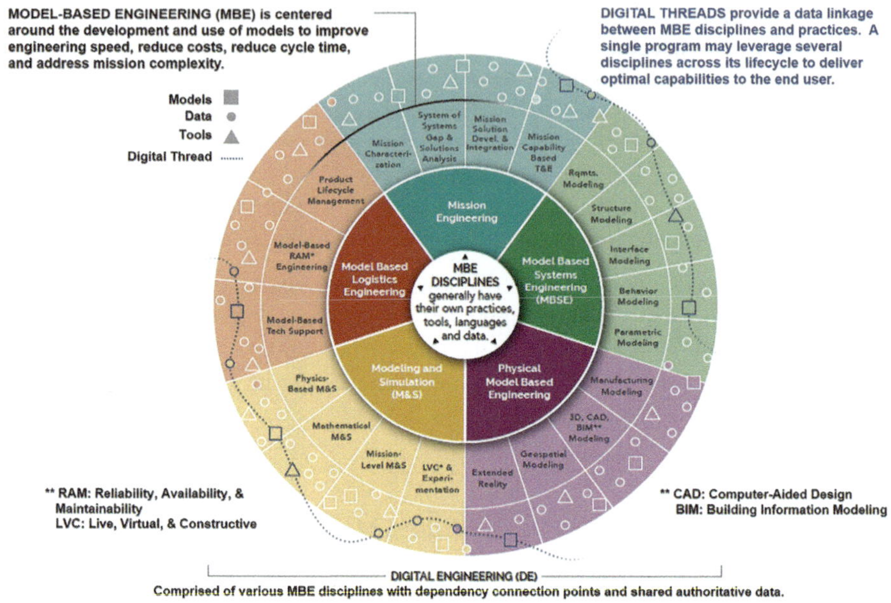

Fig. 14.9 Model-based engineering disciplines

effectiveness of engineering decisions. By investing in robust software tools, processes, and coordinating across MBE disciplines, organizations can fully realize the potential of MBE to enhance MEI, delivering more capable and resilient systems to meet mission needs as illustrated by digital threads in Fig. 14.9.

A potential future state scenario begins with analyzing critical mission threads through MEI and modeling and simulation (M&S) methods, such as mathematical, physics-based, and mission-level simulations. These methods allow engineers to evaluate mission effectiveness and identify gaps or interoperability challenges early in the process. If gaps or challenges are identified, M&S can be utilized to conduct trade studies, assess alternative solutions, and obtain preliminary performance estimates for the systems involved. Refining performance and interoperability assessments is possible through Live, Virtual, and Constructive (LVC) events. These events offer an opportunity to interact with real systems, operators, and virtual environments, creating a realistic testing ground for mission scenarios. Scenarios developed during the mission analysis phase inform the design of LVC events, which in turn may reveal additional gaps or challenges that need to be addressed, contributing to an iterative improvement cycle. To ensure alignment and interoperability across systems and programs, authoritative architectures such as the mission architecture products at the different levels of operational execution play a crucial role. These mission architectures ensure that the constituent systems within the SoS capability align with higher level requirements and work seamlessly within the broader context of critical mission threads.

The development of these requirements is supported by MBSE techniques, which enable architects and engineers to create architecture models that integrate seamlessly with the larger MEI framework. The enterprise architecture products also incorporate key technology concepts from various reference architectures, such as zero trust, unified capabilities, cybersecurity, digital platforms, and automation. These concepts ensure that integration of future state technologies and mission capabilities is coordinated and optimized to meet demands of a rapidly evolving operational environment. By combining MEI, M&S, LVC events, and authoritative architectures, DE provides a robust framework for addressing challenges and advancing system development. This approach ensures that the engineering lifecycle is aligned with mission objectives, interoperable across systems, and optimized for the integration of cutting-edge technologies, ultimately delivering enhanced capabilities to the end user.

14.3.9 Software-Enabled System Interfaces

In the context of MEI, designing and implementing interfaces for SoS is a challenging task that demands modularity, flexibility, and adaptability for heterogeneous constituent systems and technologies. Modern solutions utilize software-enabled interfaces at system boundaries to ensure smooth integration and interaction between components while supporting incremental upgrades. Retrofitting legacy systems with lightweight web technologies and leveraging standards like WebAssembly are critical strategies for achieving these objectives. WebAssembly is a compact, low-level binary instruction format optimized for high-performance execution in web browsers and other runtime environments. As a portable and platform-agnostic technology, it operates securely within a sandboxed environment, ensuring compatibility across a variety of hardware and operating systems. Supporting multiple programming languages (e.g., C, C++, and Rust), WebAssembly compiles high-level code into efficient binaries that execute with minimal overhead. This makes it particularly well-suited for resource-constrained applications like edge computing, IoT, and serverless environments.

Retrofitting provides a cost-effective pathway to modernize older systems by incorporating low-cost sensors, microcontrollers, and data processing capabilities, enabling these systems to function as integral components within modern SoS architectures. Employing WebAssembly in retrofitting facilitates deployment of lightweight, high-performance modules that can handle data collection, processing, and transmission while maintaining scalability and modularity. WebAssembly's portability ensures consistent interface behavior across heterogeneous systems. Its modular nature supports real-time updates and task reconfiguration between systems. Peer-to-peer networks complement this flexibility by enabling direct communication between edge devices, eliminating reliance on centralized systems. This distributed architecture facilitates dynamic interface updates, over-the-air software deployment, and greater resilience to system disruptions, enhancing the overall adaptability and robustness of SoS environments.

14.4 Blockchain

Blockchain offers a highly promising solution for streamlining mission capability portfolio management, making it more efficient and cost-effective. Managing complex systems-of-systems (SoS), consisting of tens of thousands of hardware and software components spread across numerous industrial facilities, presents significant challenges. These systems are manufactured in multiple plants, configured in large-scale integration facilities, and deployed to end users. Tracking and managing their lifecycle—from manufacturing to deployment and eventual disposal—is a complex task, especially when issues arise, such as damage during transit or incorrect shipments. Siloed systems and data exacerbate these challenges, making it difficult for organizations to agree on the current state of a system. Resolving discrepancies is often costly, time-consuming, and impacts user satisfaction and deployment schedules. Blockchain technology addresses these challenges by capturing all transactions and state changes as systems move through their lifecycle. By providing a single, authoritative source of truth, blockchain ensures all stakeholders have accurate and transparent information, reducing errors, improving efficiency, and supporting seamless asset management across the entire supply chain.

14.4.1 Minimum Viable Product Approach

A Minimum Viable Product (MVP) approach to implement end-to-end system management using blockchain technology provides a tangible solution within a limited scope. Blockchain technology enables businesses to form a network that coordinates sharing of one or more ledgers. These shared ledgers differ from traditional databases in a few ways. Most important, the ledgers are decentralized. Rather than being controlled by a single administrator or company, the ledger is controlled by the parties that share the ledger for the systems within the SoS of interest. Blockchain technology has four fundamental elements: (1) a shared ledger, (2) peer consensus, (3) smart contracts (business rules), and (4) privacy. Asset management teams track systems as they move through a supply chain starting when the system is ordered, through productive use, until it is disposed. Additionally, asset management service teams perform software-related activities, including license tracking, compliance, and optimization. For MVP purposes, the process starts when a serial number is generated for an individual system by the manufacturer (serialization). From there it proceeds to the recording of the system in a financial ledger as a capital item (capitalization), which happens upon shipment and continues to the receipt of a system to an end user (receiving).

In blockchain, the status of a system is recorded in different repositories as it moves through serialization, capitalization, receiving, and installation. It is not uncommon for an issue to arise that requires the asset management team to gather information from multiple repositories to resolve it. In some cases, such as when an order received by an end user includes one or more systems with serial numbers that differ from those listed on an invoice, the issue requires the end user and the manufacturer to work together to resolve the issue. The blockchain solution captures information about the systems as it moves from party to party in the supply chain (from manufacturer to end user) and as the state of the system is modified (for example, capitalized). Capturing this information in the blockchain gives the end user and its supply-chain partners a single source of truth regarding core asset information. That alone provides value, since stakeholders have one repository where they can readily get what they need to know about a system or SoS. In the future, the value delivered by this blockchain will substantially increase when all stakeholders rely on the authoritative data in the blockchain as their official record using a more automated and streamlined process that reduces cycle times and errors. The blockchain solution captures critical asset information as it moves through the supply chain, creating a unified ledger accessible to all stakeholders. By ensuring all parties have access to accurate, near real-time data, the blockchain can improve collaboration and reduce the need for manual reconciliations.

14.4.2 Overcoming Asset Management Challenges

The management of hardware and software systems in distributed environments has grown increasingly complex, compounded by the proliferation of redundant data repositories and fragmented process flows. This complexity often leads to inaccurate tracking, delays in accessing critical information, and inefficiencies in addressing issues. Transitioning from siloed, secure data repositories to cloud-based applications with metadata aggregation and cognitive capabilities is rebuilding trust in data accuracy and accessibility. A blockchain-based system fundamentally addresses these challenges by replacing legacy processes, redundant tasks, and manual interventions with a cohesive, immutable record of truth. Blockchain transforms inconsistent and fragmented data into a single "block" of reliable information, ensuring all stakeholders have access to real-time, synchronized updates. This approach eliminates delays, reduces errors, and enhances transparency across the asset lifecycle.

In today's environment, asset management processes are often fragmented, requiring stakeholders to gather information from multiple systems to resolve discrepancies. For instance, mismatched serial numbers between invoices and shipments frequently necessitated extensive coordination between primary contractors and their suppliers. By adopting blockchain, the potential exists to streamline these processes, providing a unified source of truth accessible to all parties. The ultimate goal is to improve error resolution by enabling real-time tracking of asset status. For example, when issues are identified during the receiving process, stakeholders can instantly access accurate data to resolve them. The blockchain-based solution enhances operational efficiency and sets a foundation for more agile and effective asset management practices tailored to meet end-user needs and expectations.

14.4.3 Solution Architecture and Scalability

A blockchain-based asset management system can be designed to address the challenges of managing hardware and software systems in distributed environments while ensuring scalability and adaptability. Built with a robust architecture, the system integrates advanced features such as real-time data synchronization, immutable records, and cloud-based accessibility. These capabilities provide a unified platform for stakeholders to access accurate, up-to-date information throughout the asset lifecycle. By leveraging blockchain technology, the solution ensures seamless communication between disparate systems, enhances operational efficiency, and reduces redundancy. Designed for scalability, the architecture supports the growing demands of complex supply chains and evolving operational needs, making it a future-ready solution for enterprise asset management. The following features should be considered when building the blockchain solution:

- **Shared Ledger**: A decentralized ledger serves as the single source of truth, recording every transaction and state change in an asset's lifecycle.
- **Permissioned Network**: Only authorized stakeholders such as manufacturers, logistics providers, financial teams, and customers can access the blockchain.
- **Smart Contracts**: Business rules are implemented through smart contracts, ensuring consistent execution of processes, such as validating serial numbers and managing financial capitalization.
- **Proof-of-Existence**: Transactions and events are digitally hashed and stored on the blockchain, providing an immutable record for auditing and compliance.
- **Client Interface**: A user-friendly interface allows stakeholders to interact with the blockchain, view asset lifecycle data, and integrate with external systems via APIs.

With a focus on scalability and extensibility, the blockchain solution can be built using cloud-based blockchain services. Key architectural decisions include:

- **Decentralized Ledger**: Each node in the blockchain network should maintain a copy of the ledger, ensuring data immutability and transparency.
- **Interoperability**: The system needs to be designed to integrate with existing supply-chain applications, such as manufacturing systems and financial ledgers, via APIs.
- **High Availability**: The blockchain should be deployed on a public cloud environment, ensuring global access and reliability.
- **Forward Compatibility**: The architecture should be designed to support new asset types, business processes, and future updates to the blockchain fabric.

Imagine a completely new delivery process for tracking systems. By using the business network offered in blockchain technology, all stakeholders involved in a transaction on any key data element avoid disputes, have access to the single truth of status in near real time, and mitigate, if not eliminate, software compliance exposures and even audits. The market is moving away from reactive, mid-value service via administration and analytics to high-value, automated services. These services integrate all the necessary parties transacting on data with proactive checks in a new process that eliminates any redundancy, validations, reconciliations, and compliance documentation. Blockchain establishes a system of creating a distributed consensus in the digital online world. By creating an irrefutable record in a public ledger, participating entities know for certain that a digital event occurred. Blockchain thereby opens the door for developing a democratic, open, and scalable digital economy from a centralized one. Tremendous opportunities await in this disruptive technology, and a revolution in this space has just begun.

Chapter 14 concluded with an exploration of **technological and transformational enablers**, including artificial intelligence, digital engineering, category theory, and blockchain, which together provide the computational, analytical, and architectural foundations that enhance MEI practice. Building on these transformative technologies, Chap. 15 moves from theory and methodology to practice, demonstrating how MEI principles are applied in real-world missions. This chapter applies the 10-step MEI approach to three distinct case studies: the **Hurricane Katrina Search and Rescue mission**, **NASA's Moon to Mars (M2M) initiative**, and the **Ohio Health Information Exchange (*CliniSync*) effort**. Each case provides a unique operational perspective spanning humanitarian response, aerospace exploration, and healthcare interoperability, showing how MEI frameworks can be adapted across domains to integrate complex systems, manage uncertainty, and achieve measurable outcomes. Together, these case studies illustrate the versatility and rigor of MEI as both a strategic methodology and an operational discipline for solving complex SoS challenges.

References

1. U.S. Department of Defense. (2018). *Digital engineering strategy*. U.S. Department of Defense. Retrieved from https://ac.cto.mil/wp-content/uploads/2019/06/2018-Digital-Engineering-Strategy_Approved_PrintVersion.pdf
2. DoD CIO. (2019, August 12). DoD Enterprise DevSecOps Reference Design.
3. Garrett, R., Fairbanks, J., Loper, M., & Moreland, J., Jr. (2023). The application of applied category theory to quantify mission success. *Simulation: The Society for Modeling and Simulation International, 99*(2), 201–220.
4. Murphy, A. C., & Moreland, J. D. (2022). Integrating AI microservices into hard-real-time SoS to ensure trustworthiness of digital enterprise using mission engineering. *Journal of Integrated Design and Process Science, 25*(1), 38–54. https://doi.org/10.3233/JID-210013

Chapter 15
Mission Engineering and Integration Example Case Studies

Mission Engineering and Integration (MEI) is more than a theoretical construct. It is a discipline designed to be applied in the real world to solve complex, multi-domain challenges. While the preceding chapters have established the principles, postulates, frameworks, and processes that underpin MEI, it is through case studies that the true value of this approach becomes clear. By examining missions across defense, aerospace, healthcare, transportation, and energy sectors, we can demonstrate how MEI provides a structured, repeatable method to analyze, design, and integrate systems into effective system-of-systems (SoS) that achieve mission success.

Case studies serve three essential purposes in this book. First, they anchor abstract concepts in practical realities, showing how MEI is applied under real operational and organizational constraints. Second, they provide comparative insight across diverse contexts, highlighting how common MEI principles adapt to different mission environments and domains. Third, they serve as evidence that MEI enhances outcomes by enabling leaders, engineers, and operators to manage uncertainty, balance cost and risk, and sustain mission effectiveness over time.

The sections that follow present a range of case studies, each organized around the MEI 10-step process covered in Chap. 8. Some cases draw on historical missions such as search and rescue operations following Hurricane Katrina, while others focus on ongoing national priorities such as NASA's Moon to Mars campaign. Additional cases demonstrate how MEI applies beyond defense and aerospace, illustrating its utility in addressing healthcare information interoperability, national transportation resiliency, and energy infrastructure modernization. Together, these case studies demonstrate MEI's versatility and its growing relevance to complex mission challenges in both government and industry. Through these narratives, the reader will see how mission success criteria are defined, how mission threads and kill webs are developed, how risks and costs are managed, and how integration and interoperability are achieved across systems and stakeholders. These lessons underscore MEI as not only a discipline of engineering but also a framework for leadership, governance, and strategic decision-making in an increasingly interconnected world.

J. D. Moreland, Jr., *Mission Engineering and Integration*,
https://doi.org/10.1007/978-3-032-32132-9_15

15.1 Hurricane Katrina Search and Rescue Case Study

This case study applies the 10-step MEI approach to analyze the Search and Rescue (SAR) operations following Hurricane Katrina. The analysis includes identification of mission tasks, success factors, operational conditions, performance metrics, key operational limitations, and systems integration challenges. The study is intended to inform future disaster response planning and capability development.

15.1.1 Background

Hurricane Katrina, which made landfall on August 29, 2005, was the most destructive natural disaster in U.S. history [MOYNIHAN07] [MOYHIHAN08]. A category 3 hurricane with an immense storm surge and sustained winds, it devastated the Gulf Coast with particular impact on Louisiana, Mississippi, and Alabama. New Orleans suffered catastrophic flooding due to levee failures, submerging approximately 80% of the city. The scale of the disaster was unprecedented: over 1300 lives lost, roughly 770,000 displaced persons, and more than $96 billion in damages. The destruction crippled regional infrastructure, including power, water, and emergency communication systems. SAR missions became a critical priority, but systemic shortcomings revealed widespread operational and planning deficiencies that must be addressed in future response frameworks.

The Hurricane Katrina SAR mission required coordination between its two core components: Urban Search and Rescue (US&R) and broader civil SAR. US&R focuses on rescuing victims trapped in collapsed structures, while civil SAR encompasses maritime, aeronautical, and land-based rescue operations.

15.1.2 Problem Statement

The SAR response to Hurricane Katrina faced severe operational challenges due to lack of integrated planning, interoperability breakdowns, and insufficient MEI alignment. Insufficient MEI alignment refers to the failure to systematically connect mission objectives ("ends") with appropriate tasks ("ways") and supporting capabilities ("means") in a structured, deliberate manner.

Specifically:

- Ends-Ways-Means Disconnect: No unified framework linked outcomes to operational tasks and enabling systems.
- Lack of Predefined Mission Architectures: Teams lacked integrated concepts and protocols for catastrophic SAR.
- Minimal Use of Model-Based Systems Engineering (MBSE): Reduced optimization and adaptability in real time.

- Poor Traceability of Requirements: Goals were not effectively decomposed into measurable, actionable steps.
- Limited Scenario-Based Planning: Critical interagency exercises had not sufficiently prepared teams for full-scale integration.

While many lives were saved through heroic and resourceful efforts, systemic inefficiencies produced operational gaps preventing the response from being as effective and timely as it could have been. The absence of an integrated mission plan that aligned both US&R and SAR under a unified operational framework hindered effectiveness. From a MEI perspective, this gap illustrates the critical need to define mission success criteria, decompose complex rescue missions into coordinated tasks, and establish interoperable command structures. Without a shared concept of operations (CONOPS) or mission thread that synchronized teams and systems across domains, units operated independently by setting their own objectives, selecting search areas, and determining evacuation drop-off points without systemic coherence. The lack of integrated situational awareness and command authority led to severe consequences, including instances where rescued individuals were left on highways without adequate shelter, transportation, food, or water.

These MEI gaps translated into real-world consequences: survivor rescue delays, SAR team miscoordination, and underutilization of real-time data due to system incompatibilities. The absence of a MEI approach hindered performance—not due to lack of effort, but due to insufficient structural alignment between capabilities and mission purpose.

15.1.2.1 Step 1. Identify the Missions and Tasks

This step establishes the scope and objectives of the SAR operation. The MEI approach clarifies both Civil SAR (water rescues, rooftop extraction) and Urban SAR (collapsed structures, medical extraction) as distinct but integrated components. This improves understanding of task priorities and enables tailored resource allocation (e.g., deploying Department of the Interior (DOI) boats to flooded neighborhoods and The Federal Emergency Management Agency (FEMA) teams to structural collapse zones). By structuring the mission in this way, planners ensure that overlapping SAR responsibilities are deconflicted and mutually reinforcing during execution.

Key mission essential tasks:

- **Task 1:** Surveillance
- **Task 2:** Locate stranded survivors.
- **Task 3:** Rescue and evacuate survivors.
- **Task 4:** Coordinate joint SAR operations.

- **Task 5:** Provide immediate medical care.
- **Task 6:** Maintain situational awareness and interagency coordination.

15.1.2.2 Step 2. Define Mission Success and Desired Effect

Defining mission success (i.e., rescue of all reachable survivors within 72 hours) anchors planning to outcome-driven benchmarks. This aligns tactical execution with strategic outcomes, such as reduced mortality and increased survivor stability. It provides a clear endpoint against which to measure progress, ensuring that operational tempo and resource prioritization are governed by mission objectives rather than reactive decisions.

Desired Effects:

- Civilians are safely extracted and stabilized.
- Interagency SAR operations function cohesively.
- Infrastructure damage does not significantly delay rescue efforts.

Measure of Success (MoS):

- 95% of stranded civilians rescued within 72 hours.
- Less than 5% mortality from delayed SAR.
- Zero fatalities among SAR personnel.

15.1.2.3 Step 3. Identify Mission Success Factors

Mission success factors such as timely detection of survivors and interoperable communications are pivotal in translating high-level mission goals into actionable needs. Identifying these success factors in advance enables operations planners to proactively source critical enablers (e.g., UAVs for aerial reconnaissance or satellite radios for secure communication), ensuring smoother and more predictable SAR operations.

Mission success factors depended on:

- Timely and accurate survivor detection.
- Availability and accessibility of SAR resources.
- Functional, interoperable communication systems.
- Efficient incident command and control.
- Medical capacity near high-density survivor zones.

Failure to coordinate the Urban Search and Rescue (US&R) and SAR approaches hindered overall effectiveness, revealing a lack of holistic planning.

15.1.2.4 Step 4. Identify Conditions for Each Mission Success Factor

Each success factor depends on specific operational conditions (e.g., UAV availability, reliable Geographic Information System (GIS) data, pre-positioned triage support). Clarifying these conditions ensures that decision-makers understand the dependencies that could limit mission effectiveness, allowing for contingency planning and pre-positioning of assets in critical zones before operations commence.

- **Surveillance**: Cover broad area of operational envelope around the clock.
- **Survivor Detection**: Access to updated maps, drones, and 911 data.
- **SAR Resources**: Proximity to flooded areas, fuel supplies, clear command tasking.
- **Communication**: Unified radio frequency plans, satellite uplinks.
- **Command and Control**: Centralized leadership (e.g., Joint Field Office) with rapid decision-making authority.
- **Medical Support**: Pre-positioned triage teams and mobile treatment facilities.

Lack of a shared framework meant several agencies operated independently or improvised their missions

15.1.2.5 Step 5. Map Mission Success Conditions to Mission Tasks

This step links mission-level objectives and enabling conditions to executable field tasks. Understanding these connections allows for adaptive re-tasking during the operation in the case that UAVs fail to provide visibility; alternate detection means (e.g., ground teams or crowdsourced distress signals) can be employed, thereby maintaining mission continuity despite system-level failures (Table 15.1).

15.1.2.6 Step 6. Identify Critical Conditions for Each Mission Task

This step isolates which conditions must be maintained to ensure task success (e.g., having sufficient fuel for air assets or clear extraction routes for ground-based rescues). Knowing these critical conditions enhances operational awareness and enables real-time risk mitigation (e.g., rerouting aircraft, repositioning support) during dynamic SAR missions.

Table 15.1 Conditions mapped to mission essential tasks

Condition	Mapped task
UAV imagery and 911 input	Survivor detection
Pre-staged boats/helicopters	Execution of rescues
Comms interoperability	Agency coordination
Mobile hospitals and transport	Triage and medical care
GIS and dashboards	Situational awareness and prioritization

- **Surveillance**: Integration of diverse assets and resources, ranging from aircraft and ground vehicles to specialized equipment and communication systems.
- **Survivor Rescue**: Continuous rotorcraft ops; swift location-to-rescue timing.
- **Agency Coordination**: Real-time comms across FEMA, Department of Defense, U.S. Coast Guard, and state teams.
- **Medical Care**: Support for trauma, chronic illness, and environmental exposure.
- **Asset Allocation**: Dynamic re-tasking as needs shifted geographically.

Disparate reporting streams and broken comms networks inhibited many of these tasks early in the response.

15.1.2.7 Step 7. Map Systems into Mission Tasks

This step integrates specific systems (UAVs, GIS platforms, medevac helicopters, etc.) into defined mission tasks as illustrated in Fig. 15.1 with a functional breakdown in Table 15.2. It ensures that each capability is assigned purposefully and that redundancy is built where failure could jeopardize task execution. Understanding this mapping enhances interoperability and simplifies coordination between diverse system operators in high-tempo rescue environments. Coordination gaps left some capable systems underutilized—e.g., DOI watercraft were not formally included in Search and Rescue Task Force despite offering key capabilities.

The primary mission engineering thread illustrated in Fig. 15.1 for the Hurricane Katrina SAR mission begins with surveillance, where unmanned aerial vehicles (UAVs) conduct wide-area scans of flooded urban and rural regions. The imagery collected is processed using GIS platforms to generate real-time situational maps that highlight zones with potential survivors. Building on this, the mission moves to locate survivors, as command centers analyze UAV data in combination with civilian distress calls and geospatial overlays to pinpoint individuals in need of rescue. Once

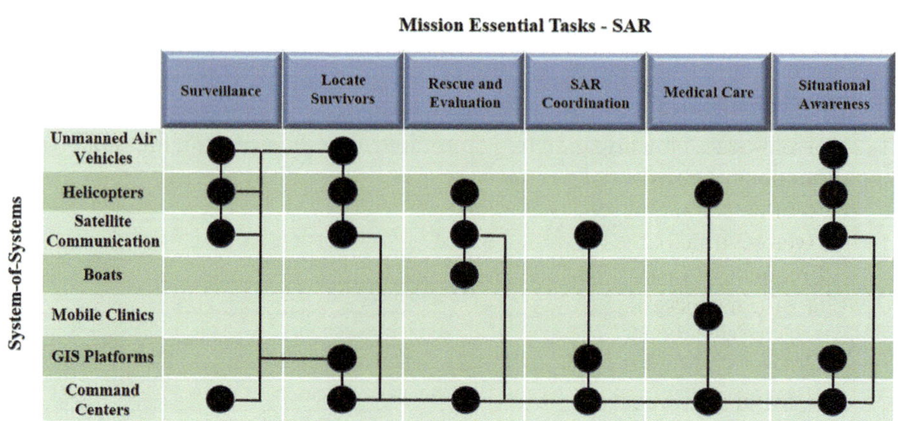

Fig. 15.1 System mapping to mission essential tasks

Table 15.2 System mapping to mission functions

System	Function
UAVs, helicopters	Broad area surveillance
UAVs, helicopters	Location and extraction of survivors
GIS platforms	Real-time mapping of needs and resources
Mobile medics (NDMS, VA teams)	Triage, transport, and stabilization
SATCOM and tactical radios	Multi-agency coordination
Boats (Coast Guard, DOI)	Shallow-water and rooftop rescues

survivors are located, the mission transitions to rescue and evacuation, with helicopters and boats dispatched through satellite communications networks to extract individuals from rooftops, attics, or isolated structures. These rescue operations are coordinated by command centers, which use SATCOM to maintain communications across federal, state, and local teams, ensuring synchronized tasking and asset allocation. Following extraction, the thread proceeds to medical care, where mobile clinics provide immediate triage and stabilization at nearby staging areas, or helicopters are used to transport more critical patients to advanced medical facilities. Throughout all phases, situational awareness is continuously maintained by feeding ISR data back into GIS-enabled command centers, which update the common operational picture (COP) and allow for real-time reassessment of mission priorities. This mission engineering thread links systems directly to mission essential tasks, creating an integrated operational path that enhances decision-making, reduces latency, and ensures the SAR mission remains focused, coordinated, and effective.

15.1.2.8 Step 8. Define Appropriate Scoring Criteria for Each Mission Task

Establishing performance metrics (e.g., time-to-rescue or percent of survivors located) allows mission managers to measure operational success in real time. It facilitates task-specific assessment and resource reprioritization, ensuring that underperforming areas are corrected before they jeopardize mission success.

- **Surveillance**:
  - % of Area of Interest scanned within specified time
  - Number of SAR targets correctly detected per flight/hour
  - % of detected objects correctly classified as people, debris, hazards, etc.
  - Average time between area scan and positive identification of victim
  - Frequency of repeat surveillance passes over a target area
  - % of time surveillance platforms are mission capable
- **Survivor Location**:
  - % located within 24, 48, 72 hours
  - Average distance between reported and actual survivor location
  - Time from initial detection to confirmed survivor location

- Probability of detection
- % of high-priority search area successfully cleared for survivors per hour
- Time to confirm identity and condition of located person

- **Rescue Execution**:
  - Time from detection to extraction
  - % of rescue attempts resulting in successful extraction without further injury
  - Time from extraction point to delivery at designated care location
  - % of planned rescues completed within mission time window
  - Time from mission assignment to task execution initiation
- **Agency Coordination**:
  - Time to unified command activation; comms uptime
  - Time delay between generation and dissemination of critical situational data
  - Average time to resolve operational or jurisdictional conflicts
  - % of attempted communications successfully delivered and understood
  - % of coordinated tasks executed without delay due to misalignment
- **Medical Support**:
  - % treated within 4 hours
  - Average time from survivor extraction to first contact with a qualified medical provider
  - % of patients correctly prioritized based on severity of injuries
  - % of required medical personnel, supplies, and equipment available during operations
  - % of rescued individuals who die before receiving advanced care
- **Situational Awareness**:
  - Frequency of Common Operating Picture (COP) updates
  - Time from incident onset to initial shared situational awareness
  - % of sensor/human inputs correctly integrated into the COP
  - % of mission area represented within the COP
  - % of relevant agencies/users with real-time access to COP and situational awareness tools

These performance measures could guide real-time prioritization and accountability.

15.1.2.9 Step 9. Apply the Scoring Criteria

Real-time application of scoring enables teams to evaluate progress and course-correct dynamically. During Katrina, this could have helped identify failing communications networks or bottlenecks in extraction zones sooner, driving more agile deployment of backup systems or alternate SAR assets.

The evaluation of mission engineering threads is conducted using the scoring criteria defined earlier, allowing decision-makers to assess how well each sequence of systems and tasks supports the desired mission outcomes. A mission engineering thread represents a direct path from enabling systems through operational tasks to the intended effects, such as locating and rescuing survivors, as shown in Fig. 15.1. By applying scoring metrics (e.g., time to complete a task, percentage of area covered, or communication latency), planners can measure the performance and effectiveness of each thread under actual operating conditions. However, real-world missions often require dynamic adaptation, which is captured through the mission engineering web, which is a network of alternative paths that offer multiple means of achieving the same objective. This web accounts for system degradations, environmental variability, or changes in resource availability by enabling rerouting or reallocation across tasks and systems. Evaluating both primary threads and their alternatives ensures operational resilience, highlights bottlenecks, and supports continuous optimization of the mission as it unfolds.

Example evaluations (illustrative):

- **Rescue Location Task:** 85% within 48 hours—*Good.*
- **Evacuation Timing:** Delays due to route inaccessibility—*Needs improvement.*
- **Coordination:** Comms gaps and fragmented ICS startup—*Major limitations.*
- **Medical Support:** High patient throughput but supply bottlenecks—*Mixed.*

This analysis reinforces the value of predefined thresholds and tolerances in mission planning as well as design challenges between systems forming the SoS for mission execution. Many times, these problems have occurred at the boundaries of the systems where the interfaces could not function adequately or are completely broken.

15.1.2.10 Step 10. Manage the Assigned Mission Areas

This step emphasizes the continuous, adaptive control of mission execution across all operational sectors. By using MEI to understand task interdependencies and success criteria, leaders can shift assets and priorities with minimal disruption, ensuring that even as conditions change, the overall mission stays on track to achieve its defined end-state.

Key management actions included:

- Dynamic reallocation of SAR assets.
- Activation of Joint Task Force Katrina to consolidate Federal military response.
- Adaptation of urban rescue protocols to aquatic conditions.
- After-action reviews to correct structural gaps (e.g., integration of SAR Task Force with DOI resources).

The absence of a unified SAR doctrine across civil and urban operations was a central coordination failure. Agencies defaulted to self-direction, which, while often heroic, lacked systematic accountability.

15.1.3 Performance Measurement Summary

Category: Measure of Success (MoS)

- 95% rescue rate in 72 hours
- <5% SAR-delay fatalities

Category: Measures of Effectiveness (MoE)

- % survivors located in 24/48 hours
- Mission coverage rate

Category: Measures of Performance (MoP)

- Sorties/day
- Tasking-to-execution latency
- Communications uptime

15.1.4 Limitations Identified Through MEI Approach

Despite operational successes, the MEI analysis revealed several integration and interoperability limitations that significantly impacted performance:

- **Communications and Information Incompatibility:** Fragmented systems led to delays and confusion and poor real-time situational awareness to include inadequate data sharing across SAR elements. This violates the MEI principle of integrated decision support and data fusion.
- **Delayed Interagency Interoperability:** Conflicting or uncoordinated rescue priorities between teams and absence of shared protocols created execution delays. This violates the MEI principle of adequate joint CONOPS and procedural alignment.
- **Command Structure Confusion:** Absence of unified or centralized command for SAR operations and disjointed chains of command delayed decisions. This violates the MEI principle of sufficient governance and coordination structure.
- **Data Fragmentation:** No centralized data hub for field and intelligence, surveillance, and reconnaissance inputs and disjointed coordination between federal, state, and local agencies as well as non-government organization and military components. This violates the MEI principle of multi-stakeholder systems integration.
- **Mission Planning Failure:** Lack of overarching plan integrating US&R with general SAR efforts. This violates the MEI principle of defining mission success and integrating mission tasks.
- **System-of-Systems Integration:** US&R and Civil SAR operated as discrete entities, leading to fragmentation and inconsistent platforms, hindering situational awareness. This violates the MEI principle of sufficient interoperability and mission thread design.
- **System Accountability:** US&R teams lacked shared tracking mechanisms for personnel, resources, or victims. This violates the MEI principle on developing mission success criteria and performance metrics with real-time feedback.

The 10-step MEI framework provides a structured lens to evaluate and enhance disaster response. The Katrina SAR case underscores that even with massive mobilization, integration and interoperability gaps can impair mission execution. Addressing these systemic issues through better engineering, planning, and coordination will be critical to future national preparedness and lifesaving efforts.

15.2 NASA's Moon to Mars (M2M) Initiative Case Study

NASA's Moon to Mars (M2M) initiative is a multi-decade strategy to establish a sustainable human presence on the Moon and to conduct the first crewed mission to Mars [LPI24]. The initiative presents complex technical, operational, and organizational challenges, involving diverse systems, international and commercial partnerships, and phased mission architectures across varying environments. To address these challenges, this case study applies the MEI 10-step approach to systematically align mission objectives, tasks, systems, and performance outcomes. This structured approach provides a framework to evaluate the effectiveness of proposed architectures, assess risks, and manage integration across system-of-systems (SoS) constructs. By defining mission success, mapping enabling conditions to specific tasks, and applying objective scoring criteria, NASA and its partners can better understand the interdependencies and performance drivers that will determine mission viability.

The analysis also reveals critical limitations and barriers with an emphasis on system integration and interoperability that could undermine the successful execution of M2M missions. The MEI process helps identify and address these issues early in the design lifecycle, improving alignment between mission needs and technical capabilities. This case study demonstrates how a mission-focused engineering framework can improve decision-making, reduce risk, and enhance NASA's ability to lead sustainable human exploration of deep space.

15.2.1 Background

NASA's M2M initiative represents a long-term strategic effort to establish a sustainable human presence beyond low Earth orbit [NASAII23] [NASAI24]. Building on the Artemis program and leveraging international and commercial partnerships, M2M seeks to enable the first crewed missions to Mars by the 2030s. The initiative begins with establishing a sustained foothold on the Moon through a series of increasingly complex missions that will develop and demonstrate the critical systems, infrastructure, and operational capabilities needed for deeper space exploration. Key elements of the M2M architecture include the Artemis missions, the Lunar Gateway, surface habitat systems, in situ resource utilization (ISRU) technologies, and the eventual development of a Mars transit system capable of supporting long-duration crewed missions [NASAI23]. The program integrates a wide range of scientific, engineering, logistical, and human factors disciplines across multiple NASA centers, international space agencies, and private industry.

The Moon serves as a critical proving ground to validate technologies, life support systems, mission operations, and risk management approaches under conditions that approximate the challenges of Mars, while remaining close enough to Earth to allow for abort contingencies. M2M is both a technical and operational transformation of human space exploration, requiring a mission-centric approach to align diverse systems, organizations, and objectives toward a coherent strategic outcome.

Initial application of the MEI approach to M2M provides valuable insights into how architectures can be better aligned with mission outcomes to include important limitations. The sheer number of stakeholders and the diversity of contributing systems are highlighted challenges in defining shared mission threads, mapping systems to tasks, and ensuring consistent interoperability across agency and industry partners. Integration remains a persistent challenge, particularly in synchronizing development timelines, data standards, and command-and-control protocols across national and organizational boundaries.

15.2.2 Problem Statement

Despite the visionary scope of NASA's M2M initiative, the program faces a range of challenges associated with the integration of complex systems, multi-agency coordination, and mission assurance across multiple phases of exploration. Key issues include:

- Diverse mission objectives spread across time and organizations with limited unifying operational threads.
- Insufficient linkage between mission-level success criteria and the performance of enabling systems and technologies.
- Gaps in structured approaches to define, analyze, and evaluate mission effectiveness across lunar and Martian environments.
- The need to manage uncertainty, long timelines, and technology maturation while maintaining alignment with national goals.
- Persistent integration and interoperability limitations, including:
  - Misalignment of interfaces and data protocols between government and commercial systems.
  - Inconsistent implementation of standards across international partners.
  - Temporal disconnects in technology readiness levels and deployment schedules.
  - Fragmented command, control, and decision-support architectures that hinder unified mission execution.

Applying the MEI 10-step approach to M2M offers a path to systematically define mission success, align architectures with mission tasks, and manage performance in a way that maximizes mission assurance and readiness for the unprecedented demands of crewed interplanetary exploration, while also identifying and addressing the integration and interoperability gaps that could otherwise delay or degrade mission outcomes.

15.2.2.1 Step 1. Identify the Missions and Tasks

The primary mission of the NASA Artemis program is to return humans to the Moon and establish a sustainable presence that will serve as a foundation for future crewed missions to Mars. This deep-space exploration endeavor integrates multiple domains (e.g., space, ground control, orbital logistics, science and technology, and human systems) into a SoS framework.

Within this overarching mission, a series of mission essential tasks is identified to achieve near-term and long-term goals. These tasks form the basis for defining mission engineering threads, enabling decomposition into executable pathways and aligned system assignments. Each task supports specific Artemis milestones such as Artemis I (uncrewed lunar orbit), Artemis II (crewed lunar flyby), Artemis III (crewed lunar landing), and Artemis IV+ (sustained lunar operations via Gateway and surface habitats).

The following mission essential tasks define the primary scope of the Artemis mission architecture:

- **T1: Launch Vehicle Delivery**
  Deliver the Orion crew vehicle and mission payloads to translunar injection (TLI) orbit using the Space Launch System (SLS).
- **T2: Orbital Navigation and Propulsion Control**
  Conduct translunar injection, mid-course corrections, lunar orbit insertion, and Earth return burns using integrated onboard and ground systems.
- **T3: Deep Space Life Support and Crew Health Maintenance**
  Sustain crew survival, health, and performance during deep-space transit through integrated environmental control and life support systems (ECLSS), radiation shielding, and medical capabilities.
- **T4: Communication and Tracking Operations**
  Maintain secure, high-bandwidth telemetry, and voice/video links between spacecraft, the Lunar Gateway, Earth ground stations, and supporting assets using the Deep Space Network (DSN) and LunaNet nodes.
- **T5: Lunar Surface Access and Mobility**
  Execute human landing and surface operations, including transfer from orbit to the Moon's surface and intra-surface mobility using Human Landing System (HLS) and pressurized/unpressurized rovers.
- **T6: Science Instrumentation and Sample Collection**
  Deploy geophysical sensors, collect lunar regolith samples, conduct in situ analysis, and return high-priority materials for Earth-based examination.

- **T7: Logistics Resupply and Habitation Infrastructure**
  Deliver and maintain logistics packages (food, oxygen, tools) and modular habitats on the lunar surface and gateway to support crew autonomy and long-duration operations.
- **T8: Mission Command and Autonomy**
  Provide hierarchical and distributed command and control architecture, integrating real-time human-in-the-loop decisions with autonomous subsystems for navigation, health monitoring, and contingency handling.

This task decomposition supports development of mission engineering threads, effects chains, and effects webs, enabling NASA and its partners (including international and commercial collaborators) to align technologies, systems, and resources with each critical capability. These threads will be further evaluated in subsequent steps for robustness, survivability, risk to mission, and overall effectiveness under various lunar mission scenarios, including nominal operations, degraded modes, and emergency contingencies.

15.2.2.2 Step 2. Define Mission Success and Desired Effect

The success of the NASA Artemis mission is defined by the ability to safely, reliably, and sustainably extend human presence beyond low Earth orbit to the lunar surface and eventually toward Mars. Success is not measured solely by the completion of a single launch or landing event, but by the ability of the integrated SoS architecture to perform as intended across a series of complex, distributed, and interdependent activities in deep space. Mission outcomes must reflect technical achievement, scientific return, international and commercial partnership execution, and crew safety, while enabling the development of long-duration space habitation and exploration capabilities.

To formalize this evaluation, success criteria are defined across strategic, operational, and tactical levels using a hierarchy of mission-level measures:

Measures of Success (MoS): *Strategic Level*

- Successful return of humans to the lunar surface and safe return to Earth.
- Establishment of sustained U.S. and allied presence on the Moon by late 2020s.
- Demonstrated readiness for Mars transit and surface mission precursors.

Measures of Effectiveness (MoE): *Operational Level*

- All mission threads achieve interoperability across NASA, international, and commercial partners (e.g., Gateway, HLS, Orion).
- Lunar infrastructure (Gateway, surface habitats) supports at least 30 continuous days of habitation.
- 95% of priority science objectives achieved across sample return, remote sensing, and in situ experimentation.

Measures of Performance (MoP): *Tactical/System Level*

- Launch vehicle (SLS) achieves TLI within required payload, trajectory, and time constraints.
- Orion maintains crew health and safety with no loss of medical, environmental, or navigation functions.
- Human Landing System (HLS) autonomously and safely delivers crew to and from lunar surface within defined delta-v and landing ellipse constraints.
- Gateway power, communication, and docking subsystems sustain nominal operations >99% of mission duration.

Desired Effect:

- Mission architecture that delivers repeatable, evolvable lunar operations with measurable impact to national space strategy, international engagement, scientific advancement, and human systems integration.

15.2.2.3 Step 3. Identify Mission Success Factors

The success of the Artemis mission depends on a set of interrelated mission success factors (MSFs) that bridge strategic vision with operational execution and technical performance. These factors define the critical characteristics of SoS performance, stakeholder integration, and environmental constraints that must be met for the mission to achieve its desired effects and support future lunar and Martian exploration.

These success factors span both engineering and enterprise domains, enabling assessment across organizational, technical, and risk perspectives. They serve as the necessary conditions for achieving the MoS, MoE, and MoP introduced in Step 2 (Table 15.3).

These ten success factors are traceable to specific Artemis phases and mission tasks and will form the basis for identifying success conditions (Step 4), critical dependencies (Step 6), and scoring metrics (Steps 8–9). They also underpin evaluation of mission engineering alternatives and the effects web that emerges from task-system interconnectivity.

15.2.2.4 Step 4. Identify Conditions for Each Mission Success Factor

Each of the 10 MSFs introduced in Step 3 depends on a specific set of *conditions* (i.e., technical, organizational, environmental, or operational) that must be present or achieved for the factor to contribute meaningfully to mission success. These conditions often span multiple systems and mission threads and must be resilient to both nominal and off-nominal scenarios, particularly given the duration, distance, and autonomy demands of lunar missions.

Table 15.4 identifies the conditions necessary to fulfill each MSF. These conditions represent the minimum enabling elements that must exist within the mission architecture or operational construct to support success.

Table 15.3 Primary mission success factors

ID	Mission success factor	Description
MSF-1	System Interoperability Across Domains	Seamless integration of systems developed by NASA, commercial partners, and international agencies (e.g., Gateway modules, Orion, HLS, rovers) enabling plug-and-play compatibility and joint operations
MSF-2	Crew Health and Safety Assurance	Reliable environmental control, radiation protection, medical readiness, and life support enabling safe human operation across deep space phases
MSF-3	Lunar Surface Accessibility and Return Capability	Precision landing, surface mobility, and safe ascent back to lunar orbit for crew and cargo, including redundancy and contingency handling
MSF-4	Sustainable Logistics and Infrastructure Support	Recurring cargo resupply and in situ support infrastructure (e.g., habitat, power, ISRU) to enable repeated and extended lunar missions
MSF-5	Robust Command, Control, and Communications (C3)	Persistent, high-bandwidth, low-latency communications and command links supporting crew operations, telemetry, navigation, and surface systems
MSF-6	Science and Technology Return on Investment	Efficient deployment of scientific instruments, collection of regolith and biological samples, and delivery of in situ and returned data to Earth
MSF-7	Autonomy, Fault Tolerance, and Contingency Management	Intelligent autonomy and fail-operational modes across critical systems, reducing Earth-reliant decision-making and enabling resilient system-of-systems behavior
MSF-8	Integrated Risk Management and Systems Engineering Discipline	Application of risk-informed decision-making, robust requirements traceability, and validation/verification (V&V) processes across the mission lifecycle
MSF-9	International and Commercial Partner Synchronization	Effective governance, interface management, and scheduling alignment across multiple stakeholders with independent deliverables and timelines
MSF-10	Launch Cadence and Payload Delivery Reliability	High-confidence transport of human and robotic payloads through launch vehicles (e.g., SLS, Falcon Heavy, New Glenn) and translunar injection stages

These conditions will be mapped to Artemis mission tasks in Step 5, enabling formal alignment between mission execution threads and the physical, procedural, and policy elements necessary for success. Importantly, these conditions also serve as critical points for verification, scoring, and risk-based analysis throughout the MEI process.

15.2.2.5 Step 5. Map Mission Success Conditions to Mission Tasks

Step 5 aligns the conditions identified in Step 4 with the mission tasks defined in Step 1, enabling a structured mapping between the enabling elements of success and the operational functions that must be performed. This mapping ensures that each task within the Artemis mission thread is fully traceable to the conditions and MSFs it supports.

Table 15.4 Conditions required for each mission success factor

Mission success factor	Required conditions
MSF-1: System Interoperability Across Domains	– Interface Control Documents (ICDs) across NASA, ESA, and commercial modules – Standardized electrical, data, and docking interfaces (e.g., IBDM, C2 protocol) – Common operating environments for robotics, sensors, and health monitoring platforms
MSF-2: Crew Health and Safety Assurance	– Fully tested Environmental Control and Life Support System (ECLSS) – Deep space medical kits, remote monitoring, and automated alerts – Dual-redundant critical life support hardware with real-time telemetry and abort logic
MSF-3: Lunar Surface Accessibility and Return Capability	– Operational Human Landing System (HLS) with integrated crew transfer – Autonomous and manual descent/ascent capability – Lunar navigation aids (e.g., surface beacons, optical terrain relative navigation)
MSF-4: Sustainable Logistics and Infrastructure Support	– Reusable cargo modules with robotic offloading capability – Surface power distribution and thermal control systems – Stowage architecture supporting modular expansion (e.g., lunar base, habitat nodes)
MSF-5: Robust Command, Control, and Communications (C3)	– Cross-platform C3 systems integrated through the Deep Space Network (DSN), Gateway relay, and LunaNet – Antennas and optical links providing lunar far-side coverage – Delay-tolerant networking (DTN) for autonomy during LOS
MSF-6: Science and Technology Return on Investment	– Sensor deployment protocols – Sample containment and biosecurity validation for Earth return – Surface analytics integrated with onboard computing for autonomous hypothesis testing
MSF-7: Autonomy, Fault Tolerance, and Contingency Management	– AI-enabled failure detection and isolation systems – Autonomous hazard avoidance on surface and in orbit – Real-time reconfiguration of system-of-systems threads (e.g., power, thermal, navigation)
MSF-8: Risk Management and Systems Engineering	– Digital engineering twins and authoritative model libraries – Continuous risk tracking across mission threads – End-to-end traceability of requirements from MoS to MoP across lifecycle reviews
MSF-9: Partner Synchronization	– Artemis Accords operationalized through partner interface reviews – Shared launch manifest and timeline reconciliation – Common verification and validation (V&V) test frameworks and certification protocols

(continued)

Table 15.4 (continued)

Mission success factor	Required conditions
MSF-10: Launch Cadence and Payload Reliability	– Redundant transport systems (SLS, Starship, Vulcan) – Ground-to-orbit integration timelines matched to lunar launch windows – Common launch processing standards across Kennedy Space Center, Wallops, and international sites

This mapping also reveals areas of concentrated dependencies where mission success is heavily reliant on one or more critical conditions. These convergence points are useful in identifying risk clusters, resilience gaps, and mission reconfiguration opportunities in later MEI steps (Table 15.5).

This task-to-condition mapping reveals the foundational role that autonomy (MSF-7), crew safety (MSF-2), and communication (MSF-5) play across nearly all mission threads. It also reinforces the layered dependency structure typical of space SoS missions, where success is not a single-point outcome but an orchestrated interaction of tightly coupled systems over extended timelines.

In Step 6, we will isolate critical conditions for each task to identify mission bottlenecks and areas requiring high reliability, redundancy, or mitigation planning.

15.2.2.6 Step 6. Identify Critical Conditions for Each Mission Task

In Step 6, the critical conditions that govern the success or failure of each mission task are isolated and assessed. Unlike broader enabling conditions identified in Step 4, these critical conditions are non-negotiable dependencies, meaning that their failure directly leads to mission degradation or catastrophic loss of function, whether technical, operational, or safety-related.

These conditions represent single points of vulnerability (if not redundantly engineered) and require special focus during mission thread analysis, risk assessment, and SoS reconfiguration (Steps 7–10). Each task's critical condition is the element that must perform to threshold levels under both nominal and contingency scenarios (Table 15.6).

This analysis highlights several recurring themes:

- Autonomy and fail-operational behavior are critical for tasks T2 and T8 due to comms latency.
- Life support and survivability threads hinge on robust environmental and power systems in T3 and T7.
- Navigation, landing, and containment risks dominate tasks T1, T5, and T6 where physical failure results in irrecoverable loss.

These critical conditions become high-priority nodes in the mission engineering thread, forming the core dependencies to be reinforced with redundancy, fault tolerance, and risk mitigation strategies in the subsequent steps.

Table 15.5 Mapping of conditions to mission tasks

Mission task (T#)	Mapped conditions
T1: Launch Vehicle Delivery	– Redundant heavy-lift capability (MSF-10) – Integrated launch manifest (MSF-9) – Ground-to-orbit handoff protocols (MSF-8)
T2: Orbital Navigation and Propulsion Control	– Precision guidance and burn planning (MSF-7) – Fault-tolerant onboard navigation (MSF-2, MSF-7) – C3 integration with ground and lunar assets (MSF-5)
T3: Deep Space Life Support and Crew Health	– Dual-redundant ECLSS with telemetry (MSF-2) – Automated radiation alert and shielding systems (MSF-2, MSF-7)
T4: Communication and Tracking Operations	– DSN, LunaNet, and Gateway comms relay (MSF-5) – Delay-tolerant networking (DTN) stack (MSF-5) – Secure, encrypted links (MSF-8)
T5: Lunar Surface Access and Mobility	– Operational Human Landing System (MSF-3) – Optical terrain nav and beacon triangulation (MSF-3, MSF-7) – Surface EVA protocols (MSF-2, MSF-6)
T6: Science Instrumentation and Sample Collection	– Regolith containment and return systems (MSF-6) – Science ops automation and AI assistance (MSF-6, MSF-7) – Sample transfer to Orion (MSF-1)
T7: Logistics and Habitation Infrastructure	– Modular cargo offloading (MSF-4) – Power and thermal conditioning for surface nodes (MSF-4) – Habitation ECLSS interoperability (MSF-1, MSF-2)
T8: Mission Command and Autonomy	– AI-enabled fault detection (MSF-7) – Distributed autonomy and human-on-the-loop models (MSF-7) – Integrated decision logs (MSF-8)

15.2.2.7 Step 7. Map Systems into Mission Tasks

In this step, assign specific systems to each of the mission tasks identified in Step 1, forming the foundation of the mission engineering thread. These systems span across NASA platforms, international partner contributions, and commercial services, and their proper assignment is essential to achieving mission effectiveness, resilience, and responsiveness.

Each system is mapped based on its primary function, integration role, and operational phase (launch, transit, orbit, surface, return). This mapping is not merely a technical allocation but a reflection of the SoS dependencies that govern how threads are sequenced, how risk propagates, and how contingency options emerge (Table 15.7).

Table 15.6 Critical conditions by mission task

Mission task	Critical condition
T1: Launch Vehicle Delivery	Reliable ignition and main engine thrust performance of SLS core stage. Failure to meet thrust curve or guidance parameters compromises translunar injection and downstream mission threading
T2: Orbital Navigation and Propulsion Control	Autonomous GN&C (Guidance, Navigation and Control) subsystem with onboard failover logic. Real-time course correction is essential due to extended communication latency with Earth
T3: Deep Space Life Support and Crew Health	Closed-loop ECLSS with dual-redundant oxygen generation and CO_2 scrubbing. Failure results in life-threatening conditions for crew beyond immediate recovery windows.
T4: Communication and Tracking Operations	Persistent link margin via DSN/Gateway relay, even during lunar far-side operations. Loss of comms results in loss of situational awareness and remote coordination
T5: Lunar Surface Access and Mobility	Throttlable, autonomous landing engine system with hazard detection and avoidance. Failure causes hard landing or abort, endangering crew and equipment
T6: Science Instrumentation and Sample Collection	Sample containment and sterilization system for Earth return. Biological or chemical compromise risks contamination and invalidates scientific payload
T7: Logistics and Habitation Infrastructure	Surface power generation and thermal regulation to sustain habitat systems. Power loss causes habitat temperature/failure cascade impacting long-duration survival
T8: Mission Command and Autonomy	Autonomous fault isolation and reconfiguration system with human-on-the-loop capability. System failure creates compounding mission thread collapse, particularly during long comm blackouts

This task-system mapping reveals a distributed yet tightly interdependent architecture forming the mission engineering thread.

For example:

- The Orion spacecraft spans multiple mission tasks (T2, T3, T4, T6, T8), acting as a multi-thread node.
- The Gateway serves as a persistent orbital outpost linking C3, logistics, autonomy, and science operations.
- International systems (ESA's ESM, JAXA modules, CSA robotics) are woven into U.S. command structures, requiring harmonized interoperability protocols.

These system assignments as shown in Fig. 15.2 will inform scoring (Step 8), degraded state assessments, and reconfiguration planning in the final stages of the mission engineering process.

Table 15.7 System assignments by mission task

Mission task	Assigned systems and platforms
T1: Launch Vehicle Delivery	– Space Launch System (SLS) core stage and ICPS – Launch Control System (LCS) at Kennedy Space Center – Ground Support Equipment (GSE) and Mission Control Centers
T2: Orbital Navigation and Propulsion Control	– Orion spacecraft: Orbital Maneuvering System (OMS), RCS thrusters – European Service Module (ESM) propulsion system – Flight software for guidance and trajectory correction
T3: Deep Space Life Support and Crew Health	– Orion ECLSS (CO_2 scrubbers, oxygen tanks, waste management) – Biomedical monitoring systems – Shielding components and crew suits for radiation protection
T4: Communication and Tracking Operations	– Deep Space Network (DSN) ground stations – Lunar Gateway communication module – LunaNet architecture including relay satellites and DTN stack
T5: Lunar Surface Access and Mobility	– Human Landing System (HLS) (SpaceX Starship or other variants) – Lander avionics and propulsion – Pressurized and unpressurized lunar rovers
T6: Science Instrumentation and Sample Collection	– Automated and crew-deployed science payloads (seismometers, spectrometers) – Sample containers with containment verification protocols – Return transfer systems for loading into Orion or cargo vehicles
T7: Logistics and Habitation Infrastructure	– Gateway Logistics Modules (delivered by SpaceX, NG, ESA) – Modular lunar habitats – Surface power systems (solar arrays, nuclear units) and thermal loops
T8: Mission Command and Autonomy	– NASA Johnson Space Center MCC for primary operations – Autonomous fault detection and response systems – Integrated flight software across Orion, Gateway, and HLS

15.2.2.8 Step 8. Define Appropriate Scoring Criteria for Each Mission Task

To objectively assess progress and mission assurance throughout the Artemis campaign, a structured scoring framework is established. This framework enables NASA, its international partners, and commercial providers to evaluate the performance of each mission essential task (T1–T8) based on measurable technical,

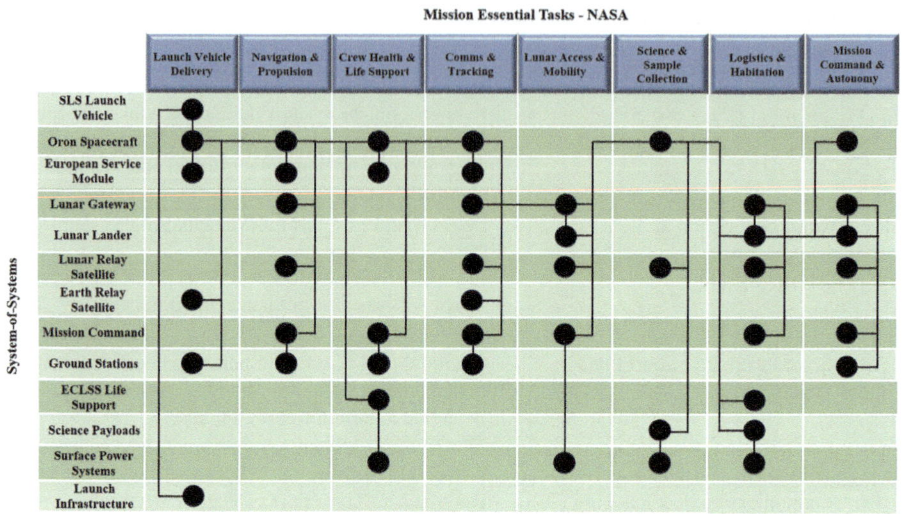

Fig. 15.2 System mapping to mission essential tasks

operational, and temporal parameters. The scoring criteria are designed to be traceable to key mission outcomes and flexible across Artemis phases (Artemis I–IV and beyond).

Each task is evaluated using a three-tiered color-coded metric system (**Green** (Fully Successful), **Yellow** (Partially Successful/Degraded), and **Red** (Unsuccessful/Failed)) based on predefined thresholds of performance, risk, and mission enablement (Table 15.8).

15.2.2.9 Step 9. Apply the Scoring Criteria

The evaluation of the Artemis mission engineering threads uses the scoring framework defined in Step 8 to assess mission performance, resilience, and adaptability. In this step, the scoring criteria are applied to actual or simulated mission scenarios to evaluate how well each mission task (T1–T8) is supported by its assigned systems under nominal, degraded, and disrupted operational conditions.

Thread-Based Evaluation

Each primary mission engineering thread (from launch to lunar surface operations and return) is modeled using a mission execution timeline and dependency logic. As operations progress, each task's score is dynamically assessed based on telemetry, autonomy logs, communication latency, system state data, and crew health

Table 15.8 Task-based scoring criteria matrix

Mission task ID	Green (G) fully met	Yellow (Y) partial/ degraded	Red (R) not met
T1: Launch Vehicle Delivery	SLS delivers Orion to target orbit w/ <5 s delay	Trajectory margin exceeded; minor anomaly	Loss of vehicle or target orbit not achieved
T2: Navigation and Propulsion	Course corrections <3%, propellant margin >10%	Minor trajectory drift; extra burns required	Critical propellant loss or misaligned orbit
T3: Crew Health and Life Support	Life support nominal; crew vitals within range	System degraded; backup systems activated	Loss of crew capability or critical failure
T4: Communications and Tracking	100% link availability; <1 s latency	Link intermittency; increased latency (>5 s)	Communication blackout >10 min or total loss
T5: Lunar Access and Mobility	Precision landing within 100 m; safe egress	Divert landing zone >500 m; mobility degraded	Abort; crash landing or failure to access site
T6: Science and Sample Collection	90%+ planned sample goals collected and preserved	<50% collection; degraded tool performance	No collection or loss of samples
T7: Logistics and Habitation	Systems fully operational; consumables stable	Resupply lag; redundancy degraded	Power, food, or thermal system failure
T8: Mission Command and Autonomy	All decision nodes operate autonomously	Autonomy degraded; mission control fallback	Autonomous command failure; manual override

monitoring. Mission outcomes are determined by whether the cumulative thread remains within mission tolerance limits.

A successful Artemis thread requires multiple mission tasks (T1–T8) to score "Green," or, in constrained scenarios, a limited number of "Yellow" scores provided contingency responses are executed without mission abort.

Evaluation of Mission Threads

For example:

Thread A: Launch → Transit → Lunar Orbit Insertion → Surface Landing → Sample Collection → Return to Earth

- If SLS launches successfully (**T1: Green**), Orion performs trajectory corrections within budget (**T2: Green**), and the lunar lander diverts due to a site anomaly but lands safely (**T5: Yellow**), the mission thread is still viable.
- However, if communications via Gateway fail during surface ops (T4: Red) and autonomy fallback does not recover control (T8: Red), the thread is considered compromised and alternate paths are triggered.

Alternative Threads and Redundancy

The Artemis mission architecture includes built-in redundancy via:

- Dual relay paths (Lunar Relay Satellite + Gateway),
- Reusable surface elements (landers and habitats),
- Autonomous re-tasking of Orion or Gateway modules.

These redundancies allow alternative mission engineering threads to be initiated in real time. For instance:

- A degraded power system on the lunar surface (T7: **Yellow**) may trigger an alternate surface operations thread leveraging remote-controlled payloads from orbit via Gateway (T6 → T8 bypass loop).

This web-based resilience modeling mirrors the Katrina SAR mission engineering web and ensures that even partial failures at the task level do not necessarily equate to mission failure.

Integrated Scoring Results

Each mission thread receives a composite score based on:

- The number and severity of degraded tasks,
- Criticality of affected systems,
- Impact on mission objectives (e.g., crew safety, sample return, comms continuity),
- Risk to mission, force, and national prestige.

The output feeds into a Mission Risk Dashboard, informing real-time decisions by Mission Command and supporting future system redesign or re-tasking strategies.

15.2.2.10 Step 10. Manage the Assigned Mission Areas

The Artemis mission requires continuous oversight, synchronization, and dynamic reallocation of mission resources across multiple SoS elements. In this step, management of the assigned mission areas ensures that system performance, task fulfillment, and mission objectives remain aligned under real-world operational variability, risk conditions, and evolving campaign priorities.

Mission Area Integration and Control

NASA's Artemis management structure incorporates:

- Mission Area Managers (MAMs) for each domain (e.g., launch ops, lunar mobility, comms, habitation),

- System owners for hardware and software elements (e.g., SLS, Orion, Gateway, landers),
- Thread coordinators responsible for end-to-end mission engineering threads.

Each manager is tasked with monitoring mission success criteria, performing real-time assessments of task scoring results (Step 9), and implementing adaptive governance through predefined contingency branches in the MEI web.

Dynamic Resource Reallocation

If a system underperforms (e.g., degraded comms via Gateway or a delay in lunar lander availability), Artemis managers can:

- Reconfigure task assignments across alternate systems (e.g., rerouting via Lunar Relay Satellite instead of Gateway),
- Shift timeline dependencies (e.g., delaying surface operations to preserve science objectives),
- Invoke autonomous replanning agents on mission command systems for decentralized adaptation.

This distributed and flexible management structure mirrors the Team-of-Teams governance model used in the Katrina SAR mission, but tailored for deep space operations and high-latency decision environments.

Campaign-Level Oversight

Artemis program control uses integrated dashboards to:

- Track mission health across threads (Green–Yellow–Red scoring)
- Visualize dependency matrices (N2 diagrams)
- Support cross-mission learning loops from Artemis I/II to Artemis III/IV

Decision authority is structured hierarchically:

- Tactical level (e.g., surface operations)
- Operational level (e.g., lunar orbit, Gateway operations)
- Strategic level (e.g., NASA HQ campaign-level objectives and national prestige outcomes)

Continuous Improvement and Feedback

After-action reviews (AARs), digital twin simulations, and mission replays are used to:

- Validate system-to-task assignments
- Refine mission scoring logic
- Identify emerging capability gaps

Lessons learned from Artemis I/II are already shaping hardware interfaces, communication protocols, and risk-tolerant architectures for Artemis III/IV, with NASA's Artemis Campaign Development Division overseeing mission area maturity and acquisition integration across the Space Administration Centers.

15.3 Ohio Health Information Exchange (CliniSync) Case Study

Healthcare delivery in the United States increasingly depends on the ability to exchange accurate, timely, and secure patient information across a complex network of providers, laboratories, and public health systems. The Ohio Health Information Exchange, known as *CliniSync*, was established to bridge fragmented data environments and create a cohesive statewide ecosystem capable of supporting clinical, operational, and strategic decision-making [LENGEL20]. As a mission, *CliniSync* embodies an SoS integration challenge which requires interoperability across hundreds of independent systems, coordination among diverse stakeholders, and alignment of policy, technical, and human factors to achieve improved health outcomes.

Applying the MEI approach to *CliniSync* enables a structured analysis of this complex mission landscape. The MEI framework decomposes the healthcare interoperability challenge into mission threads, identifies system interactions, and quantifies mission performance through measurable criteria (i.e., Measures of Success (MoS), Measures of Effectiveness (MoE), and Measures of Performance (MoP)). This allows policymakers, technologists, and healthcare leaders to understand how data interoperability contributes to statewide healthcare mission success, and how risks can be managed to maintain system resilience and trust.

15.3.1 Background

Ohio's healthcare ecosystem consists of over 150 hospitals, thousands of clinics, multiple academic medical centers, and diverse payer and public health agencies. Historically, these entities operated through siloed information systems, each optimized for local use but unable to exchange critical patient data effectively. The result was fragmented care, duplicated testing, medication errors, and inefficient coordination across providers. In 2009, supported by the Office of the National Coordinator for Health Information Technology (ONC) and the Health Information Technology for Economic and Clinical Health (HITECH) Act, Ohio initiated a statewide effort to integrate disparate systems and enable a common Health Information Exchange (HIE) infrastructure.

CliniSync, operated by the Ohio Health Information Partnership, was formed to lead this initiative. The program established data-sharing protocols, identity management solutions, and governance frameworks to support the secure exchange of electronic health records (EHRs), laboratory results, and care summaries. Over time, *CliniSync* evolved into one of the largest operational HIEs in the United States, demonstrating the benefits of shared data ecosystems. Yet, as participation expanded, new mission engineering challenges emerged which ranged from interoperability across evolving standards to cybersecurity, data quality, and alignment between technical implementation and policy objectives.

The *CliniSync* mission therefore operates at the intersection of technology, governance, and public service, much like a defense or disaster response mission. Its success depends on the coordinated performance of multiple systems and organizations; however, each has distinct objectives, authorities, and constraints. The goal is to work toward a shared mission outcome: the improvement of healthcare delivery and population health across Ohio.

15.3.2 Problem Statement

Despite significant investments in health IT infrastructure, data fragmentation and lack of interoperability remain critical barriers to effective healthcare delivery. Many healthcare organizations in Ohio continue to use different EHR platforms that cannot natively share structured data. Public health agencies and laboratories face integration challenges that delay disease reporting and response. Clinicians often lack access to a complete patient record, leading to gaps in care coordination, repeated diagnostics, and preventable adverse events.

At the enterprise level, information asymmetry prevents health systems, payers, and policymakers from fully leveraging data for population health management, cost optimization, and outcome-based reimbursement models. The resulting inefficiencies impose measurable costs such as higher expenditures, delayed treatment, and diminished health outcomes.

The problem is therefore one of mission interoperability (e.g., achieving timely, accurate, and secure information flow across a diverse SoS without compromising privacy, security, or data integrity). Applying the MEI approach enables a disciplined assessment of this problem by:

- Defining healthcare information exchange as a mission with measurable outcomes.
- Mapping participating organizations and technologies as systems within a mission-oriented SoS architecture.
- Identifying mission essential tasks such as patient identity resolution, care transition data exchange, and laboratory reporting.
- Quantifying risks to mission success to include interoperability gaps, cybersecurity threats, and governance dependencies.

By framing healthcare interoperability as a MEI problem, the *CliniSync* case illustrates how structured systems thinking can transform healthcare integration from a technical challenge into a mission-driven enterprise, aligning technology, policy, and human systems to achieve sustained improvement in public health outcomes. To operationalize this framing, the MEI 10-step approach is applied to the Ohio Health Information Exchange. This approach begins by defining the mission in clear, outcome-oriented terms and establishing the context in which interoperability contributes directly to statewide health objectives. Step 1 initiates this process by describing the specific mission and decomposing the mission into mission essential tasks for functional execution of the mission.

15.3.2.1 Step 1. Identify the Missions and Tasks

The mission of the *CliniSync* case is to ensure timely, accurate, and secure sharing of patient health information across Ohio's healthcare ecosystem to enable coordinated, high-quality, and cost-effective care. This mission serves the broader strategic goal of improving population health outcomes through seamless interoperability among hospitals, clinics, laboratories, payers, and public health agencies. At its core, this mission addresses the persistent fragmentation of health data across diverse systems. Historically, providers operated as isolated nodes with each maintaining its own records, workflows, and data standards. *CliniSync* transforms these disparate systems into an SoS that supports real-time information exchange and mission-level situational awareness of statewide health trends. As such, the mission extends beyond technical data transfer to encompass clinical effectiveness, operational coordination, and strategic resilience of the healthcare enterprise.

Mission Statement

Enable a secure, interoperable, and reliable statewide exchange of clinical information that improves care coordination, enhances patient safety, and supports evidence-based decision-making across the Ohio healthcare system.

Mission Purpose

The purpose of this mission is to create an integrated health information infrastructure capable of supporting the full continuum of care from primary diagnosis and treatment to emergency response and public health surveillance. Through interoperability, the mission seeks to eliminate data silos, reduce redundant testing, improve the timeliness of clinical decisions, and facilitate analytics that inform both operational management and public policy.

The mission also supports broader strategic outcomes:

- Improved healthcare quality through comprehensive and accessible patient data.
- Enhanced operational readiness for public health events, pandemics, and emergencies.
- Reduced costs and inefficiencies by minimizing duplicate testing and administrative burden.
- Strengthened trust and transparency among patients, providers, and governing bodies.

Mission Objectives

The Ohio HIE mission can be defined in measurable objectives that correspond to each level of mission success (Table 15.9):

Mission Tasks

The *CliniSync* mission is decomposed into a series of mission essential tasks that represent the functional activities required to achieve mission success. Each task contributes to the overall mission outcome through specific system interactions and data flows.

- **T1: Establish Data Connectivity**
  Connect hospitals, clinics, laboratories, and public health systems to the *CliniSync* Health Information Exchange (HIE) through secure, standards-based interfaces to enable comprehensive data sharing.
- **T2: Enable Patient Identity Management**
  Create and maintain a Master Patient Index (MPI) and record linkage mechanisms that ensure accurate identification of patients across disparate systems, reducing errors and duplication.

Table 15.9 Mission objectives at enterprise levels

Level	Objective	Representative measure
Strategic (MoS)	Improve population health outcomes and patient safety through integrated data systems	Reduction in preventable hospitalizations, improved chronic care metrics, lower mortality rates
Operational (MoE)	Ensure accurate, timely, and complete information flow across participating organizations	Percentage of successful data exchanges, latency in information delivery, completeness of patient records
Tactical (MoP)	Achieve consistent technical interoperability, data integrity, and system uptime	HL7/FHIR compliance rate, message success rate, network availability, cybersecurity incident frequency

- **T3: Facilitate Care Transition Exchange**
  Enable the automated exchange of discharge summaries, medication lists, and continuity-of-care documents among healthcare providers to ensure seamless transitions of care and prevent readmissions.
- **T4: Integrate Laboratory and Diagnostic Reporting**
  Automate the reporting and dissemination of laboratory results, diagnostic imaging, and pathology reports across care settings to enhance clinical decision-making and reduce delays.
- **T5: Support Public Health Surveillance**
  Provide real-time or near-real-time data feeds to the Ohio Department of Health for notifiable disease reporting, immunization tracking, and syndromic surveillance to strengthen statewide public health readiness.
- **T6: Maintain Data Quality and Cybersecurity**
  Maintain robust validation, monitoring, and security controls to ensure that shared data remain accurate, complete, and protected against unauthorized access or corruption.
- **T7: Provide Population Health Analytics**
  Aggregate, de-identify, and analyze longitudinal patient data to support predictive modeling, policy development, and population-level health interventions.
- **T8: Maintain System Governance and Stakeholder Coordination**
  Establish governance structures, participation agreements, and compliance oversight to manage shared risk, define data use policies, and ensure accountability among participating organizations.

Each mission task links directly to system functions and operational behaviors that will be analyzed in later steps. Together, they define the operational architecture that supports the health information interoperability mission and its contribution to statewide health resilience.

15.3.2.2 Step 2. Define Mission Success and Desired Effect

The success of the Ohio Healthcare Interoperability mission is defined by the seamless, secure, and actionable exchange of clinical data across diverse provider networks, payer systems, HIEs, and public health agencies. This mission supports both routine care delivery and crisis-response operations, enabling healthcare professionals, administrators, and policymakers to access timely and complete patient information to improve outcomes, lower costs, and enhance operational efficiency.

Mission success is not merely the transmission of data, but rather the conversion of shared data into improved health outcomes, higher coordination across care settings, and stronger situational awareness during both daily operations and public health events (e.g., pandemics, opioid surges, natural disasters). Therefore, success is defined in terms of mission effects across strategic (MoS), operational (MoE), and tactical levels (MoP):

Measures of Success (MoS): *Strategic Level*
At the strategic level, the desired effect is the establishment of a statewide interoperable ecosystem that improves the health of Ohio's population while ensuring the resilience of the state's healthcare delivery infrastructure. This includes:

- Reduced hospital readmissions and avoidable emergency room visits
- Improved management of chronic conditions through data continuity
- Support for public health surveillance and response

Key outcomes:

- 85% provider participation in HIE activities statewide
- Measurable population-level improvement in outcomes (e.g., diabetes, maternal health)
- Demonstrated ability to respond to statewide health crises with integrated data

Measures of Effectiveness (MoE): *Operational Level*
At the operational level, success is defined by the real-time exchange, integration, and retrieval of patient-level data across EHRs, pharmacy systems, labs, imaging centers, and payers. This ensures coordinated and informed care, especially during care transitions.

Key outcomes:

- Median response time for record retrieval <5 s
- Accurate patient matching (>95% confidence)
- Automated data quality scoring and gap identification
- Provider dashboards enabling proactive intervention workflows

Measures of Performance (MoP): *Tactical/System Level*
At the tactical level, mission success hinges on the technical performance of systems and data pipelines that enable interoperability. This includes interface stability, data integrity, and error resolution across the technical stack.

Key outcomes:

- API uptime >99.95%
- HL7 and FHIR transaction success rates >98%
- Minimal manual re-entry due to failed transactions (<2% cases)
- End-user satisfaction >80% among clinical and administrative users

Desired Effects:
- Enhanced continuity of care
- Reduced clinical errors and redundancies
- Efficient public health surveillance
- Equitable access to health services
- Stakeholder alignment across the healthcare ecosystem

The desired effect of the Ohio Healthcare Interoperability Mission is the creation of a trusted, secure, and responsive data-sharing ecosystem that enables timely, accurate, and actionable health information exchange across all healthcare entities in the state. This includes hospitals, clinics, payers, pharmacies, laboratories, emergency services, and public health agencies.

15.3.2.3 Step 3. Identify Mission Success Factors

Mission success factors for the Ohio Healthcare Interoperability Mission reflect the critical enablers that must be present to achieve the desired effects defined in Step 2. These factors represent the core capabilities, processes, and stakeholder conditions that directly support interoperability across the continuum of care and enable responsive, coordinated health outcomes at scale.

Each success factor bridges the gap between high-level desired effects and measurable conditions that must be tracked throughout implementation, operation, and sustainment phases (Table 15.10).

These 11 mission success factors collectively define the minimum viable conditions for success across Ohio's interoperable healthcare landscape. They are mapped to both mission tasks (Step 5) and mission conditions (Step 4), forming the basis of technical scoring, stakeholder engagement metrics, and risk tracking throughout the implementation lifecycle.

15.3.2.4 Step 4. Identify Conditions for Each Mission Success Factor

In this step, the mission success factors identified in Step 3 are decomposed into concrete, observable, and measurable conditions that must exist or be satisfied for those success factors to be achieved. These mission conditions are framed in operationally meaningful terms that align with the mission. They support both real-time monitoring and post-implementation assessment.

Each condition may be tied to a specific technical requirement, organizational behavior, or policy structure and may also reveal early indicators of risk to mission or system performance degradation (Table 15.11).

These conditions serve as both system validation points and early warning indicators. For example, failure to achieve C4.1 (accurate MPI) could undermine every downstream task tied to clinical data reliability. Similarly, a lapse in C9.1 (security control) creates unacceptable mission risk even if functional exchange is technically successful. Together, these conditions form the operational logic behind the upcoming mapping steps and risk-informed governance decisions.

Table 15.10 Primary mission success factors

ID	Mission success factor	Description
SF-1	Trust Framework	Stakeholders (providers, payers, public health, patients) trust the interoperability infrastructure, policies, and data governance
SF-2	Multi-Stakeholder Alignment	Shared vision and sustained commitment across health systems, HIEs, and agencies. Includes policy alignment and data-sharing agreements
SF-3	Policy and Regulatory Enablement	State-level policies and federal compliance frameworks (HIPAA, TEFCA, twenty first Century Cures Act) are integrated to reduce legal/technical friction
SF-4	Accurate Patient Identity Matching	High-fidelity patient matching across systems using MPI, demographic/biometric tools, and probabilistic algorithms
SF-5	Standards-Based Interoperability	Adoption of HL7, FHIR, Direct Secure Messaging, and IHE profiles across vendors and systems
SF-6	Data Normalization and Mapping	Unified coding systems (LOINC, SNOMED, RxNorm, ICD-10), semantic mapping tools, and structured formats for computable data
SF-7	Bi-Directional Query/ Retrieve Capability	Systems can send and receive clinical information via standard APIs with appropriate consent and security protocols
SF-8	High System Uptime and Reliability	Uptime >99.95% for exchange infrastructure, with redundancy and failover mechanisms
SF-9	Secure Identity and Access Control	Role-based access, multi-factor authentication, audit logging, and conformance with NIST standards
SF-10	Workflow Integration	Seamless integration into provider workflows via EHRs and user interfaces that support point-of-care decision-making
SF-11	Feedback and Learning Loop	Ongoing measurement, feedback, and adaptation to improve data quality, performance, and user satisfaction

15.3.2.5 Step 5. Map Mission Success Conditions to Mission Tasks

This step establishes the operational alignment between mission success conditions (identified in Step 4) and the specific mission essential tasks defined for the Ohio Healthcare Interoperability mission. This mapping ensures that each task can be evaluated against observable, measurable outcomes that directly support overall mission success.

Mapping conditions to tasks also provides the foundation for later identification of critical conditions (Step 6), system-to-task assignments (Step 7), and risk-informed governance structures to ensure mission resilience (Table 15.12).

15.3.2.6 Step 6. Identify Critical Conditions for Each Mission Task

In Step 6, the critical conditions that govern the success or failure of each mission task are isolated and assessed. Unlike broader enabling conditions identified in Step 4, these critical conditions are non-negotiable dependencies, meaning that their

Table 15.11 Conditions required for mission success factors

Mission success factor	Condition ID	Condition description
SF-1: Trust Framework	C1.1	Formal governance body with multi-stakeholder representation is chartered and meets quarterly
SF-2: Stakeholder Alignment	C2.1	>80% of major provider systems have signed data use and participation agreements
SF-3: Policy Enablement	C3.1	Alignment with federal interoperability rules (e.g., TEFCA QHIN participation, twenty first Century Cures Act provisions) is documented
SF-4: Accurate Patient Matching	C4.1	Master Patient Index (MPI) with matching accuracy >95% in pilot transactions
SF-5: Standards-Based Exchange	C5.1	All participating systems support HL7 v2.x and FHIR R4 APIs
SF-6: Data Normalization	C6.1	>90% of lab and medication data is mapped to LOINC and RxNorm terminologies
SF-7: Bi-directional Exchange	C7.1	Systems can query and retrieve clinical data via FHIR within 5 s in 95% of test cases
SF-8: High Uptime	C8.1	All HIE nodes and integration layers report system uptime >99.95% over a 30-day period
SF-9: Secure Access	C9.1	All systems use role-based access control (RBAC) and multi-factor authentication (MFA) for user sign-on
SF-10: Workflow Integration	C10.1	User interface integrates seamlessly into EHR platforms (e.g., Epic, Cerner) with no more than two clicks to access external data
SF-11: Feedback Loop	C11.1	Performance dashboard is live, with automated reports on transaction volume, latency, and error rates distributed monthly

failure directly leads to mission degradation or catastrophic loss of function, whether technical, operational, or safety-related.

These conditions represent single points of vulnerability (if not redundantly engineered) and require special focus during mission thread analysis, risk assessment, and SoS reconfiguration (Steps 7–10). Each task's critical condition is the element that must perform to threshold levels under both nominal and contingency scenarios (Table 15.13).

This matrix serves multiple functions:

- It clarifies which mission tasks are responsible for satisfying each critical condition.
- It reveals dependencies that will influence scoring and risk analysis in later steps.
- It informs system-to-task mapping by showing where technology and data infrastructure must be applied to meet conditions.

For instance, if T5 (workflow integration) is poorly implemented, it could cause mission degradation even if data exchange (T2) is technically successful, which highlights the importance of end-user alignment in system deployment.

Table 15.12 Mapping of conditions to mission tasks

Condition ID	Mapped task(s)	Rationale
C1.1: Governance body established	T8	Ensures coordinated oversight and stakeholder trust
C2.1: Participation agreements in place	T1, T2, T8	Required for sharing identity and clinical data across networks
C3.1: Regulatory alignment documented	T4, T8	Supports compliance and stakeholder assurance
C4.1: MPI matching accuracy >95%	T1	Enables accurate patient identity and record location
C5.1: HL7/FHIR interface support	T2, T3	Enables structured exchange and normalization of data
C6.1: Standardized vocabularies mapped	T3	Supports semantic interoperability and data usability
C7.1: Query/response time <5 s	T2, T5	Enables real-time access at point of care
C8.1: Uptime >99.95%	T7	Ensures service reliability and operational continuity
C9.1: MFA and RBAC in place	T4	Protects data access and ensures secure transactions
C10.1: Seamless workflow integration	T5	Reduces friction for clinical end users and increases adoption
C11.1: Dashboard and feedback live	T6, T8	Enables learning loops and governance transparency

Table 15.13 Critical conditions by mission task

Mission task ID	Mapped condition IDs	Condition focus
T1: Establish Data Connectivity	C2.1, C4.1	Participation agreements in place, high-accuracy MPI
T2: Enable Patient Identity Management	C2.1, C5.1, C7.1	Interface support (HL7/FHIR), participation, fast query
T3: Facilitate Care Transition Exchange	C5.1, C6.1	Standards-based interoperability and semantic mapping
T4: Integrate Laboratory and Diagnostic Reporting	C3.1, C9.1	Regulatory conformance and secure access control
T5: Support Public Health Surveillance	C7.1, C10.1	Real-time data retrieval and UI integration
T6: Maintain Data Quality and Cybersecurity	C11.1	Dashboard for reporting, analytics, and transparency
T7: Sustain System Uptime and Performance	C8.1	High system availability and reliability monitoring
T8: Maintain Stakeholder Trust and Governance	C1.1, C2.1, C3.1, C11.1	Multi-stakeholder governance, legal alignment, transparency

15.3.2.7 Step 7. Map Systems into Mission Tasks

In this step, we identify and assign specific systems, platforms, and infrastructures that perform the functional and technical roles necessary to carry out each mission essential task. These system-to-task mappings establish the foundation for mission thread construction, mission performance evaluation, and reconfiguration planning (Steps 9 and 10).

Systems may include hardware, software platforms, data services, or organizations that operate in an SoS construct. Each system's ability to fulfill its assigned tasks is influenced by technical compatibility, interoperability standards, policy constraints, and the mission environment (Tables 15.14 and 15.15).

Observations and Implications

- System S1 (HIE Network) is the most cross-cutting system, participating in nearly all tasks, which makes it mission-critical and a potential single point of failure.
- System S3 (EHRs) are central to tasks at the point of care, highlighting the need for vendor alignment and workflow harmonization.

Table 15.14 System identification and description

System ID	System name	Type	Function
S1	Health Information Exchange (HIE) Network	Digital Infrastructure	Core network enabling data aggregation, exchange, and routing
S2	Master Patient Index (MPI)	Software	Patient identity matching and deduplication
S3	Electronic Health Record (EHR) Systems (e.g., Epic, Cerner)	Application	Point-of-care systems used by providers to generate, retrieve, and consume clinical data
S4	FHIR/HL7 Interface Engines	Middleware	Translate, standardize, and enable bi-directional data flow
S5	Data Normalization and Terminology Services (e.g., Apelon, IMO)	Software Service	Map clinical codes (e.g., SNOMED, LOINC, RxNorm) for semantic interoperability
S6	Consent and Access Control System	Identity/Security	Enforce HIPAA/42 CFR Part 2 rules and patient consent models
S7	Public Health Data Warehouse	Data Platform	Aggregates longitudinal health data for analytics and public health reporting
S8	Governance Platform and Reporting Dashboard	Process System	Provides transparency, auditability, and stakeholder feedback loops
S9	Network and Cloud Infrastructure	Physical and Virtual Infra	Ensures high availability, uptime, and secure data transport

Table 15.15 System mapping by mission task

Mission task ID	Assigned systems	Rationale
T1: Establish Data Connectivity	S2, S1, S3	S2 performs MPI; S1 routes queries; S3 supplies patient demographics
T2: Enable Patient Identity Management	S1, S3, S4	S1 is the exchange backbone; S4 enables interoperability; S3 is the endpoint
T3: Facilitate Care Transition Exchange	S4, S5	S4 manages message parsing; S5 provides semantic translation
T4: Integrate Laboratory and Diagnostic Reporting	S6, S9	S6 handles policy-driven access; S9 enforces encrypted transport
T5: Support Public Health Surveillance	S3, S4	S3 interfaces with clinicians; S4 ensures data availability within EHR
T6: Maintain Data Quality and Cybersecurity	S1, S7	S1 routes data to S7; S7 stores and analyzes for agency use
T7: Provide Population Health Analytics	S9, S1	S9 ensures resilient architecture; S1 reports availability metrics
T8: Maintain System Governance and Stakeholder Coordination	S8, S6	S8 supports transparency and oversight; S6 provides assurance mechanisms

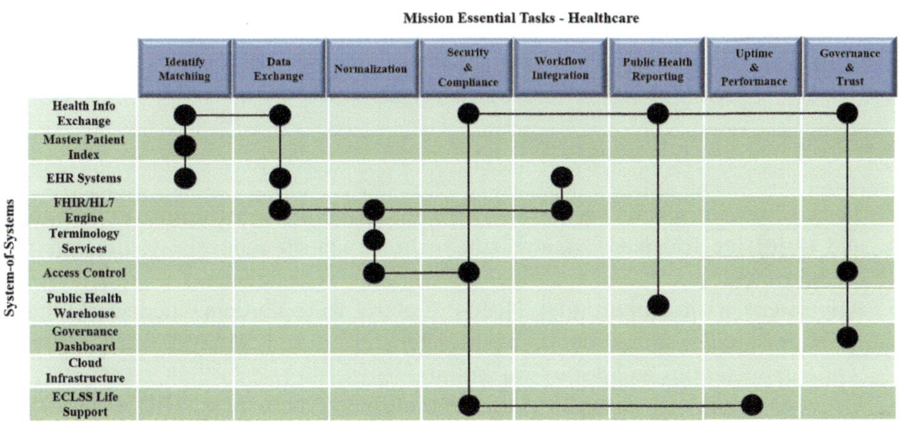

Fig. 15.3 System mapping to mission essential tasks

- Security systems (S6 and S9) are foundational enablers, whose failure would pose systemic risk even if functional data exchange continues.
- Governance (S8) ensures stakeholder trust, supporting long-term mission sustainment rather than tactical operations alone.

These system assignments as shown in Fig. **15.3** will inform scoring (Step **8**), degraded state assessments, and reconfiguration planning in the final stages of the mission engineering process.

Observations

- S1 is used in T1 and T2, highlighting its core network role.
- S3 (EHR Systems) and S4 (FHIR/HL7) support most point-of-care tasks.
- S6, S8, and S9 support security, governance, and uptime—essential backend and compliance functions.
- T4, T7, and T8 rely on fewer, more specialized systems, often requiring redundancy or high assurance design.

15.3.2.8 Step 8. Define Appropriate Scoring Criteria for Each Mission Task

To evaluate the effectiveness and performance of the Ohio Healthcare Interoperability mission and its constituent SoS, a set of quantitative and qualitative scoring criteria has been developed for each mission task. These scoring measures align with the healthcare mission's operational goals of accurate identity resolution, seamless data exchange, standards-based interoperability, secure compliance, and responsive public health reporting. Each criterion is framed in terms of measurable attributes that can be used to assess the MoPs, MoEs, and the overall contribution to MoS (Table 15.16).

Approach and Considerations

- **Tiered Metrics**: Each criterion includes both **threshold (minimum acceptable)** and **objective (desired)** values, supporting trade-off analysis during mission thread execution.
- **Alignment with Stakeholder Needs**: Criteria were co-developed with Ohio's HIEs, hospital systems, public health agencies, and IT vendors to reflect real-world expectations and policy constraints.
- **Cross-System Accountability**: Where multiple systems (e.g., HIE, MPI, FHIR engine) contribute to a task, joint performance metrics are defined to enable shared accountability and coordinated root cause analysis.

15.3.2.9 Step 9. Apply the Scoring Criteria

In this step, the scoring criteria established in Step 8 are applied to evaluate the performance of the mission engineering threads across each of the eight mission tasks (T1–T8). The evaluation process considers how well the assigned systems support the execution of their respective tasks and whether their combined performance meets the threshold and objective values for mission success.

Table 15.16 Mission tasks and evaluation criteria

Mission task ID	Scoring criteria	Metric type(s)
T1	• ≥99% uptime of health data interfaces (HIE/FHIR/API) • Link setup time ≤24 h • Interface error rate <1%	MoP
T2	• Identity match rate ≥95% • False positive rate ≤1% • Record linkage latency ≤5 s	MoP, MoE
T3	• Successful continuity of care document (CCD) exchanges ≥98% • Message acknowledgment within 30 s • HL7/FHIR compliance ≥95%	MoP, MoE
T4	• Structured lab result compliance ≥90% • Reporting latency ≤24 h • Rejected transactions <0.5%	MoP, MoE
T5	• Reporting timeliness ≤48 h from trigger • ≥99% schema compliance with state/federal reporting • Real-time dashboard availability ≥98%	MoP, MoE, MoS
T6	• Vocabulary/terminology conformity ≥95% • Policy enforcement ≥98% • Zero confirmed breaches-MTTR ≤30 min	MoP, MoE
T7	• Query response time ≤5 s • Analytics coverage ≥90% of available data • ≥85% user satisfaction with dashboard utility	MoP, MoE
T8	• ≥95% audit trail completeness • Policy update cycle ≤30 days • Stakeholder confidence rating ≥90% in governance meetings	MoE, MoS

Mission Engineering Thread Assessment

Each thread is a combination of systems supporting one or more tasks. This thread is evaluated using the mission scoring framework:

Thread Examples:

- Thread A: HIE Network + FHIR Engine + MPI supporting T1–T3
- Thread B: Lab Systems + Data Pipeline + Public Health Portal supporting T4–T5
- Thread C: Cybersecurity Gateway + EHR Platforms + Governance Portal supporting T6–T8

These threads are scored individually against their defined MoPs, MoEs, and MoS, using real or simulated operational data (e.g., system logs, data latency, user surveys, audit reports).

Alternative Path Analysis

Given the modular and distributed nature of the Ohio health IT ecosystem, alternative paths exist for achieving mission effects, especially when certain systems are degraded or unavailable. For instance:

- If the primary HIE interface is offline (affecting T1), a secondary secure FHIR bridge can be activated.
- If the MPI fails (affecting T2), the EHR-native identity linking algorithm can partially maintain continuity at a lower performance level.
- If public health dashboards are temporarily inaccessible (T5), manual CSV export paths can meet reporting thresholds, albeit with degraded timeliness and confidence.

These alternative paths are traced through the system-task-effect chain to evaluate whether the mission effects can still be delivered within tolerable bounds of risk and delay.

Decision-Support Outcomes

- Real-time monitoring (e.g., automated alerts for scoring drops)
- Resilience planning (e.g., pre-authorized system substitutions)
- Future design improvements (e.g., investing in redundant identity resolution)
- Operational planning (e.g., whether surge capacity is required for T3 during flu season)

This structured evaluation provides the foundation for Step **10** where governance, resources, and system lifecycles are actively aligned to mission threads.

15.3.2.10 Step 10. Manage the Assigned Mission Areas

The management of mission areas in the Ohio Healthcare Interoperability context involves continuous oversight, coordination, and governance across the interconnected SoS supporting healthcare delivery, public health response, and health information exchange. This final step formalizes the mechanisms by which mission threads are sustained, adapted, and improved over time in response to operational dynamics, evolving threats, stakeholder priorities, and regulatory updates.

Oversight of Mission Threads

Each mission thread is managed through assigned lead organizations, often aligned with existing state-level health governance bodies. These include:

- Ohio Health Information Partnership (OHIP)—Leads T1, T2, and T3 tasks via HIE infrastructure.
- Ohio Department of Health (ODH)—Oversees T4, T5, and T7 through lab pipelines and analytics platforms.
- Regional Health Systems and HIEs—Jointly manage task execution and continuity across clinical domains.
- Cybersecurity and Policy Offices—Ensure Task T6 and T8 compliance, trust, and resilience.

Each organization maintains responsibility for its respective mission areas while coordinating with others through joint governance boards and interoperability councils.

Monitoring and Feedback Loops

Ongoing mission thread performance is monitored through dashboards, audit logs, and surveillance indicators, linked back to the scoring criteria established in Step 8. These include:

- Real-time HL7/FHIR transaction metrics
- Identity match accuracy reports
- Public health data latency trackers
- Governance cadence compliance logs

Feedback mechanisms ensure that degradation in any part of the system (e.g., MPI accuracy, delayed lab results, security incident) is detected and acted upon. Threads are evaluated quarterly using these indicators, with scorecards reported to governance bodies.

Flexibility and Reconfiguration

Given the distributed nature of Ohio's healthcare ecosystem, the mission management structure must allow for thread reconfiguration in response to:

- System failures or cyber incidents (e.g., rapid failover to secondary HIEs)
- Policy changes (e.g., CMS mandates on patient data access)
- Epidemic surge events (e.g., real-time reallocation of public health surveillance resources)

Reconfiguration paths are pre-validated during Step 9 evaluations and encoded in contingency playbooks managed by assigned authorities. Each path is assessed for mission continuity, risk to mission, and data fidelity.

As we conclude this journey through the Mission Engineering and Integration (MEI) approach, it is both exciting and rewarding to reflect on the comprehensive exploration of the strategies, techniques, and enabling disciplines that shape effective mission execution. From the foundational concepts of creative thinking, critical thinking, and innovation to the guiding principles and operational context, this book has outlined how to think beyond traditional methods to tackle the complexities of SoS. By emphasizing forward-thinking, transformative strategies like artificial intelligence, digital engineering, category theory and blockchain, suitable solutions are driven and push the boundaries of what is possible. The MEI approach is trailblazing with the ultimate outcome of empowering leaders to think mission-first and optimize SoS capabilities before diving into the potential suboptimization design and development of the individual, constituent systems. By leveraging the enablers of mission architectures, risk assessments and mitigation, cost estimating, and mission capability portfolio management, outcomes are at a reduced risk within tolerable ranges, within economic constraints, and effectively managed for optimal impact. The path forward is paved with possibilities, and the tools and principles discussed in this book will empower future leaders to shape the operational readiness and resilience of tomorrow's missions, ensuring a sustainable and dynamic global landscape.

References

1. Lengel, A. J. (2020). Identifying barriers that prevent the usage of health information exchanges: A case study of *CliniSync* in Ohio. *Journal of Healthcare Information Management, 34*(2), 29–41. https://doi.org/10.1177/8755122520924607
2. Lunar and Planetary Institute (LPI). (2024). International Partnerships and NASA's Moon to Mars Architecture. Retrieved from https://www.lpi.usra.edu/lunar/strategies/resources/M2M_ACR2024_InternationalPartnerships.pdf
3. Moynihan, D. P. (2007) *Case studies of incident command systems: The Katrina case and others*. IBM Center for the Business of Government. Retrieved from https://www.businessofgovernment.org/sites/default/files/MoynihanKatrina.pdf
4. Moynihan, D. P. (2008). *The response to Hurricane Katrina*. IRGC Risk Governance Working Paper Series No. 2. International Risk Governance Council. Retrieved from https://irgc.org/wp-content/uploads/2018/09/Hurricane_Katrina_full_case_study_web.pdf
5. National Aeronautics and Space Administration. (2023). *Moon to Mars Architecture*. National Aeronautics and Space Administration. Retrieved from https://www.nasa.gov/moontomarsarchitecture/
6. National Aeronautics and Space Administration. (2023). *NASA's Moon to Mars Strategy and Objectives Development (NASA/SP-2023-423)*. National Aeronautics and Space Administration. Retrieved from https://www.nasa.gov/wp-content/uploads/2023/04/m2m_strategy_and_objectives_development.pdf
7. National Aeronautics and Space Administration. (2024). *Moon to Mars Architecture executive overview* (ACR executive summary). Retrieved from https://www.nasa.gov/wp-content/uploads/2024/01/2023-moon-to-mars-architecture-executive-overview.pdf

Glossary

Adaptability The ability of an individual, organization, system, or process to adjust, evolve, or change in response to new conditions, challenges, or environments involving the flexibility, resiliency, and openness to change, allowing for continuous improvement and the ability to thrive in dynamic or uncertain circumstances.

Artificial Intelligence The field of computer science that focuses on creating systems capable of performing tasks that typically require human intelligence to include reasoning, learning, problem-solving, perception, language understanding, and decision-making.

Assumption A specific supposition of the operational environment that is assumed true, in the absence of positive proof, essential for the continuation of planning.

Attack Vector A path, method, or means by which a threat actor exploits a vulnerability or weakness to gain unauthorized access to, disrupt, manipulate, degrade, or compromise a system, network, data, or other resource, potentially affecting the performance of the broader system-of-systems and its ability to accomplish the mission.

Authoritative Source of Truth The reference point for models and data across the system lifecycle. The authoritative source of truth provides traceability as the system evolves, capturing historical knowledge and connecting configuration-controlled versions of models and data.

Axioms Fundamental statements or self-evident truths that are accepted without proof and serve as the foundational basis for logical reasoning, mathematical systems, and theoretical frameworks.

Behavior Used in engineering science is a characterization that captures the range of actions or mannerisms carried out by a computer, computer application, or computer code controlling a system in response to stimuli from user inputs, environmental conditions, or an external agent.

Blue Force (BLUFOR) Combination of all friendly forces of United States.

J. D. Moreland, Jr., *Mission Engineering and Integration*,
https://doi.org/10.1007/978-3-032-32132-9

Capability The ability to complete a task or execute a course of action under specified conditions and level of performance.

Category Theory A branch of mathematics that provides a high-level, abstract framework for understanding mathematical structures and their relationships. It focuses on objects (such as sets, spaces, or algebraic structures) and morphisms (arrows) that define how these objects interact.

Command and Control The processes, systems, and structures that enable decision-makers to direct and coordinate military or organizational operations involving the management of resources, communication, and information to achieve mission objectives efficiently.

Commercial-off-the-Shelf Software, hardware, or other products that are readily available for purchase from commercial vendors and are not customized for a specific user or organization.

Complexity The characteristic of a system, problem, or environment that arises from the interactions of multiple interconnected and interdependent components, leading to unpredictable or emergent behavior.

Concept of Operations (CONOPS) A verbal or graphic statement that clearly and concisely expresses what the commander intends to accomplish and how it will be done using available resources.

Condition Refers to the specific circumstances, factors, or states that exist at a particular time or in a particular environment, influencing the way actions or processes unfold.

Configurability The ability of a system, product, or software to be customized or adjusted by users or administrators to meet specific needs or preferences without altering its core structure, allowing the user to modify certain settings, features, or components to optimize performance, functionality, or appearance based on varying requirements.

Cost Estimating The process of predicting the financial resources required to complete a project, task, or initiative involving the calculation of the anticipated costs of materials, labor, equipment, and overheads, as well as any potential risks or contingencies.

Creative Thinking The ability to generate original, innovative, and valuable ideas by approaching problems or situations from new perspectives.

Critical Thinking The ability to analyze, evaluate, and synthesize information objectively and logically to form reasoned judgments.

Cyber Kill Chain A framework used to describe the stages of a cyberattack, outlining the sequence of events that typically occur from the initial reconnaissance to the final exploitation of a system.

Cyberattack Any deliberate attempt by an individual, group, or organization to compromise the confidentiality, integrity, or availability of a computer system, network, or digital asset.

Cybersecurity The practice of protecting computer systems, networks, devices, and data from digital attacks, unauthorized access, damage, or theft involving a range of technologies, processes, and measures designed to safeguard information and ensure the confidentiality, integrity, and availability of data and systems.

Data Fusion The process of integrating and analyzing data from multiple sources to produce a unified, accurate, and more comprehensive understanding of a situation.

Deep Learning A subset of machine learning that uses artificial neural networks with multiple layers (deep neural networks) to model complex patterns in data to enable computers to automatically learn hierarchical representations from large datasets, making it particularly effective for tasks such as image recognition, natural language processing, and autonomous decision-making.

Design of Experiment A systematic approach to planning, conducting, and analyzing controlled tests or experiments to investigate the relationship between factors (independent variables) and outcomes (dependent variables) with the goal of identifying how different variables influence a process or system, optimize performance, and improve decision-making.

Developer An individual or team responsible for creating, designing, coding, and maintaining software applications, websites, or other digital products.

Digital Artifact A product or output, in computer (i.e., digital) format, created within or generated from the digital engineering ecosystem. Digital artifacts provide data for alternative views to visualize, communicate, and deliver data, information, and knowledge to stakeholders.

Digital Engineering Capability The ability to develop, validate, use, curate, and maintain technically accurate digital systems and models of systems, subsystems, and their components, at the appropriate level of fidelity to ensure test activities adequately simulate the environment in which a system will be deployed.

Digital Engineering A means of using and integrating digital models and the underlying authoritative source data to support the development, test and evaluation, and sustainment of a system.

Digital Model A digital (i.e., in an electronic form, able to be read and manipulated by computer) representation of an object, phenomenon, process, or system. The representation can include form, attributes, and functions and may be depicted visually or described via mathematical or logical expressions.

Digital Thread An extensible and configurable analytical framework that seamlessly expedites the controlled interplay of technical data, software, information, and knowledge in the digital engineering ecosystem, based on the established requirements, architectures, formats, and rules for building digital models. It is used to inform decision-makers throughout a system's lifecycle by providing the capability to access, integrate, and transform data into actionable information.

Digital Transformation The process of integrating digital technologies into all aspects of an organization, fundamentally changing how it operates, delivers value, and interacts with customers or stakeholders.

Digital Twin A computerized representation (integrated set of models) that serves as the real-time digital counterpart of a physical object or process.

Digital The use of discrete, binary data (represented by ones and zeros) to encode, store, process, and transmit information.

Distributed Test A testing approach where the testing activities, tools, or resources are spread across multiple locations, systems, or devices, often over a network.

Effects Web A dynamic, interconnected network of heterogeneous systems that collaboratively generate and deliver coordinated effects across multiple domains.

Enterprise Architecture A strategic framework used to align an organization's business processes, information systems, technologies, and resources with its goals and objectives by providing a comprehensive blueprint that defines the structure and operation of an organization, helping to ensure that all components work together efficiently and effectively.

Environmental Context The surrounding conditions, factors, or circumstances, both physical and non-physical, that influence or affect a system, decision, or process.

Fidelity A measure of the accuracy, precision, and statistical confidence to which the data, result, etc. represents the state and behavior of a real-world object or the perception of a real-world object, feature, condition, or chosen standard in a measurable or perceivable manner.

Governance Structure The system of rules, processes, and practices through which an organization or entity is directed, controlled, and held accountable.

Government-off-the-Shelf Software, hardware, or other products that are developed or acquired by government agencies for internal use, rather than being commercially available.

Human Factors Engineering The field of study and practice focused on designing systems, products, and environments that optimize human performance and safety by considering human capabilities, limitations, and behaviors.

Human Systems Integration An interdisciplinary approach that focuses on optimizing the interaction between humans and systems to improve overall system performance, safety, and user effectiveness.

Information Exchange Requirements The description, in terms of characteristics, of the requirement to transfer information between two and more end users/system elements. The characteristics described include source, recipients, contents, size, timeliness, security, and trigger. These requirements are defined to be independent of the communications medium.

Integration Readiness Levels A systematic measurement of the interfacing of compatible interactions for various technologies (components) and the consistent comparison of the maturity between integration points.

Integration The composition of a capability by designing and assembling elements in a way that allows them to work together to achieve an intended purpose.

Intelligence The ability to acquire, process, and apply knowledge and skills in various contexts, including problem-solving, learning, reasoning, and decision-making. It also refers to the collection, analysis, and dissemination of information about potential threats, adversaries, and environmental factors, helping decision-makers make informed choices.

Interoperability The ability of two or more systems or components to exchange information (syntactic), mutually understand that information (semantic), and to effectively use the information (pragmatic) that has been exchanged.

Issue Management The process of identifying, analyzing, and resolving problems or challenges that arise within a project, organization, or system before they escalate into significant risks.

Issue A problem, challenge, or concern that has already occurred and requires immediate attention or resolution.

Kill Chain A mission thread with a kinetic outcome. Dynamic targeting procedures often referred to as F2T2EA by air and maritime component forces; and Decide, Detect, Deliver, and Assess methodology by land component forces.

Kill Web An inclusive set of multiple integrated mission threads consisting of mission essential tasks for the applicable scenario or vignette of interest.

Lifecycle Cost Estimate The process of estimating the total cost of owning, operating, maintaining, and disposing of a system or product over its entire life span including all costs from initial development, acquisition, and implementation to ongoing maintenance, support, and eventual decommissioning or disposal.

Live, Virtual, and Constructive Different types of training or testing environments that vary in their level of realism and the use of technology but often used together in integrated exercises to provide a more comprehensive approach to training, testing, and evaluation, combining realism with cost-effective and flexible simulation techniques.

Machine Learning A subset of artificial intelligence that focuses on developing algorithms and statistical models that enable computers to learn patterns from data and make predictions or decisions without being explicitly programmed.

Measure of Effectiveness Measurable military effects or target values for success that are derived from executing mission essential tasks and activities to achieve the Measure(s) of Success.

Measure of Performance Measurable performance characteristics or target parameters of systems or actors used to carry out the mission essential tasks or military effect applied to key performance parameters and key system attributes.

Measure of Success Measurable attributes or target values for success within the overall mission in an operational environment. Measures of Success are typically driven by the mission objectives of the blue force.

Measure The empirical, objective, numeric quantification of the amount, dimensions, capacity, or attributes of an object, event, or process that can be used for comparison against a standard or similar entity or process.

Metric A unit of measure that coincides with a specific method, procedure, or analysis (e.g., function or algorithm). Examples include mean, median, mode, percentage, and percentile.

Minimum Viable Product The simplest version of a product that allows a team to collect the maximum amount of validated learning with the least amount of effort and contains just enough features to meet the basic needs of early adopters or users, allowing the product team to test assumptions, gather feedback, and improve the product iteratively.

Mission Architecture A view or representation that depicts the ways and means to execute a specific end-to-end mission, with relationships and dependencies among mission elements. This includes elements such as mission activities, approaches, systems, systems of systems, organizations, and capabilities.

Mission Characterization The aggregate of factors associated with military objectives and operations; this includes the mission to be accomplished in a specific time and place, the Measure(s) of Success, the threats, and constraints. Changes in any factors or operational conditions of the mission characterization may cause the mission to be redefined.

Mission Context The elements that describe who, what, when, where, and why elements of the mission to be accomplished. Changes in any elements or operational conditions of the mission context may cause the mission to be redefined.

Mission Engineering Thread Mapping of the capabilities, technologies, systems, and organizations to the mission essential tasks (mission thread) required to execute the mission to form the resulting system-of-systems.

Mission Engineering An interdisciplinary process encompassing the entire technical effort to analyze, design, and integrate current and emerging operational needs and capabilities to achieve desired mission outcomes. The process is focused on capturing and translating the intent of the customer's mission needs into implementable system, software, hardware, firmware, data, or other requirements.

Mission Essential Tasks Tasks considered essential to the accomplishment of a mission. The mission essential tasks defined in the mission essential task list include associated conditions and standards and may identify command-linked and supporting tasks.

Mission Success Criteria Predefined conditions, performance thresholds, and objectives that must be met for a mission to be considered successful.

Mission Thread A sequence of end-to-end mission essential tasks, activities, and events presented as a series of steps to achieve a mission with iterations between the tasks.

Mission The mission essential tasks, together with the purpose, that clearly indicate the action to be taken and the reasoning behind the mission.

Mission-Based Reference Architecture An architecture that consists of a list of functions and some indication of their interfaces or information exchange requirements and interactions with each other and with functions located outside the system's boundary. It facilitates a shared understanding across multiple products, organizations, and disciplines with respect to mission needs.

Mitigation The actions taken to reduce or minimize the severity, impact, or likelihood of a risk, issue, or potential problem involving the implementation of strategies, measures, or controls to lessen negative outcomes or prevent adverse situations from escalating.

Model A physical, mathematical, or otherwise logical representation of a system, entity, phenomenon, or process.

Model-Based Systems Engineering The formalized application of modeling to support system requirements, design, analysis, verification, and validation activities beginning in the conceptual design phase and continuing throughout development and later system lifecycle phases.

Modular Open Systems Approach An acquisition and design strategy consisting of a technical architecture that uses system interfaces compliant with widely supported and consensus-based standards (if available and suitable). The strategy supports a modular, loosely coupled and highly cohesive system structure that allows severable system components at the appropriate level to be incrementally added, removed, or replaced throughout the lifecycle of a system platform to afford opportunities for enhanced competition and innovation.

Monitoring The ongoing process of observing, tracking, and reviewing the performance, progress, or status of a system, project, or process to ensure it is operating as intended.

Multidisciplinary Analysis and Optimization An engineering approach that integrates multiple disciplines to analyze, evaluate, and optimize complex systems.

Multi-Domain Operations A military strategy that integrates and synchronizes operations across multiple domains (land, air, sea, space, and cyber) to achieve a unified objective.

National Security The protection of a nation's citizens, institutions, and interests from threats that could undermine its sovereignty, stability, or well-being.

Objective Quality Evidence Any statement of fact pertaining to the quality of a product or service based on observations, measurements, or tests that can be verified by valid quantitative or qualitative methods. OQE provides the verifiable evidence that deliberate steps were taken to comply with stakeholder requirements.

Ontologies Formal representations of knowledge within a specific domain, consisting of a set of concepts, their properties, and the relationships between them.

Open Standards Publicly accessible and universally accepted specifications or guidelines for technologies, protocols, or processes that ensure interoperability, compatibility, and data exchange between different systems, platforms, or organizations.

Operational Planning The process of developing detailed, short- to mid-term plans that outline how an organization or force will execute specific tasks, missions, or strategies to achieve defined objectives.

Operational Anything related to the day-to-day activities, functions, or procedures required to achieve specific objectives within an organization or system.

Opportunity Management The structured process of identifying, evaluating, and capitalizing on potential opportunities to achieve strategic or operational advantages involving proactive planning and decision-making to maximize benefits while balancing risks and resource constraints.

Opposing Force (OPFOR) Adversarial forces to the United States and Allied Partners.

Performance Measures Quantifiable indicators used to assess the effectiveness, efficiency, and overall success of a system, process, or organization in achieving its objectives.

Physiological Measures Objective, quantifiable assessments of biological functions and physical responses that indicate an individual's physiological state.

Platform A foundational system, framework, or environment that supports the development, deployment, and operation of applications, services, or other systems like a physical or digital system that serves as a base for mission execution.

Portfolio Management The process of managing a collection of projects, investments, or assets in a way that aligns with an organization's strategic goals and objectives.

Postulates Fundamental assumptions or principles that are accepted as true without direct proof, serving as the foundational basis for reasoning, argumentation, or further development of theories.

Principles Fundamental truths, rules, or guidelines that serve as the foundation for reasoning, decision-making, and system design.

Reference Architecture Authoritative sources of data about a specific subject area that guides and constrains the instantiations of multiple architectures and solutions.

Resilience The ability of a system, organization, or individual to withstand, adapt to, and recover from challenges, disruptions, or adverse events involving maintained or quickly restored functionality in the face of stress, changes, or crises.

Risk Assessment The process of identifying, analyzing, and evaluating potential risks that could impact an organization, project, or system involving the likelihood and consequences of threats, vulnerabilities, and uncertainties to determine their potential impact on objectives.

Risk Attitude An individual's or organization's approach to assessing and responding to risk, based on their willingness to accept, avoid, or mitigate uncertainty in decision-making.

Risk Category The classification or grouping of risks based on their nature, source, or impact to help in organizing and prioritizing risks to ensure appropriate mitigation strategies and responses are applied.

Risk Management The systematic process of identifying, assessing, and mitigating risks to minimize their impact on an organization's objectives.

Risk The potential for an event or condition to occur that may have negative or positive effects on objectives, typically measured in terms of likelihood (probability) and impact (consequences).

Rough Order of Magnitude An early-stage, high-level estimate of the cost, time, or resources required to complete a project or initiative providing a broad approximation based on limited information, typically generated in the initial planning phases when detailed data is not yet available.

Scalability The ability of a system, process, or infrastructure to handle a growing amount of work, or its potential to accommodate growth without compromising performance or efficiency. As demand increases, the system can expand or adapt in capacity, size, or output without requiring a complete redesign or major changes.

Scenario Specifies a projected force structure; context justifying the need for operational action(s); the nature of the mission; assumptions about the geographic, temporal, military, diplomatic, and legal conditions under which the scenario occurs; and the desired end-state.

Semantic Interoperability The ability of different systems or organizations to exchange and interpret data in a way that the meaning of the information is preserved and understood accurately across contexts.

Semantics Study of meaning in language, focusing on how words, phrases, sentences, and texts convey meaning.

Stakeholder An individual or organization affected by the system and/or who has a direct or indirect influence on the system requirements.

Syntactic Interoperability The ability of different systems or software to exchange data in a format that is mutually recognizable and structurally compatible, ensuring that data can be transmitted without errors or loss of information.

System A set of interacting or interdependent components forming an integrated whole.

System-of-Systems A collection of independent, interacting systems that work together to achieve a higher level objective that no single system could accomplish alone; however, each constituent system in an SoS maintains its own functionality and management.

Systems Engineering An interdisciplinary approach and means to enable the realization of successful systems. This structured development process integrates all the disciplines and specialty groups into a team effort that proceeds from system conception through disposal.

Systems Theory An interdisciplinary framework that studies complex systems by analyzing the relationships, interactions, and behaviors of their components as part of a unified whole.

Threat Evaluation and Weapon Assignment A decision-making process used in military and defense systems to assess threats and allocate appropriate weapons to neutralize them effectively.

Uncertainty The lack of certainty or knowledge about future events, outcomes, or conditions arising when there is insufficient information, variability, or unpredictability in a given situation, making it difficult to accurately assess or predict what will happen.

User A person or entity that consumes or uses, maintains, trains, and supports the equipment, system or service, often with some technical expertise.

Validation The process of determining the degree to which a system, model, or simulation and its associated data are an accurate representation of the real world from the perspective of the intended uses of the model.

Verification The process of determining that a system, model, or simulation implementation and its associated data accurately represent the developer's conceptual description and specifications.

Vignette A narrow and specific ordered set of events, and behaviors and interactions for a specific set of systems, to include blue force capabilities and opposing force (threats) within the operational environment. Vignettes can represent small, ideally self-contained parts of a scenario.

Zero Trust Architecture A cybersecurity model based on the principle that no entity, inside or outside the network, should be trusted by default.

Index

J. D. Moreland, Jr., *Mission Engineering and Integration*,
https://doi.org/10.1007/978-3-032-32132-9

D

N

O

P

Q

R

S

W

Z

GPSR Compliance

The European Union's (EU) General Product Safety Regulation (GPSR) is a set of rules that requires consumer products to be safe and our obligations to ensure this.

If you have any concerns about our products, you can contact us on ProductSafety@springernature.com

In case Publisher is established outside the EU, the EU authorized representative is:

Springer Nature Customer Service Center GmbH
Europaplatz 3
69115 Heidelberg, Germany

Batch number: 10761750

Printed by Printforce, the Netherlands